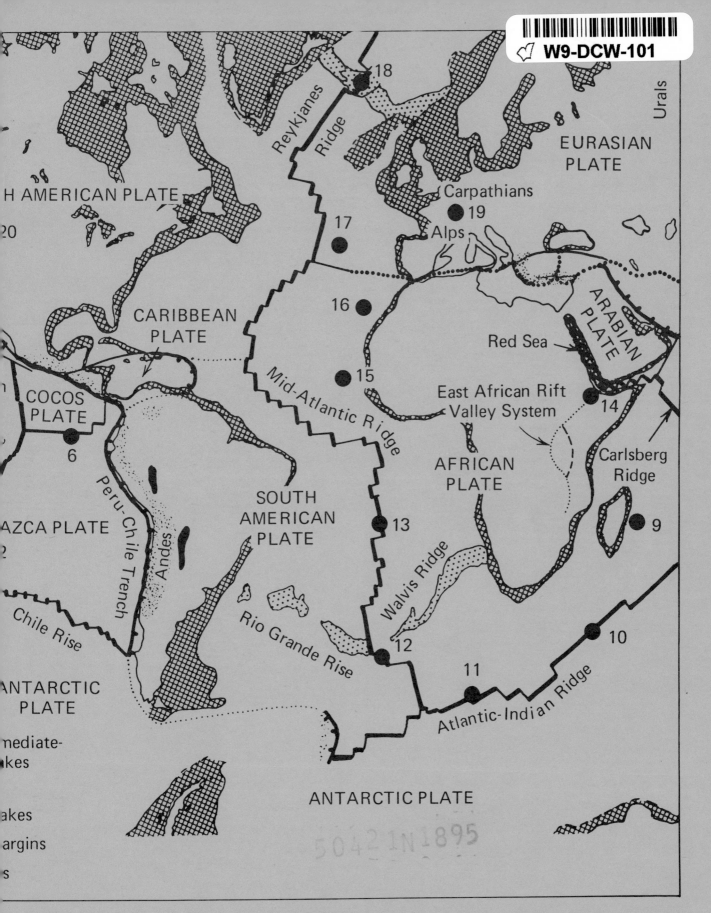

EURASIAN PLATE

Urals

Reykjanes Ridge

18

H AMERICAN PLATE

20

Carpathians

19

Alps

17

CARIBBEAN PLATE

16

ARABIAN PLATE

Red Sea

COCOS PLATE

6

15

East African Rift Valley System

14

Mid-Atlantic Ridge

AFRICAN PLATE

Carlsberg Ridge

Peru-Chile Trench

AZCA PLATE

2

Andes

SOUTH AMERICAN PLATE

9

13

Chile Rise

ANTARCTIC PLATE

Rio Grande Rise

Walvis Ridge

12

10

11

Atlantic-Indian Ridge

mediate-
akes

akes

argins

s

ANTARCTIC PLATE

5042 1N 1895

Legend:

1. Hawaii	6. Galapagos Islands	11. Bouvet Island	16. Canary Islands
2. Easter Island	7. Cobb Seamount	12. Tristan da Cunha	17. Azores
3. Macdonald Seamount	8. Amsterdam Island	13. St. Helena	18. Iceland
4. Bellany Island	9. Reunion Island	14. Afar	19. Eifel
5. Mt. Erebus (Antarctica)	10. Prince Edward Island	15. Cape Verde	20. Yellowstone

OCEANOGRAPHY

SECOND EDITION

oceanography
a view of the earth

M. GRANT GROSS

Chesapeake Bay Institute
The Johns Hopkins University

Prentice-Hall, Inc.
Englewood Cliffs, New Jersey 07632

Library of Congress Cataloging in Publication Data

GROSS, MEREDITH GRANT (date)
    Oceanography.

    Includes bibliographies and index.
    1. Oceanography. I. Title.
GC16.G7 1977   551.4′6   76-41821
ISBN 0-13-629675-0

Frontispiece

A satellite's view of the Atlantic Ocean from more than 37,000 kilometers
(22,300 miles) above Brazil. Four continents can be seen:
South America (center), North America (upper left quadrant),
Europe (upper right), and Africa (center right). Antarctica (bottom of photograph)
is obscured by clouds, the white fluffy masses often forming swirls
as part of the large storms in the lower atmosphere.
Note the large fraction of the earth's surface obscured by clouds.
(Photograph courtesy NASA.)

PRENTICE-HALL INTERNATIONAL, INC., *London*

PRENTICE-HALL OF AUSTRALIA PTY. LIMITED, *Sydney*

PRENTICE-HALL OF CANADA, LTD., *Toronto*

PRENTICE-HALL OF INDIA PRIVATE LIMITED, *New Delhi*

PRENTICE-HALL OF JAPAN, INC., *Tokyo*

PRENTICE-HALL OF SOUTHEAST ASIA PTE. LTD., *Singapore*

WHITEHALL BOOKS LIMITED, *Wellington, New Zealand*

contents

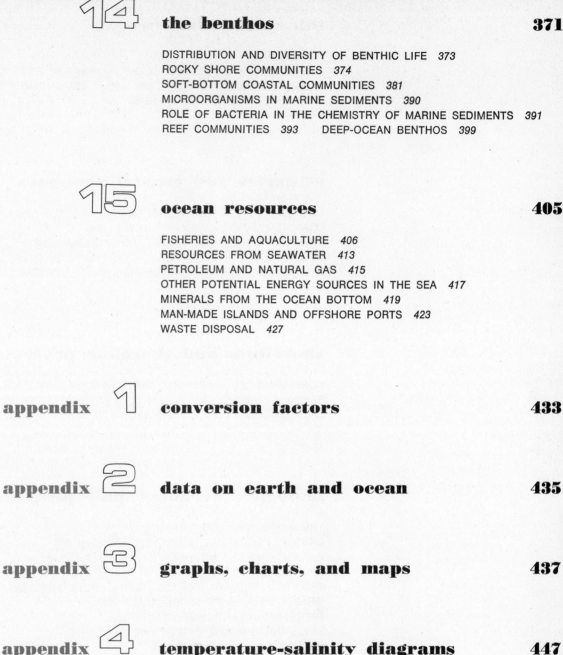

first edition preface

Oceanography—the scientific study of the ocean—presents a view of the earth that is new and useful to us. We have traditionally viewed the earth from the land and from a human frame of reference. But study of the ocean, the earth's most distinctive feature, shows us that continents are only large islands surrounded by a single body of water. Our study reveals an earth with a finite surface area and a limited capacity for production or abuse. We see an earth whose weather and ocean currents are powered by energy from the sun; we see that humans have had little success in either controlling or modifying these forces. We see that marine processes active in any single area are eventually felt throughout the world ocean.

This book examines the world ocean and those processes that control its major features and the life in it. It is intended for a general survey course, either for beginning science students or for students who may never take another course in science. The primary objective of the book is to investigate the major features of the ocean and to present some of the problems that have occupied oceanographers during the century since the beginnings of the science.

A second objective of the book is to illuminate the workings of modern science. Oceanography is still evolving at a rapid rate. Because of this rapid development, the student must recognize that many of the "facts" given in this book are actually hypotheses—ideas on trial. Some will stand the test of careful examination, further observation, and more sophisticated analyses of available data. These ideas will survive. Many will not, however, and they are then discarded or replaced by other hypotheses that explain the observations more fully or make more accurate predictions. Hence, it should be remembered that our view of the earth represents a momentary glance at a rapidly changing picture.

A book of this nature necessarily omits or condenses many aspects of oceanography. In general the ocean areas and processes selected for discussion are those most likely to be seen by land dwellers or to affect our lives. For instance, those processes that affect the coastal ocean are emphasized—beach formation, coastal currents, and sea-ice formation are just a few examples. Other interesting areas have been condensed, such as the discussion of the deep structure of the earth's crust or the early history of the earth and evolution of presently living organisms. References to more detailed treatments of these topics have been included at the end of the chapters, to aid the student in finding further material on specific subjects.

The view offered here of the coastal ocean and its limitations should be useful beyond the bounds of a college course. As citizens we are increasingly called on to make decisions or to evaluate recommendations about utilization of the coastal ocean and the coastline at its margins. Should a salt marsh be used for a housing development? A sanitary landfill? A marina? Or perhaps left in its natural state? Should waste disposal be permitted in a bay or on the continental shelf? Reasonable decisions in these areas require a knowledge of basic environmental processes. With an increased awareness and improved understanding of the processes affecting ocean and atmosphere, perhaps we can eventually break out of the cycle of buying time for today's problems by creating new problems for tomorrow. This is especially important, for the use of the ocean involves not just our own coastline but the entire planet.

second edition preface

In preparing the second edition, several objectives were foremost. First, the revision reflects the continuing integration of oceanography with the rest of science as our knowledge becomes more quantitative and less descriptive. This has been particularly evident in geological oceanography as the concept of sea floor spreading continues to develop. Chapters 2 and 3 have been extensively revised to reflect recent discoveries.

Ecological studies of marine biological processes have greatly expanded our understanding of the factors that control the productivity of the ocean. This ecological approach integrates several important processes involving physical and chemical as well as biological oceanography. So Chapters 12, 13, and 14 have been rewritten to provide a more ecological approach and to address more closely such important problems as amount and distribution of fish production.

A history of oceanography has been included as Chapter 1. The history of British and American oceanography is emphasized, but any account of an international undertaking like the study of the world ocean must include other countries.

To reduce the number of chapters for accommodation to shorter college terms. The material on coastal and estuarine processes has been condensed. The separate appendix on fishes has been eliminated, with most of it being included in Chapter 13.

Any book reflects the contributions and help of a great many people. A brief and incomplete list of people who over the years have helped to make this book possible includes: A. C. Barnes, R. H. Fleming, J. S. Creager, K. Banse, T. S. English, J. I. Tracey, Jr., H. S. Ladd, P. K. Weyl, J. L. McHugh, R. B. Williams, M. C. Glendening, W. R. Taylor, D. W. Pritchard, E. D. Goldberg, J. V. Byrne, W. C. Boicourt, E. D. Stroup and my students who taught me more than they realized. To each goes my thanks.

Owing to illness, Roger Hayward was unable to participate actively in this revision. Many of his original drawings appear in this edition with newly lettered labels; some were redrawn. The new figures were drawn by Prentice-Hall staff artists.

To my wife, Nancy, who made this book possible with her uncounted hours as typist, editor, research assistant, and general sup-

M. GRANT GROSS

porter goes the acknowledgement of a debt too great to be satisfied by a few words. And to the rest of my family and to my mother who typed this edition, I want to express my appreciation and thanks for help during the many crises of putting together the manuscript.

OCEANOGRAPHY

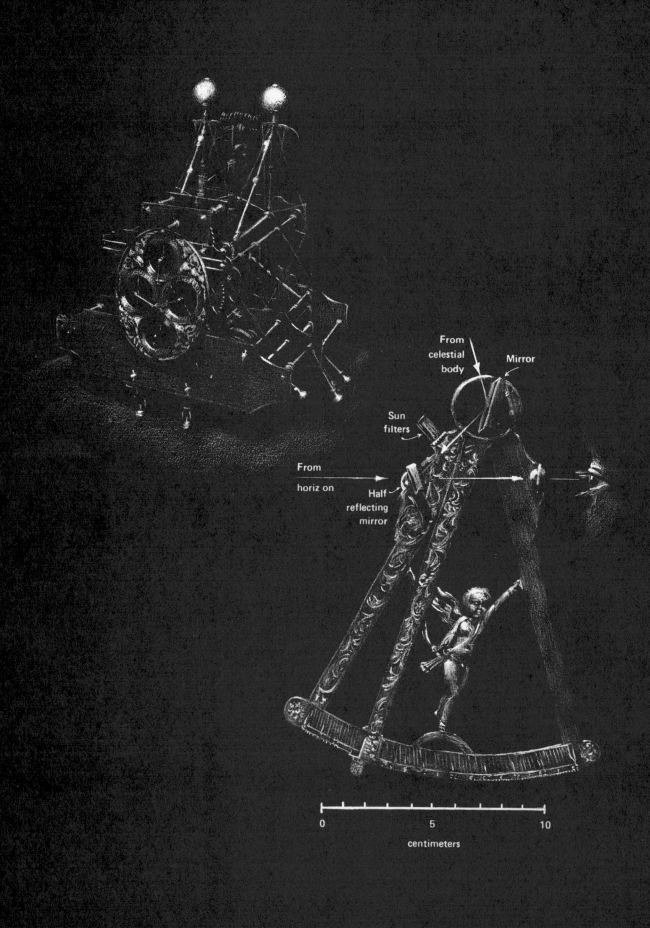

From
celestial
body

Mirror

Sun
filters

From
horiz on

Half
reflecting
mirror

0   5   10

centimeters

Eighteenth-century navigational instruments: octant and marine chronometer. By means of the two mirrors on an octant, one-half-silvered and one mounted on a movable arm, the image of a celestial body (sun, moon, or star) can be seen superimposed on an image of the horizon. The 45° of motion of the movable arm covers the angles of altitude from horizon to zenith. The diagonal lines of the calibration suggest that the required angles could be measured to within about 10 minutes of arc, equivalent to about 10 nautical miles (18.53 kilometers). This example, made of gilt bronze, is in the Altonaer Museum, Hamburg-Altona, Germany.

The chronometer shown here (left), property of the National Maritime Museum of Greenwich, England, was built by John Harrison in 1736. It is one of several clocks he invented in competition for a prize offered by Parliament to the scientist who could design a mechanism whereby a ship could determine its longitude within specified limits after six weeks at sea. The chronometer is set for Greenwich time and longitude is calculated from the precise difference between Greenwich noon and local noon. The device illustrated here features self-correcting, counter-rotating spring pendulums. Any angular acceleration causing one of them to move faster also retards the other by the same amount. Harrison won the prize with a watch-sized chronometer that achieved a precision of 5 seconds during a voyage from England to Jamaica.

1

history of oceanography

From earliest times, study of the ocean was aimed primarily at solving practical problems. The first contributions were made by fishermen, merchants, and traders, as they explored ocean margins in search of new trade routes or fishing grounds. Some basic principles of oceanography were understood by ancient sicentists. Aristotle, for instance, noted in the fourth century B.C. that the sea neither dries up nor overflows; thus the amount of rainfall is equal to evaporation over the earth. Ptolemy (A.D. 150) developed a scheme of latitude and longitude for charting the known world, and made surprisingly accurate maps of the Eastern Hemisphere.

The Venerable Bede (673–735) knew that the moon controls the tides, and in the fifteenth century, Cardinal Nicholas of Cues tried to devise a method of calculating the sea's depth without using a line. He suggested measuring the rate of rise of a buoyant object in "diverse waters" of known depth, then dragging such an object to the bottom of the ocean, releasing it, and timing its return. During the later Renaissance, explorers circled the earth and made charts as they went. And such seventeenth century scientists as Sir Robert Boyle (1627–1691) studied ocean temperatures, water pressure, sea salts, and the effects of storm waves.

FOUNDATIONS OF MODERN OCEANOGRAPHY

Since oceanic processes extend over large areas, and working at sea is generally expensive and time-consuming, most major advances have been sponsored by governments or by very wealthy individuals. Long before oceanographic research was thought of, governments supported ocean exploration. The voyages of Vasco da Gama, Columbus, Balboa, Magellan, and Captain Cook mapped the major outlines of land and ocean. Much, however, was unknown. Polar regions were still inaccessible to ships. Oceanic depths had never been charted. Only surface marine life was known. Open ocean phenomena were largely the province of myth and mystery.

Inevitably, however, ocean exploration led to studies of ocean processes. One of the most successful early explorer-scientists was a British naval officer, Sir James Clark Ross, (1800–1862). As a young man he had sailed under his uncle, Sir John Ross, in search of a Northwest Passage from the Atlantic to the Pacific, and had tried to reach the North Pole. At twenty-nine, on another Arctic expedition, he reached the magnetic pole. Later he was assigned by the British Admiralty to head an Antarctic expedition that made him famous. From 1839 to

4

1843 his ships *Erebus* and *Terror* made three separate voyages, circling Antarctica and charting hundreds of miles of its coastal waters. Scientific parties aboard collected biological specimens from great depths, though most of the samples were either lost or allowed to deteriorate through inexpert handling. The scientists also dredged to then-incredible depths of 5000 meters and tried to measure temperatures, but thermometers of that time could not give accurate readings at great pressures, so the data were wrong and misleading.

The primitive state of early nineteenth century measuring and sampling techniques caused serious misconceptions about deep-ocean conditions. Often, these caused scientists to pursue fruitless studies while ignoring fields that might have led to valuable discoveries. For example, it was thought that the deepest ocean waters did not circulate but lay stagnant at the bottom of ocean basins. This meant that no supply of incoming oxygen would be available to support life there. Furthermore a respected naturalist, Edward Forbes (1815–1854) claimed to have proven the absence of life below 600 meters. He had done some collecting in the Aegean Sea to compare findings with those of Aristotle nearly 2500 years earlier, and had noticed that fewer and fewer species were caught as he let his nets sink farther below the surface. From this he concluded that most of the ocean was "azoic," or devoid of animal life. This idea gained wide acceptance, so much so that occasional reports of animals taken from below 600 meters tended to be ignored or discounted.

Forbes's azoic theory was not refuted until the 1860's, when ocean basin topography was being studied in connection with the laying of transoceanic telegraph cables. A cable from 2000 meters depth in the Mediterranean Sea was raised for repairs, and on it were encrusted living corals. This discovery came at a time when interest in deep-sea exploration had been aroused in connection with Charles Darwin's (1809–1882) theories about the origin of species through evolution. Scientists became interested in finding stable environments, where conditions had not changed over long time periods. It was hoped that the deep ocean floor might prove to be inhabited by ancient species or "old generic types," the ancestors of modern organisms. Primitive animals on stalks, called *crinoids,* had recently been discovered by a Norwegian vicar and amateur scientist, Michael Sars, and this seemed a promising sign.

Thus by the 1860's there seemed to be several compelling reasons for studying the deep ocean. There was widespread interest in setting up large-scale research projects, for which the personal funds of even a wealthy scientist would be inadequate. Fortunately, members of the Royal Society of London, a group of eminent citizens devoted to the cause of philosophical and scientific progress, provided an important source of financial support. Another source was the British Admiralty, which provided two ships for North Atlantic deep-sea studies during the late 1860's. As a result of these research efforts, some of the old prejudices and misconceptions were overcome. It was demonstrated that waters can move through deep ocean basins. Scientists dredged to depths of 1200 meters and discovered new organisms. And the Royal Society was persuaded to sponsor the most ambitious and innovative project that had ever been attempted. This expedition, aboard H.M.S. *Challenger* (1872–1876), established a tradition of large-scale, interdisciplinary cooperation that has been an important component of major efforts in oceanography since that time.

The *Challenger* was a full-rigged corvette with an auxiliary 1200-horsepower steam engine (Fig. 1-1). The scientific party under Sir Wyville Thomson (1830–1882) had been assigned to investigate "everything about the sea"; they planned to study physical and biological

Figure 1-1 *(right)*

The *Challenger* crew attempting to land small boats at St. Paul's Rocks in the mid-Atlantic. Strong currents and high waves made this a difficult and dangerous operation.

Figure 1-2 *(below)*

This sketch shows a deep-sea dredge aboard the *Challenger* being lowered by stages to the bottom. A weight slid down the line (G, G', G'', etc.), drawing the heavy chain dredge bag downward to land on its side. This put the dredge in position to scoop up samples as the ship dragged it along.

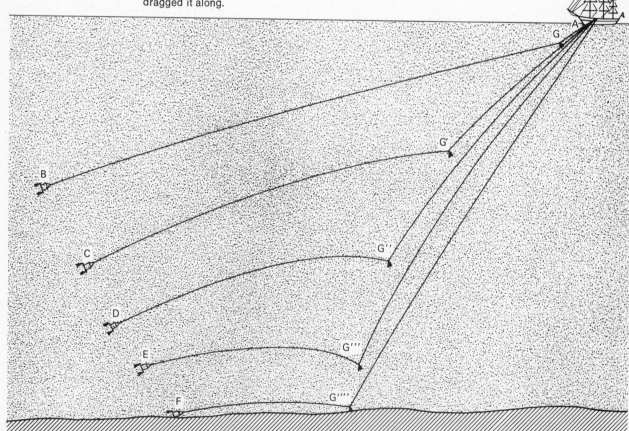

conditions in every ocean, recording everything that might influence the distribution of marine organisms. This included taking water samples and temperature measurements of both deep and surface waters, recording currents and barometric pressures, and collecting bottom samples (see Fig. 1-2) in order to study sediments and identify new species. Fish were caught in nets dragged behind the ship.

Before returning to England the *Challenger* had traveled 68,000 miles, hauled on board some 133 loads of rock and sediment from the ocean floor, and taken 492 soundings using a weighted hemp line. All oceans except the Arctic had been sampled, and 4717 new animal species had been collected and classified. Circling the globe, the ship brought back data that eventually filled fifty large volumes. The reports were written by 76 authors over a 23-year period.

Many delusions about the ocean's mysteries were swept away by the *Challenger* results. The most famous, perhaps, was the debunking of *Bathybius,* a supposed "primordial slime" that had been thought to represent preevolutionary life on the ocean floor. Bathybius had been "discovered" by Thomas H. Huxley (1825–1895), a popular biologist whose staunch support of Charles Darwin had helped to make evolution respectable. In 1868 Huxley found a slimy substance in a jar containing a biological specimen several years old. Perhaps he was overly anxious to provide new evidence for his friend's theory, which was at that time under heavy fire from supporters of the Biblical Creation. At any rate, Huxley jumped to the conclusion that he had identified the protoplasmic substance from which life originated in the deep ocean. It was a fortuitous idea, in that Bathybius stimulated much interest in marine research. But while at sea, a *Challenger* chemist revealed that the slime was made from calcium sulfate, precipitated by mixing preservative alcohol with seawater. Fortunately the theory of evolution survived this setback.

A government department, the Challenger Expedition Commission, was established to analyze and publish the results of the voyage. It was to have been headed by Wyville Thomson, but when he died suddenly that position went to Sir John Murray (1841–1914), a Scottish naturalist who had served as Thomson's assistant. Murray was a brilliant scientist whose career by no means ended with publication of the *Challenger* Reports. He is generally given credit for laying the foundations for modern submarine geology, and he is famous as the author of an important work on marine sediments. He also did valuable work on coral reefs and plankton, the microscopic drifting organisms that are the ocean's major food source. After persuading the British government to annex uninhabited Christmas Island in the Indian Ocean, Murray made his fortune there in phosphate mining. This wealth he used in part to promote oceanographic research; for example, he paid the expenses of a Norwegian expedition in the North Atlantic aboard the *Michael Sars* in 1910. In collaboration with the director of that expedition, Dr. Johann Hjort (1869–1948), Murray wrote his most important work, *The Depths of the Ocean,* which was for many years the most widely read and authoritative oceanography text available in English.

EARLY AMERICAN OCEANOGRAPHY

In the United States, governmental support for oceanography grew out of a need to solve practical problems. These included insuring the safety of passengers and goods aboard U.S. ships, maintaining adequate coastal defenses, and developing marine resources such as fisheries in U.S. waters. In 1830 the Navy created a Depot of Charts and Instruments to supervise use of navigational instruments on government vessels. Matthew Fontaine Maury (1806–1874), a brilliant naval officer, was appointed Superintendent in 1842.

Maury (Fig. 1-3) was interested in making ocean transportation safer and speedier, and he had written a navigation textbook before coming to the Depot. Once in charge, he began to organize the vast amount of wind, current, and seasonal weather data stored in ships' logbooks. To gather further information, Maury furnished ships' crews with blank charts so that they could make daily records of weather and ocean conditions, showing the time and place of each instrument reading. The Wind and Current Charts he compiled in this way revolutionized navigation and cut weeks off transoceanic runs. In addition, Maury's sounding and bottom-sampling projects were used to make simple maps of the Atlantic Ocean floor.

In 1853, Lieutenant Maury organized the first international meteorological conference, which led to international cooperation in collecting weather information at sea. His popular and influential book, *The Physical Geography of the Sea* (1855), was the first major oceanographic work in English. It stimulated widespread public interest in ocean currents, though many scientists were skeptical of Maury's rather florid and unscientific writing style, as well as his often unsupported theories about the causes of oceanic phenomena. He is sometimes called the father of physical oceanography, though today he would probably be categorized as a gifted navigator, a popularizer of science, and a dynamic administrator.

While the Navy accepted responsibility for safeguarding American lives and property on the high seas, the Treasury Department was interested in maritime commerce. Charting of harbors, ports, and coastal waters was urgently needed, and Thomas Jefferson created a civilian agency within the Treasury Department to take charge of that program. But in 1807, scientific research under government sponsorship was a new idea. Internal disorganization kept the newly formed Coast Survey (now the National Ocean Survey) from undertaking significant mapping or charting projects during its first 25 years.

Finally, Alexander Bache (1806–1867), who was trained in physics and chemistry, took over the Survey in 1843 and set up a comprehensive hydrographic program. His people worked in East Coast waters, mapping hazards to navigation, taking depth measurements, collecting bottom samples, measuring currents, and monitoring tides. They also made pioneer studies of the Gulf Stream, charting its temperatures, its depths, and changes in its velocity. In this, Bache was carrying on the work of his great-grandfather, Benjamin Franklin, who had mapped the Gulf Stream in the 1770's (Fig. 1-4). Another of his innovative ideas was to invite famous scientists to join Coast Survey cruises. This policy greatly expanded the scope of federally sponsored research and provided support for basic oceanographic science at little public expense.

In 1871 a Fish Commission was established, with a distinguished ornithologist and natural history professor as unpaid Commissioner. Spencer Fullerton Baird (1823–1887) was even more interested in establishing a center for basic research in coastal ecology than he was in regulating commercial fish stocks. For his laboratory he chose a tiny fishing village on Cape Cod—Woods Hole near Falmouth, Mass., where university-based scientists could come to work during the summer. This soon became an important center for biological research; eventually the Marine Biological Laboratory and the Woods Hole Oceanographic Institution were set up there (Fig. 1-5). Thus, two of America's foremost institutions in marine sciences began as a result of Baird's dedication and foresight and the very practical need to understand fisheries better in order to protect known stocks and to explore for new ones.

Figure 1-3

Matthew Fontaine Maury grew up on frontier farms in Virginia and Tennessee. At nineteen he was appointed a midshipman, and within the next 10 years he had sailed around the world on Navy vessels. When a stagecoach injury ended his seagoing career, Maury embarked on the administrative duties that established a foundation for government-sponsored oceanographic research.

Figure 1-4

Before the American Revolution, Benjamin Franklin noted that transatlantic mail ships made the voyage to England more speedily than the voyage home. He correctly deduced that a current flowing northward along the U.S. coast and eastward into the North Atlantic was responsible for the difference in sailing time. In 1770, he published this map, with instructions on how "to avoid the Gulph Stream, on one hand, and on the other the Shoals that lie to the Southwest of Nantucket and St. George's Banks."

Figure 1-5

Woods Hole Station, about 1890. The Fish Commission's laboratory is at center, the residence to the right, and the *Albatross* is moored at the left. (Photograph courtesy of National Marine Fisheries Service.)

The first specially equipped research ship to sail under a U.S. flag—the 234-foot steamer *Albatross*—was built in 1882 under Baird's direction. Research parties studied ways in which changing water conditions—temperature, salinity, and currents—seemed to affect the abundance of fish in a particular area. At the same time, scientists collected specimens and conducted their own studies on the way fish and invertebrate animals grow and develop in marine environments.

Scientists were glad to be invited on Fish Commission cruises. Probably the most famous was a biologist, Alexander Agassiz (1835–1910), whose father Louis Agassiz had sailed with Alexander Bache. Agassiz, Jr., was a wealthy man. He personally funded more than one Pacific cruise, outfitted the *Albatross* with special equipment, and in other ways subsidized U.S. oceanographic research. Late in the nineteenth century a record ocean depth of 7632 meters was dredged under his direction, on the Pacific floor near Japan. This was not surpassed until 1951. In addition, Agassiz maintained a private lab at Newport, R.I., where he made comparative studies of the corals and other specimens he had brought back from his travels.

Many American oceanographers have worked part-time on their own research while being paid to solve practical problems, as have scientists in every field. Some who were in administrative positions, such as Baird and Bache, encouraged this practice by supporting basic research in government laboratories. Many of the advances that kept the U.S. in the forefront of scientific discovery were made in this way. More recently, Congress has accepted the idea of paying scientists to do research that they find interesting, and assumed that the nation and the world will benefit.

NATIONAL EXPEDITIONS

Between the *Challenger* Expedition and the Second World War, small but important oceanographic expeditions were sent out from various countries. Notable among these was the work of Prince Albert I of Monaco (1848–1922). This diligent investigator had abundant financial resources and a life-long interest in marine studies. He equipped four successive private schooners as up-to-date research vessels to carry out all kinds of creative physical and biological projects. In one lengthy study Prince Albert set adrift over 1500 bottles and barrels containing messages in ten languages. Enough of these were recovered to explain the relation of the Gulf Stream to the basic North Atlantic current system. Using cleverly designed equipment he collected many new kinds of mid-depth fishes, including several specimens vomited by a dying whale. Later in his life the Prince established a museum and an oceanographic institute which continue in use today.

As we have seen, research is often stimulated by practical needs. The Scandinavian countries, whose economies are so dependent on the sea, have long been in the forefront of progress in oceanography. As investigators began to realize that the abundance of commercial fishing stock is affected by currents and the distribution of water masses, research in those areas received increasing government support. Improved equipment was developed to study movements of ocean waters, both near the surface and at great depths.

Scandinavian scientists realized that temperature and salinity differences drive currents, a fact which had already been suspected by Lt. Maury and others earlier in the century. This led to rapid developments in the field of chemical as well as physical oceanography. A surge of interest in theoretical studies went hand in hand with the development of better deep-sea *"reversing"* thermometers (Fig. 1-6) and methods of determing salt concentrations. Presently another Scandinavian, Bjorn Helland-Hansen (1877–1957), developed techniques to

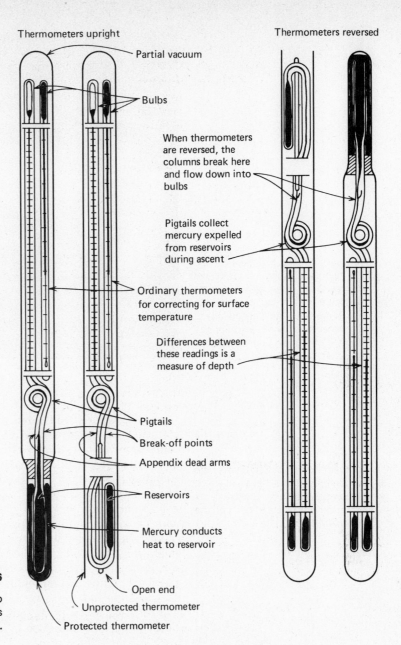

Thermometers upright

Partial vacuum

Bulbs

When thermometers
are reversed, the
columns break here
and flow down into
bulbs

Pigtails collect
mercury expelled
from reservoirs
during ascent

Ordinary thermometers
for correcting for surface
temperature

Differences between
these readings is a
measure of depth

Pigtails

Break-off points

Appendix dead arms

Reservoirs

Mercury conducts
heat to reservoir

Open end

Unprotected thermometer

Protected thermometer

Thermometers reversed

Figure 1-6

Reversing thermometers were used to
obtain precise temperature measurements
in the ocean.

calculate current direction and speed from precise observations of
temperature and salinity. This was a particularly outstanding contribution to modern oceanographic technique. Another advance was made
by V. W. Ekman (1847–1954), who showed how the earth's rotation
affects wind-driven currents.

Plankton studies (Fig. 1-7) followed the discovery that the physical and chemical properties of water masses determine which organisms
can live in them. Soon, techniques were developed for identifying water
masses by their biological communities. Since North Sea circulation is
tied to that of the Atlantic as a whole, local studies inevitably led to
more extensive work that could not be effective if carried on by one
nation alone. An influential Swedish scientist, Otto Pettersson (1848–
1941), correctly foresaw that measurements made by several vessels
simultaneously, at exactly predetermined stations, would be the most
efficient method for studying ocean properties on a large scale. He
enlisted support from Sweden's King Oscar II, and thus the plan for an
International Council for the Exploration of the Sea (ICES) was born.

Figure 1-7

Plankton (microscopic marine plants and animals) are collected by towing a cone-shaped net of fine meshed cloth through seawater. Organisms unable to swim faster than the net are captured for study. (Photograph courtesy Woods Hole Oceanographic Institution.)

Figure 1-8

South Atlantic crossings made during the *Meteor* Expedition (1925–27). Each of the tiny marks represents a station at which temperature and salinity measurements were made at various levels to get a profile of conditions throughout the entire basin.

Meteor tracks

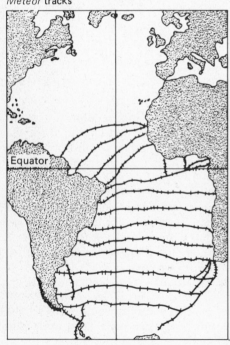

Germany, Russia, Great Britain, Holland, and the Scandinavian countries participated in ICES from its beginnings. The Council was founded in 1902 to foster international cooperation in learning how best to rationally utilize marine resources, especially fisheries. It has functioned since then as a coordinating body, and as a sponsor for joint research enterprises.

Some scientific teams working within ICES followed Pettersson and Ekman in studying currents and water masses that move with them. Others tagged fish to study their life cycles and the unpredictable migrations that had frustrated generations of fishermen. Johann Hjort was eventually able to demonstrate that fish abundance is largely determined by the amount and kind of food available to them at the time they are hatched. This was a highly significant finding in the general field of fisheries research, and Victor Hensen (1835–1924) followed it up by important studies on how larval and juvenile (planktonic) fishes are nourished. This turned out to depend again on ocean water movements, demonstrating that physical and biological processes were invariably linked within the total framework of oceanographic research. Always, the pooling of knowledge and resources at international levels promoted communication between scientists and their cooperation in solving common problems.

Word of ICES methods reached a young Harvard zoologist, Henry Bryant Bigelow (1879–1967), who had originally been trained in the naturalist tradition under Alexander Agassiz. Bigelow, encouraged by the visiting John Murray, undertook a comprehensive study of biological and physical oceanography in the Gulf of Maine. Beginning in 1912, he spent fifteen years working on detailed, comprehensive investigations, using methods pioneered by Swedish and Norwegian scientists. His work provided incentive for other Americans to plan their research around the new techniques. This meant studying the geology, biology, and chemistry of a single body of water over a long enough period to get a complete picture of its oceanographic properties.

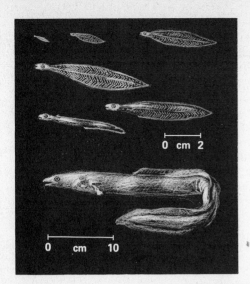

Figure 1-9

The anguillid "fresh-water" eels, 0.5 meter or longer, inhabit both sides of the Atlantic. They live in rivers and lakes, but at sexual maturity (5–8 years old) they begin a migration to the Sargasso Sea. After a journey of thousands of miles, and many months without food, they spawn at depths of around 500 meters, and then they die. Their eggs float at the surface and soon hatch into tiny larvae that resemble little fish. Each begins the long journey to fresh water first described by Johannes Schmidt. European eels take 3 years to complete the journey, but the American variety requires only 1 year. There is a physical transformation at that time; the slender young eels ("elvers"), about 7 centimeters long, travel up the streams to live there until maturity.

Figure 1-10

A geophysicist determines the velocity of sound in a sediment core recovered by modern deep-ocean drilling techniques. Such information aids in interpreting other geophysical observations on thickness and structure of rocks beneath ocean basins. Study of fossils in sediment layers may provide information on ocean conditions over hundreds of thousands of years, and paleontologists study such fossils as clues to the evolution of marine animals and plants. (Photography courtesy Deep-Sea Drilling Project, Scripps Institution of Oceanography under contract to the National Science Foundation.)

Nationally sponsored expeditions during the early part of this century produced outstanding results. We have already mentioned the 1910 Murray and Hjort collaboration for biological and physical studies on the *Michael Sars*. Another highlight was the German Antarctic Expedition of 1911–12, during which temperature and salinity distributions were charted from surface to ocean bottom and from north to south across the entire Atlantic Ocean. Basic studies of Antarctic waters completed the expedition. In 1925–27 the German research vessel *Meteor* crossed the Atlantic fourteen times between 20°N and 65°S, charting temperature–salinity profiles and mapping the ocean floor (Fig. 1-8). This was the most comprehensive study of water masses ever made for such a vast area. Studies such as these were finally beginning to provide a total oceanographic picture of the great ocean basins.

Two other government-sponsored projects warrant a brief mention. Sailing under the Danish flag, Johannes Schmidt spent twenty years trying to locate the spawning grounds of American and European eels. Finally, after being interrupted by World War I, he identified the birthplaces of these long-distance swimmers far out in the mid-Atlantic (Fig. 1-9). In 1929–1930, aboard the *Willibrord Snellius,* Dutch scientists explored Indonesian seas. Invention of the "Piggot gun" explosive bottom sampler enabled them to bring a "core" of sediment six feet long aboard the ship. By analyzing its successive layers of sediment, the history of an ocean basin is revealed (Fig. 1-10). Over the years, of course, longer and longer cores have been retrieved.

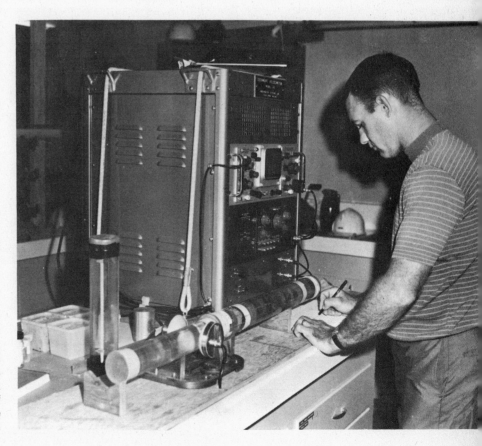

In a class by itself was the remarkable polar voyage of a Norwegian scientist and adventurer, Fridtjof Nansen (1861–1930). Convinced that winds and current drift caused pack ice to flow between the North Pole and Franz Josef Land, from the Siberian Artic to eastern Greenland, Nansen persuaded his government and the British Royal Geographical Society to help him test his theory. A special ship, the *Fram,* was built to carry a scientific party eastward from the Barents Sea toward the Pole, frozen in an ice floe (Fig. 1-11). This expedition was an outstanding success. From September, 1893, to August, 1896, the *Fram* drifted with the ice, from 70°50′ to an extreme north latitude of 85°57′, closer to the Pole than any surface ship has ever been before or since. The crew then broke her free during warm weather and sailed home. Nansen, with a companion, left the ship at 84°N and set out toward the Pole by dogsled. The two men spent fourteen months alone on the Artic ice, but made only 86°14′ before movements of the ice pack forced them to turn back toward Franz Josef Land.

Figure 1-11

Scenes from the voyage of the *Fram:*
(a) A crew member tests the depth of a
melted pool on the ice pack.
(b) Fridtjof Nansen at the helm.

(a)

(b)

The remainder of Nansen's career was equally remarkable. He worked on current theory with Ekman and Helland-Hansen, and published several books on the oceanography of the North Sea. He also designed a standard water-sampling bottle which bears his name (Fig. 1-12). From 1905 onward, Nansen was active in politics, and was awarded the Nobel Peace Prize in 1923 for contributions to the relief and repatriation of World War I refugees; he later became an expert on Soviet agriculture economics. He was a lifelong athlete and a famous skier and did much to popularize winter sports in Switzerland and elsewhere outside Scandinavia.

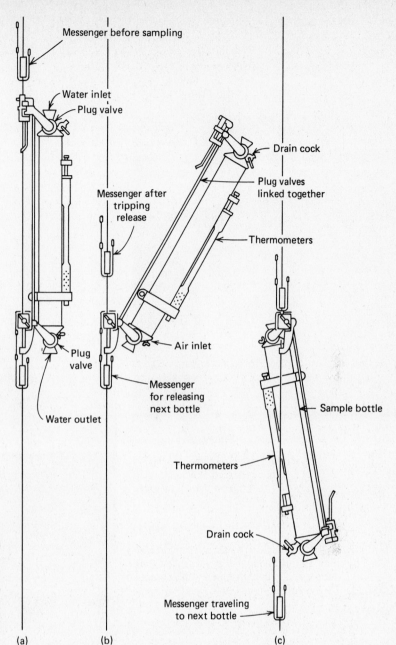

Messenger before sampling

Water inlet

Plug valve

Messenger after tripping release

Drain cock

Plug valves linked together

Thermometers

Air inlet

Plug valve

Messenger for releasing next bottle

Water outlet

Sample bottle

Thermometers

Drain cock

Messenger traveling to next bottle

(a)   (b)   (c)

Figure 1-12

Nansen bottles attached to a wire line for lowering into the water. The messenger (a brass weight on the line above the bottle) releases the trigger, permitting the bottle to invert and take a sample. Precision reversing thermometers in the two cases mounted on the bottle record the temperature at the time the sample is taken.

One important series of research projects followed from a celebrated tragedy, the sinking of the *Titanic*. She hit an unusually southerly iceberg in 1912, and the International Ice Patrol was set up to guard against future mishaps. Its primary function is to monitor the hundreds of icebergs that break from the Greenland ice cap every spring to create a hazard to North Atlantic shipping. However the Patrol soon made additional contributions to oceanography by demonstrating that the best way to predict current movements is by the so-called dynamic technique worked out by Bjorn Helland-Hansen. The method had never before been tested on a large scale. Its successful application was a triumph for those who believed that theoretical studies could be the key to man's increasing mastery of the ocean.

15

Figure 1-13 *(above)*

A modern depth recorder provides a continuous record of water depths below the ship's track.

Figure 1-14 *(right and below)*

(a) The *Atlantis* sailing under auxiliary power. (b) *Atlantis II,* a vessel designed for oceanographic research. (Photographs courtesy Woods Hole Oceanographic Institution.)

New research fields opened up in the war years of the early 1920's. Efforts at submarine detection led to development of the *echo sounder,* a device which emits sounds toward a solid object, such as a ship, an iceberg, or the sea floor. It then records the time elapsed before the returning echo is detected. Previously, ocean depths had been sounded using a weighted wire; the echo sounder made continuous depth readings possible, and ocean basin mapping took a giant step forward (Fig. 1-13). The *Meteor* expedition was one of the first to take full advantage of this method.

By the beginning of World War II, foundations for modern American research in oceanography had been laid. A new source of funds, coming from nonprofit institutions set up by industrial millionaires, was beginning to provide some solid financial support. Several million dollars from the Rockefeller Foundation supported the Woods Hole Oceanographic Institution. The Oceanographic's ketch *Atlantis* (Fig. 1-14) provided new data on the Gulf Stream and Western Atlantic Ocean. From Scripps Institution of Oceanography, also a recipient of Rockefeller moneys, *R/V E. W. Scripps* explored the nearby California Current and Gulf of California. *Catalyst,* owned by the University of Washington, worked off the Northwest Coast and in the Gulf of Alaska. For more distant studies, scientists from various institutions cruised on Coast Guard and Navy ships in the Bering Sea, through the North Pacific, and off Central America.

POSTWAR OCEANOGRAPHY

Most research came to a sudden halt with the outbreak of World War II. Scientific resources were mobilized for the war effort, and fields that had been previously neglected assumed new importance. As usual, projects initiated to attack specific problems led into theoretical studies, and basic science was advanced in many cases. Intensive acoustical studies were undertaken to help track submarines by their sounds transmitted through water. Because sharp temperature boundaries affect the acoustic properties of seawater, this led to new studies of temperature distribution in the ocean. The *bathythermograph,* which records a profile of temperatures as it is lowered through the water, was perfected in the 1930's and widely used during the 1940's. Wave research was intensified by a need to plan safer amphibious troop landings. In the same way, coral reef studies resulted from the war in the Pacific because Allied troops needed the information for landings. Sea-floor mapping was done to locate safe routes for submarines through narrow straits and other hazards.

U.S. government funding for such projects was generous, in comparison with the meager support Congress had given to prewar research. But for scientists, possibly the most significant outcome of the war was an increased confidence in their potential for contributing to the nation's welfare. After the war, as defense-related research was followed up by further studies, and prewar efforts were resumed, government grants became available for projects of generally recognized importance. This is the pattern for much research funding today.

Defense-related projects have been mainly underwritten by the Office of Naval Research, and basic science by the National Science Foundation. The Energy Research and Development Agency (formerly the Atomic Energy Commission) has been another important source of funding for marine science, outside of government laboratories.

National security, mineral and petroleum resources, living resources such as fisheries, and environmental protection are some major areas of publicly sponsored research. Basic research has also been well supported. During the postwar decades several major projects undertaken by international research teams have revolutionized our understanding, not just of ocean processes, but of the entire earth's structure.

Following directly from wartime research and postwar nuclear weapons testing, a detailed and comprehensive study was undertaken of two Pacific atolls, Bikini and Eniwetok. It was perhaps the largest oceanographic project up to that time, both in money and in personnel. The study was directed by Roger Revelle and included scientists from government, universities, and all major oceanographic research institutions in the United States. The islands were to be used as test sites for atomic bombs directed at ships, and researchers wanted to study what such operations might do to the islands themselves. Once the testing was over, further studies followed. Finally a comprehensive, detailed picture of coral atolls was obtained for the first time, including their structure, their characteristic animal and plant life, and the oceanographic processes that form them. Once again, basic scientific discoveries had come out of a problem-oriented research project.

By this time many scientists and government officials realized that small, nationally-based projects probably could not continue to make major scientific contributions. Large-scale international cooperation was clearly the way of the future. A major step in this direction was taken by the International Council of Scientific Unions (ICSU) when it organized the International Geophysical Year of 1957–58.

The IGY team was made up of loosely coordinated scientific groups from 67 nations. They examined many kinds of physical phenomena during a period of intense sunspot activity. In oceanography several areas were emphasized. For mapping currents, a recent technological development, the *Swallow float,* was used to study water movements at any desired depth. This device is preset to maintain its position in water of given density, and it emits sounds by which it may be tracked from a surface ship. Subsurface currents can be mapped by following the float's path and averaging its movements over many days.

Tide gauges were used to study variations in average sea level, which may be due to seasonal effects, temperature, or winds. Other investigations included one on carbon dioxide content of deep ocean waters, in connection with atmospheric pollution, and another on movements of deep-ocean waters using radioactive carbon-14 as a tracer.

But the most exciting oceanographic result of the IGY turned out to be exploration of the sea floor. Intensive work on the topography of

Figure 1-15

Dr. W. Maurice Ewing with Lamont's research vessel, *Vema.* (Photograph courtesy of Lamont-Doherty Geological Observatory of Columbia University.)

Figure 1-16

As navigation officer aboard a World War II troop transport, Dr. Harry Hess took almost continuous echo soundings across large regions of the Pacific Ocean floor and discovered numerous flat-topped volcanoes. Later, as he continued to study these ancient "guyots," data on their age helped convince him that the sea floor moves and is continuously consumed at ocean margins. (Photograph courtesy Department of Geology, Princeton University.)

ocean basins had already been going on for some years. At Columbia University's Lamont-Doherty Geological Observatory, for instance, geologists had been making detailed maps of ocean basins; with Lamont's director, Maurice Ewing (1906–1974) (Fig. 1-15), they had developed theories about the great submarine mountain ranges, called *mid-ocean ridges,* that occur in major ocean basins. This work was compared with important new research on the earth's magnetism, its gravitational field, and the heat that flows from its interior. Data collected from all these studies seemed to relate to a problem that had intrigued scientists for hundreds of years. This was the possibility that the continents have changed their relative positions during geologic time, as suggested by the apparent "picture-puzzle" fit of the African continent against Central and South America.

In 1962, Harry H. Hess (1906–1969) of Princeton (Fig. 1-16) formulated a theory to explain how movements in the earth's mantle, a plastic layer below its rocky crust, can cause continents to move. Robert Dietz, a government geologist, separately arrived at a similar conclusion. Data from many fields of geological science, often previously confusing and apparently conflicting, have been found to fit this theory, known as *sea-floor spreading.* It states that the ocean floor is not a permanently fixed feature, but exists in a dynamic relationship with the mantle. Oceanic crust is continually being formed by submarine volcanic action at mid-ocean ridges, and destroyed at deep trenches near ocean basin margins.

One way to test the validity of this theory was to drill into the ocean floor for samples of deep fossils and crustal rock, and to determine the age of the material recovered. Scientists from Woods Hole, Lamont, Scripps, and the University of Miami agreed to cooperate on such a project, called JOIDES, for Joint Oceanographic Institution for Deep Earth Sampling. Later, other groups joined them. The National Science Foundation provided funds, and a specially built vessel, the *Glomar Challenger,* was outfitted to do the drilling (Fig. 1-17).

Figure 1-17

The *Glomar Challenger* was specially designed to maintain position and to drill in very deep water. Position is maintained by means of a sonar beacon; dropped to the ocean floor, its emitted signals are picked up by hydrophones below the ship's hull. Computers calculate the ship's position with respect to the beacon and automatically hold her on station while drilling is in progress. Samples of sediment and rock have been recovered from more than 1000 meters below the ocean bottom, and the vessel can operate in waters nearly seven kilometers deep.

Analyses of cores from this deep-sea drilling project have provided important evidence supporting the theory of sea-floor spreading. Absence of sediment over mid-ocean ridges and progressive thickening of sediment toward ocean basin margins suggest that ridges are sites of new-crust formation. Increasing age of the crust in direct proportion to its distance from the ridge is indicated by its increasingly thick sediment cover. (Photograph courtesy Global Marine, Inc.)

By dating the sediments that had accumulated on newly formed oceanic crust, the JOIDES team confirmed that the newest crust lies adjacent to mid-ocean ridges, and its age increases with distance from them. Since the 1960's, certain modifications have been made in details of this theory, but basically it is still similar to what Hess and Dietz proposed. Continuing work, such as the 1970's French-American exploration of deep trenches using submersible research vessels, is expected to cast new light on details of the mechanism by which oceanic crust is recycled.

MODERN TRENDS   What kinds of oceanographic research will be carried on in the final decades of this century? A look backward at the past ten or twenty years yields some clues. In general, the trend has been increasingly toward study of specific problems, especially involving recognized human needs, with experimentation replacing straight description of ocean processes.

In the U.S. there have been steady increases in marine science funding (Fig. 1-18). And during the decade 1965–74 there was a dramatic change in the activities supported by these funds. National security was the largest single category in the national program in the 1960's, replaced by oceanographic research in the 1970's (Table 1-1).

TABLE 1-1

U.S. Marine Science Program, 1974*

CATEGORY	FUNDING (millions of dollars)	PERCENT OF TOTAL FUNDING
Oceanographic Research	116.1	18.1
National Security	102.4	16.0
Mapping, Charting	89.3	13.9
Coastal Zone Development	88.8	13.8
Living Resources	82.1	12.8
Nonliving Resources	22.3	3.5
Other	140.8	21.9
Total	641.8	100.0

* Data from Annual Report of the President to the Congress, 1973. *The Federal Ocean Program,* Washington, D.C.

Throughout this period, the relative amount of funding for charting oceanic depths and shorelines remained nearly constant, as did the relative amount spent for research on living resources (fisheries). Most rapid increase in funding has been for support of coastal zone development and conservation (a 2.3-fold increase) and nonliving resources, such as sand and gravel production (a 3.2-fold increase).

On a global scale, an Intergovernmental Oceanographic Commission (IOC), with close to 100 member nations, has sponsored general oceanographic expeditions similar to those of the past. An example was the International Indian Ocean Expedition of 1962–63, intended to measure and describe physical and biological conditions in a poorly-known ocean area.

In the 1970's a series of Arctic-to-Antarctic cruises through the major oceans were made by two modern research vessels, but here the goal was to study a specific process rather than to survey an area. Scientists took more than 100 series of 50 samples each from surface to depth, to study vertical mixing processes in open oceans. One objective was to understand the ways in which plant nutrients move from deep

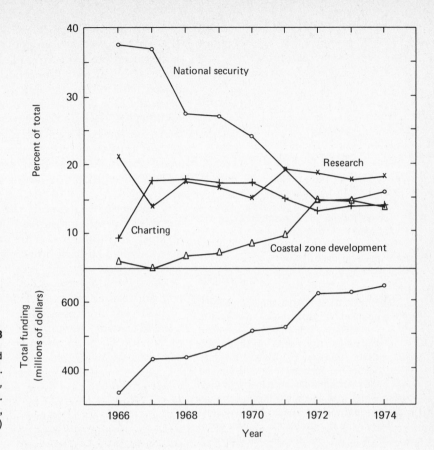

Figure 1-18

Funding of U.S. marine science programs and percent of total for some major categories. (Source: *Marine Science Affairs, 1967–1971, Federal Ocean Program 1972–1974.* Government Printing Office, Washington, D.C.)

to surface waters, thus insuring the sea's continued fertility. Distribution of radioactive substances and their behavior in the ocean was studied at the same time. This is significant because radioactive materials serve as tracers for measuring the rate and direction of diffusion in the oceans and they can be used to predict the effects of man-made radioactivity and other pollutants. Elaborate and sophisticated equipment (Fig. 1-19), both for sampling and for shipboard laboratory analyses, made this program a landmark in modern physical research.

Figure 1-19

The Wood's Hole Oceanographic Institution's *R.V. Knorr,* and Scripps Institution of Oceanography's *R. V. Melville* have deployed this underwater sensor package—a rosette of Niskin sampling bottles surrounding devices to measure conductivity (salinity), temperature, and transmission of light through water. (Photograph courtesy National Science Foundation.)

An International Decade of Ocean Exploration, endorsed by the U.N. and administered in the U.S. by the National Science Foundation, sponsors problem-oriented research. For example, in the early 1970's there were failures of important fisheries, such as the rich anchovy fishery off Peru. Changing climatic conditions caused an abrupt change in vertical water circulation, known as *upwelling,* so that surface water became depleted in nutrients and could not produce enough plant plankton to feed a large fish population. The IDOE study was directed toward predicting coastal upwelling and its effect on fish stocks. If a poor year is anticipated, it may be possible to prevent the fishery's collapse by prohibiting fishing in order to protect the small surviving stock.

Interest in large-scale climatic variation has been accelerated by the threat of widespread food shortages. Global food supplies could be seriously disrupted by fluctuations in temperature and rainfall, so that accurate prediction of future climatic variations is recognized as a vital need. To that end, the World Meteorological Organization's GARP (Global Atmospheric Research Programme) studies processes that control weather and cause long-term climate changes.

Oceans and atmosphere interact, forming a single, worldwide climate control system (Fig. 1-20). To learn the most important causes of climatic variation, scientists construct a continuous picture of global atmospheric conditions using weather satellites, deep-sea buoys, and computers to collect and correlate the data. Past climatic variations are analyzed statistically, using deep-ocean fossil records up to hundreds of thousands of years old to map ocean surface temperatures during the last Ice Age and earlier. From an understanding of past variations, scientists formulate numerical models of simulated climatic conditions to test hypotheses and predict how future climates will develop.

In the future we can probably expect more experimentation in the ocean. Recently, oceanographers have experimented with pollutants, adding controlled amounts to small, confined areas to measure their effect on biological communities. In one such study, huge (on the

Figure 1-20

Schematic illustration of the components of the coupled atmosphere–ocean–ice earth climate system. Heat, moisture, and momentum are exchanged at the sea surface; transport of heat and moisture occurs in atmospheric circulation and ocean currents. Changes in any of their relationships may cause significant variation in other parts of the system. Mathematical models are constructed to explore these adjustment mechanisms quantitatively. (From U.S. Committee for the Global Atmosphere Research Program, 1975. *Understanding Climatic Change: A Program for Action,* National Academy of Science, Washington, D.C.)

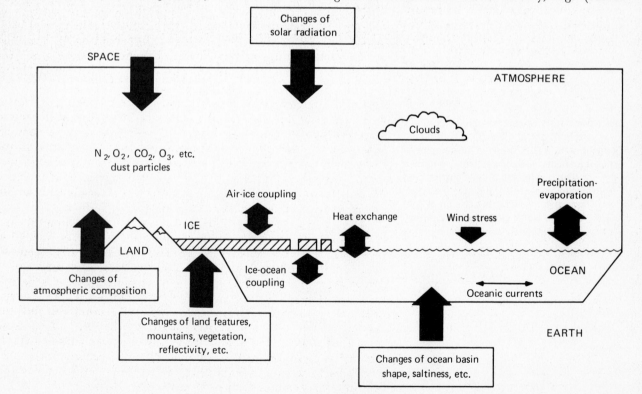

Figure 1-21

Plastic enclosures, several meters in diameter and tens of meters in depth, were collapsed on the bottom, then raised through the water to contain living communities in their natural state. Controlled experiments in pollution were performed in one cylinder, while another remained uncontaminated for comparison. (Illustration courtesy National Science Foundation.)

order of 10 meters by 30 meters) plastic cylinders (Fig. 1-21), open to the atmosphere but closed at the bottom, were suspended in the water at Saanich Inlet, Vancouver, B.C., where currents are weak, pollution levels low, and water is deeper than the levels at which plants can grow. Water and natural communities of organisms were trapped within the enclosures, so that as pollutants were added and their effects on the system assessed, scientists could judge what oils, heavy metals, or other pollutants can do to natural ecosystems.

Development of new tools and techniques will undoubtedly increase research possibilities. The Navy-developed Remote Underwater Manipulator (Fig. 1-22), for example, is an unmanned, submersible tractor with sampling devices and cameras to observe events on the deep-ocean floor. RUM is towed to an operation site by a bargelike vessel, from which it can be monitored and controlled. Difficult experiments are possible, such as measuring the respiration of an individual fish more than 1000 meters below the surface. RUM's cameras showed that many more kinds of animals exist at great depths than had previously been suspected. Entire deep-water communities can be studied with the aid of devices such as this, and samples can be taken with the same precision as if done in person on land.

23

Figure 1-22

This remote underwater manipulator moves about on the deep ocean floor and can perform televised experiments to be viewed from the towing vessel stationed above. (Photograph courtesy U.S. Navy.)

SUMMARY OUTLINE

History of oceanography—ocean studied to solve problems
> Ancient scientists studied the sea
> Explorers made charts of the ocean

Foundations of modern oceanography—early explorers knew little about the ocean
> James Clark Ross—explorer–scientist; charted waters near Antarctica
> Misconceptions delayed scientific progress—e.g., deep waters stagnant; no life below 600 meters
> Laying of transatlantic cable led to discoveries
> Voyage of the *Challenger*—Sir Wyville Thomson; beginning of modern oceanography
>> Studied marine life, ocean floors, ocean water properties
> Sir John Murray—published *Challenger* Reports, studied submarine geology

Early American oceanography—need to protect trade, passengers, aid shipping
> Matthew Fontaine Maury—Dept. of Charts and Instruments
>> Began program of Wind and Current Charts
> Alexander Bache—Coast Survey
>> Studied U.S. coastal waters, Gulf Stream
> Spencer F. Baird—Fish Commission
>> Founded Woods Hole laboratory; *Albatross* made scientific cruises; studied fish populations

National expeditions
> Prince Albert of Monaco—used personal fortune for oceanographic research; established oceanographic institute in Paris

> Scandinavian research—studied currents, temperature and salinity of water masses, in connection with fisheries research
> International Council for the Exploration of the Seas—studies of plankton, currents, water masses
> ICES-inspired research by Henry Bryant Bigelow—oceanography of Gulf of Maine
> German Antarctic Expedition of 1911–12; later *Meteor* Expedition
>> Studied temperature, salinity of Atlantic Ocean; sounded ocean floor
> Fridtjof Nansen drifted with *Fram* in pack ice
> International Ice Patrol—established after sinking of *Titanic*
> American research promoted with support from foundations, e.g., Rockefeller

Postwar oceanography
> War-related research in 1940's—development of bathythermograph, research on waves and coastal conditions, sea-floor mapping
> Government sponsored research for national security, mineral and fuel resources, fisheries
> Bikini–Eniwetok studies of coral atoll structure and formation
> International Geophysical Year—mapped currents, studied tides
> Sea-floor spreading theory—coordinated studies of earth's heat flow, gravity, magnetic field, ocean-basin topography
> Deep-Sea Drilling project drilled deep-ocean sediments, corroborated sea-floor spreading theory of moving continents, continuous formation of ocean floor

Modern trends

International Indian Ocean Expedition (1962–1963)—similar in scope to previous expeditions; studied an entire ocean in detail

Systematic chemical oceanographic studies of vertical mixing processes, to understand fertility of oceans, behavior of radioactive materials

International Decade of Ocean Exploration—problem-oriented research, e.g., coastal upwelling and fisheries studies

Global climate studies—interactions of ocean and atmosphere

Experiments—e.g., enclosing ocean areas to study effects of pollutants; remotely controlled underwater experiments

SELECTED REFERENCES

DEACON, MARGARET. 1971. *Scientists and the Sea, 1650–1900.* Academic Press, London. 445 pp. History of oceanography up to 20th century.

DIETRICH, GUNTER. 1968. *General Oceanography: An Introduction.* Interscience, London. 588 pp. Comprehensive, technical-level treatise, emphasizing physical oceanography.

GROEN, P. 1967. *The Waters of the Sea.* Van Nostrand, London. 328 pp. Physical oceanography, especially good section on sea ice.

PICKARD, GEORGE L. 1964. *Descriptive Physical Oceanography: An Introduction.* Pergamon Press, Elmsford, N.Y. 199 pp. Elementary level text, emphasizing physical oceanography.

SCHLEE, SUSAN. 1973. *The Edge of an Unfamiliar World: A History of Oceanography.* Dutton, New York. 398 pp. A readable history of oceanography since the nineteenth century, with emphasis on American oceanography.

SCIENTIFIC AMERICAN. 1971. *Oceanography.* W. H. Freeman, San Francisco. 417 pp. Selected articles from *Scientific American.*

SVERDRUP, H. V., MARTIN W. JOHNSON, AND RICHARD H. FLEMING. 1942. *The Oceans; Their Physics, Chemistry, and General Biology.* Prentice-Hall, Englewood Cliffs, N.J. 1087 pp. Classic oceanography text; detailed and scholarly.

WEYL, PETER. 1970. *Oceanography: An Introduction to the Marine Environment.* John Wiley, New York. 535 pp. Intermediate level treatment of oceanographic principles, with emphasis on physical processes and relationship to climate controls.

MOST RECENT TEN PERCENT OF EARTH'S HISTORY

Time Scale	Geo-logic Era	Fossil Record	Continental History	Time Scale	Continental History
Present	CENOZOIC	Rise of man	See detailed timetable	Present	
		Rapid rise of mammals			Arctic ice forms; major Rocky Mountain uplift
		First primates			Baja California separates from mainland, forming Gulf of California
50			Land bridge between Americas submerges		Antarctic ice sheet at maximum; Isthmus of Panama rising
	MESOZOIC	Flowering plants	North America and Europe separate	10	Mediterranean isolated, dries up; rapid growth of Himalayas
100		Continental flooding	India breaks from Australia and Antarctica		
			Africa and South America separate		
150		Age of dinosaurs	Andes form	20	Red Sea opens; Antarctic ice sheet forms
			Africa and North America move apart, splitting Pangaea		Africa moves northward; Alps rise
200		First mammals	Ocean between Europe and Asia closes; Urals form	30	Pacific Plate touches North America; alters western mountains
250	PALEOZOIC		Closure of ancient Atlantic		
300		Fern forest; coal deposits formed		40	
		Reptiles			Volcanic activity in western North America
350		Amphibians come on land	Appalachian Mountains develop		
		Age of fishes		50	India in contact with Asia
400		First vertebrates			
450			Ancient Atlantic shrinks	60	Australia separates from Antarctica

MILLIONS OF YEARS AGO

Ocean basin formation and movements of continental land masses have resulted from sea-floor spreading over the past 450 million years. Fossil records and geologic formations record the earlier positions of our present-day continents. (After Walter Sullivan, 1974. *Continents in Motion,* McGraw-Hill, New York.)

the earth and its crust

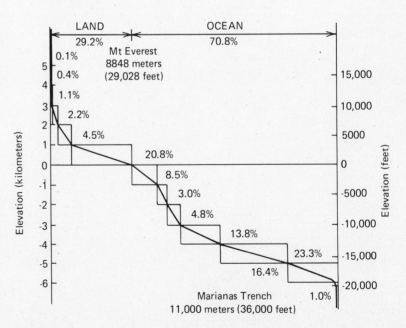

The ocean is Earth's most distinctive feature, covering the greater part of its surface in both Northern and Southern Hemisphere (Fig. 2-1, Table 2-1). Most of the land is in the Northern Hemisphere, with the Pacific, Atlantic, and Indian ocean basins extending northward as three "gulfs" from the ocean-dominated area around Antarctica. Sometimes a fourth—the Antarctic or Southern ocean basin—is designated. Its northern limit is set at 40°S latitude, where opposing currents cause a downward flow of surface waters known as the *subtropical convergence*.

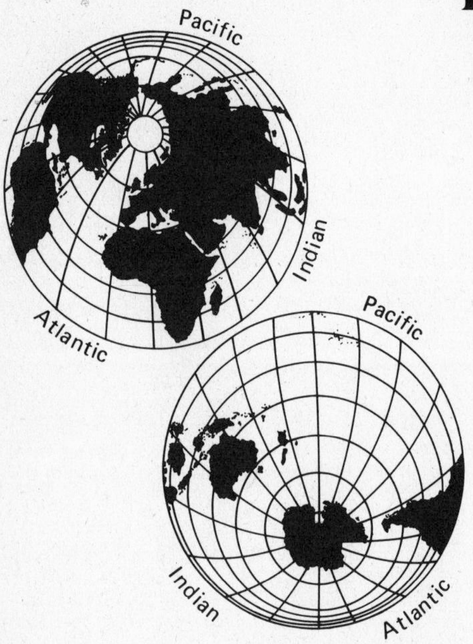

Figure 2-1 *(above)*

Land and water are unequally distributed; most land occurs in the Northern Hemisphere while the Southern Hemisphere is dominated by water.

TABLE 2-1

Distribution of Land and Water On the Earth's Surface*

HEMISPHERE	LAND (percent)	OCEAN (percent)
Northern	39.3	60.7
Southern	19.1	80.9

* After Sverdrup, Johnson, and Fleming, 1942. p. 13.

Figure 2-2 *(right)*

A hypsographic curve shows the relative proportions of the earth's surface at elevations above and below sea level. The length of each bar shows the relative proportion within each depth zone. (From Arthur N. Strahler, 1960. *Physical Geography*, 2nd ed. John Wiley, London.)

The continents form four great land masses—Eurasia–Africa, the Americas, Antarctica, and Australia—isolated in the midst of the world ocean. Though they may be joined by "land bridges," such connections tend to be narrow and easily flooded by even small rises in sea level. For example, land connected Alaska to Siberia during the last Ice Age more than 20,000 years ago, but the connection was broken when glacial melting freed the North American and Eurasian continents of ice and raised the sea level so that continental margins were flooded.

Relationships between oceanic depths and land elevations can be conveniently visualized as a *hypsographic curve* (Fig. 2-2). If the earth were a smooth sphere with the land planed off to fill the basins, it would be covered with water to a depth of 2430 meters. On the average, land projects about 850 meters above sea level and ocean basins lie an average 3730 meters below it. But despite the great depth of the oceans and the enormous height of the mountains when compared to the height of a human being, differences in elevation or depth for the earth as a whole are truly insignificant when compared to the dimensions of the earth. Viewed in this way, the ocean is a thin film of water on a nearly smooth globe.

STRUCTURE OF THE EARTH

While water covers two-thirds of the earth's surface, it accounts for only a small fraction of its total mass (see Appendix 2). Basically, the planet consists of concentric spheres (Fig. 2-3):

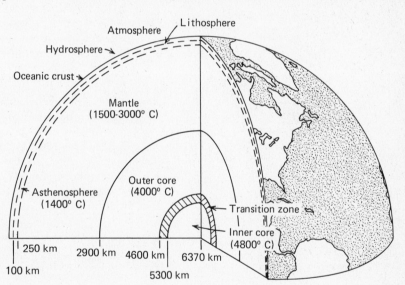

Figure 2-3

Simplified view of the earth's layered structure. (After P.J. Wyllie, 1975. "The Earth's Mantle," *Scientific American*, 232(3), 50–63.)

Core rich in iron and nickel, it is magnetized and very dense. The inner core is solid, the outer core liquid, with a transition zone between

Mantle a less dense layer of rock. The lower mantle is thought to be essentially rigid, but the upper mantle or *asthenosphere* is softer and can flow very slowly.

Lithosphere a less dense, rigid outer shell, 60–100 kilometers thick, including granitic *continental crust* as well as sediment-covered, basaltic *oceanic crust*. The crust is underlain by and apparently fused with a layer of heavier basalt, presumably solidified upper mantle material. These layers are separated at a boundary known as the *Mohorovičić discontinuity* (or *Moho*) which occurs under continents at a depth of about 25–40 kilometers and under ocean basins at 5–10 kilometers, typically 7 kilometers (Fig. 2-4).

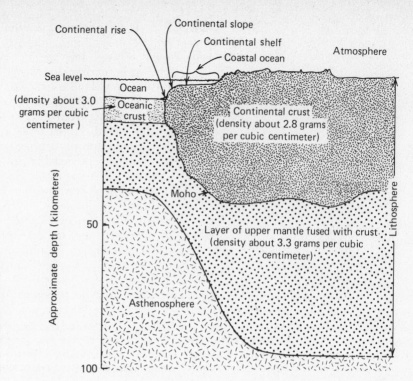

Figure 2-4

Schematic representation of the relationships between continental blocks, oceanic crust, lithosphere, and upper mantle. Note that the more buoyant continental mass is supported by a thickened lithosphere below and a corresponding depression of the more dense but plastic mantle layer beneath.

Hydrosphere all free water (Fig. 2-5) including the water that has been removed from the ocean for millions of years by incorporation in sediments and sedimentary rocks. Eventually this water also returns to the ocean, which contains about 98 percent of the earth's free water (Table 2-2).

Atmosphere primarily nitrogen, oxygen, and inert gases, mixed with variable quantities of water vapor and carbon dioxide.

Continental crust occurs as large blocks, which do not end at shorelines (see Fig. 2-4). Continental surfaces slope gently from about 1 kilometer above sea level to about 200 meters below it. The submerged portion or *continental shelf* underlies a narrow *coastal ocean* zone adjacent to the continents (Fig. 2-6). Beyond that, there is a sharp

TABLE 2-2

*Volume of Major Water Reservoirs of Earth's Crust and Mantle ***

RESERVOIR	VOLUME (millions of cubic kilometers)	(percent)
Earth's surface		
Ocean	1370	98.0
Sedimentary rocks (waters actively exchanged)	4	0.3
Glaciers	24	1.7
Lakes	0.23	0.02
Atmospheric water vapor	0.014	0.001
Rivers	0.0012	0.00009
Total	1398.245	100
Mantle (0.5% water)	13,000	

* Source: M. I. Lvovich, 1973. *The World's Water,* Mir Publishers, Moscow.

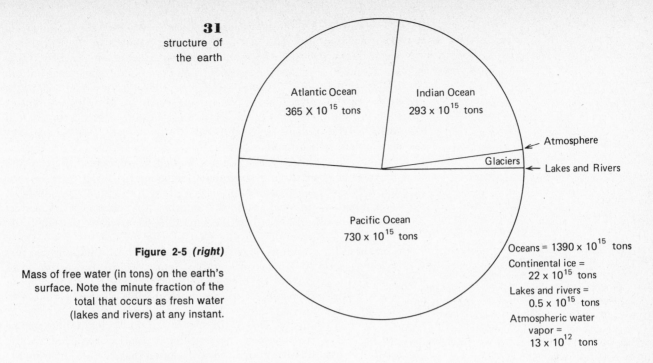

Atlantic Ocean
365 X 10^{15} tons

Indian Ocean
293 x 10^{15} tons

Atmosphere

Glaciers

Lakes and Rivers

Pacific Ocean
730 x 10^{15} tons

Oceans = 1390 x 10^{15} tons

Continental ice =
22 x 10^{15} tons

Lakes and rivers =
0.5 x 10^{15} tons

Atmospheric water
vapor =
13 x 10^{12} tons

Figure 2-5 *(right)*

Mass of free water (in tons) on the earth's surface. Note the minute fraction of the total that occurs as fresh water (lakes and rivers) at any instant.

Figure 2-6 *(below)*

Map showing relationship of land masses to the 2000-meter depth contour which marks the approximate outer limits of the continental blocks, and the 4000-meter contour that generally outlines the deep-ocean basins and the mid-ocean ridge system. (From J. E. Williams, ed., 1963. *Prentice-Hall World Atlas,* 2nd ed., Prentice-Hall, Englewood Cliffs, N.J.)

dropoff known as the *continental slope,* ending at about 2 kilometers depth (Fig. 2-4), the outer limit of continental blocks. The juncture between continental mass and ocean floor is covered by a thick apron of sediment eroded from the surface of the continent. This is the *continental rise,* which slopes gently to about 4 kilometers below sea level. Most deep *ocean basins* range in depth from about 4 to 6 kilometers, with some narrow trenches near Pacific ocean basin margins dropping to as much as 11 kilometers below sea level.

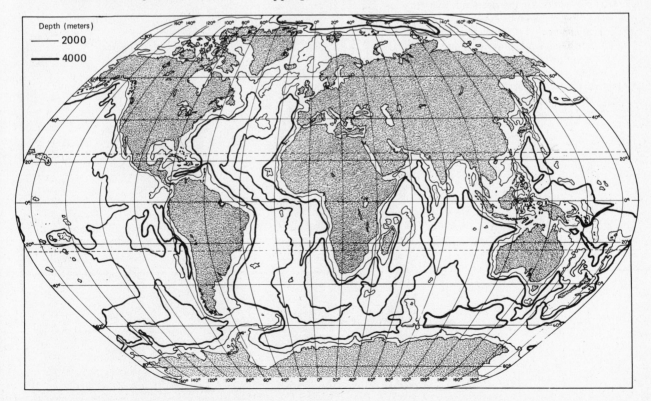

Depth (meters)
2000
4000

Continents are thick accumulations (35 to 40 kilometers) of granite containing abundant silicon and aluminum and having a mean density of about 2.8 grams per cubic centimeter. Granitic rocks are formed of material released from the mantle during mountain building. Because of their relatively low density, they tend to remain at the earth's surface. Some continental rocks are known to be nearly four billion years old. Wind and water constantly erode continental surfaces, transporting soils and silts to the ocean floor where they accumulate as sediment deposits.

Oceanic crustal rocks contain a higher proportion of iron and magnesium than do the continents; some of its basaltic rocks have densities on the average of about 3.0 grams per cubic centimeter. A layer about 7 kilometers thick underlies the ocean basins. Direct sampling has been limited by the hundreds of meters of sediment and thousands of meters of seawater that overlie the oceanic crust. Deep-ocean drilling, however, has supplied valuable information; samples have been taken from considerable depths below the ocean floor (Fig. 2-7). Long cores of oceanic sediment provide information on ocean basin history and composition comparable to that available to geologists working on land.

The lithosphere "floats" in the less rigid asthenosphere below (see Fig. 2-4). The surface of the continental blocks extends above the oceanic crust because granite is less dense than basalt. Continents have their "roots" deeply submerged in the heavier upper mantle, just as an iceberg floats in equilibrium with the slightly denser water

Figure 2-7

A core of altered igneous rock (solidified from a melt) overlain by white marble (an altered sedimentary rock). The core was obtained by drilling operations in water 4700 meters deep in the western North Atlantic Ocean. This rock sample, taken after drilling 450 meters of sediment almost 100 million years old, provides a direct means of investigating ocean-bottom structure and ocean-basin history. (Photograph courtesy Deep-Drilling Project, Scripps Institution of Oceanography, under contract to the National Science Foundation.)

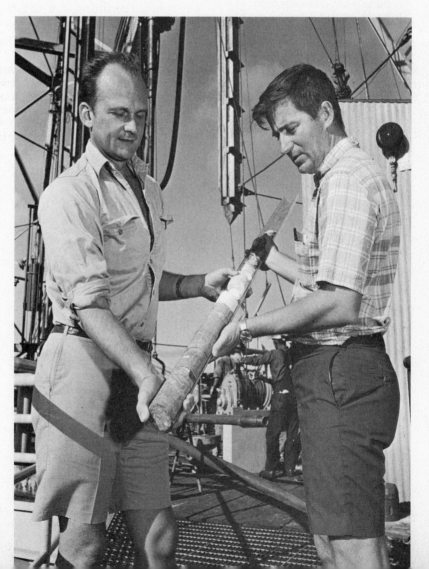

medium by having most of its bulk below the water line. A concept of gravitational equilibrium known as *isostasy* predicts that the earth's crust behaves like a series of large blocks, each of which "floats" in isostatic equilibrium with the plastic or deformable deeper layer. Therefore the shape of the Moho below a continent is a mirror image of the surface relief, exaggerated several times. Furthermore, isostatic adjustment takes place whenever a portion of the crust becomes heavier or lighter. If an appreciable mass is removed from part of a continent or ocean basin, that part rises to maintain a constant ratio of exposed to submerged material. For example, parts of Scandinavia and North America are rising as much as one meter per century, due to melting of glaciers within the past several thousand years that had previously weighted down the crust. Conversely, whenever a volcano builds a mountain on the ocean floor its great mass achieves isostatic equilibrium by sinking deeper into the underlying material. Many volcanoes that once formed islands have disappeared below sea level through this process.

METHODS OF PROBING THE EARTH

The earth's interior is studied using primarily indirect techniques. One of the most useful is *seismology,* in which the speed and direction of waves from earthquakes or explosions are studied as they pass through the earth's layers. Earthquake-generated waves change their speed and direction according to the density, elasticity, and flow properties of the rocks through which they pass. Certain types of waves (*compressional waves*) cause the medium to compress and expand; these pass through liquids and solids. Others, known as *shear waves,* distort the shape of a material but do not change its volume. These can pass through high-strength, nondeformable substances, but not through liquids. Both types of waves are slowed down when passing through rocks of lowered density, through hotter layers, or through low-strength materials which can flow or be very slowly deformed. High pressures increase wave speed, but tension—the stress that occurs when materials are being pulled apart—lowers it.

Waves are partly bounced back or *reflected* (Fig. 2-8(a)) from a *discontinuity* (boundary) between layers, and partly transmitted. If wave velocity is different in the new layer, the waves are bent or *refracted* (Fig. 2-8(b)). The Moho, for example, was recognized from seismic observations; compressional waves from earthquakes abruptly change speed there, from 6.1 to 6.7 kilometers per second in the crust, to about 8.1 kilometers per second in the heavier lithosphere below.

Two other physical properties are used to study the earth's structure and composition—its gravity and the flow of heat from its interior. Variations of mass anywhere on the earth cause disturbances in the normal gravity field known as *gravity anomalies.* More dense materials exert a locally greater gravitational force, while less dense rocks weaken the gravity above them. The size and depth of a subsurface anomaly can be calculated if its density is known; for example, a relatively dense volcanic mass that appears as a mountain on the ocean floor may have a deep root in the oceanic crust that is detectable by a gravity meter as a positive anomaly. A negative gravity anomaly occurs where an area of thick crust provides buoyancy below mountains that are not sinking, but "floating" in isostatic equilibrium. The anomalous negative value is caused by displacement of heavier mantle material by the less dense crustal root.

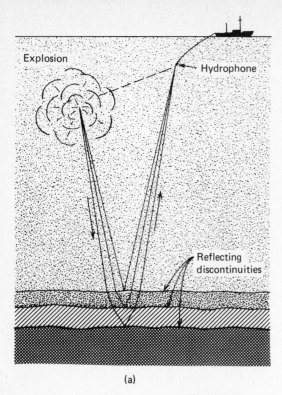

Shooting ship | Recording ship

Explosion
Hydrophone
Sediments
Volcanic layer
Mantle

(b)

Figure 2-8

(a) In seismic-reflection surveys, a ship drops an explosive charge at some preset point. Sound pulses reflected from different layers are picked up by sensitive microphones—*hydrophones*—towed by a ship. These signals are recorded and analyzed to determine depths to the reflecting rock units.
(b) Seismic-refraction studies commonly use two widely separated ships, one to release the explosive charge, the second to receive and record the signals. Sound waves from the source travel along the layers and their boundaries. They move up through the overlying layers to be received at a second ship.

Heat flows from continental blocks and ocean basins at an average of about 1.5 microcalories per square centimeter per second (about 1/3000 the average amount of energy from the sun at the ocean surface). Heat flow is measured by inserting sensitive thermometers below the surface and measuring the rate of increase in temperature with depth. This value, multiplied by the thermal conductivity of the rock, gives the outward heat flow per unit area and unit time. Heat is produced in rocks by decay of radioactive isotopes of thorium, uranium, and potassium, and this may also be the source of the earth's high interior temperatures. But although the basalts of the oceanic crust contain less radioactive material than continental granite does, and oceanic crust is only about one-fourth as thick as continental crust, the average heat flow from ocean basins is about the same as from continents.

But note that oceanic heat flow is highly variable (Table 2-3). It is almost twice as great as average near mid-ocean ridges (see Fig. 2-6) and below average at the deep trenches that border the Pacific basin. Hot material rising through the asthenosphere from deep within the mantle may cause locally high heat flows, for instance at mid-ocean ridges. Very high flows can be explained as indicating pockets of molten rock close beneath the ocean floor. The molten rock might be due to locally high concentrations of radioactive elements or to local upward movements of hot material. Such areas, particularly where they occur in inhabited regions such as Iceland or Central America, are potential sources of geothermal power.

Clues as to when and how a surface feature formed are found in iron-rich rocks, which become magnetized in the direction of the

TABLE 2-3
Heat Flow Through the Earth's Crust

AREA	HEAT FLOW (microcalories per square centimeter per second)
CONTINENTS	
Old continental areas (older than 600 million years)	0.9 ± 0.2
Old mountain areas (from 600 million to 230 million years)	1.2 ± 0.4
Younger mountain belts (younger than 230 million years)	1.9 ± 0.5
Undeformed areas (no deformation in past 600 million years)	1.5 ± 0.4
OCEANS	
Trenches	1.0 ± 0.6
Basins	1.3 ± 0.5
Oceanic ridges	1.8 ± 1.6

After W. H. K. Lee and S. Uyeda, 1965. "Review of Heat Flow Data," in *Terrestrial Heat Flow*, Geophysical Monograph, American Geophysical Union, Washington, D.C., no. 8, p. 87.

Figure 2-9

Time scale for reversals of the earth's magnetic field. The polarity (direction of the magnetic field) is shown for the past 4 million years. (After A. Cox, R.R. Doell, and G. B. Dalrymple, 1964. "Reversals of the Earth's Magnetic Field," *Science*, 144, 1537–43.)

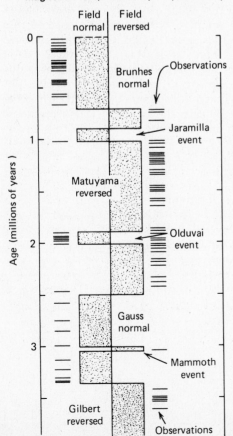

earth's magnetic field at the time of their formation. Oceanic crust, for example, is formed when iron-rich lavas flow upward from the earth's interior and solidify at the surface. As molten rock cools below the 600°–700°C temperature range, iron minerals become permanently aligned with magnetic North.

The earth's magnetic field, created by circulation of core materials, has reversed itself many times in the past. This has occurred at intervals on the order of half a million years (Fig. 2-9), with each reversal probably taking a few thousand years. During periods of so-called normal magnetic orientation the north-seeking end of a compass needle would behave as it does now. But during periods of reversed magnetism, the north-seeking end would point toward our present south. The permanent magnetism of a single rock can be determined in the laboratory. On a larger scale, instruments can be towed by aircraft or ship to map the earth's magnetic field over wide areas.

Such a survey over a portion of the Atlantic Ocean basin revealed the striped pattern shown in Fig. 2-10. Each stripe corresponds to a period of normal or reversed magnetism, as illustrated on the chart in Fig. 2-9, and each stripe on the ocean floor can be dated according to the chart. Thus the ocean basin floor acts like a gigantic tape recorder, the tape being the newly formed crust and the signal being the earth's magnetic field at the time of crustal formation. Data for interpretation of the pattern has been extracted from lavas of known ages on the continents. Comparison of the pattern in the ocean's magnetic stripes with this dated pattern in terrestrial lavas permits ages of various parts of the ocean bottom to be estimated. On this evidence it is estimated that about 50 percent of the deep-ocean bottom has formed during the past 70 million years.

35

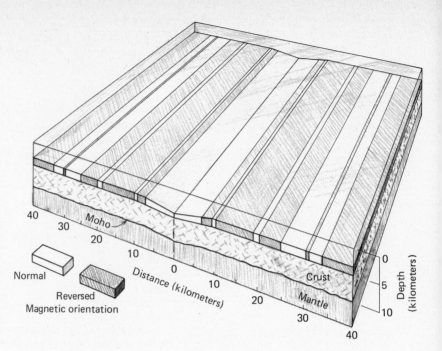

Figure 2-10

A striped pattern caused by reversals of the earth's magnetic field appears in this survey of the North Atlantic Ocean southwest of Iceland over a portion of the Mid-Atlantic Ridge. Areas of positive anomalies (unusually high intensity) are shown in white; areas of low intensity are shown in black.

SEA-FLOOR SPREADING

Interpretation of geophysical data has changed radically in recent years. Before 1960, geologists thought that the earth's crust was a rigid, stationary layer surrounding a nearly molten interior. But as data on the heat flow, topography, and magnetic patterning of ocean basins were collected and coordinated in the 1950's and 1960's, a different picture began to emerge. According to the new, dynamic view, the rigid lithosphere and soft mantle below are linked and slowly moving, giving rise to complex processes that shape continents and ocean basins.

The lithosphere may be compared to an eggshell cracked into six major plates and several smaller ones (Fig. 2-11). Divisions between plates occur at three kinds of boundaries (Figs. 2-12 and 2-13): at *mid-ocean ridges* and *rises* (see Fig. 2-12); at *trenches* that ring the Pacific Ocean basin; and at cracks in the crust known as *faults*. The plates move away from each other (*divergence*), toward each other (*convergence*) or past each other at rates shown in Fig. 2-12, in response to very slow movements in the mantle below.

Figure 2-11

An artist's "front" (a) and "rear" (b) view of the earth, divided into its major plates, showing how their boundaries cut across ocean basins and continents: (1) Pacific Plate, (2) Antarctic, (3) Nazca, (4) South American, (5) African, (6) North American, (7) Cocos, (8) Caribbean, (9) Eurasian, (10) China, (11) Australian.

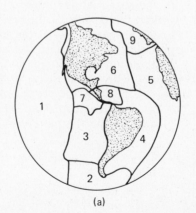

(a)

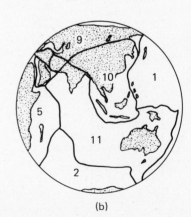

(b)

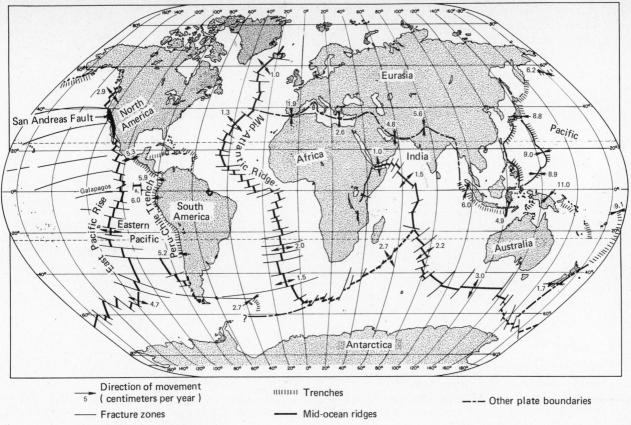

Direction of movement
5 (centimeters per year)

||||||| Trenches

--- Other plate boundaries

—— Fracture zones

—— Mid-ocean ridges

Figure 2-12

Connected oceanic rises form a *mid-ocean ridge system* that generally bisects ocean basins. Deep *trenches* occur mainly around Pacific margins. Lithospheric plates including continental and oceanic crust move generally away from mid-ocean ridges and rises, and toward trenches where subduction takes place. Faults occur where plates move past each other.

New crust is formed when molten rock or *magma* moves upward from the earth's interior and reaches the surface at mid-ocean ridges and rises. Through continuing volcanic activity, about 12 cubic kilometers of new oceanic crust are formed each year as magma fills the cracks formed when plates are pulled or pushed apart. Reduced pressure near faults or cracks permits rocks to melt at lower temperatures, and the faults provide routes for lava to move to the surface. Surveys of magnetic patterning in rocks on either side of the Mid-Atlantic Ridge show that rock age increases with distance from the ridge crest (Fig. 2-14). Rate of crustal generation is determined from the amount of material created per unit time. The process does not proceed at a uniform rate at all points along a ridge, and consequently the sections of a rise are offset at intervals as shown in Fig. 2-12. Where sections have moved past each other, a *transform fault* connects the ends of the offset (Fig. 2-13).

Since the earth is not expanding, oceanic crust must be destroyed at about the same rate as it is formed. Crust is destroyed by being *subducted*, or drawn down, and resorbed into the mantle at the trenches. Continental crust as such is not formed at mid-ocean ridges and rises, nor is it subducted at trenches, but the relative position of continental masses changes as a result of lithospheric plate movements. Over the entire earth, the average rate of ocean floor movement from ridges and rises toward trenches is about 2.5 centimeters per year. For example, the Atlantic Ocean has only two small trench systems but an extensive mid-ocean ridge. Consequently it is growing wider by about 4 centimeters per year. In the process, Europe, Africa, and the Americas are being dragged farther from the Mid-Atlantic Ridge. Plates move at rates as fast as 10 centimeters per year in the Pacific basin and as slow as 1–2 centimeters per year in the Atlantic (Fig. 2-12).

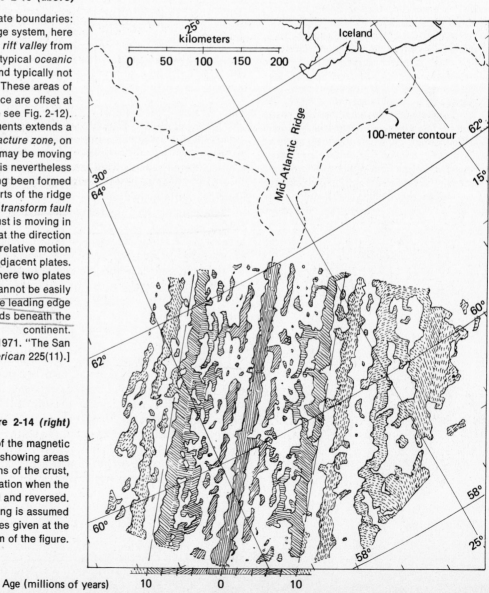

★ ★ ★ ★ ★ = Earthquakes

Ocean

Oceanic crust

Rift Valley
(area of divergence)

Mid-Ocean
Ridge or Rise

Plate

Plate

Fracture Zone

Oceanic crust

Continent at forward
edge of plate

Transform Fault

Plate

Plate

Oceanic crust

Oceanic crust

Trench (area of convergence)

Magma

Figure 2-13 *(above)*

Three types of plate boundaries:
(a) The mid-ocean ridge system, here
shown with a steep central *rift valley* from
which magma issues. The typical *oceanic
rise* is gentler in relief and typically not
topped by a pronounced rift. These areas of
spreading or divergence are offset at
intervals (also see Fig. 2-12).
(b) Beyond the offset segments extends a
type of fault known as a *fracture zone,* on
each side of which material may be moving
in the same direction but is nevertheless
discontinuous, not having been formed
at the same time. The two parts of the ridge
crest are connected by a *transform fault*
where newly formed crust is moving in
opposite directions. Note that the direction
of a fault indicates the relative motion
of adjacent plates.
(c) A trench occurs where two plates
converge. Continental crust cannot be easily
subducted, so the plate whose leading edge
is of oceanic crust descends beneath the
continent.
[After D.L. Anderson, 1971. "The San
Andreas Fault," *Scientific American* 225(11).]

Figure 2-14 *(right)*

Diagrammatic representation of the magnetic
field at the Mid-Atlantic Ridge showing areas
of different magnetic orientations of the crust,
resulting from crustal formation when the
earth's magnetic field is normal and reversed.
A constant rate of spreading is assumed
in estimated crustal ages given at the
bottom of the figure.

kilometers

0   50   100   150   200

Iceland

Mid-Atlantic Ridge

100-meter contour

Age (millions of years)   10   0   10

Figure 2-15

An artist's concept of recent continental movements that have given rise to their present distribution over the globe: (a) *Pangaea,* the single continental land mass that existed around 250–450 million years ago. The earth's other side was wholly ocean. (b) About 200 million years ago, Pangaea began to break up. Antarctica and Australia were one land mass, at the south end of the map. India was detached, as a large island. (c) As the Atlantic Ocean developed, it was wider in the north because that was where it had begun to form. Note that India is moving toward Asia. (d) Sixty million years ago most of the land was still in the Northern Hemisphere. (e) The earth at present. (f) Predicting from present plate movements, fifty million years from now a small part of Eastern Africa will have broken off and the Red Sea will be wider, both due to spreading along the East African Rift Valley (see Figure 2-17). (g) Northwestward Pacific Plate movement will have detached Baja California from the rest of the North American continent.

Several prominent types of global surface features show how generation and movement of lithospheric plates have shaped present ocean basins. The six great continental land masses, now widely separated over the globe, are each surrounded on all sides by ocean, but they have not always been so. Apparently about 225 million years ago, there was only a single land mass or supercontinent, called *Pangaea* after the Greek earth goddess (Fig. 2-15 (a)). There is considerable evidence for this theory, even without recent geophysical data. The fit of the present continental blocks against one another is excellent, like pieces of a jigsaw puzzle, the few overlaps or gaps are attributable to well-understood processes; and the distribution of certain types of rocks in South Africa appears to be identical with similar rocks of comparable ages found in South America.

About 180 million years ago, Pangaea began to break up, initially forming a long, thin, east-west trending basin, like the Red Sea or the Gulf of California (Fig. 2-15 (b)). Two subcontinents, designated *Laurasia* and *Gondwana,* were separated by this Sea of Tethys. Then Antarctica and Australia moved southward, perhaps connected to each other for a while. Around a hundred million years ago North America, followed by South America, was split off by the development of a north-south ridge system under Laurasia, then Gondwana. The Americas moved westward as the Atlantic Ocean opened between them and the remaining continental mass; this process is still happening. Finally, India separated from Africa, and moved rapidly northward to collide with Southern Asia. That collision caused India to move under part of Asia, buckling the land above and creating the Himalayas, the world's highest mountains. Today the Pacific Ocean and the Mediterranean Sea are closing, the Atlantic and the Red sea are becoming wider and Africa, underlain by a developing rise system, is apparently going to break into two parts (Fig. 2-15 (f)).

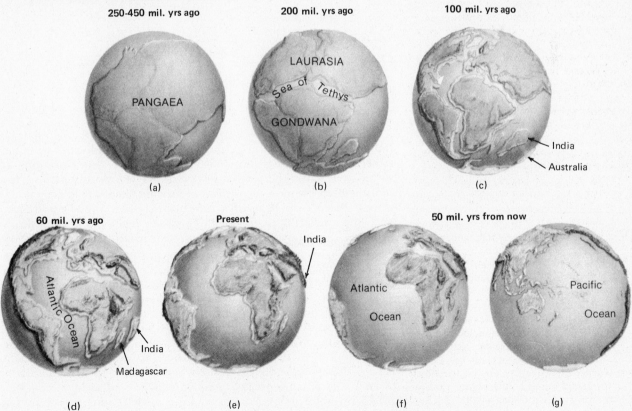

250-450 mil. yrs ago — PANGAEA (a)

200 mil. yrs ago — LAURASIA / Sea of Tethys / GONDWANA (b)

100 mil. yrs ago — India, Australia (c)

60 mil. yrs ago — Atlantic Ocean, India, Madagascar (d)

Present — India (e)

50 mil. yrs from now — Atlantic Ocean (f)

Pacific Ocean (g)

No part of the present ocean floor is older than 225 million years (Fig. 2-16). But there is no reason to assume that movement of crustal plates has occurred only in the past 225 million years. Evidence indicates that an ancient Atlantic Ocean separated Europe and America during the early Paleozoic era, more than 450 million years ago, and was later closed only to be replaced by the present Atlantic (see Chapter 2 frontispiece). Many other major rearrangements of land masses also must have taken place in the more than three billion years that continental material has floated at the earth's surface.

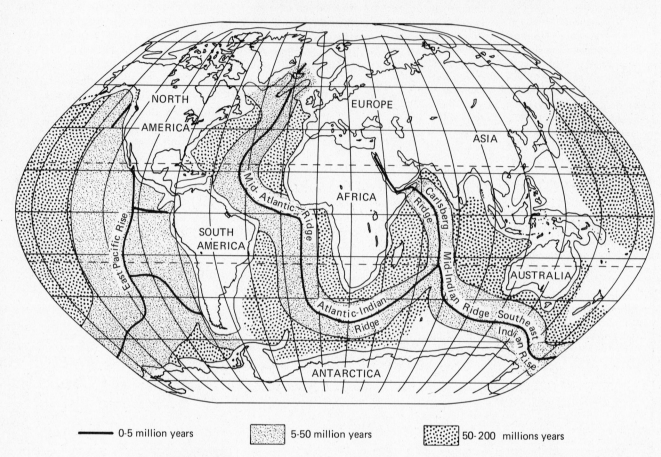

——— 0-5 million years   ▦ 5-50 million years   ▦ 50-200 millions years

Figure 2-16

The ages of ocean floors were determined on the basis of magnetic anomaly profiles, made by towing magnometers across ocean basins behind ships or airplanes. Points in the magnetic reversal sequence were then dated by Deep-Sea Drilling Project teams aboard *D/V Glomar Challenger* and found to coincide closely with the magnetic data. Note that in the Atlantic the age of the ocean floor is symmetrical on either side of the Mid-Atlantic Ridge, but that a large area of the Eastern Pacific has been subducted. Compare with Fig. 2-12, where rates of motion are indicated, and with Fig. 2-17. There is a recent rift at the Galapagos volcanic center in the equatorial Pacific. In the Indian Ocean, three spreading centers join to form an inverted Y whose northern branch causes rifting in the Gulf of Aden and the Red Sea.

What causes sea-floor spreading? It seems probable that heat from the inner mantle gives rise to convective movements within the asthenosphere, causing hot material to rise in some locations and spread out across the top of the plastic layer. As this moving layer cools, it becomes more dense and eventually sinks. Some geophysicists believe that as the edge of a lithospheric plate sinks at a trench it pulls the rest of the plate along. This is analogous to a towel floating on water, one end of which has lost the surface tension that kept it afloat so that it sinks and drags the rest of the towel after it.

Another possibility is that upward moving, hotter, and therefore less dense material from deep layers pushes up the lithospheric plates at mid-ocean ridges and rises. This would explain the buoyancy of the mid-ocean ridge system, for if the rocks beneath had the same density as ordinary oceanic crust, presumably the huge mountain range system would attain isostatic equilibrium by sinking or flattening out to a depth uniform with the rest of the deep-ocean floor. At the leading edges of lithospheric plates, on the other hand, the thick accumulations of land- and ocean-generated sediments that lie adjacent to continents may depress the plates. According to this model, young crust is elevated

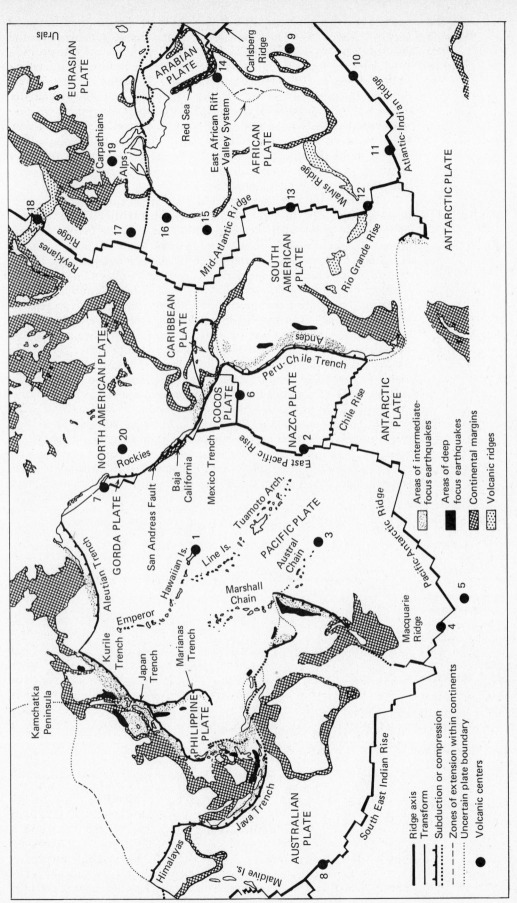

Figure 2-17

Worldwide lithospheric plate system, showing some centers of volcanic activity. Three prominent Pacific seamount chains, most of which do not now appear above sea level, demonstrate Pacific Plate movement away from thermal centers, or "hot spots." Submarine volcanic ridges, extending northeast and northwest from Iceland and Tristan da Cunha, and northwest from Reunion Island, were produced by long continued volcanic eruptions at the "hot spots" now marked by those islands. Note that transform faults, oriented in the direction of plate movement, extend perpendicular to ridge axes. Deep and intermediate depth earthquakes are associated with trenches. (After J. F. Dewey, 1972. "Plate Tectonics," *Scientific American*, 226(5), 56–58.)

Legend:

1. Hawaii
2. Easter Island
3. Macdonald Seamount
4. Bellany Island
5. Mt. Erebus (Antarctica)
6. Galapagos Islands
7. Cobb Seamount
8. Amsterdam Island
9. Reunion Island
10. Prince Edward Island
11. Bouvet Island
12. Tristan de Cunha
13. St. Helena
14. Afar
15. Cape Verde
16. Canary Islands
17. Azores
18. Iceland
19. Eifel
20. Yellowstone

Ridge axis
Transform
Subduction or compression
Zones of extension within continents
Uncertain plate boundary
Volcanic centers

Areas of intermediate-focus earthquakes
Areas of deep focus earthquakes
Continental margins
Volcanic ridges

relative to older portions, and gravity therefore acts to drag the lower edges of plates "downhill" to sink at the trenches that occur in the deepest parts of ocean basins.

Another line of reasoning suggests that a "plume" of magma rises from deep within the mantle at each of twenty or more major centers of prolonged volcanic activity (Fig. 2-17) and spreads radially beneath the lithosphere. Many such volcanic centers occur at or near the present ridge system. Other active areas may mark plumes whose force has been insufficient to crack the crust in a line along which magma could come to the surface. When large scale rifting of plates occurs, a new ridge system forms. The sea-floor spreading continues, but with plate movements occurring in different directions.

Evidence for this theory is found in the chains of volcanic seamounts illustrated in Fig. 2-17. In each chain, the islands become progressively older with increasing distance from the active volcanic center. To picture their formation, go back about 70 million years. Imagine a "hot spot" or plume of rising molten basaltic rock in the mantle under Hawaii, under Easter Island, and under the Macdonald Seamount, a submerged mountain (Fig. 2-17). Then picture the Pacific Plate moving almost due north over those "hot spots" for tens of millions of years. During this time, periodic volcanic eruptions have built several submerged volcanic chains, the Emperor Seamounts, the Line Island chain and the Marshall-Ellice Islands (Fig. 2-18). But during that period, and subsequently, erosion has worn them down above sea level, and isostatic adjustment has caused them to sink. Now only the tops of some of the largest volcanoes appear as tiny islands.

After these northward-trending island chains were built, imagine that the movements of the Pacific Plate shifted direction. Since then it has moved generally northwestward, as indicated by the arrows in Fig. 2-12, perhaps because of tension where the corners of plates collide. Subsequently, the Hawaiian, Tuamoto, and Austral Islands formed. Figure 2-18 shows these chains in detail, plus a smaller group

Figure 2-18

Pacific seamount and island chains. Most of the volcanoes do not reach sea surface; only a few form islands. Many volcanoes in these chains occur in pairs. This evidence suggests that the processes causing the volcanic center or plume affects an area some 300 kilometers in diameter. (After G. B. Dalrymple, E. A. Silver, and E. D. Jackson, 1973. "Origin of the Hawaiian Islands," *American Scientist*, 61(3) 294–308.)

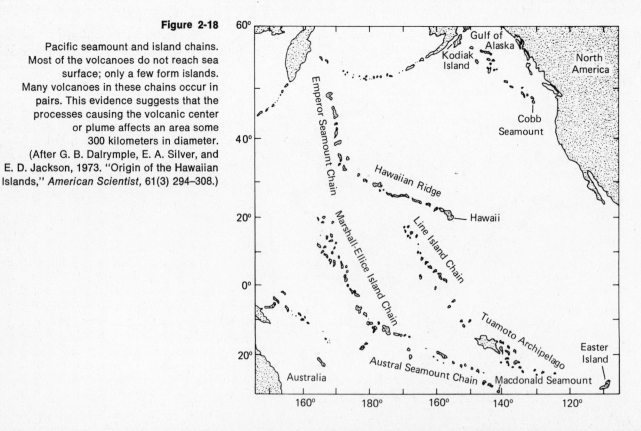

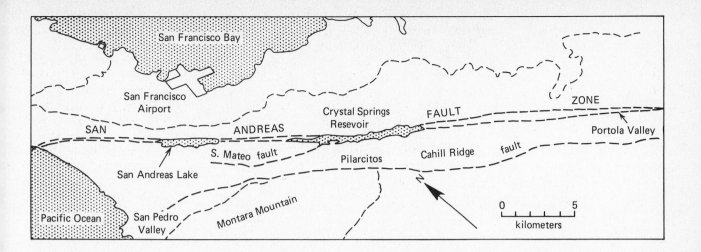

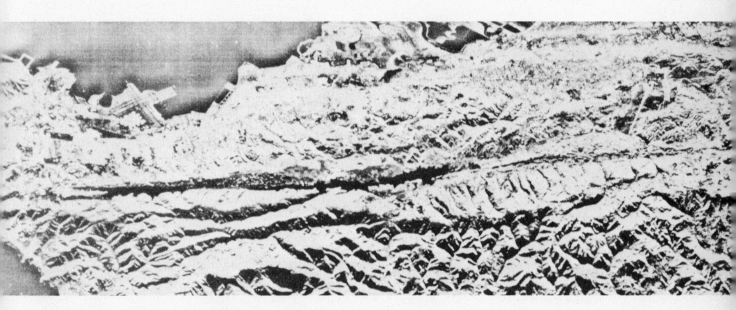

Figure 2-19

Radar image of the San Andreas Fault (California). Along the California coast, the Pacific Plate (see Fig. 2-17) moves northwestward at its boundary with the westward-moving North American Plate. Baja (lower) California is part of the Pacific Plate and is being dragged away from the rest of the continent. Strain builds up steadily along the San Andreas; eventually one side moves with respect to the other and an earthquate results. This fault is an example of a *transcurrent fault.* (Photograph courtesy of U.S. Geological survey.)

that arises from the Cobb Seamount thermal center. Spreading beneath the East Pacific Rise, which causes Pacific Plate motion, results in earth movements along California's well-known San Andreas Fault (Fig. 2-19). Baja California, being part of the Pacific Plate, will eventually be drawn away from the California mainland as the Gulf of California extends northward to join the Pacific Ocean (Fig. 2-17).

Another prominent feature of deep ocean basins, also attributable to volcanic activity in conjunction with sea-floor spreading, are the *volcanic ridges.* Figure 2-17 shows the Rio Grande Rise and Walvis Ridge extending on either side of the Mid-Atlantic Ridge at Tristan de Cunha. Another such feature centers at Iceland and a third trends northwestward from Amsterdam Island; its extension on the extreme left of the figure, the Maldive Island group, lies south of India. As in Pacific Island chains, these ridges form when a great amount of lava is erupted from a thermal center. The difference is that submarine ridges originate at centers of sea-floor spreading, where lava erupts fairly continuously and in large quantities, instead of sporadically as with volcanic islands. Figure 2-20 shows diagrammatically how ridges or island chains develop through time, and how they record the direction of plate movements.

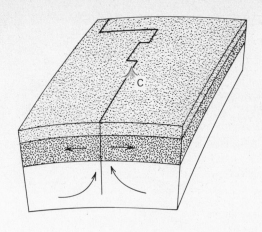

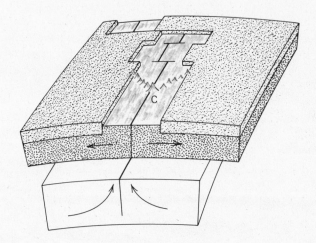

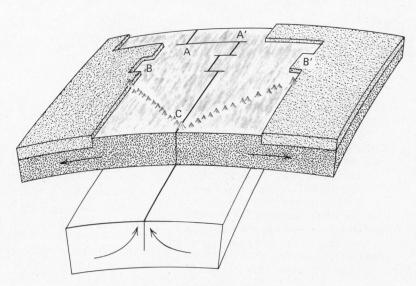

Figure 2-20

Formation of a submarine ridge or volcanic chain. The Walvis and Rio Grande ridges have probably been developing for around 140 million years as the African and South American plates moved apart. There has apparently been little or no change of direction during that time. [After R. S. Dietz, and J. C. Holden, 1970. "The Breakup of Pangaea," *Scientific American*, 224(10).]

Volcanic ridges and mid-ocean rises separate the deep ocean into several small basins (Fig. 2-21), so that bottom waters cannot flow freely between different parts of the world ocean. In the Southeast Atlantic Ocean, the Walvis Ridge extends from the Tristan da Cunha–Gough Island area to southwest Africa. This sharp ridge comes to within 2 kilometers of the surface and prevents waters moving along

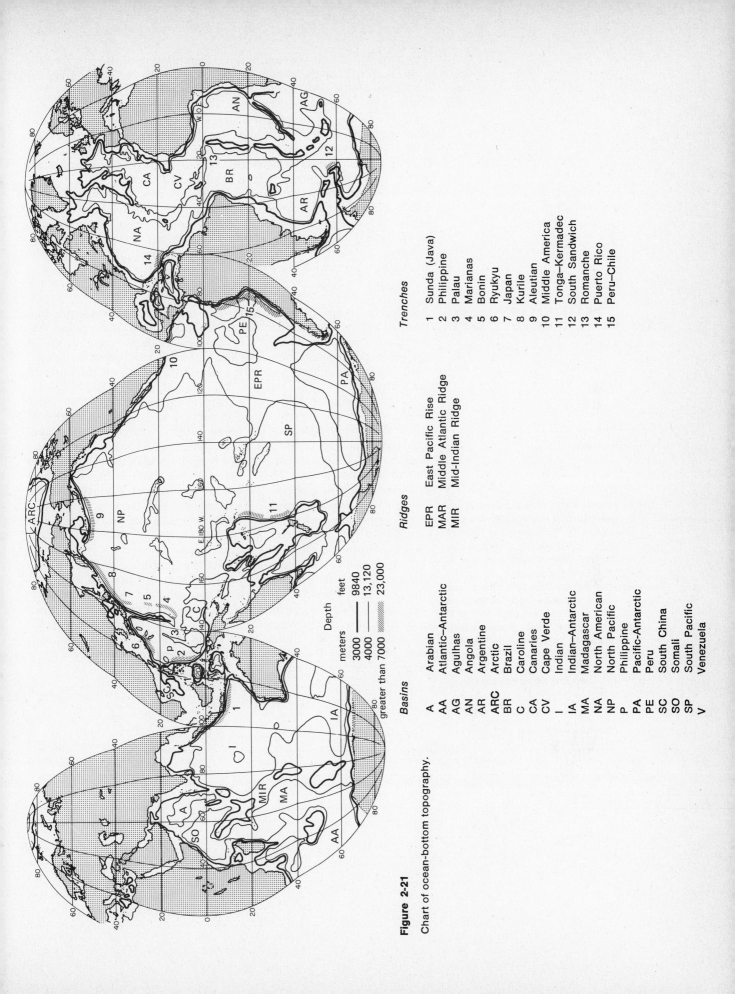

Figure 2-21

Chart of ocean-bottom topography.

Depth

meters	feet
3000	9840
4000	13,120
greater than 7000	23,000

Basins

A	Arabian
AA	Atlantic–Antarctic
AG	Agulhas
AN	Angola
AR	Argentine
ARC	Arctic
BR	Brazil
C	Caroline
CA	Canaries
CV	Cape Verde
I	Indian
IA	Indian–Antarctic
MA	Madagascar
NA	North American
NP	North Pacific
P	Philippine
PA	Pacific–Antarctic
PE	Peru
SC	South China
SO	Somali
SP	South Pacific
V	Venezuela

Ridges

EPR	East Pacific Rise
MAR	Middle Atlantic Ridge
MIR	Mid-Indian Ridge

Trenches

1	Sunda (Java)
2	Philippine
3	Palau
4	Marianas
5	Bonin
6	Ryukyu
7	Japan
8	Kurile
9	Aleutian
10	Middle America
11	Tonga–Kermadec
12	South Sandwich
13	Romanche
14	Puerto Rico
15	Peru–Chile

the bottom from the Antarctic region from flowing northward into the eastern portion of the deep Atlantic basin. In the North Atlantic, the massive volcanic ridge stretching from Greenland to Iceland and eastward to the Faeroe and Shetland Islands inhibits the flow of Arctic waters into the North Atlantic basin. East–west circulation in the deep ocean is generally inhibited because the location of continents and mid-ocean ridges favors north-south circulation. However in certain areas, east-west trending ridges also restrict the flow of near-bottom water, for instance in the Atlantic and Indian Oceans.

Furthermore, world climate is strongly affected by the position of great land masses with respect to each other and to the geographic poles. Major climatic changes affect water movements in every part of the world ocean. Thus, although the oceans themselves are very ancient, the patterns in which they circulate are only as old as the basins through which they flow.

PLATE TECTONICS AND THE SHAPING OF CONTINENTS

The forces that create ocean basins also shape continents. Continental margins adjacent to trenches are particularly subject to alteration, but anywhere a continent overlies a plate boundary there is activity within the earth that causes changes at the surface. Earthquakes, volcanoes, uplifting of mountains, and rifting of continents are all evidences of such activity.

Most earthquakes occur at plate margins (see Fig. 3-8). They result from strains that build up in rocks, and are suddenly released when the rocks finally break. Along ridges and faults this seldom happens below 70 kilometers depth, but at subduction zones intermediate and deep-focus earthquakes typically originate at depths from 100 to 700 kilometers. This is because the plate being subducted descends under the overriding plate for hundreds of kilometers before losing its identity as a result of extreme heat and pressure (Fig. 2-22). Deep earthquake foci are distributed along inclined planes known as *Benioff zones,* as shown also in Fig. 2-13. Figure 2-17 shows this trench-related earthquake distribution throughout the world.

Studies of earthquake wave behavior indicate that lithospheric plates are under tension where they bend at trench margins, but pressure on them increases as they descend through the asthenosphere. This suggests that plates meet increased resistance as they penetrate farther into the mantle. Earthquakes do not occur below 700 kilometers depth, indicating either that lithospheric plates do not descend below this level, or that they become too soft from heating to sustain the stresses that generate earthquakes.

Figure 2-22

A simplified view of plate subduction and the generation of new basaltic crust, as envisioned by Professor Harry Hess, of Princeton University. Vertical scale throughout is highly exaggerated. Note that inland mountain ranges are built by release of material from both lithospheric plate and mantle, because the plate descends to great depths at an angle of about 45° and volcanic activity occurs along its length. At left, sunken volcanoes, their tops planed off by erosion that took place while they were above sea level, are borne into the trench.

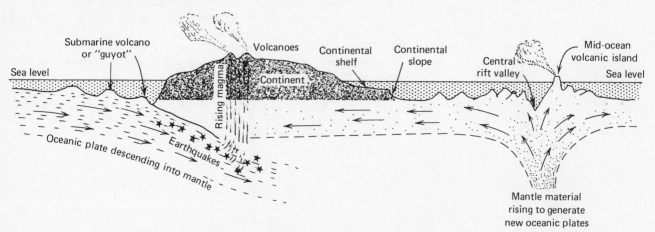

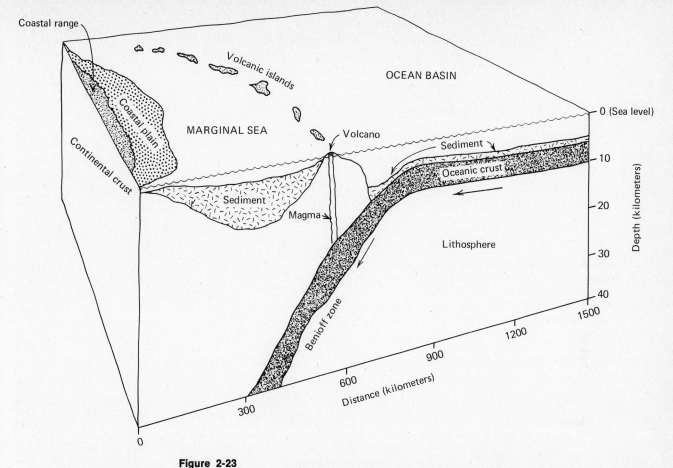

Coastal range

Volcanic islands

OCEAN BASIN

Coastal plain

MARGINAL SEA

Volcano

Sediment

Continental crust

Oceanic crust

0 (Sea level)

Sediment

10

Magma

20

Lithosphere

30

40

Benioff zone

1500

1200

900

600

Distance (kilometers)

300

0

Depth (kilometers)

Figure 2-23

Schematic diagram of trench and island
arc separating marginal basin from
open ocean.

When a continent rides on the leading edge of an advancing plate and overrides an oceanic plate, a coastal trench and parallel mountain ranges are formed (Fig. 2-22). An example is the Peru–Chile Trench, which parallels the western coast of South America. Close beneath it lies an earthquake zone which extends under the continent at an angle, to a depth of about 600 kilometers. About 300 kilometers inland is a range of volcanic mountains, the Andes, which parallel the west coast of South America for more than 6000 kilometers. The Andes, and other trench-related mountains, are made of *andesite,* a type of volcanic rock intermediate in silica and iron-magnesium content between granites and basalts. Thick sediment layers overlying the ocean floor near a continental margin travel into the trench and are reworked, along with the oceanic crust itself. Altered by the heat and pressure of the melting region, and possibly also incorporated partially molten mantle rocks, both plate and sediment contribute material to the building of andesitic coastal ranges. The Sierra Nevada in California is another example.

In the western Pacific, relationships are more complex. Trenches occur several hundred kilometers from the Asian continent and beyond them are arcs of volcanic islands. Shallow marginal seas, filled with sediment washed from the adjacent continent, separate these island arc systems from the mainland (Fig. 2-23). The Aleutians, Indonesia, and Japan (Fig. 2-17) are examples.

Island arcs are built by much the same mechanism as andesitic coastal ranges. Basaltic sea floor, together with a thick layer of wet, silica-rich sediment, descends diagonally and melts, forming an andesitic magma as it reaches the asthenosphere. Made buoyant by heat

47

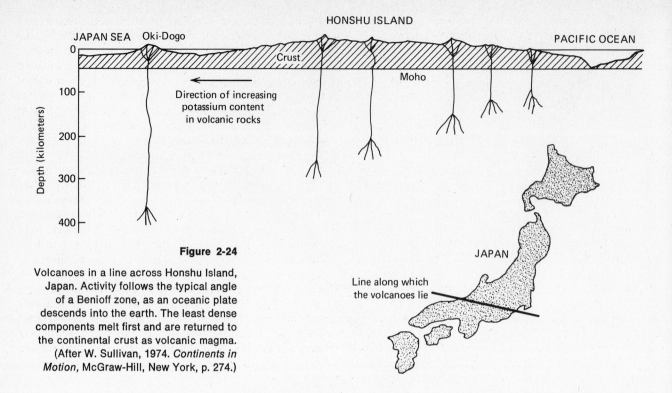

Figure 2-24

Volcanoes in a line across Honshu Island, Japan. Activity follows the typical angle of a Benioff zone, as an oceanic plate descends into the earth. The least dense components melt first and are returned to the continental crust as volcanic magma. (After W. Sullivan, 1974. *Continents in Motion*, McGraw-Hill, New York, p. 274.)

and the pressure of expanding volatile materials, magma rises and erupts as gases and lava at an overlying volcanic arc. Across a large island, such as Honshu in Japan, volcanoes have sometimes formed in a line from the trench or ocean side to the marginal sea (Fig. 2-24). The shallowest seem to originate at about 80 kilometers depth, where low density constituents in the rocks begin to melt and move upward. As the lithospheric plate descends to greater depths with increased distance from the trench, heavier materials begin to melt and undergo *metamorphosis,* or changes in mineral composition and texture. For example, potassium content of the extruded lava consistently rises with increasing depth of melting. This means that changes in potassium content of rocks can be used as "footprints" to mark the direction of a descending plate, even in continental mountain ranges which were formed during earlier eras of sea-floor spreading.

Very hot water may collect in deep cracks or fissures in rock, for instance in contact with rising magma at an area of plate subduction or divergence. Certain metals in the rock dissolve in water at high temperatures and pressures, so that the resulting *hydrothermal solution* is highly concentrated in metal ions. When it rises toward the surface through faults or fissures, the water cools and metals are selectively deposited as mineral precipitates or metal ores in *hydrothermal veins.* Major deposits of hydrothermal sulfide ores, especially iron, lead, zinc, mercury, and copper are located at ancient or modern convergence zones. These include the Philippine Islands, Japan, the Coast Ranges of North and South America, and a belt extending from the eastern Mediterranean Sea to Pakistan. Gold deposits also seem to be associated with convergence zones. There is evidence that different metals may rise from a descending plate at different depths along the Benioff zone; for instance, iron seems to come to the surface from shallower layers than copper.

Deep basins containing metal-rich brines and oxide or sulfide precipitates sometimes occur at zones of divergence or rifting where there is high heat flow. Along the axis of the Red Sea, two kilometers below sea level, intense heat is generated at an active spreading center (Fig. 2-17). Layers of ore, penetrated during the early 1970's by the Deep-Sea Drilling Project, occur there in sediment 20 to 100 meters thick. They were apparently formed when seawater, or water released from rising magma, entered cracks caused by rifting. Hydrothermal iron and copper sulfide ores have also been found in basins on the Mediterranean island of Cyprus, surrounded by metal-rich sediment deposits. The findings suggest that Cyprus was once a site of plate divergence, perhaps a mid-ocean ridge, now elevated above sea level.

Continental crust, being less dense and more buoyant than oceanic basalt, is not subducted at trenches. Instead, continental margins are deformed by the intense pressure that builds up when continental blocks meet head-on. This can happen, for example, if a trench beside one continental mass is approached by another continent, after the oceanic crust that separated the continents has been completely subducted. Collisions between continental masses result in folding, thickening, or thrusting of the continental crust as shown in Fig. 2-25. Tectonic plate movements have been going on perhaps fifteen times longer than the age of present ocean basins, and the same continental materials must have collided many times, in different relative positions. Complex continental structures, such as gigantic, upthrust, folded mountain ranges containing mile-high sections of former ocean floor, exist as a result of billions of years of deformation.

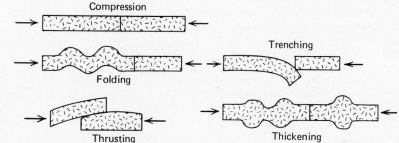

Figure 2-25

Mountains form as continents are pressed together. Most mountains were formed by a combination of these processes, often acting together with volcanism.

The Himalaya Mountains, built about 50 million years ago when India encroached on Asia (see Chapter 2 frontispiece, and Fig. 2-15(c)), provide a dramatic example. India underthrust the Asian land mass, creating a double continent 70 kilometers thick (Fig. 2-26). A former continental slope, the Himalayan plateau rises sharply to a height of four kilometers above the region south of it, and folded, thickened mountains extend as far northward as the Russia–China border in Central Asia (Fig. 2-17). Deep-sea drilling in the Indian Ocean has provided data indicating more than 50 million years of pressure at that contact, and the Himalayas are still adjusting, as evidenced by the earthquakes there.

The Appalachian and Allegheny mountain ranges of eastern North America probably formed hundreds of millions of years ago when the ancient Atlantic Ocean closed as Africa and Europe met the North American continent (Fig. 2-27). When the supercontinent of Pangaea began to break up about 150 million years ago (see frontispiece), the continents separated a few hundred kilometers east of their former boundaries. Sections of the former African continent remained in the Western Hemisphere to form parts of the present North American coastal plain.

Continents grow when materials from the ocean floor are added to their margins. Evidence of this process can often be observed in mountains built over a present or former trench area. These often

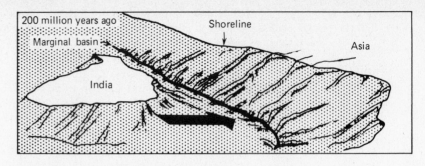

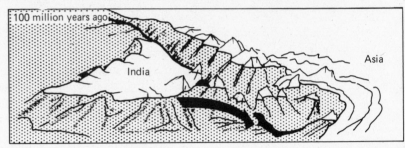

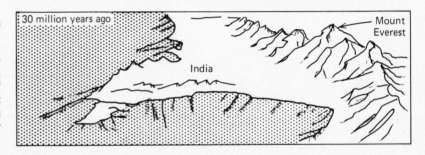

Figure 2-26

About 200 million years ago the northern edge of the Indian subcontinent began to underthrust Asia, eventually creating a double continental thickness with folding of the crust that forms the Himalayas, highest mountains on earth. (After J. Weisberg and Howard Parish, 1974. *Introductory Oceanography*, McGraw-Hill, New York.)

contain *ophiolitic suites,* or complexly faulted assemblages of deep-sea sediment, submarine lava, and oceanic crust that were once part of the sea floor adjacent to a continent. In the process of trenching under extreme pressure, ophiolites have broken off a descending plate and been thrust to great heights. These layers of ancient ocean floor—volcanic basalts and the sedimentary rocks that formed on top of them through compression over millions of years on the sea floor—are wedged high over much younger rocks in the world's greatest mountain ranges. For instance, ophiolites occur along the axis of the Urals (Fig. 2-17), far from any present ocean, and in the more recent mountains behind San Francisco Bay. A clearly defined series of bands marks the junction of Switzerland, Italy, Southern Yugoslavia, Albania, and Greece with the rest of Europe. Mountains were formed in these regions when the northern portion of what is now Africa collided with present Northern Europe, closing the ocean that formerly separated the continental masses.

As oceanic plates descend into a trench, thick layers of sediment eroded from the continent may be scraped off against the trench wall by the overriding plate. These layered sediments are folded, sliced, and altered by extreme pressure while being carried to depths of thirty kilometers or more. But instead of descending farther toward the mantle, this relatively buoyant material is sometimes returned to the surface. This type of a deposit, known as a *mélange,* is characteristic of present and ancient subduction zones. It typically contains ophiolitic suites and additional layers of sediment originally eroded from a continental mass and deposited at the base of a continental slope.

50

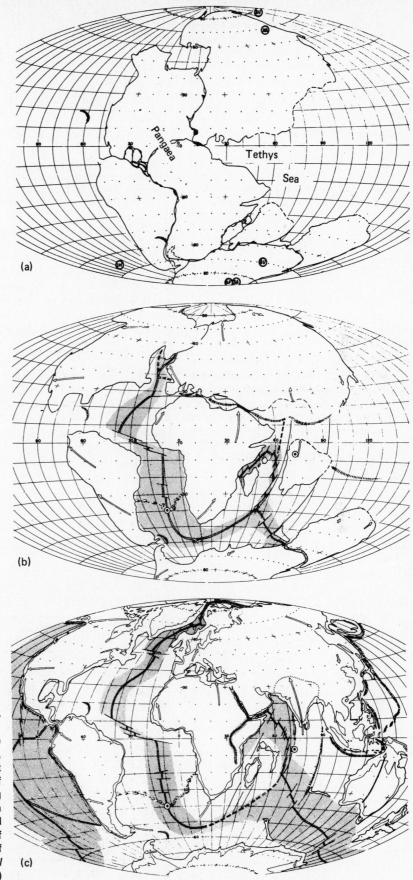

Figure 2-27

Movement of continental blocks during the past 225 million years of earth history, following the breakup of the ancient continent, Pangaea (a). Note the closing of Tethys, an ancient marginal sea; the opening of the Atlantic Ocean (b); and the formation of the Indian Ocean (c). (R. S. Dietz and J. C. Holden, 1970. "Reconstruction of Pangaea; Breakup and Dispersion of Continents, Permian to Present," *Journal of Geophysical Research,* 75, 4939–56.)

Figure 2-28

An island arc or coastal range is commonly
formed from reworked marine sediments
and basalts in areas of subduction.
Seaward from the volcanic feature there
may be a mélange deposit consisting of
sediment originating on the continent and
deposited near shore, plus deep-ocean
sediment carried toward the trench on a
descending plate, and also incorporating
ophiolitic suites of sedimentary and basaltic
rock broken from the ocean floor during
subduction. The drawing is not to scale.
(After W. R. Dickinson, 1971.
"Plate Tectonics in Geologic History,"
Science, 174, 107.)

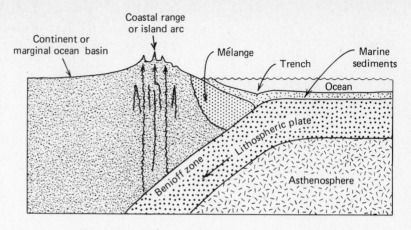

Mélanges commonly occur seaward of andesitic mountain chains or
island arcs that have been built by volcanic processes associated with
a descending plate (Fig. 2-28).

As the continents grow by progressively adding material, the
ocean basins have gradually become deeper. It is estimated that the
oceans have deepened at a rate of about one meter per million years,
and the continental blocks have accumulated at a similar rate.

ORIGIN OF OCEAN AND ATMOSPHERE

The ocean is an ancient feature of the earth. But how the ocean and
atmosphere formed and at what rate, and what controlled their
composition, are questions still actively debated by scientists; thus the
answers offered here may well be challenged and even displaced by
new hypotheses within even a few years. Satellite observations of the
earth, powerful microscopes for studying ancient microfossils, and
the recovery of rocks from the moon all have contributed to our present
understanding.

Evidence for the age of the earth is indirect, coming from radiometric dating of the decay products of uranium and thorium isotopes
which gradually accumulate after billions of years. Several lines of
evidence suggest that the earth formed more than four billion years
ago, but there is no known record of the first billion years of earth's
history. The earth probably formed from an accumulation of cold
particles, especially silicon compounds and oxides of iron and magnesium. Gravity caused the heaviest materials to fall toward the center
so that the mass became increasingly compressed. This compaction
caused interior heating and finally melting, while heat released by
the decay of radioactive materials also raised interior temperatures.
Three concentric layers were formed—the heaviest material, primarily
iron and nickel, became the core, heavy silicate rocks formed the
mantle, and the lightest materials rose to the surface as crust. Minerals
from ancient rocks of Europe, North America, and the Southern Hemisphere continents represent the earliest evidence of a solid crust on
the earth. These rocks are 3.5 to 3.6 billion years old.

If the primordial earth had an atmosphere, it apparently was
lost, perhaps as a result of the same heating that caused the segregation
of core, mantle, and crust. At any rate, there is relatively much less
rare noble gas (krypton, neon, xenon) in the present atmosphere than
in the universe as a whole. This indicates that these gases were not
retained in the abundance that one would expect from the retention
of an ancient atmosphere. Thus, the present ocean and atmosphere
must have been derived from later alteration of the earth rather than

from cosmic sources. Both the water and gases must have been combined in rocks since they were not lost in the early escape of gas from the earth's surface.

Based on present knowledge of geologic processes in the ocean and on the continents, it seems probable that the hydrogen and oxygen that now form water were originally bound in minerals. As the earth became hotter, parts of the upper mantle or crust melted and the more volatile, water-rich portions of these molten rocks escaped as lava from volcanic eruptions or intruded overlying rocks. In the process a part of their original water and gas content was lost to the atmosphere. In time, the water condensed to fill depressions in the crust—the ocean basins—and the gases remained in the atmosphere. This process apparently continues today, although the amount of new water is too small to be detected with confidence.

The ocean and atmosphere thus seem to have originated as a result of the same reworking of crustal materials that formed the continents and ocean basins, but their early history is poorly known. The earliest evidence for the existence of the ocean is the presence of recognizable water-laid sediments formed about 2.5 to 3 billion years ago. While this indicates the presence of abundant water on the earth's surface, it does not tell us about the chemical composition of this primordial ocean. Carbonate sediments about 2 billion years old suggest that ocean water has been salty for at least half of the earth's history. Very likely the major constituents of sea salt were not greatly different from those now present in the ocean. Fossils about 0.6 billion years old suggest that since that time the chemical composition of seawater has changed little.

Evidence suggesting the composition of the early atmosphere is mostly indirect. It probably contained water vapor, hydrogen, hydrogen chloride, carbon monoxide, carbon dioxide, and nitrogen. Free oxygen almost certainly did not occur. We know this because some ancient sediments originally formed in the ocean contain minerals that are not stable in the presence of oxygen. Further evidence is provided by the existence of iron-rich deposits that have been extensively mined as iron ore. These could not have been formed in the past 2 billion years, because iron is highly insoluble in seawater containing dissolved oxygen, and today does not form large-scale marine sedimentary deposits. Ancient ocean waters were essentially devoid of dissolved oxygen, so that large quantities of iron were transported and precipitated to form extensive deposits.

Substantial amounts of free oxygen were apparently not introduced to the atmosphere, nor to the ocean, until after the evolution of *photosynthesis.* That is the process by which green plants such as algae combine carbon dioxide and water in the presence of sunlight to form organic carbon compounds and free oxygen. Fossil evidence suggests that this may have begun to happen about 3 billion years ago. The subsequent accumulation of oxygen in air and ocean permitted the evolution of *respiration,* whereby animals combine oxygen with organic carbon compounds to release energy.

Small amounts of free oxygen apparently began to be used for respiration about 2.7 billion years ago. But oxygen and organic carbon levels continued to build up until about 500 million years ago, when higher animals evolved. The oxygen accumulated in ocean and atmosphere, and the organic carbon was incorporated in sediments, from which a small fraction is now recovered as fossil fuel. Relationships between oxygen and carbon in photosynthesis and respiration, showing the former excess production of organic carbon and oxygen, are shown in Fig. 2-29.

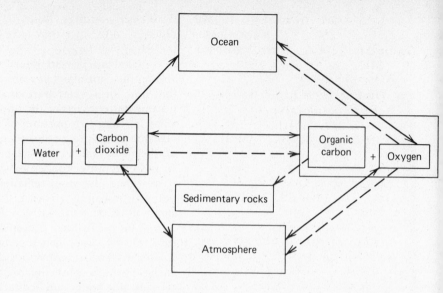

Figure 2-29

Before about 500 million years ago, an excess of photosynthesis over respiration built up surplus organic carbon and oxygen. The former was retained in sedimentary rocks, and the latter in atmosphere and ocean. Today oxygen and organic carbon are used by animals in respiration at about the same rate as they are produced by plants in photosynthesis.

← — — — — — — — — Surpluses building up prior to 500 million years ago

The Red Sea (on the left) and the Gulf of Aden extending into the Indian Ocean on the horizon. This seaway formed about 20 million years ago when Africa (bottom) split from Eurasia (upper left) and subsequently drifted apart. Note the parallelism of the coasts of the Gulf of Aden. Clouds occur over the land, especially just above the spacecraft in the lower portion of the photograph. (Photograph courtesy NASA.)

SUMMARY OUTLINE

Ocean covers most of the earth
 Ocean in three "gulfs"—mostly in Southern Hemisphere
 Continents are four land masses
 Hypsographic curve—two levels
  Continents—0 to 1 kilometer above sea level
  Ocean basin—4 to 6 kilometers below sea level

Structure of the earth
 Water is a small fraction of total mass
 Earth in three concentric layers: core, mantle, lithosphere
  Core—dense, rich in iron and nickel
  Mantle—outer asthenosphere capable of flow
  Lithosphere, 60 to 100 kilometers thick—includes continental, oceanic crust fused with layer of rigid material derived from mantle
   Mohorovičić discontinuity separates crust from rigid mantle layer
  Hydrosphere—includes water in sedimentary rocks; most water in ocean
  Atmosphere—includes water vapor
 Continental crust—35 to 40 kilometers thick, granitic; extends out to 2000 meter depth contour
 Oceanic crust—7 to 10 kilometers thick, basaltic; ocean basins 4 to 6 kilometers below sea level
 Lithosphere floats on asthenosphere: isostatic equilibrium of crust maintained by rising or sinking with respect to deeper layers

Methods of probing the earth—indirect techniques
 Behavior of earthquake waves demonstrates layered earth structure
 Gravity anomalies indicate differences in mass
 Heat flow varies—high from ocean basins, especially mid-ocean ridges; low at trenches
 Magnetic "striping" reveals age of ocean floor, due to periodic reversals of earth's magnetic field that caused minerals in molten rocks to align themselves with magnetic North

Sea-floor spreading
 Lithosphere divided into plates—boundaries at mid-ocean ridges, trenches, faults
 Generation of crust by magma rising at ridges and rises; transform faults connect offset sections of ridges
 Subduction at trenches destroys crust
 Pangaea—ancient supercontinent; broke up 200 million years ago to form present continents, Atlantic Basin
 Sea-floor spreading caused by convection in mantle; younger crust elevated relative to older crust due to buoyancy, lack of sediment cover
 Plumes of magma rise at volcanic centers; seamounts form in linear progression as crust moves; volcanic ridges form similarly, divide ocean into basins.

Plate tectonics and the shaping of continents
 Deep-focus earthquakes along inclined plane beneath continents
 Andesitic mountain chains, island arcs form from reworked sediments, plate, mantle
 Hydrothermal deposits precipitate from metals dissolved in water under heat, pressure at zones of subduction, rifting
 Continental crust deformed when continents converge—mountains built by folding, thickening of crust; contain layers from ocean basins
 Mélange deposits add to continental margins

Origin of ocean and atmosphere
 Primordial atmosphere lost when earth heated, formed layers
 Water and gas released from interior during volcanic activity; free oxygen absent
 Oxygen produced during photosynthesis during past 3 billion years
  Organic carbon stored—now used for fuels
  Respiration, photosynthesis now maintain oxygen, organic carbon balance

SELECTED REFERENCES

JUDSON, SHELDON, KENNETH S. DEFEYES, AND ROBERT B. HARGRAVES. 1976. *Physical Geology*. Prentice-Hall, Englewood Cliffs, N.J. 551 pp. Elementary; includes discussion of sea-floor spreading.

PRESS, FRANK AND RAYMON SIEVER. 1974. *Earth*. W. H. Freeman, San Francisco. 945 pp. Comprehensive treatment of earth science.

SCIENTIFIC AMERICAN. 1972. *Continents Adrift*. W. H. Freeman, San Francisco. 172 pp. Reprinted articles from *Scientific American*. Elementary and intermediate in difficulty.

SULLIVAN, WALTER. 1974. *Continents in Motion, The New Earth Debate*. McGraw-Hill, New York. 399 pp. Well-written treatment of development of plate tectonic theory.

A satellite's view of the Atlantic Ocean from more than 37,000 kilometers (22,300 miles) above Brazil. Four continents can be seen: South America (center), North America (upper left quadrant), Europe (upper right) and Africa (center right). Antarctica (bottom of photograph) is obscured by clouds, the white fluffy masses often forming swirls as part of the large storms in the lower atmosphere. Note the large fraction of the earth's surface obscured by clouds. (Photograph courtesy NASA.)

the ocean floor

In the previous chapter we studied the processes that shape continents and ocean basins. This chapter will be devoted to a description of ocean floor features beginning with the major provinces: continental margin, oceanic rise, and deep-ocean basin. Within the ocean basin proper we will concentrate particularly on volcanic features including reef and atoll formation. The major ocean basins—Pacific, Atlantic, and Indian Oceans and the Arctic Sea—will be examined briefly, and we will note some differences in their water properties and in the amount of sediment they receive. Lastly there will be a look at the marginal oceans that lie close to continents and are strongly affected by them.

CONTINENTAL MARGIN

The continental margin consists of the *continental shelf, slope,* and *rise* (see Fig. 3-1). The continental shelf is the submerged top of a continent's outer edge, and it begins at the shoreline. This is the part of the continent that has been shaped by the ocean, except where mountains occur at the edge of the continent. There the shaping force has been the processes that formed the mountains. From the shoreline the shelf slopes gently toward the *continental-shelf break* typically at a depth of about 130 meters. Off Antarctica, however, the shelf break

Figure 3-1

Schematic representation of the continental margin showing the continental shelf and its relationship to the coastal plain, the continental slope, and continental rise. Note that a submarine canyon cuts across the shelf break and extends to the base of the continental slope, where it joins the continental rise.

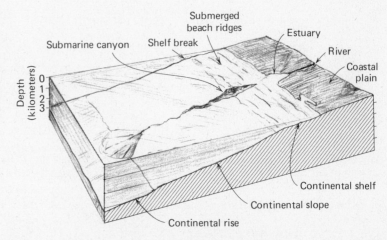

is more than 350 meters deep. At the shelf break the gently sloping shelf gives rise to the much steeper continental slope.

On the average the continental shelf is about 70 kilometers wide. On much of the Pacific Coast it is relatively narrow, only a few tens of kilometers across. The widest shelves occur in the Arctic Ocean. A broad continental shelf is the sculptured build-up of sediment shed by a continent, particularly where there has been no major mountain building at the margin over hundreds of millions of years.

Topographically, we may consider the *coastal plain* to be a terrestrial analog of the continental shelf. Where the coastal plain is absent and mountains form the coastline, as in Southern California, the continental margin is rugged, broken by submarine ridges, banks, and basins. Continental shelves near glaciated coasts are also typically rugged. These shelves tend to be wide and contain basins or deep troughs more than 200 meters deep, generally paralleling the coast. Broad ridges often divide these basins. The Gulf of Maine off the New England coast is such an area.

Some troughs cut across the shelf and connect with flooded, deep indentations called *fjords*. Such troughs can be quite large, such as the trough at the entrance to the Strait of Juan de Fuca connecting the Pacific Ocean with Puget Sound and the Strait of Georgia. Topography of these shelf areas was formed by glaciers during periods of lowered sea level, and have not been modified greatly by presently active marine processes.

On unglaciated continental shelves, the bottom topography tends to be much smoother, commonly gently rolling and subdued. Elongate low sand ridges (sometimes containing gravel) generally parallel the shore. The ridges, formed by infrequent, violent storms that stir waters to the bottom and redistribute sediments, are up to 10 meters thick.

The continental-shelf break is remarkably uniform in depth, averaging about 130 meters over most of the world ocean. The shelf break is thought to have formed some 20,000 years ago, when sea level stood at its lowest during Pleistocene glacial times. At this time the shoreline was at the edge of the present continental shelf, which was then a low-flying coastal plain. Estuaries and lagoons occurred along the edge, the estuaries occupying the sites where present submarine canyons indent the shelf break. Traces of ancient lagoons are not readily discernible although they must have been common in many areas. After that time, as glaciers melted, the ocean began to rise. The present level was reached about 3000 years ago; since that time there have been relatively few changes in sea level. In some formerly glaciated areas, the land has risen in response to the melting of glaciers.

Relatively shallow, canyon-like shelf valleys extend from the mouths of estuaries across the continental shelf, except in areas where sediment discharged from the estauary has obliterated them. These valleys, a few tens of meters deep, were formed by tidal currents that flowed in and out of the estuary during the migration of sea level.

In tropical oceans there are often coral reefs on the continental shelf. These commonly occur on the outer portion of the shelf and are separated from the land by a navigable channel paralleling the coast.

Continental slopes are the relatively narrow, steeply inclined submerged edges of continental blocks. On an ocean-free earth these would be the most conspicuous boundaries on the earth's surface. Although their total relief is substantial, ranging from 1 to 10 kilometers (0.6 to 6 miles high), continental slopes are not precipitous. They average about 4°, ranging from 1° to 10° of slope, either quite straight or gently curving. The most spectacular slopes occur where a continental block borders a deep trench. On the western coast of South

America, the peaks of the Andes Mountains rise to elevations of about 7 kilometers, while the adjacent Peru–Chile Trench is 8 kilometers deep. This adds up to a total relief of 15 kilometers (9 miles) within a few hundred kilometers. Continental slopes with trenches at their base are common in the Pacific and account for about half the world's continental slopes.

Faulting on the slope often forms small basins that trap sediment, changing the local topography from a complex of steep-walled basins to small, flat-floored basins that gradually become filled and smoothed over, eventually disappearing. With accumulation of a sufficient thickness of sediment, the original outlines of the continental slope are completely buried. Margins of many continental blocks have numerous buried ridges and ancient coral reefs, now covered by sediment deposits. Many of these filled basins are prime locations for petroleum accumulation.

Thick accumulations of sediment cover many of the world's continental slopes, except along the Pacific Ocean. When sediment deposits become too thick, and the slopes too steep, the deposits may slump. This leaves large, spoon-shaped scars on the edge of the slope, and a mass of deformed sediment at the base of the slump. This probably is fairly common, and may account for many of the hills on the lower parts of continental rises.

Submarine canyons are large features that occur commonly on continental slopes. They are typically V-shaped, follow curved paths, and have tributary channels leading into them. Some are as large as the Grand Canyon of the Colorado River. A few submarine canyons are obviously associated with large rivers on the continent—for instance, offshore from the Hudson and Congo Rivers. In these cases, the upper portions of the valley could have been cut during times of lowered sea level when a river flowed through the canyon, so their origin posed no problem. How, then, does one explain the portion of a canyon that swept out to the edge of the continental slope where it merged into much shallower channels with low levees on each side? These channels often run for tens or hundreds of kilometers across the fan of sediment that forms the base of the slope. Many canyons have no obvious connection to present-day rivers, although they may have been associated with rivers no longer in the region. Some canyons were cut through sedimentary rocks that are relatively soft. Others were found to be cut through harder granite. Some sediment commonly occurs on canyon floors.

These canyons, it is now generally agreed, are cut by *turbidity currents,* flows of sediment-laden water that move along the ocean bottom carrying material from the continental shelf to the deep ocean floor. The process of canyon cutting is still active in many areas, especially where the continental shelf is relatively narrow and sediment can be readily transported to the head of the canyon. Such canyons have been studied especially closely in Southern California, where their upper portions lie quite close to the beach.

Continental rises (Fig. 3-1) are smooth-surfaced accumulations of sediment at the base of continental slopes. They cover the transition between the continent and the ocean basin. Low hills are present, perhaps formed by slumping of sediment from the continental slope. Channels a few meters deep, apparently routes of sediment movement, cross the rise. These channels are often connected to submarine canyons.

Continental rises are apparently formed by sediments carried by turbidity currents and deposited at the base of the slope. Deposition is caused by the reduction in current speed when it flows out onto the gently sloping ocean floor. Initially each canyon has its fan-shaped deposit of sediment at its mouth, similar to mountain streams flowing onto a flat desert floor. Eventually individual fans build sideways and coalesce to form a continuous sediment wedge.

Typically, the continental rise is made of sediment deposits several kilometers thick, and tens of kilometers wide. One estimate of the total volume of sediment in the world ocean gave the following results:

Continental shelf	1.2 billion cubic kilometers
Continental slope and rise	1.6 billion cubic kilometers
Deep-ocean bottom	0.8 billion cubic kilometers

These data indicate that about 40 percent by volume of the sediment in the ocean lies in the continental rises at the base of continental slopes. Altogether, deposits on the continental margin account for about 78 percent of the total recognized sediment deposits in the world ocean.

OCEANIC RIDGES AND RISES

The deep ocean floor, which we take to begin at about 4 kilometers below sea level (see Fig. 2-4) is shown very simply in Fig. 3-2. Mid-ocean ridges and rises stand out clearly above the deep basins, covering 23.1 percent of the earth's surface (Table 3-1). Compare this with the land above sea level, which covers 29.2 percent of the earth. The mid-oceanic ridge and rise system includes high-relief segments, as shown schematically in Fig. 3-3, and low-relief segments, as in Fig. 3-4.

TABLE 3-1

Physiographic Provinces of the Ocean *

OCEAN †	SHELF AND SLOPE (percent)	CONTI-NENTAL RISE (percent)	OCEAN BASIN (percent)	VOLCANOES AND VOL-CANIC RIDGES (percent)	RISE AND RIDGE (percent)	TRENCHES (percent)
Pacific	13.1	2.7	43.0	2.5	35.9	2.9
Atlantic	19.4	8.5	38.0	2.1	31.2	0.7
Indian	9.1	5.7	49.2	5.4	30.2	0.3
World ocean	15.3	5.3	41.8	3.1	32.7	1.7
Earth's surface	10.8	3.7	29.5	2.2	23.1	1.2

* After H. W. Menard and S. M. Smith, 1966. "Hypsometry of Ocean Basin Provinces," *Journal of Geophysical Research*, 71, 4305.

† Includes adjacent seas; for example, Arctic Sea included in Atlantic Ocean.

The rugged high-relief Mid-Atlantic Ridge with its many fracture zones and volcanoes, stands 1 to 3 kilometers above the nearby basin floor. It is 1500 to 2000 kilometers across and has a prominent steep-sided rift valley (Fig. 3-5). Rocks dredged from the rift valley floor have evidently solidified from lavas fairly recently, within the past few million years. The valley is usually 25 to 50 kilometers wide, and 1 to 2 kilometers deep. It is bordered by steep mountains, also formed by faulted blocks; these peaks come to within 2 kilometers of the ocean's surface. The closest terrestrial analog to conditions on the Mid-Atlantic Ridge is found in the East African Rift Valley shown in Fig. 3-6.

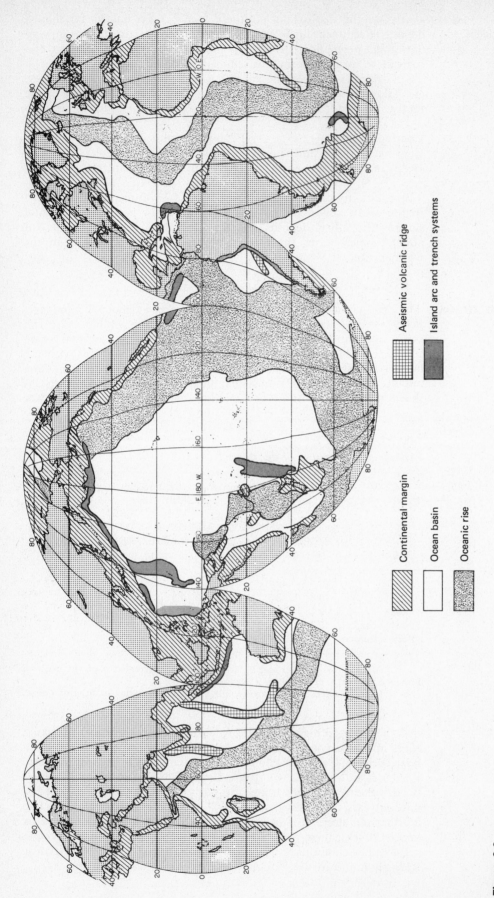

Figure 3-2

Ocean-bottom provinces showing relationships between submerged
continental margins, mid-oceanic rises, ocean basins, island arcs,
and volcanic (aseismic) ridges. Note that the trenches, the deepest parts
of the ocean bottom, occur along the basin margins.

Continental margin

Ocean basin

Oceanic rise

Aseismic volcanic ridge

Island arc and trench systems

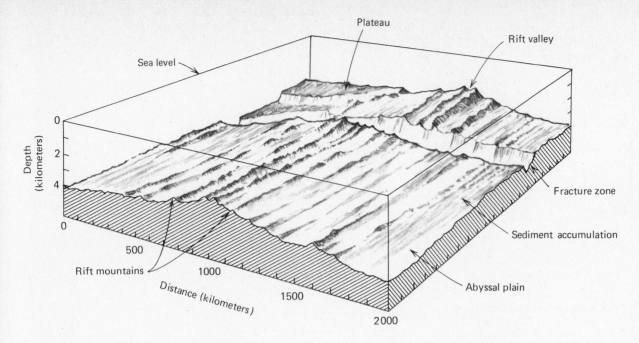

Figure 3-3

Schematic representation of ocean-bottom topography of a portion of the Mid-Atlantic Ridge, typical of a rugged, faulted mid-ocean rise. Note the rough topography along the fracture zone where two segments of the rise are offset. The fault along which movement occurred is a transform fault. Sediment forming the abyssal plain is deposited on the margin of the rise.

Figure 3-4

Schematic representation of ocean-bottom topography along a portion of the East Pacific Rise, an example of a broad, low rise with relatively subdued relief.

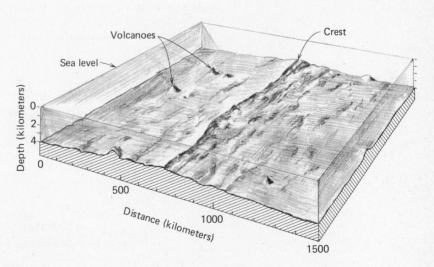

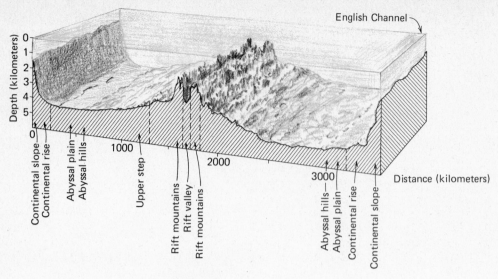

Figure 3-5

Diagrammatic view of the North Atlantic ocean floor between Canada and Great Britain. Note the great exaggeration in vertical relief.

Away from the central valley of the Mid-Atlantic Ridge, bottom topography is still quite rugged. There is a great deal of faulting and insufficient sediment accumulation to bury the irregular terrain. Faults cut across the ridge and offset it in a series of segments, as indicated in Fig. 3-7. One of these transverse faults is the Romanche Trench, an important link for the flow of deep-ocean water from the western to the eastern Atlantic basin. There are numerous and frequent shallow-focus earthquakes (occurring within 70 kilometers of the earth's surface) along the Mid-Atlantic Ridge as well as volcanoes and volcanic islands, including the Azores, Iceland, Ascension, and Tristan de Cunha; the distribution of these features is indicated in Fig. 3-8. The Mid-Atlantic Ridge extends into the Arctic Ocean.

The *East Pacific Rise* lies near the eastern margin of the Pacific Basin. It is not rugged like the Mid-Atlantic Ridge, but is a vast, low bulge on the ocean floor, approximately equal in size to North and South America combined. The rise stands about 2 to 4 kilometers above the adjacent ocean bottom and varies from 2000 to 4000 kilometers in width. Intersecting the North American continent in the Gulf of California, its continuation reappears off the Oregon coast and extends into the Gulf of Alaska. The two segments of the rise system are connected by the San Andreas fault in the state of California. Movement along this fault caused the San Francisco earthquake of 1906.

Figure 3-6

Profiles of the rift valley of the Mid-Atlantic Ridge and the Tanganyika rift of East Africa. Note the similarity in form and size of the two features. (After Holmes, 1965.)

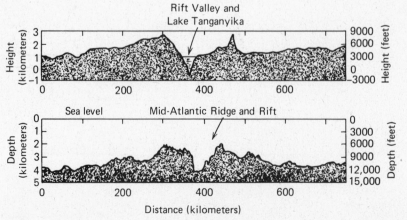

In the Indian Ocean, the high-relief Mid-Indian Ridge intersects Africa-Asia in the Red Sea area (see Fig. 2-17), and the active system apparently extends under the East African Rift Valley. On the south, the Mid-Indian Ridge system branches to join the system of active ridges and rises that circle Antarctica. Near Madagascar a low section of the ridge system permits deep water to move between the deeper parts of the Indian Ocean and the Atlantic Ocean. The Mid-Indian Ridge separates two distinctly different ocean areas.

A narrow belt of frequent shallow-focus earthquakes runs along the crests of oceanic rises (Fig. 3-8). In the North Atlantic, where the ridge is about 1500 kilometers wide, the belt of earthquake *epicenters* (point on the earth's surface above the earthquake focus) is about 150 kilometers wide. Detailed surveys show that these earthquakes coincide rather closely with the axis of the rift valley. Indeed, earthquakes have been used to pinpoint active ridge systems in little-known ocean areas. Such information is especially useful where other information is lacking or where relationships are unclear, as in the Arctic Ocean.

Figure 3-7

Artist's rendition of the bottom of the North Atlantic Ocean. Elevations and depths are noted in feet. (Painting by Heinrich Berann, courtesy Aluminum Corporation of America.)

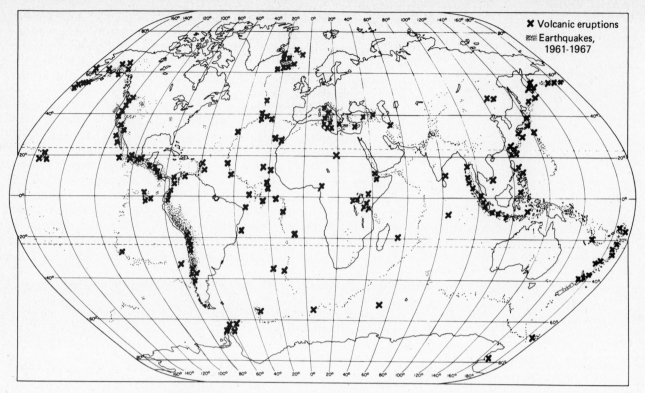

Figure 3-8

Distribution of earthquakes (1961–1967) and active volcanoes. Note the correspondence between earthquake epicenters and the boundaries of the large crustal blocks shown in Fig. 2-17. (Compiled from Bryan Isacks, J. Oliver, and L. R. Sykes, 1968. "Seismology and the New Global Tectonics," *Journal of Geographical Research,* 73, 5855–5900, and from A. Holmes, 1965.)

Mid-oceanic ridge systems are also sites of high heat flow from the earth's interior—about 1 to 3 microcalories per square centimeter per second (μcal cm^{-2} sec^{-1}). These "hot spots" do not correspond as closely to the location of rift valleys as do the sites of earthquakes. Instead, areas of unusually high heat flow (up to 8 μcal cm^{-2} sec^{-1}) tend to form narrow bands, suggesting narrow intrusions of molten rocks.

The shallowness of mid-ocean areas affects sediment accumulation. Distinctive types of carbonate-rich sediment, not found in deep-ocean basins, tend to accumulate in shallow areas, particularly in the small, fault-bounded basins characteristic of mid-oceanic rise systems. Furthermore, the presence of barriers to the movement of sediment along the bottom prevents continental sediment from reaching many areas of the ocean bottom. For instance, the segment of the East Pacific Rise off the coast of Washington and Oregon blocks the movement of sediment along the bottom so that none of it reaches the deep North Pacific ocean floor. Thus the accumulation of sediment on the landward side of the ridge has caused it to be hundreds of meters shallower than the seaward flank of the rise.

OCEAN BASIN The ocean basin proper occupies about 29.5 percent (Table 3-1) of the earth's surface. Its floor is covered by sediment deposits about 300 meters thick, much of which has accumulated particle by particle so that sediment layers are draped over preexisting topography as a snow-fall blankets and blurs the sharp features of the land. Near continents, on the other hand, heavy turbidity currents can completely bury original topography.

Oceanic features have a much longer existence than land features, because there is no deep-ocean analog to the vigorous erosion by glaciers, wind, or rivers that takes place on land. Nor is there the chemical alteration of rock by water, atmospheric gases, and the plant

66

action of a terrestrial environment. Submarine volcanoes formed more than 25 million years ago are still very much in evidence—for instance, as foundation for the atolls at the northwestern end of the Hawaiian Island chain. Rocks have been recovered from submarine volcanoes that apparently formed about 80 million years ago, older than any known feature on land.

Ocean-bottom topography is dominated by the striking fracture zones that extend for hundreds or thousands of kilometers on either side of mid-ocean ridges and rises. In the Pacific Ocean, these bands of volcanoes and mountainous terrain are 100 to 200 kilometers wide. Each consists of individual ridges and troughs that are several hundred kilometers long and a few tens of kilometers wide. Some have cliffs or ridges up to a few kilometers above the nearby ocean floor. The greatest depths in the central Pacific Ocean (more than 6 kilometers) occur along these fracture zones.

In some areas, fracture zones offset continental margins. The sudden bend in the coast between Cape Cod and New Jersey is thought to be caused by a major fracture zone. This zone can be traced through the deeper ocean basin where it is marked by the New England Seamounts.

Low *abyssal hills* (less than 1000 meters high) cover about 80 percent of the Pacific Ocean Basin floor, and about 50 percent of the Atlantic, including portions of the Mid-Atlantic Ridge. They are probably also abundant over much of the Indian Ocean Basin, although data there are more sketchy. These hills have an average relief of about 200 meters with diameters of about 6 kilometers, and are the most common topographic feature on the earth's surface. Many appear to be small volcanoes covered with a thin layer of sediment that is slightly thicker in the valleys than on the hills themselves. Some abyssal hills may be formed by intrusions of molten rock that push up the overlying sediment. Some are probably formed by faulting of oceanic crust associated with movements of the sea floor and the mantle below.

Some immense areas of exceedingly flat ocean bottom, called *abyssal plains*, lie near continents (Fig. 3-9). These are among the flattest portions of the earth's surface. They are defined as having slopes of less than 1 in 1000, equivalent to a slope of 1 meter per kilometer. Areas of comparable flatness are found in the High Plains area of the mid-continental United States, where the surface slopes about 1.5 in 1000. Abyssal plains commonly occur at the seaward margin of the continental rise. In fact, there is usually a gradual merging of the deep-sea fans that make up the continental rise with the adjacent abyssal plain.

Figure 3-9

Block diagram showing the Atlantic Ocean floor in the North American Basin.

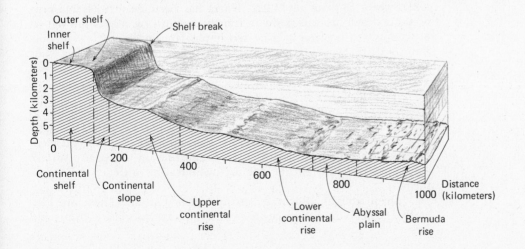

Most abyssal plains are heavily covered with sediment derived from the continents, to which we attribute their nearly flat topography. Deep-sea channels, a few meters deep, with levee-like margins, cut across them. These are extensions of deep canyons across continental margins, through which sediment is carried into the deep ocean. For instance, Cascadia Channel in the Northeast Pacific, extends more than 500 kilometers from the base of the continental slope near the mouth of the Columbia River and crosses through a gap in the mid-oceanic rise, finally extending out onto the adjacent Tufts Abyssal Plain. Similar channels have been described in the Atlantic and Indian Ocean.

Abyssal plains are especially common in marginal seas, such as the Gulf of Mexico, Caribbean, and the many marginal basins of the Pacific borderland. The Sigsbee Abyssal Plain has been explored extensively; it lies at the base of the Mississippi Cone, built by sediment coming from the Mississippi River. Sediment that escapes deposition within the river delta or on the cone is deposited on the abyssal plain along with a minor amount coming from the Campeche Bank to the south. During the last glacial period (Wisconsin stage), sediment accumulated on the plain at the rate of about 60 centimeters per thousand years. In the past 10,000 years (since the retreat of the ice), sediment has accumulated more slowly, about 8 centimeters per thousand years.

Small hills, first discovered by echo soundings and subsequently studied by deep-drilling techniques, contain thick layers of salt that have been compressed upward to form cylindrical masses. The presence of such *salt domes* indicates that lenses of salt, tens to thousands of meters thick, lie at some depth below the abyssal plain. They apparently formed in areas that were at one time covered by shallow seas, for example at sites of present or former continental rifting like the Red Sea or the Gulf of Mexico. Alternate periods of flooding and evaporation in these shallow basins caused heavy brines to accumulate, from which thick deposits of salt eventually precipitated. If such a salt lens is later covered by 600 meters or more of sediment, which compacts and becomes more dense than the salt beneath, the relative buoyancy of the salt layer causes it to be deformed and to flow upward as a narrow cylindrical mass. Plugs of salt up to 2 kilometers in diameter may rupture through several kilometers of overlying sediment and sedimentary rock as they rise to the surface. Domes of rock, distinctively faulted and folded, mark their presence under the sea floor. Salt domes are particularly common on continental shelves, about 350 having been identified in the Gulf of Mexico alone.

The greatest ocean basin depths occur in trenches (Table 3-2). Deepest is Challenger Deep, a part of the Marianas Trench (see Fig. 2-21), where a depth of 11,022 meters has been recorded. Obtaining accurate depth measurements in trenches is not easy, and depth figures are subject to revision as new techniques are developed or old ones improved. In general, the trenches are hundreds of kilometers wide and about 3 to 4 kilometers deeper than the surrounding ocean bottom. Individual trenches have lengths of thousands of kilometers.

The sides of trenches often consist of a series of steep-sided steps, suggesting that extensive faulting has taken place and caused blocks of the earth's crust to subside. The result is a narrow, deep, V-shaped profile which is almost invariably flat-bottomed due to sediment deposits. The largest negative gravity anomalies on the earth's surface are associated with trenches, due to downbuckling of the crust beneath them.

Deep troughs also occur on the ocean bottom. These are generally less deep than the trenches and not always associated with island arcs. Some deep areas are apparently associated with fracture zones that cut the mid-ocean rises and ridges, such as the Romanche Trench in the Atlantic.

TABLE 3-2
Characteristics of Trenches *

TRENCH	DEPTH (kilometers)	LENGTH (kilometers)	AVERAGE WIDTH (kilometers)
Pacific Ocean			
Kurile–Kamchatka Trench	10.5	2200	120
Japan Trench	8.4	800	100
Bonin Trench	9.8	800	90
Marianas Trench	11.0	2550	70
Philippine Trench	10.5	1400	60
Tonga Trench	10.8	1400	55
Kermadec Trench	10.0	1500	40
Aleutian Trench	7.7	3700	50
Middle America Trench	6.7	2800	40
Peru–Chile Trench	8.1	5900	100
Indian Ocean			
Java Trench	7.5	4500	80
Atlantic Ocean			
Puerto Rico Trench	8.4	1550	120
South Sandwich Trench	8.4	1450	90
Romanche Trench	7.9	300	60

* After R. W. Fairbridge, 1966. "Trenches and Related Deep Sea Troughs," in R. W. Fairbridge, (ed.), *The Encyclopedia of Oceanography,* Reinhold Publishing Corporation, New York, pp. 929–38.

VOLCANOES, REEFS, AND ATOLLS

Volcanoes and volcanic islands are among the most conspicuous features of ocean basins. Projecting usually one kilometer or more above the surrounding sea floor, they are usually cone-shaped, as on land. It is estimated that there are at least 10,000 volcanoes in the Pacific alone, often grouped into provinces covering up to 10 million square miles. Some, like the Hawaiian Islands, form long chains; other groups are more circular. The location of these volcanic provinces often corresponds to deep faults, through which molten rock can move up to the ocean bottom. Considering that the Pacific is about half the world ocean, it is likely that there are 20,000 or more volcanoes scattered across the ocean bottom. We have no data to indicate how many former volcanoes are now buried by later flows of volcanic material or covered by sediment.

Once formed, volcanoes persist for a long time. They are alternately active for extended periods during which a volcanic cone builds up, then dormant. Between eruptions, volcanoes often subside slightly as crust and mantle adjust to the added load of volcanic rock. Periods of activity last for millions to tens of millions of years. Certain areas of the southwestern Pacific Ocean appear to have undergone an extended period of volcanic activity during which immense volumes of lava poured out onto the ocean floor, beginning about 10 million years ago and lasting perhaps 30 million years. After cessation of activity, the region gradually sank. In such areas of subsidence there is often a compensating low rise of the ocean bottom beyond the depression. Many extinct volcanoes are known to occur in other ocean basins—for instance the Bermuda Islands shown in Fig. 3-10.

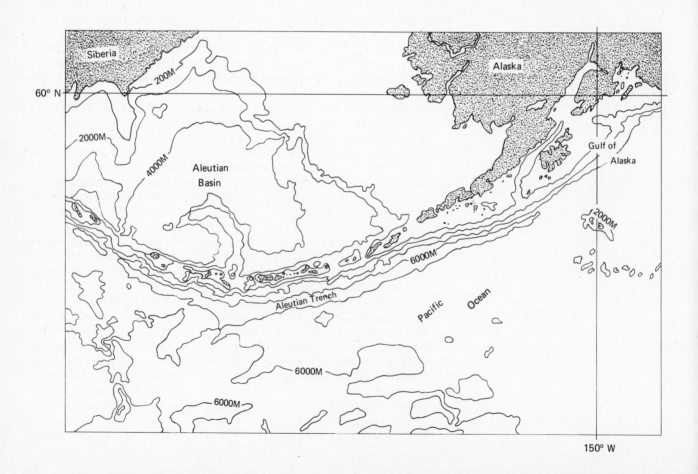

The andesitic island arcs associated with subduction zones (see Fig. 2-23) are sometimes relatively simple, and sometimes quite complex. The Aleutian Islands southwest of Alaska, for example, are a single chain with a trench on the outward, or seaward side, as shown in Fig. 3-11. West of the Aleutians, the Kamchatka Peninsula consists of a group of similar volcanoes but there a platform of older, greatly altered rocks is exposed above sea level, as it is in the Japanese Islands to the south. Probably such older, deformed rock underlies many of the world's island arcs but is submerged and not directly observable.

The complexity of some island-arc systems is illustrated by the Indonesian area, as shown in Fig. 3-12, where several chains of active or recently active volcanoes form complexly twisted arcs. The Java (Indonesian) Trench has several associated island groups including the island of Timor, one of the few places on earth where deep-ocean sediments are exposed on land. The West Indies of the western Atlantic is another example of a complex island-arc system. Deep-ocean sediments are exposed on the West Indian island of Barbados.

The Fiji Islands east of Australia provide a case history of changes in volcanic island composition that occur after an island arc has moved away from a trench system. The Fiji group was formerly associated with the Kermadec–Tonga trench system, and it was therefore andesitic in composition. Subsequently the islands have been carried westward, away from the subduction zone. Volcanism since that time has become increasingly basaltic, like that of an ocean island chain formed at a volcanic center.

Volcanic lava commonly erupts at temperatures between 900° and 1200°C. It flows for a while as molten rock, then solidifies first at the surface and finally throughout the flow. Further cooling turns the lava into volcanic rock.

Not all volcanic activity forms volcanoes. There are large areas on the ocean bottom where smooth plains of volcanic rock occur. Apparently lava flowed out in large volumes, covering all previous topography over extensive areas of the ocean bottom and leaving a bare rock surface, now slightly covered by sediment. Probably these lavas were at first quite fluid. There may have been several centers of eruption, perhaps long breaks forming linear fissure eruptions. These *archipelagic plains* (or *aprons*), as they are called, surround volcanic groups or islands and extend tens or hundreds of kilometers from the volcanoes themselves. About 80 percent of the volcanic rock in the Pacific is thought to occur in these vast archipelagic plains. Only about 15 percent occurs in volcanic ridges and large volcanic island groups.

There are two modes of volcanic eruption in the ocean. Some lavas form tranquil flows in which the surface cools first, forming an insulated cover for the still-molten material in the interior of the flow. When this type of lava flows into the ocean it forms "pillow lavas," rounded masses of volcanic rock. High pressure at oceanic depths seems to favor such tranquil flows, for they appear to be quite common in the deep ocean. The other type of marine volcanic eruption is explosive, producing large amounts of volcanic ash, small pieces of volcanic rock, and glass formed by quenching of molten rock (Fig. 3-13).

Numerous ancient Pacific islands were eroded to sea level and then submerged to depths of one to two kilometers by subsidence of the ocean floor. These are sometimes known as "guyots." The fact that they were once at sea level is demonstrated by the remains of shallow-water organisms and wave-rounded cobbles on their flat tops. Slow substance permits the growth of massive reefs and atolls around volcanic islands. These wave-resistant structures are built by carbonate-secreting organisms, mainly in open, subtropical oceans where average annual water temperatures are around 23° to 25°C. They are rare or absent where water temperatures in the coldest months fall below 18°C (Fig. 3-14).

Figure 3-10 *(facing page, top)*

The Bermuda Islands are solidified sand dunes built on a wave-eroded extinct volcano. The shallow top of the volcano (now forming a bank) was exposed during periods of lowered sea level, and carbonate sands were blown across the bank forming the dune now changed to soft, crumbly limestone. Reefs built primarily of calcareous algae can be seen in the lower portion of the photograph. (Photograph courtesy Bermuda News Bureau.)

Figure 3-11 *(facing page, bottom)*

Aleutian Islands, a simple island-arc system with trench on the convex, seaward side. These islands are built of volcanic rock from ancient eruptions. Some of the volcanoes are still alive.

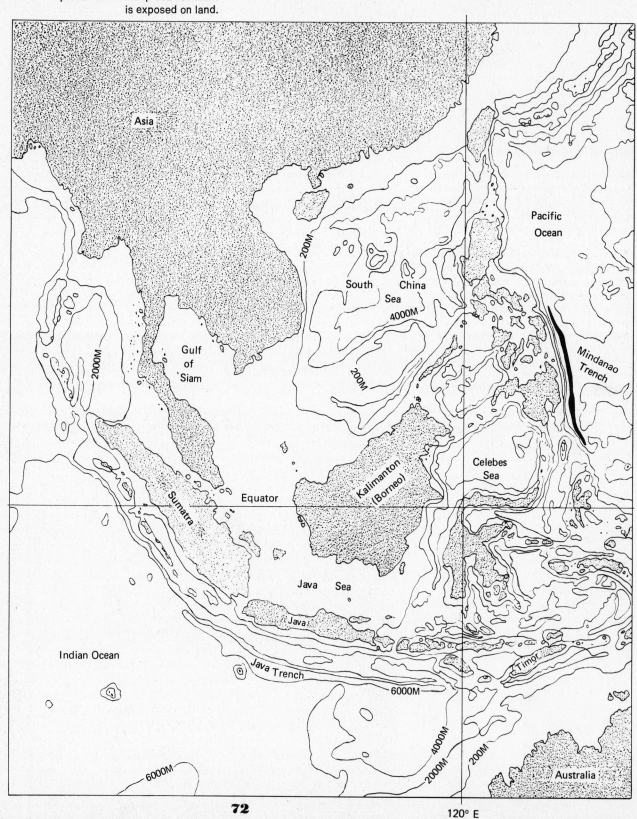

Figure 3-12

Indonesia, a complex island-arc system. Active volcanoes occur on the larger islands. Some smaller islands near the trench are built of deformed sediment. The island of Timor in the group is one of the few places where deep-sea sediment is exposed on land.

(a)

Figure 3-13

Eruption of Kovachi, a submarine volcano in the Solomon Islands, South Pacific, in October, 1969, caused shock waves and water eruptions approximately 30 seconds apart (a). Explosions ejected water and steam 60 to 90 meters into the air. The sea surface was discolored for 130 kilometers (80 miles). (Photograph courtesy Dr. R. B. M. Thompson and Smithsonian Institution Center for Short-lived Phenomena.) A new island (b) was built by Kovachi in March, 1970. A similar island formed by an eruption in 1961 has since been eroded by waves and is now completely submerged. (Photograph courtesy Smithsonian Institution Center for Short-Lived Phenomena.)

(b)

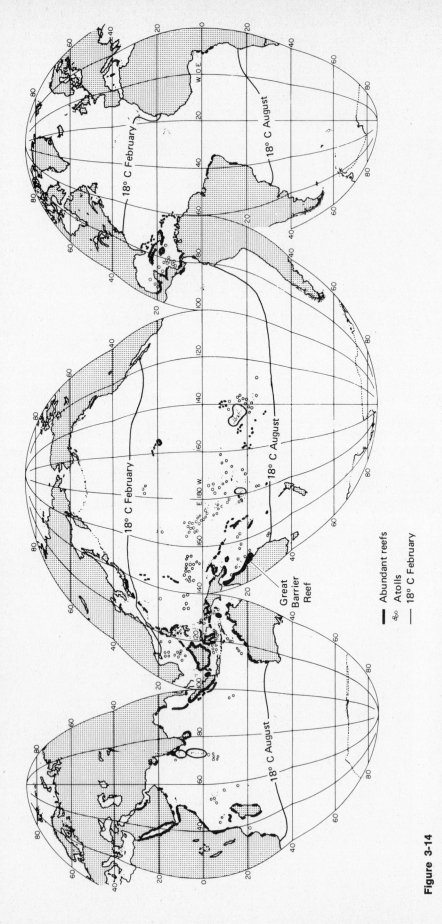

Figure 3-14

Distribution of coral reefs and areas of abundant atolls. The contour lines show the limits of waters that never get colder than about 18°C.

Carbonate-secreting animals and calcareous algae grow on the shallow bottom, eventually depositing a thick layer of calcareous skeletal material. If the platform on which they are growing slowly sinks, the organisms can grow upward and thus keep the upper part of the reef in shallow waters. Where this has occurred, a characteristic sequence of reef forms can be found. First, a *fringing reef* (Fig. 3-15(a)) grows along the shore of an island with gaps at river mouths. As the island sinks, the land area becomes smaller. Rain, wind and wave action erode it, further reducing the size of the island. But while the coastline retreats, the fringing reef offshore grows upward becoming a *barrier reef* (Fig. 3-15 (b)), and a lagoon separates the reef from the island. The Great Barrier Reef off Australia is a well-known example.

Figure 3-15

Stages in the transition from fringing reef
to barrier reef to atolls.

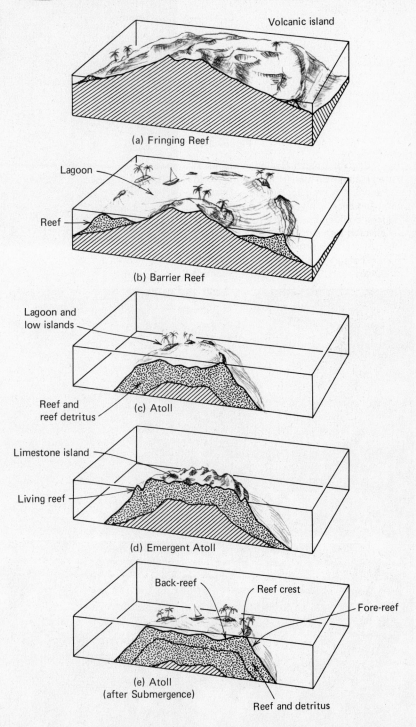

Figure 3-16

Six atolls in the Tuamoto Archipelago, part of French Polynesia, in the Central South Pacific Ocean (16°S, 145°W). The white bands are reefs surrounding lagoons. Islands occur on the reefs, often at the sharp bends in the atoll, and can be recognized because they appear darker owing to the vegetation on them. Faint lines transverse to the reef mark the deep-water passes through which ships enter the lagoon. The large white masses are clouds. (Photograph courtesy NASA.)

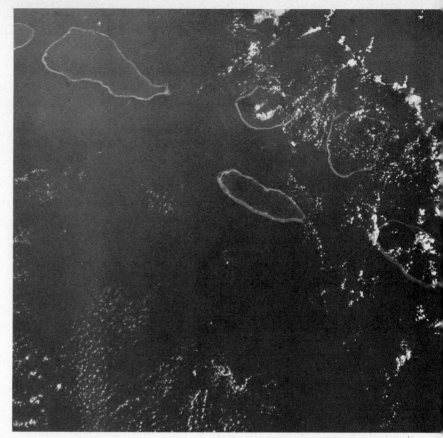

Figure 3-17

Edge of a patch reef in the lagoon at Midway Island, an atoll in the Central North Pacific. The irregular surface consists of coral and encrusting calcareous algae. The lagoon floor is about 5 meters deep and is covered with carbonate sand and larger pieces broken off the patch reef. (Photograph courtesy U.S. Geological Survey—J. I. Tracey, Jr., photographer.)

Finally the island may totally disappear to leave only the reef, now called an *atoll,* surrounding the lagoon (Fig. 3-15 (c)). Atolls are commonly irregular in shape, often generally elliptical in outline (Fig. 3-16). The lagoon is typically about 40 meters deep and connects to the ocean through passes—interruptions in the reef large enough for ships to navigate. There are often small islands on top of the reef or near it within the lagoon, made of sand or reef debris. Small *patch reefs,* their tops covered with coral and algae, grow up from the lagoon floor (Fig. 3-17).

A typical reef, for instance off the West Indian island of Jamaica, can be divided into 3 distinctive zones: back-reef, reef crest, and fore-reef (Fig. 3-15 (e)). The *back-reef* region consists of the lagoon and the protected inner side of the reef. This rises to a high point, the *reef crest,* which is exposed to the air at low tide. It extends from about sea level to depths of a few meters and includes a buttress zone which is commonly cut by a series of grooves alternating with spurs of reef growing seaward. This *spur-and-groove* structure absorbs wave energy and protects the reef from damage by breakers. Carbonate sediment and debris, swept by waves from the reef crest, slide down the channels of the steep reef-front to collect in deeper, more protected waters.

Reef crest and buttress zone are near-surface features, extending to depths of 7 to 10 meters. Below this the seaward slope or *fore-reef* is typically subdivided into a gently sloping *fore-reef terrace* (about 7 to 15 meters depth) bordered by a steep, pinnacled *fore-reef escarpment* shown in Fig. 3-18. At about 30 meters depth the reef front levels off to a seaward dipping, sandy *fore-reef slope* studded with coral pinnacles. Below about 50 to 60 meters the reef drops off sharply to depths of several hundred meters.

Figure 3-18

Sketch of fore-reef, fore-reef slope, and deep fore-reef from the perspective of a diver at 50 meters depth. Fore-reef terrace and escarpment are seen at upper left, merging with fore-reef slope. Coral pinnacles dot the fore-reef slope and large pinnacles rise above it to become promontories of the deep fore-reef. (Sketch by T. F. Goreau.)

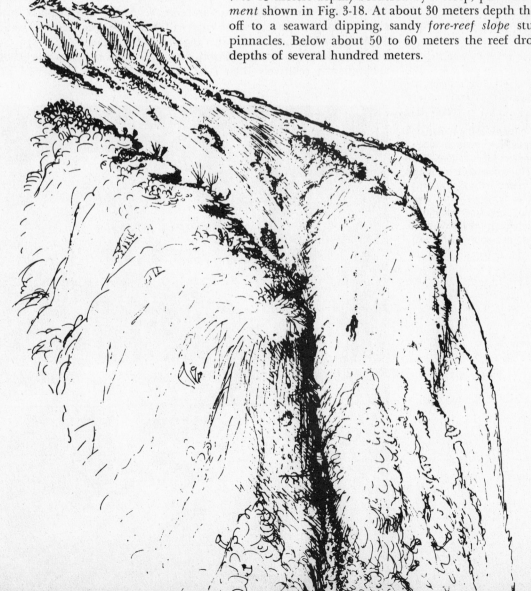

The reef itself consists of a carbonate framework built of coral skeleton bound together by encrusting calcareous algae. This forms an open, cavernous structure. Loose carbonate sediment collects in the internal spaces and is itself cemented, forming a nearly solid carbonate mass.

Reefs usually grow for millions of years. Vertical accretion of calcium carbonate on the fore-reef terrace and slope of a typical West Indian reef proceeds at a rate of 0.5 to 1 meter per thousand years, and horizontal accretion of material on the deep fore-reef has averaged about 0.2 meter over the same period.

THE MAJOR OCEAN BASINS

Up to this point we have studied the ocean floor in general. In this section we examine individual ocean basins, including some of the continental features that border them. The rivers that drain into an ocean basin play an important role in its oceanographic characteristics, because they discharge water and sediment. Mountains, on the other hand, can prevent rivers from reaching the ocean and cause them to drain into the continental interior.

Stretching about 17,000 kilometers (10,000 miles) from Ecuador to Indonesia, the Pacific Basin occupies more than one-third of the earth's surface (Fig. 3-19) and contains more than one-half of the earth's free water (see Fig. 2-5). As we have seen, mountain building is a dominant feature of Pacific shorelines, the mountains creating a barrier between inland areas and ocean. Along the Pacific's entire eastern margin (the western coasts of the Americas from Alaska to Peru) rugged young mountain ranges parallel the coast.

All these mountains inhibit direct access of rivers to the Pacific; consequently relatively little fresh water or sediment is carried to it

Figure 3-19

Satellite view of parts of the Indian and Pacific Oceans taken from an altitude of about 10,000 kilometers (6568 miles). Much of Asia (upper left) is obscured by clouds except for portions of the China coast. Australia is clearly visible in the lower portion of the photograph, a consequence of dry atmospheric conditions prevailing there. (Photograph courtesy NASA.)

from the interiors of the continents. Continental margins are narrow because the steep slope of the mountains tends to continue below sea level. As.a result, the central Pacific is little affected by the continents around it.

The Pacific is the deepest, the coldest, and by far the largest of the ocean basins (see Table 3-3). Paradoxically, it is also the least salty (see Table 3-4), even though it receives the least river water of any major ocean (see Table 3-5 and Fig. 3-20).

TABLE 3-3

*Surface and Drainage Areas of Ocean Basins and Their Average Depths ***

OCEAN †	OCEAN AREA (millions of square kilometers)	LAND AREA DRAINED ‡ (millions of square kilometers)	RATIO OF OCEAN AREA TO DRAINAGE AREA	AVERAGE DEPTH † (meters)
Pacific	180	19	11	3940
Atlantic	107	69	1.5	3310
Indian	74	13	5.7	3840
World ocean	361	101	3.6	3730

   * From H. W. Menard and S. M. Smith, 1966. "Hypsometry of Ocean Basin Provinces," *Journal of Geophysical Research,* 71, 4305.
   † Includes adjacent seas. Arctic, Mediterranean, and Black Seas included in the Atlantic Ocean.
   ‡ Excludes Antarctica and continental areas with no exterior drainage.

TABLE 3-4

*Average Temperature and Salinity of World Ocean ***

OCEAN	TEMPERATURE (°C)	SALINITY (parts per thousand)
Pacific	3.36	34.62
Indian	3.72	34.76
Atlantic	3.73	34.90
World	3.52	34.72

   * From R. B. Montgomery, 1958. "Water Characteristics of Atlantic Ocean and of World Ocean," *Deep-Sea Research,* 5, 146.

TABLE 3-5

*Water Sources for the Major Ocean Basins (centimeters per year) ***

OCEAN	PRECIPI-TATION	RUNOFF FROM ADJOINING LAND AREAS	EVAPO-RATION	WATER EXCHANGE WITH OTHER OCEANS
Atlantic	78	20	104	6
Arctic	24	23	12	35
Indian	101	7	138	30
Pacific	121	6	114	13

   * From M. I. Budyko, 1958. *The Heat Balance of the Earth's Surface,* trans. N. A. Stepanova, Office of Technical Services, Dept. of Commerce, Washington.

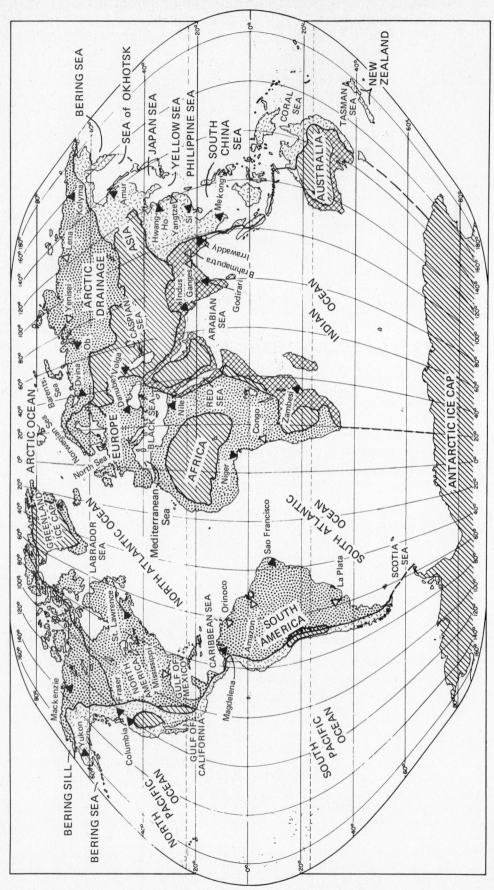

Figure 3-20

Map of the earth's surface showing the arbitrary boundaries drawn for the major oceans. Areas of the continents whose rivers drain into each of the ocean basins are shown as well as some of the major rivers. Note the large area that drains into the Atlantic and Arctic Oceans.

Legend:

▽ Rivers discharging more than 15,000 cubic meters (525,000 cubic feet) per second

▼ Rivers discharging more than 3000 cubic meters (100,000 cubic feet) per second

Runoff to Atlantic Ocean and Arctic Sea

Runoff to Pacific Ocean

Runoff to Indian Ocean

No runoff to oceans

The largest Pacific islands, occurring along the western margin of the basin, are not volcanic but continental in structure. They stretch from New Zealand in the south to Japan in the north, the largest being New Guinea. Consisting of continental-type rocks, these large islands are generally richer in industrial minerals and other natural resources than the smaller volcanic or carbonate islands.

The Atlantic Ocean basin was the first to be explored by Europeans, and it has been studied more thoroughly than any other. This ocean is relatively narrow, about 5000 kilometers wide (3000 miles). Its shorelines form an S-shaped basin which is the major connection between the two polar regions of the world ocean (Fig. 3-20).

The boundary between the Atlantic and Indian Oceans is arbitrarily drawn from the Cape of Good Hope at the southern tip of Africa, and extends southward along longitude 20°E to Antarctica. The boundary between the Atlantic and Pacific Oceans is drawn between Cape Horn at the lower tip of South America to the northern end of the Palmer Peninsula of Antarctica. When the Arctic Ocean is included as part of the Atlantic, the northern boundaries of both the Atlantic and the Pacific are made to meet at the Bering Strait between Alaska and Siberia—a shallow area known as the Bering Sill forms an easily recognizable boundary between the two oceans. (Although oceanographers draw such boundaries with ease, the ocean pays little attention to them. There is substantial movement of water across all these artificial boundaries, including the Bering Sill.) Defined this way, the Atlantic Ocean extends from the shore of Antarctica northward across the North Pole to about 65°N on the other side of the earth. Neither of the other two major ocean basins extends so far in a north–south direction (Fig. 3-21).

Figure 3-21

Distribution of land and water in each 5° latitude belt. (Redrawn from M. Grant Gross, 1976. *Oceanography*, 3rd ed., Charles E. Merrill, Columbus, Ohio.)

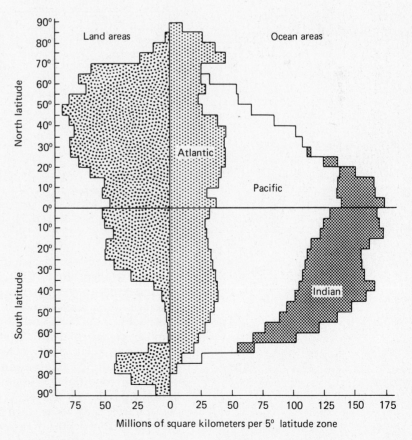

The Atlantic is relatively shallow, having an average depth of only 3310 meters (10,800 feet), the result of an abundance of continental shelf in the ocean, and shallow marginal seas and the presence of the mid-Atlantic Ridge. The Atlantic also receives a relatively large amount of sediment from numerous rivers, which covers the ocean floor and further reduces its depth. There are relatively few islands in the Atlantic Ocean. Greenland, the world's largest island, is actually a part of the North American continent, isolated by the relatively narrow channel formed by Davis Strait and Baffin Bay. Most other Atlantic islands are tops of volcanoes, many of them located on the mid-Atlantic Ridge.

The Atlantic Ocean receives a large amount of fresh water from river discharge. The Amazon and the Congo, the world's two largest rivers (see Fig. 3-20 and Table 3-6) flow into the Atlantic equatorial region, together discharging about one-quarter of the world's river discharge to the ocean. Other large rivers flow into the marginal seas that are common around the North Atlantic. For example the Mississippi River, draining much of the North American continent, discharges into the Gulf of Mexico, and the St. Lawrence flows into another small gulf. Also, fresh water running from several large rivers into the Arctic Ocean eventually enters the Atlantic.

TABLE 3-6
River Discharge to Ocean

	DISCHARGE		
RIVER	(cubic kilometers per year)	(cubic miles per year)	OCEAN BASIN
Amazon	5550	1330	Atlantic
Congo	1250	300	Atlantic
Yangtze	688	165	Pacific
Ganges	588	141	Indian
Mississippi	555	133	Atlantic
St. Lawrence	446	107	Atlantic
Mekong	350	84	Pacific
Columbia	178	42.7	Pacific
Yukon	165	39.6	Pacific
Nile	117	28	Atlantic
Colorado	5	1.2	Pacific
Total World	30,000	7400	

After H. G. Deming, 1975. *Water: The Fountain of Opportunity,* Oxford University Press, New York, p. 28.

Despite its connections to both polar regions and the large amount of fresh water discharged into it, the Atlantic is the warmest (average temperature 3.73°C) and saltiest (average salinity 34.90 parts per thousand—usually written as 34.90 $^o/_{oo}$) of the world ocean basins (average ocean temperature is 3.52°C, average salinity 34.72 parts per thousand). This apparent paradox is probably a result of the large number of marginal seas adjoining the Atlantic. The Mediterranean Sea in particular exerts a strong influence on the entire Atlantic. Waters in the Mediterranean are warmed and exposed to dry atmospheric conditions. Substantial amounts of water are lost by evaporation from the sea surface, so that the sea water becomes more saline. This warm, salty water then flows out into the Atlantic and is detectable at mid-depths over a wide area extending many thousands of kilometers from the Strait of Gibraltar.

The Arctic Ocean is a landlocked arm of the Atlantic including all waters north of Eurasia and the North American continent. It is unique in several respects: continental shelf forms one-third of its floor, it is almost completely surrounded by land, and—especially characteristic—much of it is covered by sea ice during part or all of the year.

The Arctic as a whole is quite shallow, largely because of its wide continental shelves. On North America, between Alaska and Greenland, the shelf is only 100 to 200 kilometers (60 to 100 miles) wide, but offshore from western Alaska and Siberia it is much wider, 500 to 1700 kilometers (300 to 1000 miles) across, as shown in Fig. 3-22. During much of the year it is unnavigable because its surface is covered by *pack ice* (sea ice in thick chunks that freeze together, break apart, and freeze again to form a solid surface) to a depth of 3 to 4 meters (10 feet or more). In the central Arctic this cover is permanent, though it may melt to a minimum thickness of about 2 meters by the end of August.

Figure 3-22

Map of the Arctic Ocean showing submarine topography.

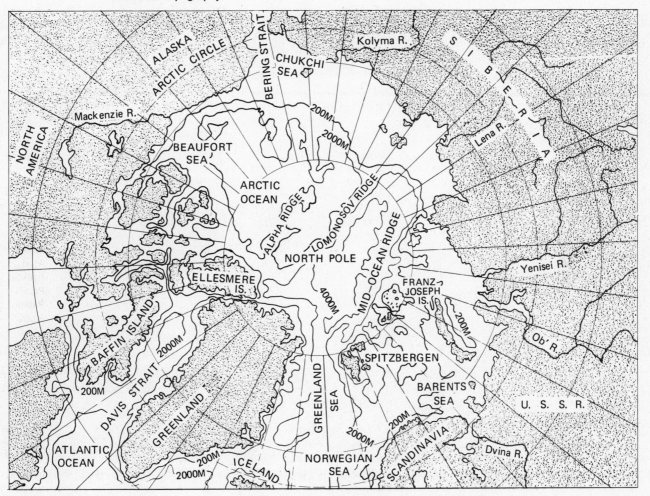

Being surrounded by continents, the Arctic Ocean is greatly influenced by river discharge and has low salinity. Large amounts of sediment on the bottom of Arctic basins testify to extensive erosion of surrounding continents by glaciers during the past several million years. Much of this material was deposited as glaciers melted and lost the sediment they had accumulated while advancing over the land. Rivers also discharge considerable sediment into the Arctic region.

The Indian Ocean lies primarily in the Southern Hemisphere, extending from Antarctica to the Asian mainland at about 20°N, as shown in Fig. 3-20. Its Pacific boundary runs through the Indonesian Islands and extends from Australia, through Tasmania and southward to Antarctica along longitude 150°E. It is the smallest of the three major ocean basins, triangular in shape, with a maximum width of about 15,000 kilometers between South Africa and New Zealand. The continental shelves around this ocean are relatively narrow, but the large amount of sediment that has been deposited in the northern part of the basin makes it intermediate in depth.

Three of the world's major rivers (Ganges, Indus, Brahmaputra) discharge into the Indian Ocean (see Table 3-6) along its northern margin, also the location of its only marginal seas. On these the Red Sea and Persian Gulf have the greatest effect on the Indian Ocean as a whole, because they are areas of intensive evaporation. Most of the

Figure 3-23

The subcontinent of India and the island of Sri Lanka (formerly Ceylon) as seen from Gemini XI in September, 1966, from an altitude of 815 kilometers (507 miles). A well-developed sea breeze circulation has suppressed clouds along the coast, although clouds obscure much of the land. (Photograph courtesy NASA.)

water discharged into the Indian Ocean drains from the Indian subcontinent (shown in Fig. 3-23); little comes from Africa.

There are relatively few islands in the Indian Ocean, Madagascar being the largest one. It is continental in composition, as are several shallow banks in the ocean basin—all apparently fragments of an ancient continental block. There are a few volcanic islands and several groups of carbonate islands, including atolls.

MARGINAL-OCEAN BASINS

Marginal-ocean basins are large depressions in the ocean bottom lying near continents. They are separated from the open ocean by submarine ridges, islands, or parts of continents. Despite their proximity to continents, marginal basins are usually more than 2 kilometers deep, so that their bottom waters are partially isolated from ocean waters at comparable depths outside the basin.

These small seas are, in a sense, transitional between continents and deep-ocean basins, as their location would suggest. Studies of their crustal structure has shown that they commonly have a typical oceanic crust overlain by thick sediment deposits. Being near the coast, they receive the discharge of many large rivers. The sediment brought by these rivers is deposited in the basins, resulting in smooth sediment deposits many kilometers thick. These deposits contain about one-sixth of all known ocean sediments.

Nearly all marginal-ocean basins are located in areas where crustal deformation or mountain building is active. They may eventually be changed and incorporated into the adjacent continental block. Three kinds of associations are typical of marginal-ocean basins. The most common is that of a basin associated with island arcs and submarine volcanic ridges which partially isolate surface waters and may completely isolate the subsurface waters. The many small basins around Indonesia and along the Pacific margin of Asia are examples of this association.

A second type of marginal sea lies between continental blocks— for instance, the Mediterranean Sea between Europe and Africa, the Black Sea (now almost completely landlocked), and the Gulf of Mexico and Caribbean Sea between North and South America (see Fig. 3-20).

The third type is the long, narrow marginal sea formed where oceanic rises are breaking up continental blocks. The Red Sea (shown in Fig. 3-24) and the Gulf of California are examples of such basins.

Marginal seas are subjected to more changeable weather conditions than occur in the open ocean, owing to the proximity of large land masses. Marginal seas in high latitudes, such as the Sea of Okhotsk, are cold enough during the winter months to freeze over. In mid-latitudes, where the world's deserts are located, marginal seas such as the Mediterranean Sea and Persian Gulf are strongly evaporated, so that their waters are much warmer and saltier than the world-ocean average.

Because of their marked response to seasonal weather changes, water circulating through marginal seas has a recognizable influence on adjacent ocean basins. As we have already mentioned, this is especially true for the Atlantic Ocean, which has many marginal seas and is a relatively small ocean basin.

In Europe and North America, shallow seas occupy depressed areas at the edge of continental blocks, such as Hudson Bay in Canada. The North Sea of northern Europe is another such flooded continental area, leaving only England and her associated islands above sea level at the present time.

Figure 3-24

The Red Sea with the Gulf of Aqaba and the Gulf of Suez. The dark band on
the right is the Nile Valley and the river can be seen as the light meandering line.
Note the straight coastlines on either side of the Red Sea. The absence of
clouds is an indication of the region's aridity. The photograph was taken from
Gemini XII at an altitude of 280 kilometers (175 miles); portions of the
spacecraft are visible on the right side. (Photograph courtesy NASA.)

SUMMARY OUTLINE

Continental margin
    Continental shelf—average 70 kilometers wide; rough if region was glaciated, smooth if adjacent to smooth coastal plain
        Features formed by recent migration of shoreline
        Shelf break averages 130 meters depth
    Continental slope—narrow, steeply inclined edge of continents
        Formed by faulting and slumping of rocks in many places
        Cut by submarine canyons probably formed by turbidity currents
    Continental rise—sediment accumulations at base of continental slope contain about 45 percent of identified sediment in ocean

Oceanic rises—comprise a single system throughout all ocean basins
    Mid-Atlantic Ridge—rugged, faulted, with rift valleys
    East Pacific Rise—low bulge, little faulting, no rift valleys
    Mid-Indian Ridge—resembles Mid-Atlantic Ridge
    Characteristics of oceanic rises:
        Comparable in size to continents
        Many volcanoes and shallow-focus earthquakes
        High heat flow, usually in narrow bands
        Near continents, they block sediment transport to deep ocean basin

Ocean basin
    Sediment cover typically accumulated particle by particle, covering older topography
    Fracture zones cross ocean basins
    Abyssal hills abundant
    Abyssal plains—among flattest portions of earth's crust
        Occur at seaward margin of continental rise
        Crossed by deep-sea channels, avenues of sediment dispersal
        Common in marginal seas, e.g. Gulf of Mexico, Caribbean
    Salt domes—in areas of evaporation, flooding; may trap fossil fuels
    Trenches—greatest depths in ocean

    Negative gravity anomalies
    Faulted sides, sediment in bottom

Volcanoes, reefs, atolls
    Volcanoes—1 kilometer or more in height, tops may form islands
    Andesitic island arcs—simple or complex
    Archipelagic plains—flat, smooth lava flows; 80 percent of volcanic rock in Pacific
    Guyots—subsided, flat-topped seamounts
    Coral reefs—wave-resistant platforms built in shallow water by carbonate-secreting organisms
        Corals, encrusting calcareous algae form framework, sediment is trapped
        Grow in shallow, warm, sunlit waters of near-normal salinity
    Atolls result from submergence of platform, most common in Western Pacific
        Back-reef—inner surface, lagoon
        Reef crest—exposed at low tide, spur-and-groove structure below
        Fore-reef—coral pinnacles

Major ocean basins—terrain of shoreline governs water, sediment discharge
    Pacific Ocean—largest basin, contains more than 50 percent of surface water
        Bordered by mountains; land isolated from ocean
        Many marginal seas; volcanoes and volcanic islands; large continental islands
    Atlantic Ocean—long, narrow; connects North and South Polar regions
        Relatively shallow, many marginal seas
        Receives about 68 percent of world's drainage
    Arctic Ocean—nearly landlocked arm of the Atlantic
        Shallow, wide continental shelves
        Low salinity, covered by ice most of the year
    Indian Ocean—Southern Hemisphere; smallest of three major basins
        Major rivers discharge into northern section

Marginal-ocean basins
    Usually formed by partial isolation resulting from tectonic processes
    Usually have distinct oceanographic characteristics—highly variable

SELECTED REFERENCES

HOLMES, ARTHUR. 1965. *Principles of Physical Geology*, 2nd ed. Ronald Press, New York. 1288 pp. Advanced treatment of geological principles.

MACDONALD, GORDON A. 1972. *Volcanoes*. Prentice-Hall, Englewood Cliffs, N.J. 510 pp. Comprehensive study of volcanic activity.

SHEPARD, FRANCIS P. 1973. *Submarine Geology*, 3rd ed. Harper & Row, New York. 517 pp. Advanced discussion of marine geology.

Small free-fall gravity corer ready for sampling deep-ocean sediment. (Photograph courtesy Woods Hole Oceanographic Institution.)

Small free-fall gravity corer ready for sampling deep-ocean sediment. (Photograph courtesy Woods Hole Oceanographic Institution.)

4

sediments

M ost of the ocean bottom is covered by sediment deposits—unconsolidated accumulations of particles brought in by rivers, glaciers, and winds and mixed with shells and skeletons of marine organisms. But a few areas of ocean bottom are devoid of sediment cover; some are swept clean by strong currents, others are newly formed and have not accumulated much sediment. Sediment is usually thickest near continents, especially in marginal-ocean basins where barriers interfere with sediment transport into the main ocean basin, and thinnest on newly formed mid-oceanic ridges.

Throughout its history, the ocean has been the final repository for debris from the land. Recoverable sediment deposits record ocean basin history during the past 200 million years. Earlier deposits have apparently been destroyed by crustal subduction. Deep-sea deposits also tell much about oceanic processes, such as nutrient supply to organisms and the effect of climatic changes on oceanic circulation. In short, the study of oceanic sediments and the fossils in them provides us with a time dimension for ocean studies.

ORIGIN AND CLASSIFICATION OF MARINE SEDIMENT

Sediment particles are classified in two different ways—by origin and by size. Classified by origin, sediment particles are divided into the following groups:

Lithogenous particles (derived from rocks): primarily silicate mineral grains (listed in Table 4-1) released by the breakdown of silicate rocks on the continents during weathering and soil formation; in the open ocean, volcanoes are locally important sources of lithogenous particles.

Biogenous particles (derived from organisms): the insoluble remains of bones, teeth, or shells of marine organisms. Some typical particles are illustrated in Fig. 4-1, 4-2, and 4-3.

Hydrogenous particles (derived from water): formed by chemical reactions occurring in sea water or within the sediments; manganese nodules illustrated in Fig. 4-4 are examples.

Sediment particules are also classified according to size, as in Table 4-2. Particle size is an important property of sediments because it determines the mode of transport and also determines how far particles travel before settling to the bottom. Typical sizes of particles from various sources are indicated in Fig. 4-5.

TABLE 4-1

Common Minerals In Marine Sediments and Their Chemical Composition

MINERAL	CHEMICAL COMPOSITION
Silicates—primarily lithogenous, derived from silicate rocks:	
Quartz	SiO_2
Opal	$SiO_2 \cdot H_2O$, glasslike substance formed by diatoms, radiolaria
Feldspars	Alumino-silicates with Na, K, and Ca in variable proportions
Micas	Primarily muscovite (a K, Al, hydrous silicate)
Clays	Similar to micas but with more variable composition and structure
Illite	Primarily altered, fine-grained muscovite
Montmorillonite	A group of minerals formed commonly from volcanic glass
Chlorite	Fe, Mg, Al hydrous silicate, commonly formed during weathering
Kaolinite	Al silicate formed during tropical weathering
Gibbsite	Al oxide, formed during tropical weathering
Zeolites	Hydrous alumino-silicates, similar to feldspars in structure
Oxides, hydroxides—occur primarily in Fe–Mn nodules:	
Hematite	Fe_2O_3, probably red coloring on deep-sea clays
Goethite	$FeO(OH)$, constituent of Fe–Mn nodules
Lepidocrocite	$FeO(OH)$, another mineral form of Fe in nodules
Pyrolusite	MnO_2, occurs in Fe–Mn nodules
Carbonates—primarily formed by marine calcareous organisms:	
Calcite	$CaCO_3$, forms foraminiferal shells, coccoliths, mollusk shells
Aragonite	More soluble form of $CaCO_3$, in pteropod shells, corals
Phosphates—rare in sediments, common in bones and teeth:	
Apatite	$Ca_5(PO_4)_3F$
Sulfates:	
Barite	$BaSO_4$, found in certain primitive organisms, and sediments rich in organic matter
Celestite	$SrSO_4$, found in shells of certain radiolaria, rarely in sediment

Biogenous minerals in boldface. Hydrogenous minerals are in italics. All others are lithogenous in origin.

TABLE 4-2

Classification of Sediment Particles By Size
(Wentworth Scale)

SIZE FRACTION	PARTICLE DIAMETER (millimeters)
Boulders	>256
Cobbles	64–256
Pebbles	4–64
Granules	2–4
Sand	0.062–2 (62–2000 μ *)
Silt	0.004–0.062 (4–62 μ)
Clay	<0.004 (<4 μ)

* μ is a *micron,* one millionth of a meter, or 39.4 millionths of an inch.

91

If a sediment consists of single-sized particles, it is *well sorted*. A *well-sorted sediment* is one in which a limited size range of particles occurs, the other-sized particles have been removed, usually by mechanical means. Beach sand is a typical well-sorted sediment, in which particles have been segregated by size as a result of wave or current actions. A *poorly-sorted sediment* contains different sized particles, usually indicating that little mechanical energy has acted to sort the particles. Glacial marine sediments, derived from mechanical wearing down of rocks by glaciers, are poorly sorted. Commonly they contain particles ranging in size from boulders to fine silt (see Table 4-2).

Sands usually occur along coastlines where waves and currents remove the finer particles, depositing them in deeper water. (Effects of currents are shown in Fig. 4-8.) Sands are rare in the deep ocean, although, as we shall see, they do occur there. *Mud*—a mixture of silt- and clay-sized particles—is the most common type of sediment in the deep ocean, as indicated in Table 4-3.

TABLE 4-3

Abundance and Composition of Deep-Ocean Sediment *

		Siliceous Muds		
CONSTITUENTS	CALCAREOUS MUD (percent)	DIATOMA-CEOUS (percent)	RADIO-LARIAN (percent)	BROWN MUD (percent)
Biogenous:				
Calcareous	65	7	4	8
Siliceous	2	70	54	1
Lithogenous and hydrogenous	33	23	42	91
Abundance in deep-ocean sediments	48	14		38

* Recalculated after R. R. Revelle, 1944. *Marine Bottom Sediment Collected in the Pacific Ocean by the* Carnegie *on its Seventh Cruise,* Carnegie Institution Publication 556, part 1.

Lithogenous constituents of marine sediment are derived from silicate rocks. Most rocks originally were formed at high temperatures and pressures in the absence of free oxygen and with little liquid water. These rocks break down due to a complex of chemical and physical processes known as *weathering*. Certain soluble constituents are released and carried to the ocean dissolved in river water. The remaining rock is often broken up by physical processes into sand- and clay-sized grains, which are also carried suspended in running water.

Some minerals—for example, quartz and some feldspars—resist chemical alteration by air and water. These usually enter the ocean virtually unaltered and commonly occur in sands and silts. Some products of rock weathering and the slightly altered remains of mica-like minerals form the clay minerals, common in fine-grained deep-sea clays.

Dissolved substances, removed from seawater, are used by organisms to make their shells and skeletons. Materials most commonly used are calcium carbonate, silica, and calcium phosphate. When the plant or animal dies the shells sink to the bottom, but during their descent they may be consumed and digested by larger animals. Only the sturdiest survive to be incorporated into sediment deposits. Fragile shells are rarely preserved.

The most common *biogenous* remains in sediments consist of calcium carbonate, primarily the mineral *calcite*. The chemically similar mineral *aragonite* is more soluble and therefore less frequently preserved in deep-ocean sediments. The most common among the larger biogenous particles are shells of *foraminifera,* which are one-celled animals, pictured in Fig. 4-1. Among the minute calcium carbonate particles, *coccoliths*—platelets formed by tiny, one-celled algae—are common in carbonate-rich sediments. Shells of *pteropods*—floating snails—form rare deposits at shallow depths, primarily in the Atlantic Ocean; at greater depths, their fragile aragonitic shells are broken up and usually dissolved.

Figure 4-1

Some calcareous, deep-sea sediments obtained by the *Challenger* Expedition. (top) *Globigerina* mud; (bottom) pteropod mud.

Siliceous shells of *diatoms, radiolaria,* and *silicoflagellates* are also common biogenous constituents. Of these, the shells of diatoms (one-celled algae) and radiolaria (one-celled floating animals) are the most important. Siliceous *sponge spicules*—the spikey skeletal elements shown in Fig. 4-2—are also found in many deep-ocean sediments, but these never dominate the deposits to the extent that diatoms and radiolaria do.

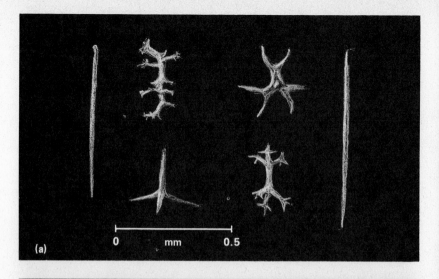

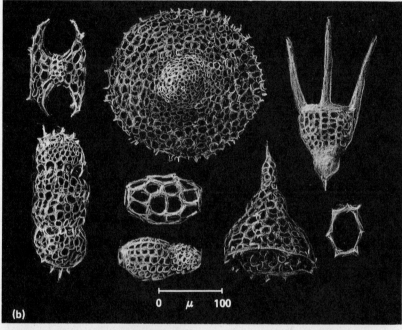

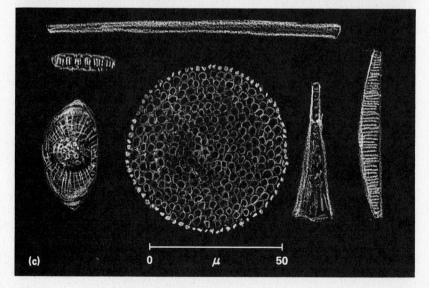

Figure 4-2

Some common siliceous biogenous
constituents in marine sediments:
sponge spicules (a) radiolaria, (b) and
diatoms (c).

Sediment containing more than 30 percent biological constituents by volume is classified as an *ooze*: a "diatom ooze" if diatoms were dominant, a "foraminiferal ooze" if shells of foraminifera were the most abundant carbonate constituent. For our purposes, we call such sediments *diatomaceous muds* if diatom shells constitute more than 30 percent by volume, or *radiolarian muds,* or *foraminiferal muds,* using the same standard. If none of these biogenous constituents dominate the sediment, then we shall refer to the deposit as a *deep-sea mud,* sometimes also called a *red clay* because of its typical red or red-brown color, especially in the Atlantic.

Other biogenous remains are less common. For example bones are rather fragile, open structures made up of minute calcium phosphate crystals embedded in an organic matrix. When bones are broken and eaten, bacteria decompose the organic matrix, leaving only tiny fragments of phosphate crystals which commonly dissolve. Only the most resistant bones, such as sharks' teeth or whales' earbones, remain intact on the ocean bottom; they are recovered by dredging, and are sometimes found in the centers of manganese nodules as those in Fig. 4-4.

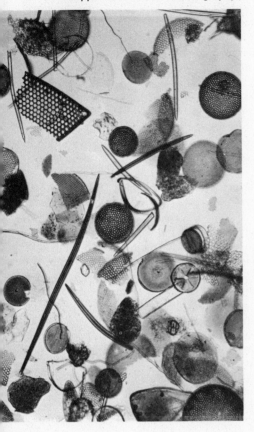

Figure 4-3 (below)

Diatoms, siliceous constituents of deep-sea sediments, are typically about the size of a sand grain (0.05 to 0.1 millimeters). (Photographs courtesy of W. R. Reidel, Scripps Institution of Oceanography.)

Figure 4-4 (right)

Manganese nodules, 8–10 centimeters in diameter, on the ocean floor in certain parts of the South Pacific. (Photograph courtesy National Science Foundation.)

Hydrogenous constituents are formed through chemical processes occurring in seawater itself. There is some indication of bacterial activity involved in formation of manganese nodules (also known as iron-manganese nodules, because of their high iron content). These elements, originally brought to the ocean by rivers and possibly from volcanic activity on the ocean bottom, apparently precipitate in the presence of the abundant dissolved oxygen typically found in deep-ocean waters. Their precipitates form tiny particles that are swept along by currents until they contact a surface. The particles tend to adhere to solid surfaces, so they form a stain or coating around a solid nucleus, such as a fragment of volcanic rock or other debris. In the deep ocean, where sediment accumulates slowly, sharks' teeth and whales' earbones are common examples of such biological residues. Through time an iron-manganese coating builds up slowly around the original nucleus, and eventually a nodule is formed.

Nodules form by accretion of layers and they contain most particles commonly found in seawater. Presumably these, too, have been incorporated in the nodules by striking the surface and adhering to it. Deep-sea nodules have high copper, cobalt, and nickel contents, which makes them attractive as a possible source of copper and other scarce metals.

Manganese nodules form at a rate of only a few atomic layers per day, one of the slowest chemical reactions known. On the Blake Plateau of the continental rise off southeastern North America, the bottom is covered with such nodules, which apparently form at rates of about 1 millimeter per million years. In other areas, nodules seem to form 10 to 100 times faster. In some isolated or nearly isolated marine basins, such as the Baltic Sea, manganese nodules form at rates of 20 to 1000 millimeters per 1000 years. The speed record for nodule growth is the manganese coating that formed on a naval artillery shell fragment dredged off Southern California; it formed at the rate of about 10 centimeters per 100 years.

Figure 4-5

Grain-size distribution of common marine sediments and sediment sources.

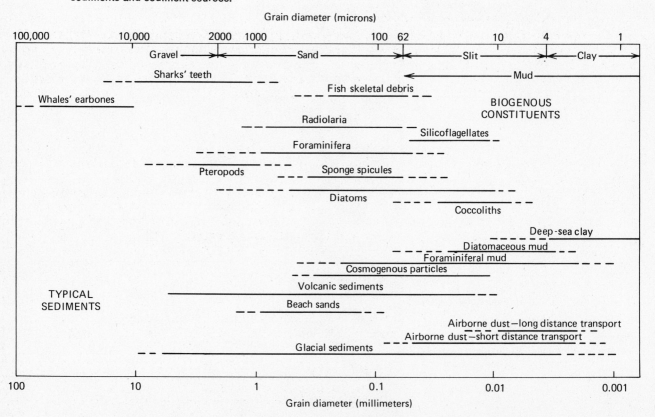

Forming so slowly, manganese nodules cannot grow to any substantial size in areas receiving large amounts of other sediments, because they will be buried too quickly. Consequently, nodules are rare in the Atlantic, which is well supplied with lithogenous sediment. In the Central Pacific where sediment accumulates slowly, nodules are quite common on the bottom. Manganese nodules are estimated to cover 20 to 50 percent of the deeper parts of the Pacific, based on analyses of deep-ocean photographs and sediment samples.

Particles from space also occur in marine sediments. Many are not recognizable meteorites but fall as dust, much of it apparently derived from meteorites. Cosmic spherules (Fig. 4-6), first recognized in deep-ocean sediment by Sir John Murray, are about 200 to 300 microns in diameter. They are magnetic, and largely composed of iron or iron-rich minerals.

Since the Industrial Revolution, man has supplied substantial amounts of sediment to the ocean through discharges of municipal and industrial wastes. Because of the high level of industrialization along their coasts, the United States and Northern Europe are large contributors of such man-made sediments. Most of these deposits appear to remain on continental shelves near coastal urban regions.

Figure 4-6

Cosmic spherules from deep-ocean deposits.
(*Challenger* Reports.)

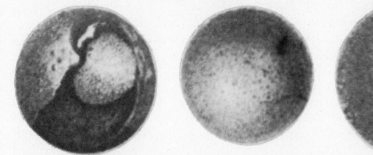

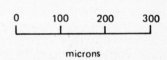

microns

SOURCES AND TRANSPORT OF LITHOGENOUS SEDIMENT

Rivers are the largest source of sediment in the ocean—about 20 billion tons each year, most of it coming from Asia as shown in Table 4-4. Ten rivers, listed in Table 4-5, contribute more than 100 million tons of sediment each year. Of these, the four largest are Asian rivers: the Ganges and Brahmaputra of India, and China's Hwang Ho (Yellow) and Yangtze. Together these supply about 25 percent of the total sediment discharge from continents.

TABLE 4-4

River Discharge of Suspended Sediment from the Continents *

CONTINENT	AREA DRAINING TO OCEAN (millions of square kilometers)	ANNUAL SUSPENDED SEDIMENT DISCHARGE $\left(\dfrac{10^9 \text{ tons}}{\text{yr}}\right)$	$\left(\dfrac{\text{tons/km}^2}{\text{yr}}\right)$ †
North America	20.4	2.0	96
South America	19.2	1.2	62
Africa	19.7	0.54	27
Australia	5.1	0.23	45
Europe	9.2	0.32	35
Asia	26.6	15.9	600
Total	100.2	20.2	

* From J. N. Holeman, 1968. "The Sediment Yield of Major Rivers of the World," *Water Resources Research*, 4, 737.

† Tons of sediment per square kilometer of drainage area per year.

TABLE 4-5

*Major Sediment-Producing Rivers ***

RIVER	DRAINAGE AREA (thousands of square kilometers)	Average Annual Sediment Discharge	
		SEDIMENT	
		$\left(\dfrac{10^6 \text{ tons}}{\text{yr}}\right)$	$\left(\dfrac{\text{tons/km}^2}{\text{yr}}\right)$
Yellow (Hwang Ho)	666	2100	2900
Ganges	945	1600	1500
Brahmaputra	658	800	1400
Yangtze	1920	550	550
Indus	957	480	510
Amazon	5710	400	70
Mississippi	3180	340	110
Irrawaddy	425	330	910
Mekong	786	190	480
Colorado	630	150	420

* After J. N. Holeman, 1968. "The Sediment Yield of Major Rivers of the World," *Water Resources Research*, 4, 737.

Except for the Ganges–Brahmaputra system, and the Amazon and Congo rivers (which drain into the equatorial Atlantic), the world's major sediment-producing rivers drain into marginal seas. Consequently, most of the sediment discharged by rivers into the ocean can not reach the deep ocean floor directly.

Much lithogenous sediment entering the ocean is derived from semiarid regions, usually from mountainous areas, as indicated in Fig. 4-7. Rainfall in such areas is usually inadequate to support an erosion-resisting cover of vegetation, but sufficient to erode soil. In Asia and

Figure 4-7

Major source areas of lithogenous sediment. The rivers shown discharge more than 100 million tons of suspended sediment each year. (Data from J. N. Holeman, 1968. "The Sediment Yield of Major Rivers of the World," *Water Resources Research*, 4, 737.)

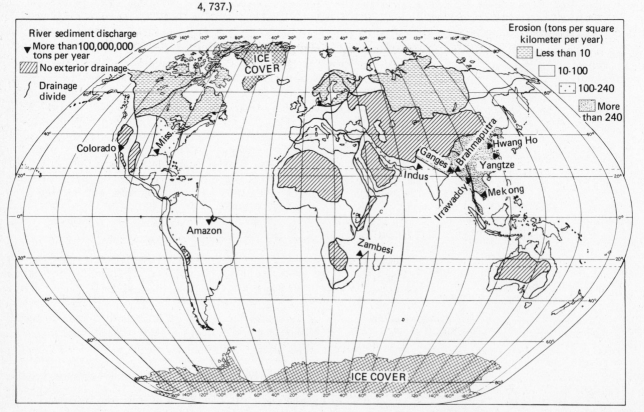

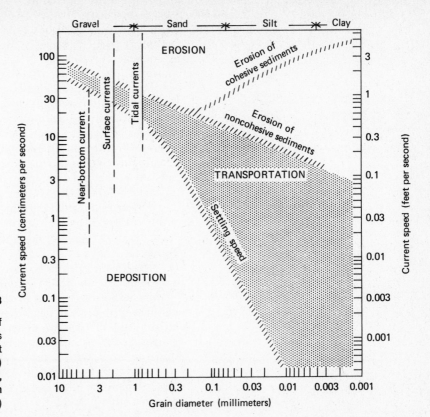

Figure 4-8

Current speeds required for erosion of sediments and transport of grains as compared to settling speeds for sediment grains. (Note use of logarithmic scale.) (After B. C. Heezen and C. D. Hollister, 1964. "Deep-Sea Current Evidences from Abyssal Sediments," *Marine Geology*, 1, 141.)

India, long-continued intensive agricultural development has further increased sediment yield. Low-lying tropical areas usually supply relatively little sediment per unit of drainage basin because the heavy vegetation inhibits particle movement. Also, intensive weathering of rocks in tropical climates tends to break them down completely, forming soluble components, rather than to leave discrete mineral particles. In polar regions, low temperatures retard chemical decomposition of rocks because water, of major importance in weathering, is frozen for much of the year. Furthermore, the glaciation of these high-latitude regions during the last Ice Age scoured out many lakes which act as traps, inhibiting sediment transport to the ocean. Transport of sediment particles is determined primarily by physical characteristics. Particle size (see Fig. 4-5) and current speed (see Fig. 4-8) are most important. Vertical transport (sinking of particles through a column of water to the ocean floor) is easiest to visualize.

The settling speed of a particle is controlled primarily by its diameter; large particles fall faster than small ones. Settling times for different sized grains are:

Particle	Settling velocity (centimeters per second)	Time to settle 4 kilometers
Sand (100 μ)	2.5	1.8 days
Silt (10 μ)	0.025	185 days
Clay (1 μ)	0.00025	51 years

From this it is obvious that sands could settle out near the point where they enter the ocean, silts could be transported moderate distances, and clays could theoretically be transported throughout the world ocean before settling through the 4 kilometers of sea water.

But let us consider another aspect of sediment movement: lateral transport. Figure 4-8 summarizes available estimates about the current

99

speeds necessary to suspend sediment particles and to transport them. Note that for noncohesive sediments, the speed necessary to move sediment particles steadily decreases as particle size decreases, so that even weak currents can erode and transport fine-grained sediment. If, however, particles are cohesive (stick together), the force required to erode the deposit first decreases as the grain size gets smaller and then increases, so that erosion of cohesive silts and clays may require currents just as strong as those necessary to move sands and gravels. Once eroded and set in motion, particles tend to settle out. This tendency is opposed by the upward-directed but basically random water movements existing during turbulent flow. Even at current speed less than necessary to erode the sediment, particles may still be transported some distance before settling out of the water.

Surface and tidal currents (see Fig. 4-8) are often strong enough to erode and transport most common sediment types. Even deep-ocean currents can erode and transport noncohesive, fine-grained bottom deposits.

Sediment is initially transported by rivers either suspended in the water (usually about 90 percent of the river's sediment load), or rolling along the bottom as part of the so-called "bed load." The bed load is commonly deposited near the head of the estuary, where the downstream flow of river water encounters upstream near-bottom currents caused by the estuarine circulation (see Chapter 10). At that point the downstream current is too weak to transport the relatively large grains that constitute the bed load, so they are deposited.

Organisms also cause sediment deposition. Many marine organisms obtain food by filtering water. Mineral particles removed along with their food are compacted into fecal pellets and then excreted. Oysters, mussels, clams, and many other attached and floating animals effectively remove particles from suspension, and bind them into fecal pellets which accumulate in sediment deposits. Plants growing along the banks and on tidal flats also trap sediment by providing quiet sheltered areas where these small particles can settle out. Plants then prevent erosion of sediment deposits when tidal currents or heavy waves scour the area.

As a result of these processes (discussed in Chapter 10), most sediments are trapped and deposited near river mouths. Estuaries that have not been filled by sediment act as effective sediment traps; for example, rivers of the U.S. Atlantic coast transport an estimated 20 million tons of sediment each year to their estuaries, but virtually none reaches the continental shelf. Thus the continental shelf of eastern North America is covered primarily by sediment laid down about 3000 or more years ago when sea level was lower.

When the sediment load of a river is large enough, it may fill its estuary; then the sediment load can be carried out into the ocean. The Mississippi River is a good example: its large sediment load has not only filled the estuary but also created the Mississippi delta, a sediment deposit that extends into the Gulf of Mexico. Estimates for different rivers suggest that from 10 to 95 percent of a river's sediment load is deposited in its delta.

Perhaps as much as 100 million tons of lithogenous sediment per year are transported by winds to the ocean, mainly from deserts and high mountains having sparse vegetation and strong winds. Wind transport of sediment to the deep ocean seems to be especially important in middle latitudes, where there are few rivers owing to the arid climate. It seems to be less important in the Southern Hemisphere, where there is less land.

Particles less than 20 microns in diameter may be carried great distances by winds. Volcanic fragments smaller than about 10 microns may be carried around the world, if an eruption is powerful enough to inject fine ash into the stratosphere. In 1956, for instance, a large volcanic eruption on the Kamchatka Peninsula, northeast of Japan, was later detected in the atmosphere over England at heights of 18,000 meters.

Particles eroded by winds from soils in arid or mountainous regions generally remain in the lower atmosphere, forming a "plume" of such material downwind. This is particularly noticeable off Africa, where Sahara sand is often blown hundreds of kilometers to sea. Precipitation quickly removes particles from the lower atmosphere; they serve as condensation nuclei for raindrops and snowflakes.

TURBIDITY CURRENTS

A turbidity current is a dense mixture of water and sediment that flows along the bottom, transporting sediment from continental margin to continental rise and even into the deep ocean basin. Such flows have never been directly observed in the ocean, although their occurrence in lakes is well documented. An earthquake may set one off, as happened in November 1929, when a turbidity current triggered by the Grand Banks earthquake hurtled down the continental slope off Newfoundland, breaking telegraph cables along its way. Such a flow moves like a gigantic submarine avalanche at speeds of up to 100 kilometers per hour (60 miles per hour), finally depositing a thin layer of sediment over a large area of ocean bottom. A sudden large discharge of riverborne sediment may trigger a turbidity current, with a force strong enough to erode rock.

Deposits formed by turbidity currents have distinctive properties that occur widely, not only in present day deep-ocean sediments but also in rocks thought to have originally been deposited in ancient deep-ocean areas. Among these features is *graded bedding,* in which the base of a single unit contains large sediment grains—commonly sand overlain by progressively smaller-grained deposits until the top of the unit is silt- and clay-sized particles. These units are often repeated many times in a single sediment core.

When the current first passes, it is moving at high speeds and carrying sand-sized particles. Later sections of the flow move progressively more slowly, transporting finer particles which are deposited on top of the heavy ones. The result is a rapidly formed sediment deposit showing distinct gradation in particle size, changing from coarsest grained at the bottom to the finest at the top.

A distinctive aspect of transport by turbidity currents is that they frequently carry shells of shallow-water organisms and fragments of land plants onto the deep ocean floor. Plant fragments—some still green—have been dredged from the ocean bottom following cable breaks. Turbidite layers containing remains of shallow-water organisms often alternate with layers of sediment that accumulated slowly particle-by-particle and contain organisms or exhibit other features typical of deep-ocean sediments.

Turbidity currents are most active where continental shelves are narrow, least active on wide continental shelves. In the Atlantic, turbidity current deposits have buried most of the older topography so that currents flow from the edge of the continents almost to the mid-ocean ridge. Likewise, sediment brought into the northern Indian Ocean can flow unimpeded over great distances of the deep-ocean bottom. But in the Pacific, trenches and ridges along the margins of the ocean basin now prevent movement of turbidity currents to the deep ocean floor.

Besides carrying sediment and burying ocean-bottom topography, turbidity currents are also thought to cut the submarine canyons that indent the margins of most continents. Whether turbidity currents are the sole agent responsible for these large features has been debated extensively, but there is good reason to suppose that they serve in many areas to remove the large volumes of sediment that would otherwise fill these canyons completely. Evidence of this comes from frequent breakage of submarine cables off the mouths of large rivers, such as the Congo.

BIOGENOUS SEDIMENTS

More than half of the deep-ocean bottom is covered by biogenous sediments, either calcareous or siliceous (see Table 4-3). Formation of these deposits is governed by three fundamental processes: *productivity* of overlying surface waters, *destruction* of the shells before burial, and *dilution* by other kinds of sediment. Where nutrients (phosphate and nitrate) are resupplied to the surface layers and there is sufficient sunlight to support photosynthesis, marine life is usually abundant and biogenous constituents of sediment are generated.

Diatomaceous sediments predominate in the high latitudes both in the North Pacific and surrounding Antarctica. Absence of predominantly diatomaceous sediment in the North Atlantic is presumably caused by mixing of diatom shells with abundant lithogenous sediment from neighboring continents. Furthermore, surface waters of the North Atlantic are unproductive. In the equatorial Pacific, sediment accumulations may be markedly thicker than elsewhere on the deep ocean floor—that is, in excess of 600 meters and in some areas (120° to 170°W) over 1 kilometer thick. Accumulation rates of about 15 millimeters per thousand years are typical of the most productive parts of the equatorial Pacific, compared to about 1 millimeter per thousand years elsewhere.

A second factor controlling distribution of biogenous sediment is *destruction* of skeletal remains. Phosphatic and siliceous constituents apparently dissolve at all depths in the ocean, as indicated in Fig. 4-9, so their survival depends in large measure on how sturdy the fragments were initially. Only the most resistant bones and most heavily silicified shells of radiolaria and diatoms are common sedimentary constituents.

Figure 4-9

General relationship between depth and weight loss in particles due to dissolution after 4 months of submergence at various depths in the Central Pacific Ocean. (After W. H. Berger, 1967. "Foraminiferal Ooze: Solution at Depths," *Science,* 156(3773), 383–85.)

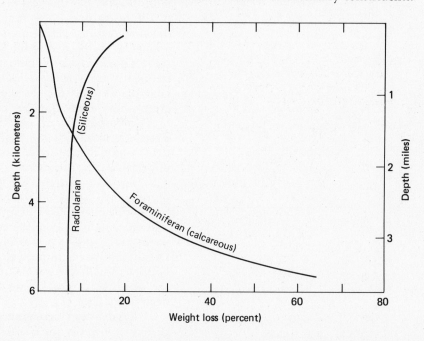

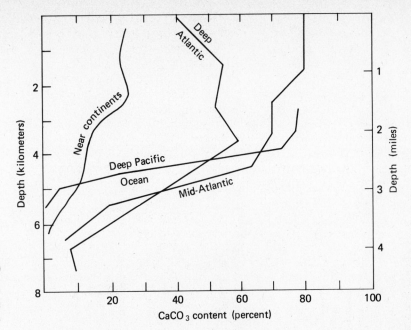

Figure 4-10

Distribution of calcium carbonate in
sediment deposited at various depths in
the deep ocean and near the continents.

Thin or slightly silicified shell fragments are easily broken down and
readily dissolved; they do not survive the long settling time and the
long period of exposure at the sediment surface.

   Effects of particle destruction (see Fig. 4-9) can most readily be
seen in the distribution of calcareous deep-ocean sediment. Surface
ocean waters are usually saturated with calcium carbonate, so that
calcareous fragments there are not dissolved. At mid-depths, the lower
temperatures and higher CO_2 contents of seawater cause dissolution
of calcareous fragments, but rather slowly. Below about 4500 meters,
waters are rich in dissolved CO_2 and able to dissolve calcium carbonate
much more readily. This depth relationship is markedly shown in the
calcium content of deep-ocean sediments when plotted as a function
of depth, as in Fig. 4-10. Carbonate-rich sediments are common at
relatively shallow depths in the ocean but almost completely absent
below about 6000 meters, as Fig. 4-11 indicates. For instance, aragonitic

Figure 4-11

Depth distribution of deep-ocean sediment.
Minimum, maximum, and average depths
are shown for each major sediment type.
Frequency of occurrence of various depth
zones is also shown.

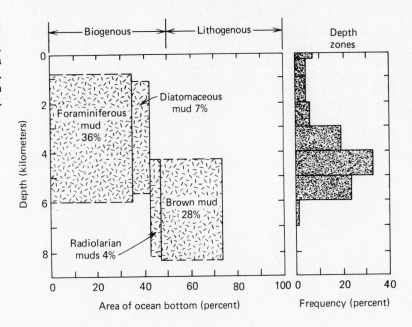

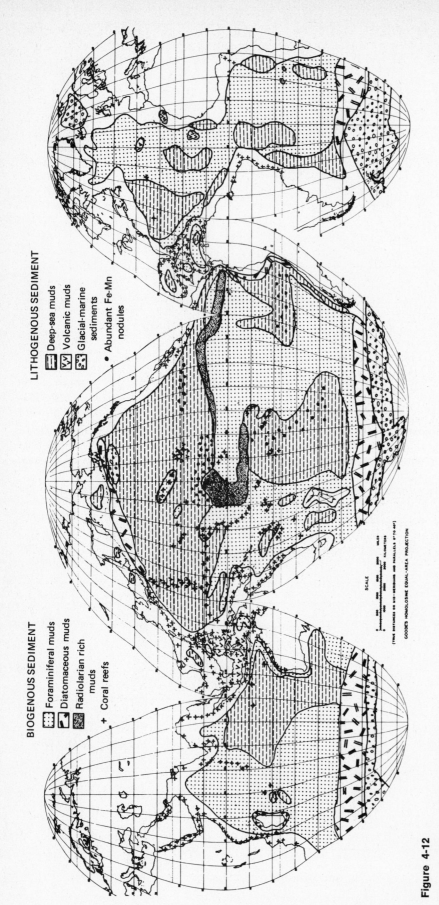

BIOGENOUS SEDIMENT

☷ Foraminiferal muds

᯼ Diatomaceous muds

Radiolarian rich muds

+ Coral reefs

LITHOGENOUS SEDIMENT

▤ Deep-sea muds

⩗ Volcanic muds

⩲ Glacial-marine sediments

• Abundant Fe-Mn nodules

SCALE

0   500   1000   1500   2000   MILES

0   500   1000   1500   2000   2500   3000   KILOMETERS

(TRUE DISTANCES ON MID-MERIDIANS AND PARALLELS 0° TO 40°)

GOODE'S HOMOLOSINE EQUAL-AREA PROJECTION

Figure 4-12

Distribution of deep-ocean sediments.

pteropod shells are found only on the tops of submerged peaks in the Atlantic. Distribution of carbonate-rich foraminiferal and coccolith muds reflects this depth zonation. Note that these sediments are most common on the shallower parts of the ocean bottom: on the Mid-Atlantic Ridge, the East Pacific Rise, and the Mid-Indian Ridge (see Fig. 4-12).

A third factor controlling distribution of biogenous sediment is *dilution* by lithogenous sediment. Remember that we defined biogenous sediment as consisting of more than 30 percent by volume of a biogenous constituent. We thus change the classification of such a deposit if large amounts of lithogenous sediment are added. Near continents where material derived from soil erosion is supplied relatively rapidly, the probability of biogenous sediment formation is slight except on the tops of banks or ridges where turbidity currents cannot cover them.

In order to estimate the rate at which biogenous sediment is formed, we assume a steady state—in other words, that conditions in the ocean are not changing through time. If the chemical composition of seawater is constant, then the amount of dissolved calcium, silica, and phosphate brought to the ocean each year must equal the amount removed during the year by organisms and incorporated into the sediment. Remember that this takes into consideration only the amount actually reaching the sediment, not the amount removed by organisms from seawater, most of which is returned to the water by dissolution of particles as they sink to the bottom. This approach indicates that about 1.5 billion tons of calcium carbonate is deposited in sediments each year; most of this is probably removed by foraminifera and coccoliths. About 0.5 billion tons of siliceous sediment forms each year. In total, this amounts to about 10 percent of the total sediment load brought to the ocean each year, ignoring the apparently small amount of phosphate added to sediments.

CLAY MINERALS

Deep-sea muds cover most deeper parts of the ocean basins. This very fine-grained deep-ocean sediment is especially interesting because it contains a distinctive suite of layered-silicate minerals (clays) derived from different climatic zones (see Table 4-6). These mineral grains are too small to be studied visually and require examination by X-ray techniques which differentiate them on the basis of their internal crystal structure. While present in the muds on continental shelves and rises, the clay mineral fraction is most important in the deep ocean, where it constitutes 60 percent or more of the nonbiogenous fraction of these sediments, as indicated in Fig. 4-13.

Clay minerals are also interesting because they are small enough (less than 2 microns) to be transported by all sediment-moving processes—winds, rivers, and glaciers. Because of their distinctive composition, clay minerals permit us to determine the importance of these transport mechanisms in various parts of the ocean.

Tropical rivers contribute distinctive clay minerals (*kaolinite, gibbsite*) that typically form in the highly weathered tropical soils. These clays are widely distributed, as Fig. 4-14 shows, across the tropical ocean, carried by the equatorial-current system. *Illite,* a group of minerals derived from slightly altered micas that occur commonly in many different kinds of rocks, is also a common constituent of deep-ocean sediment. River-transported illite is abundant in the Atlantic Ocean basin, as Fig. 4-15 indicates. In the Central Pacific, where river-transported sediment does not occur, illite is also a common constituent. Presumably it is carried there by winds.

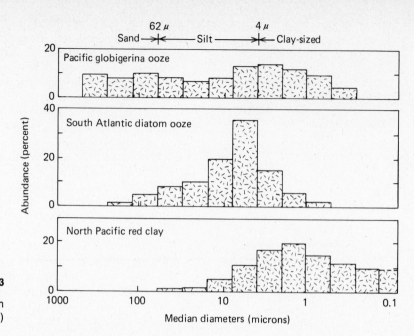

Figure 4-13

Grain-size distribution in typical deep-ocean sediments. (After Sverdrup et al., 1942.)

Figure 4-14

Distribution of kaolinite in the fine-grained portion (less than 2 microns) of deep-ocean sediment. (After J. J. Griffin, H. Windom, and E. D. Goldberg, 1968. "Distribution of Clay Minerals in the World Ocean," *Deep-Sea Research,* 15, 433–59.)

Another clay mineral, *montmorillonite,* comes in part from certain soils but is also formed in the ocean from the in-place alteration of volcanic glass from eruptions. In the Central Pacific where submarine volcanism is common and other sediment sources far removed, this mineral occurs in great abundance, often to the exclusion of other clay minerals. The relatively high concentrations of montmorillonite near the Mid-Indian Ridge may also be due to the high incidence of volcanism in that region. Other common silicate minerals in deep-ocean sediments include the *zeolites (phillipsite),* which also form from altered volcanic glass.

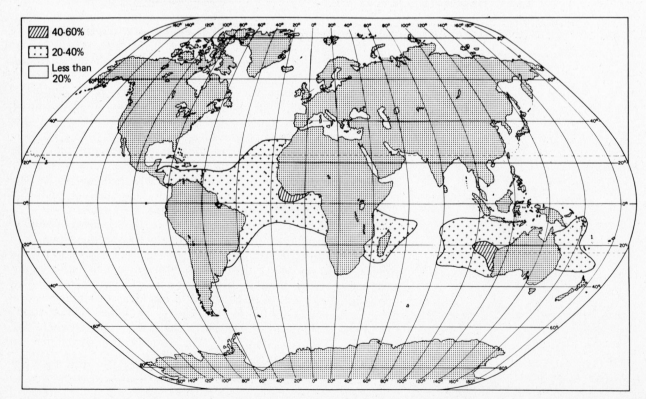

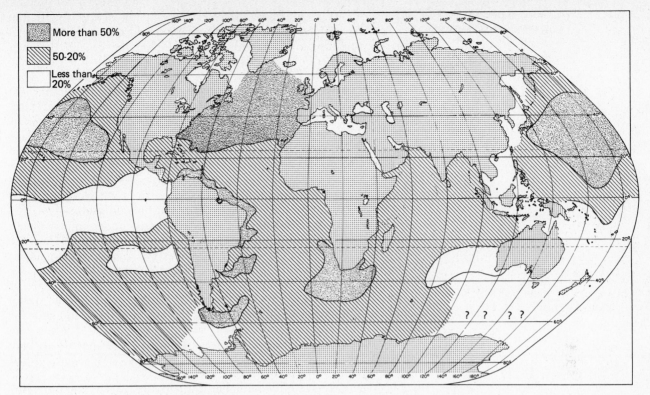

Figure 4-15

Distribution of illite in the fine-grained portion (less than 2 microns) of deep-ocean sediment. (After Griffin et al., 1968.)

At higher latitudes, typical clay minerals (primarily *chlorite*) are those formed in arctic soils. Because of the low temperatures, these clay minerals are much less altered than are clay minerals derived from tropical soils. Ice transport of sediment is important in the polar and subpolar regions, usually in latitudes between 50° and 90° in both hemispheres. In mid-latitudes, jet streams are important transport routes for wind-blown material from the continents.

TABLE 4-6
*Common Clay Minerals and Their Distribution In Marine Sediment ***

MINERAL	ORIGIN	AREA OF ABUNDANCE IN OCEAN
Chlorite	Abrasion of metamorphic and sedimentary rocks, especially by glaciers in polar regions	Polar and sub-polar regions
Montmorillonite group of minerals	Chemical alteration of volcanic ash; may form in place after deposition in sea water	South Pacific and Indian Oceans near mid-ocean ridges
Illite (hydro-mica group of minerals)	Chemical and mechanical alteration of mica minerals in rocks on continents	Near continents, especially common in North Atlantic
Kaolinite	Forms by decomposition of silicate minerals under intensive weathering in the formation of tropical soils	Equatorial Atlantic, and Indian Ocean near Australia

* Data from J. J. Griffin, H. Windom, and E. D. Goldberg, 1968. "The Distribution of Clay Minerals in the World Ocean," *Deep-Sea Research*, 15, 433–59.

DISTRIBUTION OF SEDIMENT DEPOSITS

Effects of various transport processes are reflected in the distribution of marine sediments. In general, river-transported sediment is restricted to continental margins. Most sediment remains on the shelf or rise except where turbidity currents are active and can carry material out into the deep ocean.

Ocean margins are usually covered by thick deposits of lithogenous sediment, much of it having accumulated fairly rapidly. Transport of this material from continents to the deep-ocean bottom is inhibited by the topography of most continental margins and the lack of modern sediment brought to the continental shelf by rivers, a consequence of the changes in sea level following the last retreat of the glaciers. Some deposits were formed by processes still active in the ocean; others, formed at some earlier time, remain exposed on the ocean bottom but do not reflect present oceanic processes.

Deposits of sand and silt that were formed under conditions no longer existing in an area are known as *relict sediments*. They are recognized by distinctive features, such as the remains of organisms that no longer inhabit the area. For example, oyster shells found far out on the U.S. continental shelf indicate deposition at a time of greatly lowered sea level when these areas were near the shore. Mastodon teeth in marine sediment, or other remains of extinct land animals, indicate that the area was once dry and that its surficial deposits date from that period. Other characteristics of relict sediments include iron stains or coatings on grains that could not have been formed under present marine conditions.

Where rivers supply sediment to the coastal ocean, relict deposits are buried. Alternatively, they may be reworked by the action of waves during periods of rising sea level so that the characteristic features of a relict deposit are destroyed or masked. About 70 percent of the world's continental shelves are now covered by relict sediments.

Most sand is deposited near the mouth of a river and moved along the coast by longshore currents. The association of sand beaches and river mouths is especially obvious on the Pacific coast of the United States, where the largest beaches and dunes in Washington and Oregon are associated with the Columbia River mouth. Sands are common in areas of strong wave action because finer-grained materials are removed.

Silts and clays are carried farther seaward and may be deposited at depths where wave action is not strong enough to stir up sediment or erode the bottom. On many continental shelves this occurs at depths of 50 to 150 meters.

Sands and silts deposited on continental shelves commonly accumulate at rates greater than 10 centimeters per 1000 years. This is relatively rapid accumulation, and particles have insufficient time to react chemically with the seawater or with near-bottom waters as they lie on the ocean bottom before being buried. Coastal-ocean sediments therefore retain many of the characteristics that they acquired during weathering. The colors of these rapidly accumulating deposits tend to be rather dark—gray, greenish, or sometimes brownish—and they contain abundant organic matter (1 to 3 percent is not uncommon).

In polar regions, glacial marine sediments are the most common type of sediment on the ocean bottom. These deposits contain all sizes of particles—from boulders to silt (see Fig. 4-5) deposited by melting icebergs from continental glaciers. Sea ice commonly contains sediment only if it has gone aground and incorporated material from the shallow continental shelf, which happens in the Arctic Sea. When such sea ice melts, its sediment load is deposited on the bottom, where it mixes with the remains of marine organisms. In the Antarctic, icebergs carry sediment picked up from the bottom, and also material trapped in the ice when it froze.

In those parts of the continental shelf where the temperature during the coldest month of the year is 18°C or warmer, reef-building corals can grow. Many continental shelves within this belt are covered by biogenous sediment usually consisting of shallow-water calcareous algae, shallow-water foraminifera, and coral. In the United States such sediments occur around the southern tip of Florida. They are also common on the shallow Bahama Banks, Bermuda, and off the Yucatan Peninsula of Mexico, and are especially extensive around the northern coast of Australia, where the *Great Barrier Reef* (150 kilometers wide) extends for about 2500 kilometers.

Near volcanic areas the ocean bottom also receives substantial amounts of volcanic rock fragments, usually the relatively large particles (greater than 50 microns in diameter) formed during eruptions. Sediments deposited in the Indonesian area, around the Aleutians off Alaska, and near the many island arcs of the western Pacific contain abundant volcanic debris. Volcanic sediments are not restricted to the Pacific Ocean but occur as ash in all ocean basins, transported by winds.

Sediments that accumulate fairly rapidly near continents (indicated in Table 4-7)—transported either by running water, turbidity currents, or ice—cover about 25 percent of the ocean bottom. Accumulations are thickest in marginal ocean basins; although these account for only about 2 percent of the ocean area, they contain about one-sixth of all identifiable oceanic sediment.

TABLE 4-7
Typical Sediment Accumulation Rates

AREA	AVERAGE (RANGE) ACCUMULATION RATE (centimeters per thousand years)
Continental margin	
Continental shelf	30 (15–40)
Continental slope	20
Fjord (Saanich Inlet, Brit. Col.)	400
Fraser River delta (Brit. Col.)	700,000
Marginal ocean basins	
Black sea	30
Gulf of California	100
Gulf of Mexico	10
Clyde sea	500
Deep-ocean sediments	
Coccolith muds	1 (0.2–3)
Deep-sea muds	0.1 (0.03–0.8)

The remainder of the ocean floor is covered by *pelagic sediments* that accumulate slowly—between one and ten millimeters per thousand years. Deposits formed in this way differ substantially from those formed by turbidity currents. They blanket the original ocean-bottom topography, faintly preserving its outlines. A comparison has often been made with fallen snow on land. Sometimes, however, as with a heavy accumulation of snow, deposits on sides of hills slump down into nearby valleys. When this happens, the effect is similar to a turbidity current.

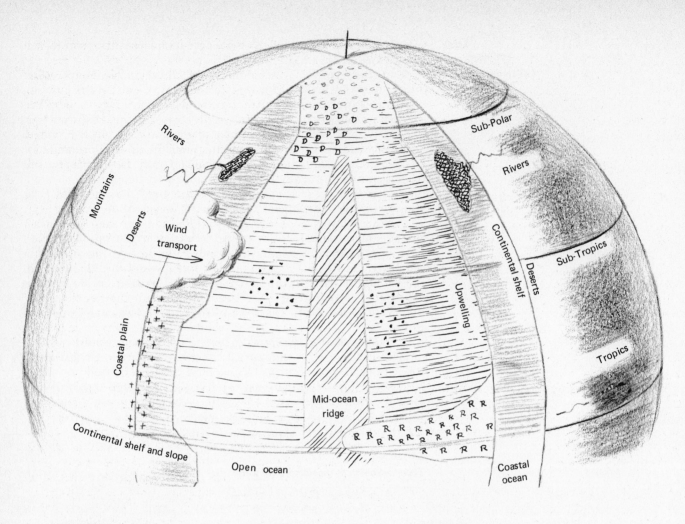

Rivers

Mountains

Deserts

Wind transport

Coastal plain

Continental shelf and slope

Sub-Polar

Rivers

Continental shelf

Deserts

Sub-Tropics

Upwelling

Mid-ocean ridge

Tropics

Open ocean

Coastal ocean

Coral reefs and mangrove swamps	Glacial marine sediments	Foraminiferal sediment (on mid-ocean ridge)
Modern lithogenous sediment	Diatomaceous sediment	Brown mud
Relict sediment (continental shelf)	Radiolarian-rich sediment	Manganese nodules

(a)

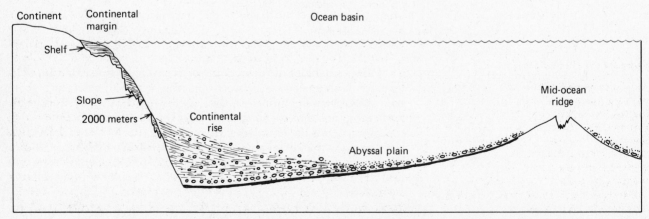

Continent

Continental margin

Ocean basin

Shelf

Slope

2000 meters

Continental rise

Abyssal plain

Mid-ocean ridge

Iron-manganese sediment   Carbonate ooze   Turbidite deposits   Pelagic deposits

(b)

Clays and very small particles that escape the river mouth can be carried great distances. Even relatively large clay particles (1 micron) require about 50 years to settle through 4 kilometers of water. Smaller particles (less than 0.5 microns), which are more typical of the particles in the deep ocean, theoretically require up to 1000 years to sink to the bottom in still water. Most of these particles are removed from suspension within less than a hundred years, probably due to filtration of seawater by planktonic organisms which ingest particles and incorporate them in fecal pellets. These relatively heavy pellets sink at rates of 10 meters or more per day.

Deep-ocean sediments accumulate very slowly, commonly at rates of less than one centimeter per thousand years. Thus the particles not only spend years suspended in ocean water, but then remain exposed on the ocean bottom for centuries before they are finally buried and sealed off from contact with near-bottom waters. Burrowing organisms further increase particle exposure by stirring up the deposits. As a result, there is usually ample time for particles in deep-ocean sediments to react with seawater. Abundant dissolved oxygen in deep-ocean waters causes iron to be converted to the ferric state (iron rust) and gives red clays their typical color. The presence of dissolved oxygen also favors utilization of organic matter by organisms or its destruction by inorganic processes; consequently deep-ocean sediments commonly contain less than one percent organic matter.

Over most of the deep ocean, the sediment cover is relatively thin—usually between 0.5 and 1 kilometer, averaging about 600 meters. Thin sediment deposits generally occur on mid-oceanic ridges or rises, where large areas may be completely swept clear of sediment, apparently by strong currents. Sediments do accumulate there in certain protected "pockets." Sediment thickness increases away from the ridges and is generally greatest over the old crust near the continents along the ocean-basin margins. In general, these sediments are flat-lying and in many areas apparently consist of alternating layers of turbidites (deposits formed by turbidity currents) and particle-by-particle accumulations. Such stratified sediments are especially common in the Atlantic and Indian Oceans, and in parts of the Pacific. There is little evidence for disturbance of these sediments since their deposition many millions of years ago.

Manganese nodules are common in the central portions of the Pacific, far from major sources of lithogenous sediment. In the Atlantic, manganese nodules are less common; an exception is the Blake Plateau off the southeastern United States, where the Gulf Stream keeps the ocean bottom free of sediment.

Some general relationships between marine processes and sediment distribution are summarized diagrammatically in Fig. 4-16. Note in Fig. 4-16 (a) that recent ice ages and the present polar ice cover have controlled the distribution of glacial marine sediments, both on the shallow and deep-ocean bottom. A less obvious effect of the present climatic regime is the formation of cold, dense bottom waters in the Antarctic that probably influence the dissolution of carbonate sediments in the deepest part of the ocean, causing dissolution in the deepest portions.

Effects of high productivity can be seen on the deep-ocean bottom and on the continental shelf. On the deep-ocean bottom (Fig. 4-16 (a)), bands of diatomaceous muds in high latitudes and radiolarian muds in equatorial regions directly reflect the high biological productivity of surface waters in these regions. On the continental shelf, the abundance of recent carbonate sediment in tropical waters is again a consequence of the locally high carbonate productivity. Figure 4-16 (b) shows that turbidity currents leave thick sediment deposits on and near

Figure 4-16 (facing page)

(a) Schematic representation of the general distribution of sediment in the deep and coastal ocean. (Modified in part after K. O. Emery, 1968. *Relict Sediments on Continental Shelves of World,* American Association of Petroleum Geologists Bulletin, 52(3), 445–64.) (b) Sediments are thickest near ocean basin margins because of heavy discharge from continents. Also, ocean basins are youngest near mid-oceanic ridges.

continental margins. Note also that sediment cover is thinnest at mid-ocean ridges, and thicker near continents where the ocean floor is older.

CORRELATIONS AND AGE DETERMINATIONS

So far we have considered sediment distributions in particular parts of the world ocean. This leads to questions about the ages of deposits and correlation of specific zones of equal age between ocean areas.

The most common techniques used to date or correlate sedimentary deposits employ biogenous remains such as those in Fig. 4-17. Fossils show changes through time that paleontologists use to assign

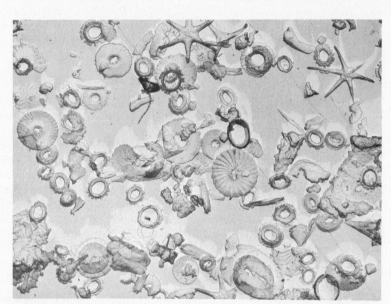

Figure 4-17

Fossil coccoliths and discoasters (calcareous nannoplankton) from sediment cores. In this electron photomicrograph, the fossils are magnified several thousand times; they range from 2 to 25 microns in diameter, equivalent to fine silt particles. The star-shaped fossils in the upper portion of the photograph are discoasters, the remains of extinct organisms thought to have been related to coccoliths. Disks and rings are coccoliths, an abundant constituent of nannoplankton now living in the ocean. (Photograph courtesy Deep-Sea Drilling Projects, Scripps Institution of Oceanography, under contract to the National Science Foundation.)

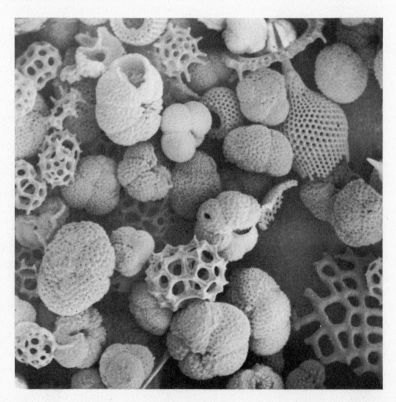

Figure 4-18

Skeletons of foraminifera and radiolaria taken from sediment in the western Pacific Ocean in waters 2660 meters deep. The core penetrated 131 meters into the ocean bottom. The fossils have been magnified about 160 times by an electron scanning microscope. (Photograph courtesy Deep-Sea Drilling Project, Scripps Institution of Oceanography, under contract to the National Science Foundation.)

ages to well-characterized assemblages of fossils. Time zones are commonly defined in terms of presence or absence of several different fossils. Microorganisms are preferred for work with marine sediments, since a large number can be recovered from even a small volume of sediment. Foraminifera, radiolaria, and coccoliths are commonly used (Fig. 4-18); even pollen grains have been used to correlate deposits.

From studies of fossil assemblages one can establish time zones and correlate deposits in widely separated areas. Eventually the history of an entire ocean basin over the past 200 million years may be worked out from fossils buried in its sediments. Where absolute ages in years are needed, radioactive substances are used as clocks. Radionuclides decay at known, unvarying rates; the time elapsed since the rock or shell formed is reflected in the amount of decay product and the amount of the original radionuclide present in a deposit. Depending on the age of the sediment, different radionuclides, depicted in Fig. 4-19, are used.

Figure 4-19

Radionuclides used for dating sediments, showing time period for which each is useful. Uranium and thorium have many decay products; during deposition, the original relationship among them is disturbed and only gradually returns to equilibrium. The relative abundance of "daughter" products for a specific radionuclide provides a measure of time elapsed since the sediment formed.

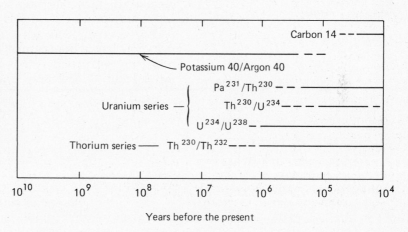

For recently deposited sediment, carbon-14 is a useful radionuclide. Cosmic rays bombard the earth, forming carbon-14 in the upper atmosphere, and all living organisms have carbon-14 atoms amidst the carbon of their tissues. When an organism dies, no further carbon is taken up or exchanged with atmosphere or ocean water. Instead, the amount of carbon-14 begins to decrease at a steady rate. One-half of it is gone in 5600 years (the *"half-life"* of this radionuclide), and after 11,200 years only one-quarter of the original amount remains. Thus by comparing the amount of carbon-14 per unit amount of carbon in a fossil with that in organisms presently living under similar conditions, it is possible to determine the time elapsed since that organism was alive. Assuming that the organism died and was incorporated in the layer when it formed, we can assign an age to the deposit. Although extremely useful for recently deposited sediment, carbon-14 has too short a half-life to permit dating of sediments older than about 30,000 years.

Dating older sediments requires the use of radionuclides with longer half-lives. Potassium-40 is extremely useful for this purpose. Most rock-forming minerals contain potassium, which includes 1.2 percent potassium-40 having a half-life of 1.3 billion years. When potassium-40 decays, 11 percent of the daughter product is argon-40, an element that is not present in the mineral when it forms. Thus, a comparison of the amount of argon-40 in a mineral with its potassium content provides a measure of the time elapsed since that mineral formed. The more argon-40 in a mineral, the greater its age. To use this radionuclide to date sediment deposits, it is necessary to date only minerals that formed when the deposit was laid down. Volcanic ash and glauconite (a micalike mineral) are often used in dating deep-ocean deposits. The age of a mineral must be further interpreted to calculate sediment age. Because of its long half-life, potassium-40 is useful only for sediments older than about 100,000 years. In general, older sediments may be dated by isotopic techniques more easily than younger ones.

Within the last decade or so, sedimentary evidence has been used for dating different parts of the ocean floor. We have seen that magnetic reversals leave laterally extending magnetic patterns in oceanic crustal rocks (Fig. 2-14). It turns out that this same record is preserved vertically in sediment cores. Particles in a sediment layer reflect the earth's magnetic orientation at the time that layer was deposited. This was first documented in the mid-1960's, when a record of the million year old Jarmilla event (Fig. 2-9) was found in a core.

A record of worldwide climatic changes is also preserved in oceanic sediment. For instance, shells of organisms known to thrive where surface waters are cold may be found in a sediment layer deposited many millions of years ago, taken from a part of the world that now has a tropical climate. Movements of continents, changed ocean current patterns, and major climatic changes have been documented and studied using fossils from deep ocean deposits.

SUMMARY OUTLINE

Origin and classification of marine sediment
Lithogenous particles—derived from silicate rocks
Biogenous particles—skeletons of various plants and animals
Hydrogenous particles—formed by Fe–Mn precipitation from seawater
Minor sources—cosmogenous debris from space; man's contribution

Sources and transport of lithogenous sediment
Rivers—supply about 20 billion tons of lithogenous sediment per year
Wind erosion and transport—important in desert and high mountain regions
Biogenous sediments—about 2 billion tons formed annually
Settling velocity—controlled by particle size
Most sediment trapped near river mouth or in estuary
Relict sediment—covers about 70 percent of continental shelves of world
Wind transport—important for small particles in mid-latitudes; removal by precipitation

Turbidity currents
Dense, sediment-laden waters move along the bottom
Form characteristic deposits—graded bedding, displaced shallow-water plant and animal remains
Important means of sediment transport across continental shelves, especially narrow shelves

Bury ocean-bottom topography forming smooth plains

Biogenous sediments
Cover more than half of ocean basin
Distribution controlled by production, destruction, dilution
Most rapid accumulations in regions of high productivity, especially in equatorial region
Destruction of fragile or soluble remains—leaves only sturdy and less soluble shells and skeletons
Dilution by lithogenous constituents occurs near the continents

Clay minerals—layered silicates indicative of climatic zone where they formed in soils
Transported by winds and rivers

Distribution of sediment deposits
Reflects effects of various transport processes
Thickest deposits near continents, cover about 25 percent of ocean bottom; accumulate at rates up to 10 centimeters per 1000 years
Manganese nodules—common in areas of slow sediment accumulation, such as North Pacific

Correlations and age determinations
Detailed studies of fossil assemblages give relative ages
Radiometric techniques give "absolute" ages—carbon-14, potassium-40

SELECTED REFERENCES

GARRELS, R. M. AND F. T. MACKENZIE. 1971. *Evolution of Sedimentary Rocks*. Norton, New York. 397 pp. Intermediate level textbook on sedimentary processes and their long-term geochemical implications.

SHEPARD, F. P. 1974. *Submarine Geology,* 3rd ed. Harper & Row, New York. 517 pp. General treatment of sediments in ocean, emphasizing continental shelf and adjacent environments; technical, descriptive.

TUREKIAN, K. K. 1976. *Oceans,* 2nd ed. Prentice-Hall, Englewood Cliffs, N.J. 149 pp. Emphasizes chemical aspects of sedimentary process; elementary.

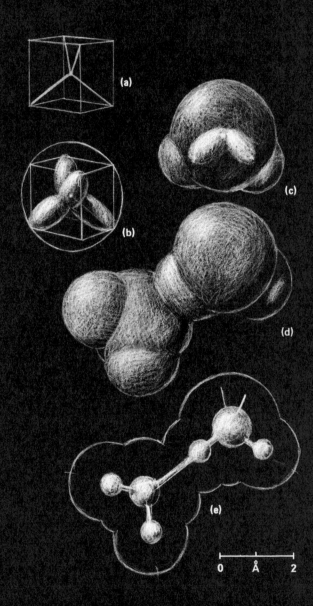

(a)

(b)

(c)

(d)

(e)

0   Å   2

Different ways of representing water molecules. To demonstrate the structure of a single molecule, bond directions (a) and (b) and the electron cloud (c) are shown. Two molecules bonded together as they would occur in ice appear as electron clouds (d) and as "Tinker-Toys" (e).

seawater

P revious chapters have discussed the earth's surface and the ocean basins. The characteristic feature of the ocean, however, is not the basin that contains it, but seawater itself. To understand the global importance of the ocean we must understand the physical and chemical properties of water and seawater. In this chapter we first consider pure water and then the salts dissolved in ocean waters.

Seawater is a complex substance. On the average, it is about 96.5 percent water containing 3.5 percent salt, a few parts per million of living things, and perhaps an equal amount of dust. Some properties of water are little affected by addition of sea salts. An example is water's capacity to absorb and give up large quantities of energy as heat with little change in temperature. Other characteristics, such as its suitability for life, are strongly influenced by the dissolved materials. Minute differences in chemical composition can drastically alter its biological properties. Processes of living organisms in turn affect the abundance of dissolved gases and salts in seawater. Finally, some important properties of ocean water, such as the way it transmits sound, are controlled in part by the structure of water and the behavior of its dissolved salts.

MOLECULAR STRUCTURE OF WATER

Two hydrogen atoms and one oxygen atom combine to form one water molecule. The oxygen atom has two unshared electron pairs and two electron pairs shared with the hydrogen atoms. The unusual properties of water result from the structure of this molecule, specifically its associated electron cloud. This cloud has a definite shape, forming the "third and fourth corner" of the molecule (see Frontispiece p. 116). It may be described as resembling a child's four-pronged jack. The oxygen atom, being relatively large, accounts for most of the molecule's volume. The two hydrogen atoms are much smaller and are partially buried in the electron cloud of the oxygen atom, forming two of the four "prongs" of the molecule (see Fig. 5-1).

The other two "prongs" are formed by the unshared electrons of the oxygen atom. The hydrogen atoms are on one side of the molecule, the unshared electrons on the other side. Thus, viewed on an atomic scale, a water molecule is essentially negative on one side (the side with the unshared electrons) and positive on the other (the side with the hydrogen atoms). Such a molecule, whose electronic charges are separated, is called a *polar molecule*. It responds to electrical charges in the vicinity. The unshared electrons on one molecule tend to form

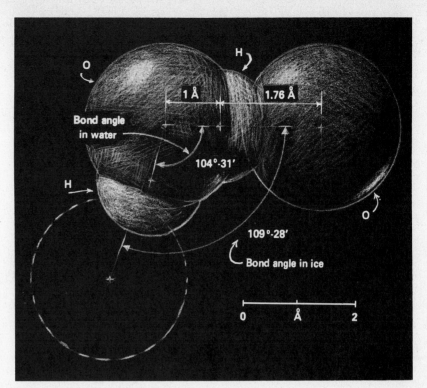

Figure 5-1

Schematic representation of a water moleclue, showing the bonding between the hydrogen atoms on the molecule and adjacent oxygen atoms in water molecules on the right and lower left. Note the separation between the side of the molecule having no hydrogen atoms and the side of the molecule having no hydrogen atom (An Angstrom, A°, is one ten-billionth of a meter.)

rather weak bonds, called *hydrogen bonds,* with hydrogen atoms of the adjacent molecule. This tendency for hydrogen bonding between water molecules gives water many of its unusual properties, including its large capacity for absorbing heat, its abnormally high boiling point and freezing point, and its capacity to dissolve substances that are held together by *ionic bonds;* some of these properties are listed in Table 5-1, in which water is compared to a liquid that does not have the strongly polar characteristics.

TABLE 5-1

*Comparison of Physical Properties of Water and A "Normal" Liquid (n-heptane) ***

PROPERTY	WATER $(H_2O)_n$	NORMAL HEPTANE (C_7H_{16})
Molecular weight	$(18)_n$	100
Density (g/cm³)	1.0	0.73
Boiling point (°C)	100	98.4
Melting point (°C)	0	—97
Specific heat (cal/g per °C)	1	0.5
Heat of evaporation (cal/g)	540	76
Melting heat (cal/g)	79	34
Surface tension, 20°C (dynes/cm)	73	25
Viscosity, 20°C (poise)	0.01	0.005

* Modified from G. Dietrich, 1963. *General Oceanography.* Interscience, New York. p. 40.

Crystals and molecules are attracted to one another and held together by various forces. The weakest of these results from fleeting electronic interactions between molecules. These so-called *van der*

119

Waals bonds form, break up, and reform easily; although present in water, they are greatly overshadowed in importance by the stronger hydrogen bonds.

Ionic bonds form between adjacent molecules (or atoms) which have lost or gained electrons thus acquiring either a positive charge, where an electron has been lost, or a negative charge, where an electron has been gained. One example of ionic bonding is the association between sodium (Na^+) and chlorine (Cl^-) to form table salt, sodium chloride. The bond is the result of strong attraction between closely spaced molecules of opposite charge. To get some idea of the strength of these bonds, consider the amount of energy (expressed in kilocalories, a measure of heat energy) required to break each type of bond in one *mole* of a substance, e.g., water. One mole of water weighs 18 grams, or about 1/2 ounce—that is its formula weight in grams, oxygen having an atomic weight of 16 and each hydrogen atom having an atomic weight of 1:

van der Waals bonds:	0.6 kilocalorie per mole
hydrogen bonds:	4.5 kilocalories per mole
ionic bonds:	10's of kilocalories per mole

At the relatively low temperatures and pressures of the earth's surface, water is one of the few substances that we commonly see in all three forms of matter—solid, liquid, and gas. *Water vapor*—the gas—is probably the easiest to understand. In water vapor, each molecule exists separately and is little affected by other molecules. Therefore, each molecule is free to move with little restriction, except for the walls of the container in which we enclose it. This freedom to move accounts for the characteristics of a gas: it has neither size nor shape but expands readily to fill any container in which it is placed.

Water molecules striking the sides of a container exert *pressure*. Pressure can be increased in two ways: we can put more material in the container or shrink it, thereby increasing the rate of molecules striking the sides of the container; alternatively, we can make the molecules move faster and strike the container walls more frequently. The first of these we do by placing more material (even another gas) in the container, the second by increasing the temperature, which makes molecules move more rapidly.

At the opposite extreme in internal order is the solid—in this case, *ice*. Solids have a definite size and shape. Most are also crystalline —that is, they have a definite internal structure. Usually solids break, or perhaps bend, when enough force is applied. This resistance to flow, or deformation, results from the orderly internal structure of the solid. Each atom, or molecule, has a position which it occupies for long periods of time. These atoms or molecules cannot move readily from position to position, nor rotate in a given position. But at room temperatures there is always some vibration of atoms in their fixed position. Some small amount of movement is possible if there is a hole nearby into which a molecule can move. Just as with water vapor, the movements of the atoms or molecules increase with increasing temperature, so that movement between positions also increases with rising temperature.

In ice, the water molecules are held together by hydrogen bonds, as shown in Fig. 5-2. The large oxygen atoms have definite positions, and the hydrogen atoms in each molecule are also oriented in a regular manner. The oxygen atoms form six-sided puckered rings which are arranged in layers, each layer a mirror image of the adjacent layer. This forms a fairly open network of atoms and thereby gives ice a somewhat lower density (approximately 0.92 grams per cubic centi-

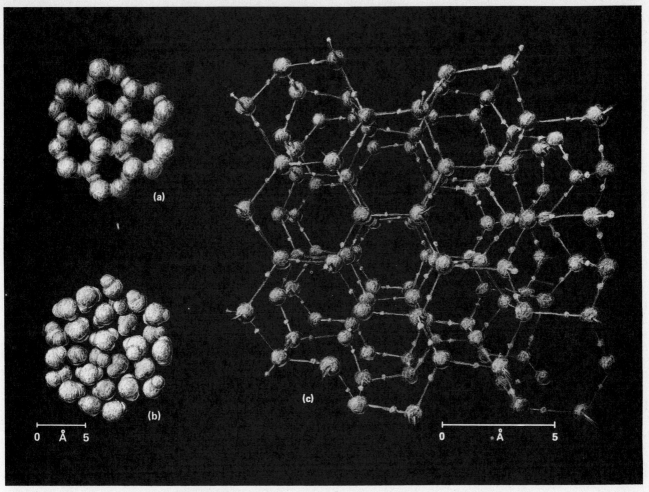

Figure 5-2

Crystal structure of ordinary ice showing the fixed positions of water molecules (a). Note the six-sided rings formed by 24 water molecules. In the same volume of liquid water, 27 molecules would be present (b). The ice lattice is shown in (c).

meter) than that of liquid water (approximately 1 gram per cubic centimeter) because the molecules are not as tightly packed together.

Despite the openness of the internal structure of ice crystals, impurities such as sea salts are not readily incorporated into the holes. Consequently salt is excluded from ice formed from seawater. The excluded sea salts remain in pockets of unfrozen liquid (brine).

Liquid water has several anomalous physical properties (see Table 5-1) intermediate between those of solids and gases. It apparently consists of two different forms of molecular aggregates (Fig. 5-3) in a state of equilibrium. The first component consists of clusters of hydrogen-bonded water molecules. These clusters form and reform very rapidly—10 to 100 times during one-millionth of a microsecond (10^{-10} to 10^{-11} second). Although the lifetime of any individual one is extremely short, clusters persist long enough to influence the physical behavior of water. These structured portions of water are less dense than the unstructured portion. In some respects, this lowered density of the structured portion arises from the same causes that make ice less dense than liquid water. It seems likely, however, that the ice structure is not exactly the same as that of the flickering clusters of molecules.

Although clusters form and reform rapidly, they persist in liquid water until broken up by external forces. The clusters apparently disappear when the pressure exceeds about 1000 atmospheres (a pressure reached only in the few deepest trenches of the ocean floor) or at temperatures exceeding 100°C (at atmospheric pressure), where liquid structures break up and molecules escape to form a gas.

The other constituent of liquid water is unstructured or "free" water molecules which surround the structured portions. These molecules move and rotate without restriction. Interactions with nearby molecules are weaker than in the structured portion. The "free" water portion of liquid water is denser than the structured portion, since the molecules fit more closely together. The relative proportion of the structured and unstructured portions of liquid water vary with changes in temperature, pressure, and salt content and composition.

Because of its unusual molecular structure, liquid water is strikingly different from hydrogen compounds of elements with chemical properties similar to oxygen, as indicated in Fig. 5.4. The melting point and boiling point of water occur at temperatures 90° and 170°C higher, respectively, than might be predicted from studying the chemical behavior of these other compounds. If water were a "normal" liquid, it would occur only as a gas at earth-surface temperatures and pressures.

Figure 5-3 *(above)*

Schematic representation of liquid water at atmospheric pressure. Note the two types of water structures: areas of bonded molecules (ordered regions) and free water molecules between them.

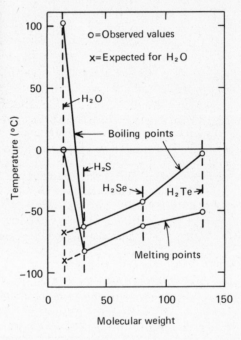

Figure 5-4 *(right)*

Melting and boiling points of water and chemically similar compounds. (After Horne, 1969.)

TEMPERATURE EFFECTS ON WATER

Temperature affects the internal structure of water and its properties. Much of the heat energy absorbed by water is used up in changing the internal structure so that water temperature rises less than other substances, after absorbing a given amount of heat. The large amount of water on earth acts as a climatic buffer preventing the wide variations in surface temperature experienced by a waterless planet. For example, surface temperatures on the moon go from about +135°C at noon during the lunar day to about −155°C during the lunar night. On earth, the highest temperature ever recorded was 57°C, at Death Valley, California (July 10, 1913), and the lowest was −68°C, at Verkhoyansk, in Russia (February, 1892).

Earth's limited temperature range is controlled primarily by the abundance of water on the earth's surface, since much of the incoming solar radiation goes into evaporating water and melting ice. Let us see what happens to water during these changes and how this affects the earth's heat balance.

First, we must define a measure of heat, the *calorie,* as the amount of heat (a form of energy) required to raise the temperature of 1 gram of liquid water by 1 degree Celsius (1°C). This means that we can change the temperature of 1 gram of water by 50°C by supplying 50

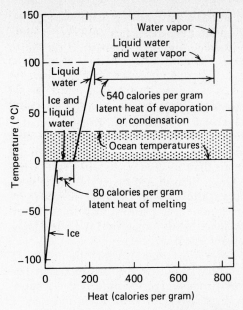

Figure 5-5

Temperature changes when heat is added or removed from ice, liquid water, or water vapor. Note that the temperature does not change when mixtures of ice and liquid water or liquid water and water vapor are present. (Redrawn from Gross, 1976.)

Figure 5-6

Effect of changes in temperature on the relative abundance of unbroken hydrogen bonds, cluster size, and the number of molecules per cluster. (Data from G. Nemethy and H. A. Scheraga, 1962. "Structure of Water and Hydrophobic Bonding in Protein," *Journal of Chemical Physics*, 36, 3382–3400.)

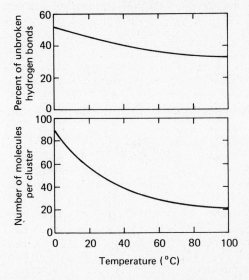

calories of heat. Alternatively, we would change the temperature of 50 grams of water by 1°C with the same amount of heat.

Breaking bonds in ice and liquid water requires energy, usually heat. Conversely, heat is released when they reform. Consider what happens when 1 gram of ice just below the freezing point is heated slowly; let us keep track of the amount of heat added.

As we add heat to ice, its temperature increases about 2°C for each calorie of heat we add (see Fig. 5-5). When the ice reaches its melting point, 0°C, the temperature remains constant and instead the ice begins to melt. As long as ice and liquid water exist together in a container, the temperature of our system remains fixed at 0°C. After we have added about 80 calories per gram, the last bit of ice melts, leaving only liquid water.

As we continue to apply heat, water temperatures rise, but now at a slightly lower rate: 1°C per calorie of added heat. This rate of temperature change holds nearly constant between 0°C and 100°C. At 100°C, the boiling point, the temperature rise again ceases and gas (water vapor) forms. At the boiling point, the same situation occurs that we experienced at the melting point. The temperature remains fixed as long as liquid and gas are present. After adding about 540 calories per gram, the last of the water evaporates. If we capture the vapor (steam) and continue heating it, we find that the temperature rises much more rapidly: about 2°C per calorie of heat added—approximately the same as we experienced when heating ice.

Heat added to the water is taken up in two forms. One is *sensible heat*, heat that we detect either through our sense of touch or with thermometers. This change in temperature results from the increased vibration of molecules or their motion in the gas state.

The other form is called *latent heat*, which represents the energy necessary to break bonds in the water structures. As we have seen, at both the melting point and the boiling point of water there was no change in temperature as long as two states of matter existed together. The added energy went into breaking the bonds in the disappearing phase. This is called latent heat because we get back exactly the same amount of heat when the process is reversed. Condensing water vapor to form liquid water releases 540 calories per gram at 100°C; freezing water at 0°C releases 80 calories per gram. The difference between the latent heats of melting and evaporation arises from the fact that only a small fraction of the hydrogen bonds are broken when ice melts. All hydrogen bonds are broken when water evaporates.

Although ice freezes at 0°C and water boils at 100°C, it is possible for molecules to go from vapor to solid or liquid at other temperatures. The processes involved are similar to those described before, but the amount of heat involved is different. For example, the latent heat of evaporation changes as follows:

Temperature (°C)	Latent heat of evaporation (calories per gram)
0	595
20	585
100	539

It takes more energy to remove a water molecule from liquid water at 0°C or 20°C than it does at 100°C. This is an important factor in the ocean since most water evaporates from the ocean surface at temperatures around 18° to 20°C. If water had to reach 100°C before evaporating, we would have no water vapor in the atmosphere.

Both cluster size and number of molecules per cluster decrease with increasing temperature, as Fig. 5-6 illustrates. It is interesting to note in this figure that a large number of hydrogen-bond water molecules remain bound together even after ice melts at 0°C. There is some evidence that subtle changes in structure may occur at other intermediate temperatures.

DENSITY

Whether a substance sinks or floats in a liquid is determined primarily by its *density* (mass per unit volume, expressed in grams per cubic centimeter). A substance less dense than its surroundings tends to move upward; in other words, it floats. If its density is greater than its surroundings, it displaces less mass than its own mass, settles out of the liquid, and sinks. The sinking rate is determined by its relative density and the resistance it experiences while moving through the water. A small object generally experiences more drag per unit mass than a large object, and thus a small particle takes much longer to sink than a large object which experiences relatively less drag.

A *density-stratified system of fluids* is one in which lighter fluids float on heavier ones. A mass of fluid may sink through materials less dense than itself until it reaches a zone where the fluid below is more dense and the fluid above less dense. At this point, there is no force acting on the fluid and it tends to remain at that level. Furthermore, vertical displacement of the fluid will be countered by forces tending to keep it at the same relative position. (An example is the pousse-café, an after-dinner drink in which colorful cordials of varying densities are layered in a glass, each floating on a heavier one below.) In the ocean, the density of fluids and solids determines in most instances whether a body of fluid or a solid remains in the upper layers or sinks to the bottom. Density differences drive currents in the ocean and atmosphere, as we shall see in later chapters.

Factors controlling seawater density include temperature, salinity, and pressure. Warming influences density by increasing the vibrations and movements of atoms and molecules, so that molecules vibrate or move more, and thus effectively occupy more volume. For a constant mass, the density decreases as the volume increases. Remember that density is a ratio; it decreases either because of a decrease in the numerator or an increase in the denominator.

Ice, water vapor, and seawater behave like most materials, becoming less dense with increasing temperature. In any stable structure of liquid water or seawater the least dense material will generally be the warmest—if salinity is constant—and will occur at the top.

Pure liquid water has an anomalous density maximum at 4°C. It is an important factor in freshwater lakes of cold regions but not in

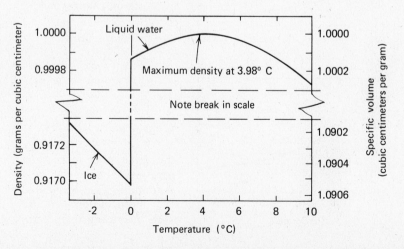

Figure 5-7

Effect of temperature on density, expressed in grams per cubic centimeter, and specific volume, expressed in cubic centimeters per gram, for both ice and pure liquid water.

the ocean. The anomalous density maximum does not occur in sea-water of salinity greater than 24.7 parts per thousand.

Ice at 0°C has a density of about 0.92 gram per cubic centimeter. Like most substances, ice becomes less dense as its temperature increases, illustrated in Fig. 5-7. Since ice is about 8 percent less dense than liquid water, it floats on water. When liquid water at 0°C is warmed, its density increases slightly until it reaches a density maximum at 3.98°C. After that temperature is reached, the density decreases as temperature increases further. In a freshwater lake, water cooled to 4°C sinks to the bottom and there is protected against further cooling. This keeps the bottom of the lake ice-free and supplies oxygen, dissolved in the water when it was in contact with the atmosphere, to bottom-dwelling animals. After the basin is filled with 4°C water, further cooling forms a less dense water layer on the lake surface, which eventually freezes if the weather remains cold long enough. If water behaved like most fluids and its density steadily increased with decreasing temperature, the coldest water would be found in the bottom of the lake and ice would also sink. This mode of freezing, with ice sinking to the bottom, would be far more efficient than freezing from the top. Most lakes at high altitudes and in cold climates would likely remain frozen solid during the summer.

As we have previously seen, temperature has a profound effect on the relative proportions of the structured and "free" portions of liquid water. Since these two constituents have different densities, the temperature of water affects its density. One explanation for the anomalous density maximum in pure liquid water is that there is a relatively rapid increase in the amount of structured portion of water below 4°C. This makes water less dense as the denser "free" portion of the mix is diminished.

Adding salt to water (as in Fig. 5-8) or increasing the pressure causes a lowering of the temperature of maximum density. The amount of salt in seawater is sufficient to eliminate this anomalous feature. Furthermore, adding salt to water increases its density. The salinity of seawater is commonly the dominant factor controlling its density near shore; temperature dominates in the open ocean.

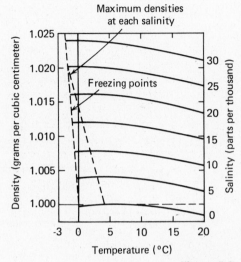

Figure 5-8

Effect of salinity on temperature of maximum density and initial freezing temperature of seawater. Note that increasing salinity from 0 to 25 parts per thousand causes the temperature of maximum density to decrease markedly, but lowers the initial freezing point only slightly.

COMPOSITION OF SEA SALT

Virtually anyone with access to seawater can make his own crude salinity determinations. The simplest way, though not the most precise, is to place a known amount of seawater in a pan and let it evaporate. The result is a gritty, bitter-tasting salt that never completely dries. The weight of the salt in a known volume of seawater provides a crude measure of salinity.

Determination of the composition of this salt residue, as shown schematically in Fig. 5-9, has occupied many excellent chemists since the first analyses were made in 1819. By now, at least traces of nearly all the naturally occurring elements have been detected in sea salts as shown in Fig. 5-10, and the job is still not finished. Most of the analyses have been made on surface waters collected near the coast. Such waters are most likely to exhibit variability in the chemical composition of their dissolved salts. Furthermore, the development of each new analytical instrument opens new horizons for additional, more detailed studies of ocean chemistry.

Six constituents—chloride (Cl^-), sodium (Na^+), sulfate (SO_4^{2-}), magnesium (Mg^{2+}), calcium (Ca^{2+}), and potassium (K^+)—comprise 99 percent of sea salts. The other elements present in sea salt add up to only 1 percent. As Fig. 5-9 shows, Na^+ and Cl^- alone make up 86 percent of sea salt.

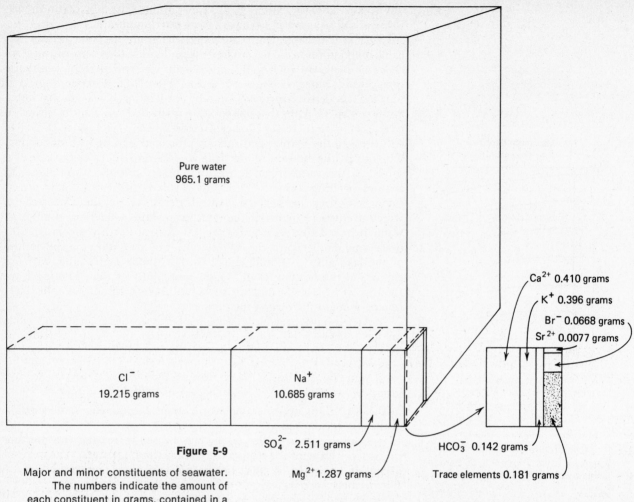

Pure water
965.1 grams

Ca²⁺ 0.410 grams
K⁺ 0.396 grams
Br⁻ 0.0668 grams
Sr²⁺ 0.0077 grams

Cl⁻
19.215 grams

Na⁺
10.685 grams

SO₄²⁻ 2.511 grams

HCO₃⁻ 0.142 grams

Mg²⁺ 1.287 grams

Trace elements 0.181 grams

Figure 5-9

Major and minor constituents of seawater.
The numbers indicate the amount of
each constituent in grams, contained in a
kilogram of seawater (S = 34.7 parts
per thousand).

To express the amount of dissolved salt present in seawater, oceanographers commonly use two concepts—chlorinity and salinity, both based on chemical determinations.

Since the oceans are well mixed, sea salts have a nearly constant composition. Therefore, we can use the most abundant constituent, chloride (Cl^-) as the index of the amount of salt present in a volume of seawater. *Chlorinity* (*Cl*) is defined as the amount of chlorine, in grams, in 1 kilogram of seawater (bromine and iodine are replaced by chlorine). Chlorinity is expressed as parts per thousand, or parts per mil, written $^0/_{00}$.

Chlorinity indicates the amount of chlorine. Oceanographers commonly convert this to salinity, a measure of the total amount of dissolved salt. *Salinity* (*S*) is defined as the *total amount of solid material, in grams, dissolved in 1 kilogram of seawater* (iodine and bromine are replaced by chlorine and all organic matter destroyed). Salinity is also commonly expressed as parts per thousand.

Salinity (*S*) and chlorinity (*Cl*) are both measures of the saltiness of seawater. This relationship can be expressed mathematically by

$$S(^0/_{00}) = 1.80655 \times chlorinity$$

For example, if a seawater sample had a measured chlorinity of 20.00 parts per thousand, its salinity would be calculated as follows:

$$S = (20 \times 1.80655) = 36.13^0/_{00}$$

Using chemical techniques, oceanographers can determine chlorinity within ±0.01 parts per thousand. Laboratory instruments (salinometers) for measuring conductivity by physical means can determine chlorinity to within ±0.005 parts per thousand or better. For this reason, among others, salinometers are commonly used for salinity determinations on oceanographic ships and at shore-based laboratories.

While the relative proportions of major elements in sea salts change little in the ocean, physical processes can change the amount of water in seawater. Therefore, salinity and concentrations of the so-called *conservative properties,* such as sodium and magnesium which are not involved in biological processes, are changed by:

1. evaporation and precipitation (rainfall, snow);
2. formation of insoluble precipitates whereby formerly dissolved elements or compounds settle out of seawater;
3. mixing of water masses having different salinities;
4. diffusion of dissolved materials from one water mass to another; and
5. movement of water masses within the ocean;
6. freezing and thawing.

Figure 5-10

Elements detected in seawater, arranged as in the periodic table. Most non-conservative elements are involved in biological processes (underlined). (Redrawn from Gross, 1976.)

Salinity of seawater is most variable near the air–sea interface, at boundaries of ocean currents, and in coastal areas. Even though the composition of seawater changes little, those slight deviations that can be detected are extremely useful to trace movements of water masses from their source to areas where they mix or otherwise lose their identity.

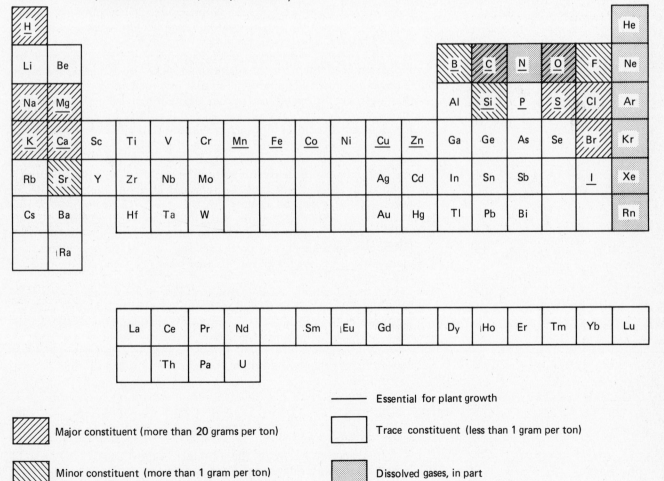

Many of the minor elements present in seawater exhibit pronounced changes in relative abundance owing to biological and chemical processes. These we call *nonconservative properties*. Some nonconservative elements are incorporated in surface-ocean dwelling organisms and settle to the bottom when the organism dies, thus depleting that element in the surface layer. When organic matter is decomposed and the element released to seawater, it goes into near-bottom waters and is then involved in the slow circulation of these waters before eventually returning to the surface.

Other nonconservative constituents are associated with inorganic particles and may be removed near their point of entry into the ocean. For example, aluminum is usually associated with rock or soil particles, most of which settle out near the mouth of the river that brought them to the ocean. If the particles are wind-transported and enter the ocean through its surface, the elements they contain are likely to be more widely dispersed than if brought in by rivers.

SALT COMPOSITION AND RESIDENCE TIMES

Each year, rivers bring about 4 billion tons (4×10^{15} grams) of dissolved salts to the ocean. Nearly all the NaCl in river water is recycled sea salt that fell on land in rain, coming from sea-salt particles derived originally from the sea surface. The remaining salts dissolved in river water come from the chemical breakdown (weathering) of rocks on land. Although this amount of salt is only about one-thousandth of the total amount of salt in the ocean, it would seem reasonable to find that seawater is getting saltier as the earth grows older.

All our data, however, indicate that the salinity of seawater has changed little. Direct measurements of salinity go back only a few hundred years and indicate no appreciable change. But this short period of observation is inadequate to decipher the ocean's billions of years of history.

Marine fossils preserved in rocks provide the best long-term information available to us. Fossils in these rocks—once marine sediment deposits—are usually found to be similar or closely related to organisms that now live in open-ocean waters. This bit of evidence, combined with some chemical data, suggest that the salinity of the open ocean has changed little, if at all, during the past half-billion years of ocean history. Salinity of seawater before the time when animals developed preservable skeletons remains essentially unknown.

Since rivers are delivering salt to the ocean while seawater salinity remains unchanged, salt must be removed at about the same rate at which it is supplied, an example of the *steady-state condition* of the world ocean which is changing little—if at all—through time. Some of these salt-removing processes apparently involve complex chemical reactions with sediments or particles suspended in the ocean. In other cases, seawater is evaporated in isolated basins to form salt deposits in arid regions.

A useful concept for analyzing substances in seawater is *residence time,* the time required to replace completely the amount of a given substance in the ocean. We can develop this concept in either of two ways, using either the rate of salt addition by rivers or the rate of removal of elements incorporated in sediments depositing on the ocean bottom. Using the second option, we can define residence time (T) as:

$$T = \frac{M}{R}$$

where
M = total amount (grams) of the substance in the ocean
R = rate of removal by sediment (grams per year).

For example, we can calculate the residence time for Na$^+$ in seawater as follows:

$$T_{Na} = \frac{M_{Na}}{R_{Na}} = \frac{1.5 \times 10^{22} \text{ g}}{5.7 \times 10^{13} \text{ g/yr}} = 2.6 \times 10^8 \text{ yr}$$

Sodium's calculated residence time of 260 million years is one of the longest residence times for an element in the ocean. Other elements, such as Al, have residence times of about 100 years.

Residence time of an element in the sea is apparently related to its chemical behavior. Elements such as Na$^+$, which are little affected by sedimentary or biological processes, generally have residence times of many millions of years. Elements used by organisms or readily incorporated in sediments tend to have much shorter residence times, ranging from a few hundred to a few thousand years.

We can even define a residence time for water. There is a net removal, due to evaporation, of a layer of water about 10 cm thick from the ocean surface each year. This water falls on the land and returns to the ocean through river runoff. Recall that the ocean has an average depth of about 4000 meters. From this we obtain an approximate residence of 40,000 years for water.

SALT IN WATER Natural waters are rarely pure, and usually contain some salt. River waters contain on the average about 0.01 percent dissolved salts, and average seawater contains about 3.5 percent (35 parts per thousand) of various salts (Table 5-2). Even rainwater usually contains small amounts of salts and gases that it has dissolved while falling through the atmosphere. In large measure these impurities are a consequence of the remarkable solvent powers of water.

Sodium chloride crystals consist of sodium ions (Na$^+$) and chloride ions (Cl$^-$) tightly bound together by ionic bonds. These bonds are the result of strong attraction of unlike charges on nearby ions. The charges arise because the ions have lost or gained an electron. Sodium readily loses an electron, forming a sodium ion. Chlorine, on the other hand, readily accepts an extra electron, froming a chloride ion, Cl$^-$. Water molecules disrupt these bonds by orienting themselves around ions and effectively shielding adjacent ions from the influence of each other. As a result, we see that the crystal has disappeared and we say that it has *dissolved*.

Dissolving salt in water affects its behavior in several ways. We have already discussed the effect on temperature of maximum density. Among the other properties affected by salts are:

1. *temperature of initial freezing*—decreased by increased salinity;
2. *vapor pressure*—decreased by increased salinity; and
3. *osmotic pressure*—increased by increased salinity (water molecules move through a semipermeable membrane from a less saline solution to a more saline solution; the opposing pressure necessary to stop this movement is called the osmotic pressure).

Consider the temperature of initial freezing. Pure water freezes at at 0°C and the temperature of the water–ice mixture remains fixed until there is only ice. In seawater, the temperature of initial freezing is lowered by increased salinity. Furthermore, salts are excluded from the ice and remain in the liquid, causing the brine to become still more salty. Hence the temperature of freezing for the remaining liquid is still lower and the temperature of the system must drop before additional ice forms. Consequently, seawater has a lower temperature of

TABLE 5-2
Elements Detected In Seawater *

	ELEMENT	CHEMICAL FORM	CONCENTRATION (ppm)	RESIDENCE TIME (thousands of years)
Ag	silver	$AgCl_2^-$	0.0003	2100
Al	aluminum		0.01	0.1
Ar	argon	Ar	0.6	
As	arsenic	AsO_4H^{2-}	0.003	
Au	gold	$AuCl_4^-$	0.000011	560
B	boron	$B(OH)_3$	4.6	
Ba	barium	Ba^{++}	0.03	84
Be	beryllium		0.0000006	0.15
Bi	bismuth		0.000017	45
Br	bromine	Br^-	65	
C	carbon	CO_3H^-, organic C	28	
Ca	calcium	Ca^{2+}	400	8000
Cd	cadmium	Cd^{2+}	0.00011	500
Ce	cerium		0.0004	6.1
Cl	chlorine	Cl^-	19000	
Co	cobalt	Co^{2+}	0.00027	18
Cr	chromium		0.00005	0.35
Cs	cesium	Cs^+	0.0005	40
Cu	copper	Cu^{2+}	0.003	50
F	fluorine	F	1.3	
Fe	iron	$Fe(OH)_3$	0.01	0.14
Ga	gallium		0.00003	1.4
Ge	germanium	$Ge(OH)_4$	0.00007	7
H	hydrogen	H_2O	108000	
He	helium	He	0.0000069	20000
Hf	hafnium		<0.000008	
Hg	mercury	$HgCl_4^{2-}$	0.00003	42
I	iodine	I^-, IO_3^-?	0.06	
In	indium		<0.02	
K	potassium	K^+	380	11000
Kr	krypton	Kr	0.0025	
La	lanthanum		0.000012	0.44
Li	lithium	Li^+	0.18	20000
Mg	magnesium	Mg^{2+}	1350	45000
Mn	manganese	Mn^{2+}	0.002	1.4
Mo	molybdenum	MoO_4^{2-}	0.01	500
N	nitrogen	organic N, HO_3^-, NH_4^+	0.5	2.5
Na	sodium	Na^+	10500	260000
Nd	neodymium		0.00001	0.3
Ne	neon	Ne	0.00014	
Ni	nickel	Ni^{2+}	0.0054	18
O	oxygen	OH_2, O_2, SO_4^{2-}	857000	
P	phosphorus	PO_4H^{2-}	0.07	
Pa	protoactinium		2×10^{-9}	
Pb	lead	Pb^{2+}	0.00003	2
Ra	radium		6×10^{-11}	
Rb	rubidium	Rb^+	0.12	270
Rn	radon	Rn	6×10^{-16}	
S	sulfur	SO_4^{2-}	885	
Sb	antimony		0.00033	350
Sc	scandium		<0.000004	5.6
Se	selenium		0.00009	
Si	silicon	$Si(OH)_4$	3	8
Sn	tin		0.003	100
Sr	strontium	Sr^{2+}	8.1	19000
Ta	tantalum		<0.0000025	0.35
Th	thorium		0.00005	0.16
Ti	titanium		0.001	
Tl	thallium	Tl^+	<0.00001	
U	uranium	$UO_2(CO_3)_3^{4-}$	0.003	500
V	vanadium	$VO_5H_3^{2-}$	0.002	10
W	tungsten	WO_4^{2-}	0.0001	1
Xe	xenon	Xe	0.0001	
Y	yttrium		0.000052	
Zn	zinc	Zn^{2+}	0.01	7.5
Zr	zirconium		0.000022	180

* After R. A. Horne, 1969. *Marine Chemistry: The Structure of Water and the Chemistry of the Hydrosphere.* Wiley-Interscience, New York. p. 153.

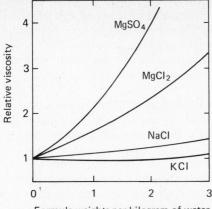

Figure 5-11

Effect of various salts on viscosity of water at 25°C. (After Horne, 1969.)

initial freezing and no fixed freezing point as observed in pure water.

Sea salts affect the internal structure of water. Some dissolved ions such as Na^+ and K^+ cause a shift in the equilibrium toward water's unstructured phase, while other ions such as Mg^{2+} "favor" the structured portions. The effect of changing the relative proportions of these constituents can be seen in changes of such properties as *viscosity*, the internal resistance of a liquid to flow, as indicated in Fig. 5-11.

After a crystal has dissolved, water molecules remain associated with ions, forming an envelope or cloud that actually moves with the ion. It is postulated that about 4 water molecules are associated with each sodium ion, about 2 molecules with each chloride ion. Ions with higher charges, such as Mg^{2+} or SO_4^{2-}, also have these so-called *hydration atmospheres* (or *sheaths*), usually containing more water molecules. Like other structured parts of liquid water, the hydration atmospheres are affected by temperature and pressure. They are apparently destroyed by pressures exceeding 2000 atmospheres, but are not completely destroyed by temperatures above 100°C.

Hydration sheaths affect the water around them. Some, such as NaCl and especially $MgSO_4$, "favor" the structured portion of water and actually increase viscosity, a property of water that responds sensitively to the relative abundance of the structured portion. Salts such as potassium cloride (KCl) seem to "favor" the unstructured portion of the water and might be called *"structure breakers."* They cause an initial reduction in the relative viscosity of water. Dissolved gases and insoluble or partially soluble substances also affect the relative abundance of structured and "free" water nearby.

Materials that dissolve completely forming separate ions in water are called *strong electrolytes*. Sodium chloride is an example. After a crystal dissolves, the ions form hydration sheaths that move more or less independently, so that ions formed by strong electrolytes behave like separate ions. In general, strong electrolytes have little effect on water structure as shown by their slight influence on water viscosity (see Fig. 5-11).

Figure 5-12

Schematic representation of hydration sheaths of ions formed by strong electrolytes (NaCl), ion pairs ($Fe^{3+}NO_3^-$). (After Horne, 1969.)

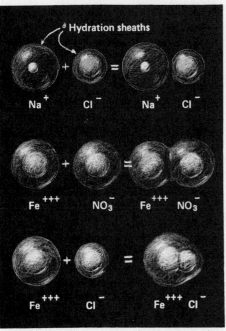

TABLE 5-3

Chemical Species of Some Major Constituents in Surface Seawater *

| CATIONS | FREE ION (percent) | Combined with: | | |
		SULFATE (percent)	BICARBONATE (percent)	CARBONATE (percent)
Na^+	99	1.2	0.01	—
Mg^{2+}	87	11	1	0.3
Ca^{2+}	91	8	1	0.2
K^+	99	1	—	—

| ANIONS | FREE ION (percent) | Combined with: | | | |
		Na (percent)	Mg (percent)	Ca (percent)	K (percent)
SO_4^{2-}	54	21	21.5	3	0.5
HCO_3^-	69	8	19	4	—
CO_3^{2-}	9	17	67	7	—

* From R. M. Garrels and M. E. Thomson, 1962. "A Chemical Model for Sea Water at 25°C and One Atmosphere Pressure," *American Journal of Science*, 260, 57–66.

Not all materials separate completely into individual ions when they dissolve, and we say that they are *weak electrolytes*. Magnesium sulfate ($MgSO_4$) is an example of a weak electrolyte. Only a fraction of the material dissociates into individual ions, as shown in Table 5-3. About 11 percent of the magnesium ions remain associated with sulfate ions. Sulfate ions are partially associated with Mg^{2+}, partially with Na^+.

Still another mode of association is possible in seawater— the formation of *ion pairs,* associations of ions that remain together much of the time even though they retain their individual hydration sheaths. In human terms we could think of it as a liaison, rather than a marriage. Ion pairs can have charges where the *valences* (amount of positive or negative charge) of individual members of the pair do not completely cancel. *Complex ions* are still more strongly attracted to each other, so much so that their hydration sheaths merge to form a single sheath, as represented in Fig. 5-12.

DISSOLVED GASES

Atmospheric gases are soluble in water, passing into a dissolved state at the air–water interface. Conversely, molecules of gas also pass through the interface in the opposite direction—back into the atmosphere. Water can only dissolve a certain amount of any substance under given conditions of temperature and pressure, and when that limiting value is reached the amount of gas going into solution is the same as that going out. At this point the water is *saturated* with the gas, which is then said to be present in *equilibrium concentration*.

Temperature, salinity, and pressure all affect the saturation concentration for a gas. In the normal range of oceanic salinity, temperature is the dominant factor, as indicated in Fig. 5-13. In general those gases such as nitrogen, oxygen, or the rare gases that do not react chemically with water become less soluble in seawater as temperature or salinity increase. Seawater at all depths is saturated with atmospheric gases. Exceptions are dissolved oxygen (O_2) and carbon dioxide (CO_2), which are involved in life processes.

Nitrogen (N_2), the most abundant gas in the atmosphere (Fig. 5-14), is also a common gas dissolved in seawater (Fig. 5-15). It is not involved in biological processes (with a few minor exceptions), and thus remains very near saturation throughout the ocean. Sometimes, in deep water, more nitrogen is present than can be accounted for by the temperature, salinity, and pressure of the water. We then say that the water is *supersaturated* with nitrogen, a condition which may be explained as follows:

The solubility of a gas in seawater is determined primarily by temperature, pressure, and salinity. Since exchange of dissolved gas occurs at the sea surface, conditions at the surface control the amount of gas dissolved in water. As a water mass moves away from the surface where it last exchanged atmospheric gases, its temperature and salinity can change as a result of mixing with other water masses. Also, depth changes are accompanied by pressure changes. The observed slight supersaturations of nitrogen and rare gases in the deep ocean can thus be accounted for by considering these changes in physical conditions of the water since it was last at the surface.

There is, to be sure, some slight production of nitrogen by bacteria living in oxygen-deficient basin waters. These bacteria break down nitrate ions ($NO_3{}^{2-}$) from seawater and release nitrogen gas (N_2). In the absence of dissolved oxygen, most biological processes proceed slowly, and the volume of ocean area deficient in oxygen is small, so this process is a minor factor in the dissolved-nitrogen picture. There is also some formation of nitrate from dissolved nitrogen by bacteria, but again it appears to play a minor role.

Figure 5-13

Solubility of oxygen and nitrogen in pure water and in seawater. Note that both gases are more soluble in pure water.

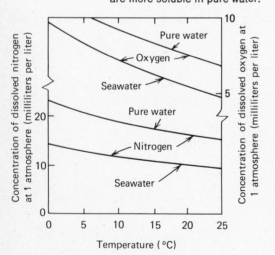

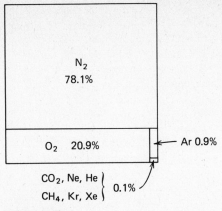

Figure 5-14

Relative abundance (by volume) of gases in dry atmosphere.

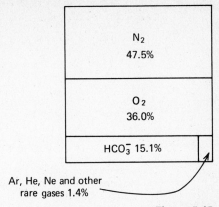

Figure 5-15

Relative abundance (by weight) of gases dissolved in surface seawater, $S = 36$ parts per thousand, $T = 20°C$. Note the abundance of HCO_3^- and rare gases compared with their abundance in the atmosphere (see Fig. 3-14).

A small amount of helium (He), a rare gas, is produced by radioactive decay in rocks and sediment at the sea floor. Otherwise, the rare gases—argon (Ar), krypton (Kr), and xenon (Xe)—behave identically to nitrogen, allowing for their slightly different solubilities.

Oxygen is one of the most variable of the dissolved gases in seawater. As all the other gases, it enters the ocean primarily through the surface. Oxygen is also produced by photosynthesis of plants in the sunlit layer of the ocean, usually restricted to a zone a few tens of meters deep just below the ocean surface. In the deep ocean, dissolved oxygen is supplied by the sinking of cold, oxygen-rich waters in the Antarctic and Arctic regions. Oxygen is used by both plants and animals at all depths in respiration.

In those parts of the ocean where the rate of oxygen consumption by marine life exceeds the rate of resupply of dissolved oxygen, certain bacteria are able to derive their necessary oxygen by breaking down compounds dissolved in seawater. First the nitrate ions are broken down, releasing nitrogen. There being little nitrate in most ocean waters, this does not last long. Next, different organisms break down SO_4^{2-} to obtain oxygen and release hydrogen sulfide, H_2S, as a metabolic by-product, causing the familiar rotten-egg smell sometimes noticeable in salt marshes at low tide.

CARBON DIOXIDE AND CARBONATE CYCLES

The carbon dioxide cycle in seawater is the most complicated of all the dissolved gases. Carbon dioxide occurs as a dissolved gas, as a weakly dissociated acid, and as the dissolved ions of carbonate (CO_3^{2-}) and bicarbonate (HCO_3^-). Its abundance is controlled by physical properties of seawater, by biological processes such as photosynthesis and respiration, and by the formation and destruction of carbonate shells by marine plants and animals.

By itself, carbon dioxide has a complicated chemistry. It dissolves in water, forming carbonic acid (Fig. 5-16); this acid ionizes slightly, yielding carbonate and bicarbonate in equilibrium. If carbonate is used in plant production, the equilibrium shifts in response (in Fig. 5-16) and causes more bicarbonate to dissociate. Carbonate precipitation as calcium carbonate ($CaCO_3$) in plant and animal skeletons represents a net CO_2 loss to the cycle.

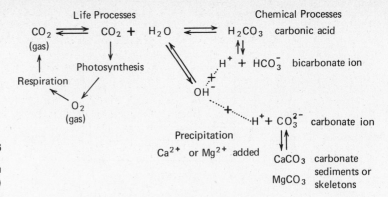

Figure 5-16

Carbonate–carbon dioxide cycle in the ocean. (After Horne, 1969.)

Another biologically significant aspect of CO_2 in the ocean is that the weak acid (H_2CO_3) acts as a *buffer*. Addition of acidic substances (H^+) causes the equilibrium in Fig. 5-16 to shift, thereby, creating more bicarbonate. This ionizes very little, so that the abundance of hydrogen ions (pH), which controls the relative alkalinity or acidity of a solution, remains fairly constant. Thus respiration and decomposition processes producing CO_2 hardly affect the pH of seawater, nor does the removal of CO_2 in photosynthesis. Hydrogen ions in seawater are freely accepted and given up by both carbonate and bicarbonate ions, creating a buffer against sudden, sharp changes in acidity.

Carbon dioxide is highly soluble in seawater and enters the ocean through the sea surface. It is also produced at nearly all depths by the respiration of plants and animals.

Calcium carbonate shells formed in surface waters dissolve in intermediate and deep waters after the organism dies and sinks below the surface. On the ocean bottom less than about 4 kilometers deep (the depth varies in different ocean basins), the rate of supply of carbonate shells is greater than the rate of dissolution. However, the rate of resupply is much too slow to prevent the dissolution of carbonate in the cold, CO_2-rich waters deeper than the 4-kilometer level. There the sediments are nearly devoid of carbonate because of its rapid dissolution; this is the red mud area of the deep-ocean basins.

Over most of the ocean, removal of calcium carbonate is accomplished by biological processes. Inorganic precipitation seems to be a relatively rare occurrence, restricted to warm, highly saline waters. At present the Bahama Banks, east of Florida, is one of the few areas of the ocean where this occurs. Seawater is warmed as it flows slowly over the Banks, causing two effects; the solubility of $CaCO_3$ in water is diminished, and the solubility of carbon dioxide decreases so that some escapes to the atmosphere. While in solution, each Ca^+ ion is balanced by the presence of two bicarbonate ions. If bicarbonate is lost by dissociation and escape of CO_2, then $CaCO_3$ (lime) precipitates as a solid coating on rounded grains (oolites), or as tiny needles of the mineral aragonite.

PHYSICAL PROPERTIES OF SEAWATER

Pressure in the ocean increases 1 atmosphere for each 10 meters of depth. Thus pressures in the deepest trenches are around 1100 atmospheres.

Viscosity—resistance of a fluid to flow—in "normal" liquids increases as pressure increases. In a simple way, we can imagine that this results from the increased resistance of atoms or molecules sliding past each other as they are pressed more closely together.

Liquid water responds in exactly the opposite way from a "normal" liquid; its viscosity decreases with increasing pressure. It reaches a minimum viscosity at pressures of about 500 to 1000 atmospheres, corresponding to depths of 5000 to 10,000 meters.

Ease of molecular movement in water (and therefore its viscosity) is related to the degree of binding of molecules and the amount of structuring of the water. The simplest explanation of the pressure effect on water is that increased pressure in the depth range to 5000 meters or more favors the destruction of the structured portions, thereby increasing the ease of movement of molecules. This decreases its viscosity. At pressures exceeding 1000 atmospheres, the structured portion of the water has been completely destroyed, and water then acts like a "normal" liquid. Its viscosity increases with further pressure increase.

Air trapped in seawater near the surface forms bubbles. These range in size from a few millimeters to a few microns in diameter and are formed by breaking of waves and impact of raindrops. Bubbles provide an important route for gases in seawater to pass in and out of solution and for the production of sea-salt particles that serve as condensation nuclei for raindrops and snow particles in the atmosphere.

Development and bursting of bubbles has been studied in detail by high-speed photography, as shown in Fig. 5-17. First a gas bubble forms (a) and rises toward the surface at rates of about 10 centimeters per second. Reaching the surface (b) the bubble has a cap or thin film of water that bursts (c), forming a thin spray of droplets (d) whose diameters are about 1 to 20 microns. Shortly thereafter a large droplet (e), about 100 microns in diameter, is formed by a jet rising rapidly from the bottom of the bubble and is ejected into the atmosphere at speeds of 10 meters per second. This rises 10 centimeters or more above the sea surface. The smaller droplets are caught by wind and transported long distances in the atmosphere. The larger droplets rapidly fall back to the sea surface.

Airborne droplets evaporate, leaving tiny salt particles that are carried by winds. The residence time of most such particles in the

Figure 5-17
Bubble bursting at the sea surface forming water droplets in the atmosphere. Many of these droplets evaporate, forming salt particles that serve as condensation nuclei for raindrops or snowflakes.

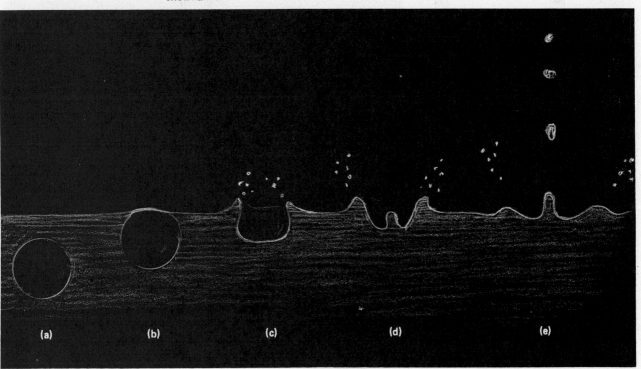

(a)   (b)   (c)   (d)   (e)

atmosphere is probably short. Some, however, are carried high into the atmosphere, where they eventually act as nuclei around which raindrops or snowflakes form. In this manner, escaped salt is carried back to the earth's surface. The total amount of salt moved by this process is about 1 billion tons per year.

Composition of the salt particles closely resembles that of normal sea salt, although there is some enrichment in the more volatile materials that evaporate from the sea surface, such as boron and iodine.

Surface tension is the force necessary to break the filmlike surface layer of a liquid. Water has an anomalously high surface tension, such that a clean needle can float on a water surface. Surface tension causes an undeformed water drop to form a sphere. This may be seen by shaking a bottle of oil and vinegar—the vinegar forms droplets in the oil during the temporary period of suspension.

In water, high surface tension seems to result from strengthened local water structure near any interface, including the air–water boundary. Near an interface there is more hydrogen bonding of the water molecules, which tend to orient themselves with their oxygens pointing out of the liquid, as shown in Fig. 5-18. This structured zone appears to extend about 10 molecules deep into the liquid. Because of the highly structured nature of interface water, certain ions tend to be excluded so that the chemical composition of this zone may differ appreciably from that of bulk seawater. This selective exclusion likely affects the chemical composition of sea salts that pass into the atmosphere from breaking bubbles and other processes. In fact, several elements are more abundant in these airborne sea salts than they are in normal seawater.

Figure 5-18

Changes in water structure near the air–water interface at the top of figure.

Surface films are sensitive to the presence of impurities. Many substances that are insoluble in water tend to collect at the interface. Many are molecules whose various components have different chemical behavior. For example, many surface-active agents—substances that concentrate at or change the properties of an interface—have one end that readily dissolves in water while the other end of the molecule (often a long, complicated structure) does not dissolve in water. The water-soluble end is called the *hydrophilic* ("water-loving") end, the insoluble end the *hydrophobic* ("water-hating") end; this relationship is illustrated in Fig. 5-19 if you look closely.

Surface-active agents change the surface tension of seawater, commonly lowering it. Surface "slicks" or films, where surface-active materials have been swept together by wind or current action, are common in coastal waters, especially where currents converge. These areas have relatively few small waves at the water surface, and their relative smoothness compared to adjacent waters has given rise to the descriptive term "slick." They commonly contain a wide variety of organic materials of an oily or otherwise hydrophobic nature and are of considerable biological significance in the ocean. The subject will be discussed in Chapter 12.

Figure 5-19

Changes in water structure in the vicinity of a nonpolar, hydrophobic material (right). Note that the polar water molecules exclude the nonpolar propane molecule from the water structure (below).

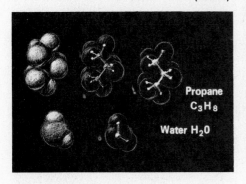

Detergents and soaps are familiar surface-active agents. Their long molecules surround dirt or grease, substances easily attached to the hydrophobic end of the surface-active molecule. The hydrophilic end is directed toward the surrounding water, and the whole aggregate moves as if it were soluble, so that the otherwise insoluble dirt or grease can "wash away" in the water.

An increasingly common method of determining the salinity of seawater is to measure its *conductivity*. Conductivity of water is controlled by ion movement through the water. The more abundant the ions, the greater the transmission of electricity and the higher the conductivity. In seawater, the relative abundance of the major ions is nearly invariant, and the conductivity provides a precise means of determining salinity. The major ions conducting electrical charges are:

	Contribution to Conductivity
Ion	(percent)
Cl^-	64
Na^+	29
Mg^{2+}	3
SO_4^{2-}	2
Other	2

In fresh waters the composition of dissolved salts is more variable, and conductivity does not provide a precise measure of salt content. In fresh water the use of this method requires careful study to insure that the changes in conductivity are not caused by changes in chemical composition but are caused by changes in their concentration.

SOUND IN THE SEA

Behavior of sound in the ocean is controlled by temperature, pressure, and salinity. Sound is a form of mechanical energy, consisting of the regular alternation of pressure or stress in an elastic substance. Of all the various forms of radiated energy, sound penetrates ocean water best, so that sounds made by organisms or physical processes can be detected at relatively great distances. Sound is also used to determine depth to the bottom, to detect other vessels or objects (including fish), and for emergency signaling using the technique known as *sonar*, an acronym for "sound navigation and ranging," an underwater equivalent of radar.

Figure 5-20

Acoustical properties of seawater: (a) Typical profiles showing variation in speed of sound with depth in various latitudes. (b) Absorption coefficients in seawater and in distilled water. (After R. J. Ureck, 1969. *Principles of Underwater Sound for Engineers*, McGraw-Hill, New York.)

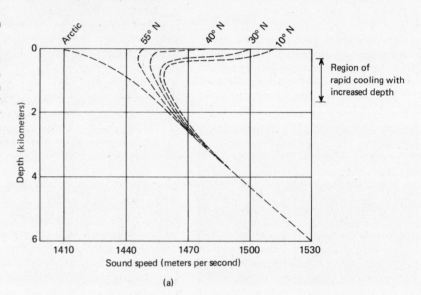

(a)

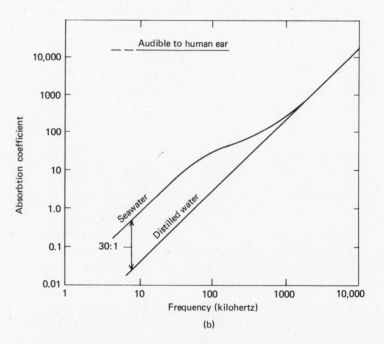

(b)

In the ocean, sound travels at a speed of about 1480 meters per second. The speed of sound in water is increased by increasing hydrostatic pressure (which increases with depth), temperature, and salinity. In most of the ocean, the effect of salinity is least important, the effect of pressure the most important. Sound in water can be slowed by bubbles or excessive concentrations of dissolved gases in some bays and harbors.

Because of the pronounced effects of temperature on sound speed, there is substantial variation in the speed of sound with depth in the ocean, indicated in Fig. 5-20(a). Near the ocean surface the temperature effects dominate. At greater depths, the pressure effect becomes appreciable.

The intensity of a sound wave traveling through water is diminished as energy is absorbed. The absorption of sound in seawater is far greater than in pure water at frequencies less than 100 kilocycles per second (kilohertz), but is essentially identical to pure water at higher frequencies, as Fig. 5-20(b) demonstrates. The presence of $MgSO_4$ is the cause of this anomaly. As previously pointed out, $MgSO_4$ in seawater is typically surrounded by a large, bulky hydrated aggregate. This complex aggregate is capable of absorbing acoustical energy over a large frequency range, which accounts for its unusual sound absorption in seawater.

Sound passing through water can be considered as traveling in paths, called *rays*. Some of these rays are direct—in other words, they pass directly from source to receiver. Others are indirect and may involve reflection from some sharp boundary such as the air–water interface or the water–sediment interface. This reflection of sound is the basis for the echo-sounding technique previously discussed.

As sound rays pass through water of varying sound speed (caused by variations in temperature, salinity, and pressure) the rays are bent toward regions of lower sound speed. Thus, sounds originating in the surface ocean tend to be bent or refracted downward into the depth zone of lowered sound speed. Sound originating below such a layer is likewise bent upward. Consequently the sound ray tends to be trapped as if it were in a channel. This *sound channel* is a typical feature of the open ocean at depths of around 1000 meters at mid-latitudes to near the surface in polar regions, and it greatly extends the transmission range of underwater sound. This sound channel has been used for long-range underwater signaling, called *sofar* ("sound fixing and ranging"). Emergency location of downed aircraft crews is just one example of its utility. Other sound channels occur near the surface and in shallow waters less than 200 meters deep, owing to temperature changes or physical boundaries.

Marine organisms interfere with the operation of sound-ranging devices in several ways. Systems that passively listen to detect characteristic noises of ships passing or of other activities are especially troubled by the noises made by fishes—especially drumfish, grunts, snappers, and croakers, whose names suggest some of the sounds they make. These sounds have been confused with ships' engines, propellers, and other equipment. The utility of these noises to the fish is not always apparent. Mammals—including whales, dolphins, and seals—also produce noises apparently used for communication and their own form of sound ranging. Even the octopus dragging shells along the bottom has interfered with sound ranging.

More active sound-ranging systems are also affected by the presence of organisms. Many marine organisms have minute cavities that are highly efficient reflectors of sound energy. Consequently, they register as much larger targets on a detection device, and can disrupt the operation of the equipment.

Seawater—a living soup, 96.5 percent water, 3.5 percent salts

Molecular structure of water

Water—a polar molecule

Many unusual properties result from formation of hydrogen bonds

Ionic bonds readily broken in water; weak van der Waals bonds also present

In water vapor, molecules move freely, exert pressure

Ice, held together by hydrogen bonds, has properties of a solid, regular crystal structure; salt is excluded from structure

Liquid water an anomalous substance, containing structured and unstructured portions in a dynamic equilibrium

Temperature effects on water

Heat energy absorbed in changes of state with addition of heat

Latent heat utilized in breaking bonds; heat is released in condensation of water vapor and freezing of water

Cluster size and number of molecules per cluster decrease with increasing temperature

Density

Sinking (or floating) is determined by density relationships

Seawater density dependent on temperature, salinity, and pressure

Density of seawater decreases with increasing temperature.

Fresh water has a density maximum at 3.98°C; thus ice is less dense than liquid water.

Salinity commonly the dominant factor controlling density of seawater

Composition of sea salt

Six ions comprise 99 percent of sea salts (Cl^-, Na^+, SO_4^{2-}, Mg^{2+}, Ca^{2+}, K^+)

Chlorinity—(grams per kilogram or parts per thousand) = amount of chlorine in seawater, a measure of salinity

Salinity—total amount of dissolved salts in parts per thousand ($^0/_{00}$) in seawater

Conservative elements are those whose relative proportions in seawater are usually constant; their concentrations may change with addition or removal of water

Nonconservative elements—variable, generally low concentrations which depend on biological activity

Salt composition and residence times

Salts dissolved from continental materials enter seawater; these are eventually removed by incorporation in sediments

Residence time of an element in seawater is related to chemical behavior

Salt in water

Water shields ions of unlike charge from one another—therefore breaks ionic bonds

Temperature of initial freezing is affected by salinity changes

Different ions affect structuring properties of liquid water, cause formation of hydration atmospheres around the ion

Strong electrolytes dissolve completely; weak electrolytes dissociate only in part

Other modes of association—complex ions and ion pairs

Dissolved gases

Equilibrium concentration depends on solubility of a gas

Saturation concentration affected by salinity, temperature, and pressure conditions at the surface

Nitrogen—near saturation throughout the ocean

Oxygen—produced by photosynthesis; utilized by plants and animals at all depths

Carbon dioxide and carbonate cycles

Carbon dioxide occurs as a dissolved gas, carbonic acid, carbonate, and bicarbonate.

H_2CO_3 acts as a buffer

Calcium carbonate removed from seawater by plants and animals

Some inorganic precipitation in warm waters

Physical properties of seawater

Pressure and viscosity increase in direct proportion in "normal" liquids

At less than 1000 atmospheres pressure, increased pressure reduces water viscosity because it "favors" the unstructured mode

Bubbles and sea salt

Bubbles promote solution and evaporation transfer of gases

Bursting bubbles release salts to the atmosphere

Surface tension

High surface tension results from strengthened local water structure near any interface

Surface-active agents change surface tension of water, cause hydrophobic substances to "dissolve" in water

Electrical conductivity can be used to measure salinity

Sound in the sea—a form of mechanical energy

Speed (1480 meters per second) increased by increasing temperature, salinity, pressure

Sound channel (approximately 1000 meters depth) used for long-range signalling

Many fish and mammals use sound

SELECTED REFERENCES

BROECKER, W. S. 1974. *Chemical Oceanography*. Harcourt Brace Jovanovich, New York. 214 pp. Intermediate level discussion of processes affecting ocean chemistry.

DEMING, H. G. 1975. *Water: The Fountain of Opportunity*. Oxford University Press, New York. 342 pp. General treatment of water; Chapter 15 discusses desalination of seawater.

GROSS, M. G. 1976. *Oceanography*, 3rd ed. Charles E. Merrill, Columbus, Ohio. 150 pp. Elementary discussion of seawater.

HORNE, R. A. 1969. *Marine Chemistry: The Structure of Water and the Chemistry of the Hydrosphere*. Wiley-Interscience, New York. 568 pp. Advanced treatment of the subject, also a reference handbook.

KUENEN, P. H. 1963. *Realms of Water: Some Aspects of its Cycle in Nature*. Science Editions (John Wiley), New York. 327 pp. Elementary to intermediate in difficulty; stresses hydrologic cycle.

MACINTYRE, FERRAN. 1970. "Why The Sea is Salt." Scientific American, 223 (Nov., 1970), 104–115.

RILEY, J. P. AND R. CHESTER. 1971. *Introduction to Marine Chemistry*. Academic Press, New York. 465 pp. Intermediate in difficulty.

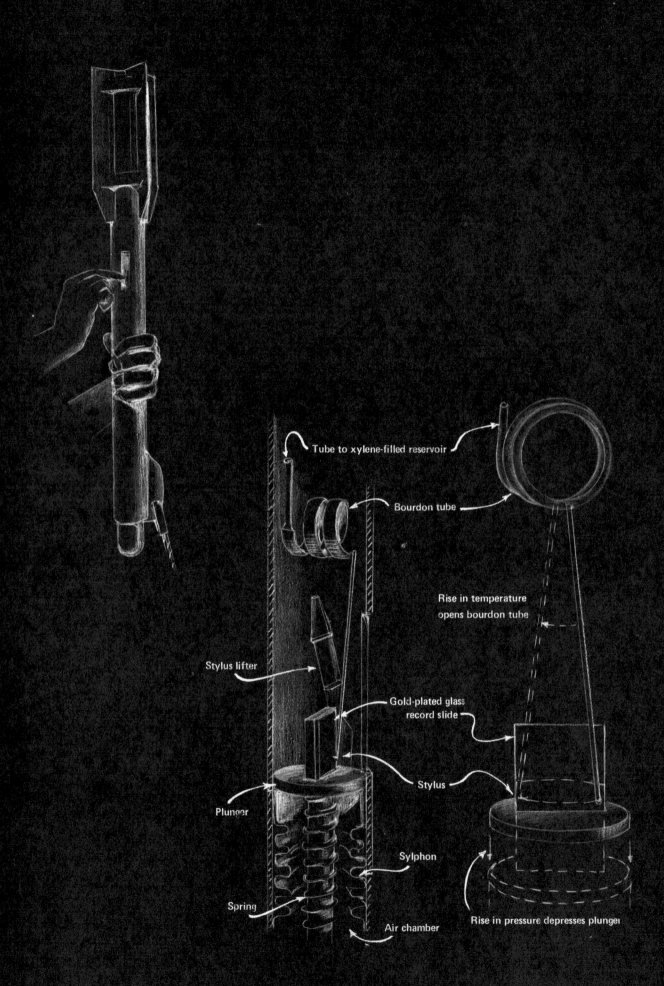

Tube to xylene-filled reservoir

Bourdon tube

Rise in temperature
opens bourdon tube

Stylus lifter

Gold-plated glass
record slide

Stylus

Plunger

Sylphon

Spring

Air chamber

Rise in pressure depresses plunger

The *bathythermograph* is a torpedo-shaped device used to measure temperature changes with depth below the surface while a ship is underway. A temperature-sensitive element causes a stylus to move. Increasing pressure compresses a bellows moving a metal-coated slide which is scratched by the stylus. A small graph is thus drawn of temperature at various depths.

temperature, salinity, and density

Having considered the physical and chemical properties of seawater, including the detailed structure of water and sea salts, we will now describe the behavior of ocean waters on a global scale.

Two different approaches are used. The first is the concept of the *budget*. Heat and water budgets are constructed, showing the major sources of each and what happens to both in the ocean. Regarding the water budget, it is important to remember that the amount of water on the earth's surface is nearly fixed. New sources need not be considered, only redistribution of water on the earth during the year (or perhaps some longer time period).

The second approach determines routes followed by ocean water as a result of the major processes. This is done by studying distributions of water temperature and salinity. Ocean and atmosphere are closely coupled systems, and a complete description of either requires detailed pictures of both oceanic and atmospheric conditions and current patterns throughout the earth over the course of a year. Despite enormous advances in oceanography in the past century, such a complete descriptive series is still not available. However, by using ships, aircraft, and/or satellites simultaneously, enough information about temperature and salinity distributions is accumulated to portray ocean conditions as they exist for a short time. Such an undertaking is called *synoptic oceanography*. Increasing successful use of satellites and buoys offers substantial hope that this may be achieved over many ocean areas in the future, providing nearly instantaneous pictures of the ocean like those available for the atmosphere.

LIGHT IN THE OCEAN Radiation from the sun striking the earth's surface is the source of virtually all the energy that heats the ocean surface and warms the lower portion of the atmosphere. Part of this incoming solar radiation is within the visible part of the spectrum and provides the energy needed by plants for photosynthesis, on land and in the ocean. After passing into the surface of the ocean, most of this energy is converted into heat, either raising water temperatures or causing evaporation.

The spectrum of radiant energy from the sun is filtered once as it passes through the atmosphere, as shown in Fig. 6-1, and is fur-

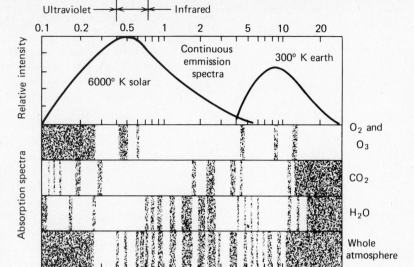

Figure 6-1

Emission spectrum of the sun and earth as compared to the absorption spectrum for the atmosphere and several important constituents. The earth radiates to space as much energy as it receives from the sun. The major difference is the shift from the relatively short-wave radiation of the sun, which we experience as visible light, to the longer-wave infrared radiation which we perceive as heat. The amount of heat received from the sun is balanced by the amount of heat radiated back to space from the earth. (After Weyl, 1970.)

Figure 6-2

Spectrum of extinction for visible light in 1 meter of seawater in areas removed from sources of fresh-water discharge. *A* is the curve for extremely pure water in the open ocean; *B* is for relatively turbid tropical–subtropical ocean water; *C* is for ocean water in mid-latitudes; *D* is for clearest coastal water; and *E, F,* and *G* are for coastal waters of increasing turbidity.

All wavelengths in the violet-to-yellow range are transmitted freely in open-ocean water. Near coasts, the presence of dissolved pigments ("yellow substance" of organic origin) strongly inhibits transmission in the violet and blue ranges. (After Dietrich, 1963.)

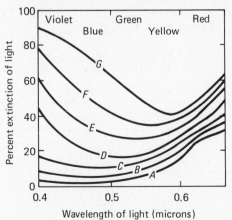

ther filtered in the surface ocean. Within the first 10 centimeters of even pure water, virtually all the infrared portion of the spectrum is absorbed and changed into heat. Within the first meter of seawater about 60 percent of the entering radiation is absorbed, and about 80 percent is absorbed in the first 10 meters. Only about 1 percent remains at 140 meters in the clearest subtropical ocean waters.

In coastal waters, the abundance of floating marine organisms, suspended sediment particles, and various dissolved organic substances causes the absorption of light to take place at even shallower depths. Near Cape Cod, Massachusetts, for instance, only 1 percent of the surface light commonly penetrates to a depth of 16 meters. In such waters, the maximum transparency shifts from the bluish region typical of clear oceanic waters to longer wavelengths, as shown in Fig. 6-2. In very turbid coastal waters, the peak transparency occurs around 0.6 micron, in the yellow range. The Thames River water near New London, Connecticut, shows maximum transparency in the red, at wavelengths between 0.65 and 0.7 micron. In highly polluted waters, absorption of all light takes place within a few centimeters of the surface.

Far from the coast, the ocean often has a deep luminous blue color quite unlike the greenish or brownish colors common to coastal waters. The deep blue color indicates an absence of colored particles or dissolved substances. In these areas, usually restricted to the central portions of the deep-ocean basins, the color of the water is thought to result from *scattering* of light rays within the water. A similar type of scattering is responsible for the blue color of the clean atmosphere.

The amount of light reflected from the ocean is controlled by the state of the sea surface and the angle at which the sun's rays strike the water. When the sun is directly overhead, only about 2 percent of incoming radiation is reflected; the remainder enters the water. When the sun is near the horizon, nearly all incoming radiation is reflected. Waves on the sea surface increase the amount of light reflected by as much as 50 percent, but when the sun is very near the horizon, waves serve to decrease the amount of reflection.

ATMOSPHERIC CIRCULATION AND HEAT BUDGET

One way to study average conditions on the earth as a whole is to construct a *budget*—of energy, or of a substance such as water. Budgets indicate where materials or energy come from *(sources)* and where they go *(sinks)*, but they often cannot give detailed information about the exact rate at which these transfers occur or the specific routes they follow. Thus our generalized picture may differ substantially from real conditions at any time or location.

First we consider the earth's *heat budget*. The earth is essentially a sphere which is exposed to the sun's radiation from a distance of 149 million kilometers (about 90 million miles). The top of the earth's atmosphere receives 2 calories per square centimeter per minute (cal cm^{-2} min^{-1}), and radiant energy is distributed over the earth as it rotates about its axis every 24 hours. Thus, on the average, about 0.5 calories per square centimeter per minute strike any given point at the top of the atmosphere, as Fig. 6-3(a) indicates, since the earth's surface is dark for part of the 24-hour day.

There is also a small amount of heat coming from the earth's interior, as we learned in Chapter 2; it may be disregarded in this chapter.

Earth's atmosphere contains constant amounts of nitrogen and oxygen, plus small and variable amounts of dust, water vapor, carbon dioxide, and ozone. Although present in minute quantities, dust,

Figure 6-3

Incoming solar radiation at the top of the earth's atmosphere. (a) shows the average solar radiation received over 24 hours at different latitudes in the Northern Hemisphere at the autumnal equinox, September 23; (b) shows the instantaneous maximum of incoming solar radiation at the earth's surface during the summer solstice when the sun is directly overhead at 23.5°N. Note the change in albedo between the equator and the pole.

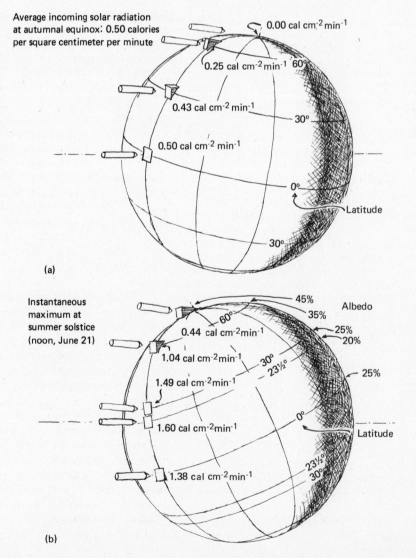

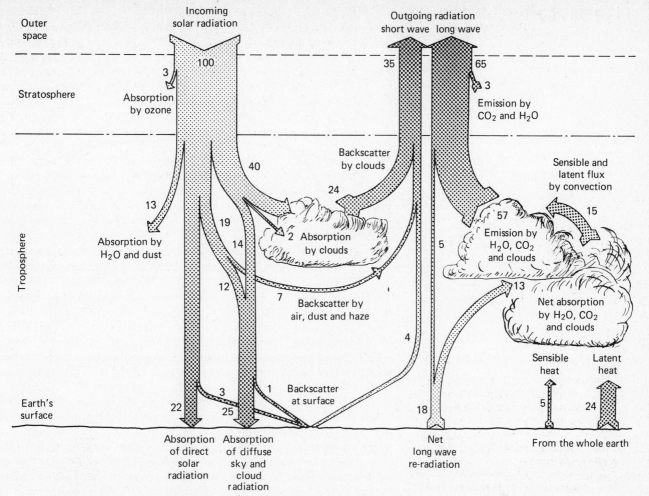

Figure 6-4

Average annual heat budget of the earth showing the major interactions with incoming solar radiation and loss of heat from earth.

carbon dioxide, and water absorb some incoming solar radiation; thus only about 47 percent of the solar radiation striking the top of the atmosphere actually reaches the earth's surface, as indicated in Fig. 6-4. The remainder is absorbed by the atmosphere or scattered and reflected back to space. Estimates of the earth's local reflectivity (albedo) range from 0.30 to 0.35. Figure 6-4 is based on an albedo of 0.35. Clouds, covering on the average 50 percent of the earth's surface, account for about two-thirds of the back-scattered radiation.

Despite the large amount of heat received by the earth from the sun each year, historical records indicate that the earth's surface-temperatures over the past few thousand years have been so constant that any significant changes are extremely difficult to document. Therefore, the earth is radiating back to space as much energy as it receives from the sun. The sun's surface, being extremely hot, radiates energy including the visible portion of the electromagnetic spectrum. Because the earth's surface is so much cooler, its radiation occurs in the *infrared* portion of the electromagnetic spectrum (see Fig. 6-1).

Of the energy lost to space, only about 5 percent is radiated back directly from the earth's surface (this is commonly called *backradiation*). About 60 percent is radiated back to space primarily from clouds (see Fig. 6-4). The atmosphere is nearly opaque to the earth's radiations, so that it acts much like a blanket in keeping the surface warmer than would be the case under a transparent atmosphere. A so-called *"greenhouse effect"* is caused by strong absorption of infrared radiation by carbon dioxide and water-vapor molecules. Without these

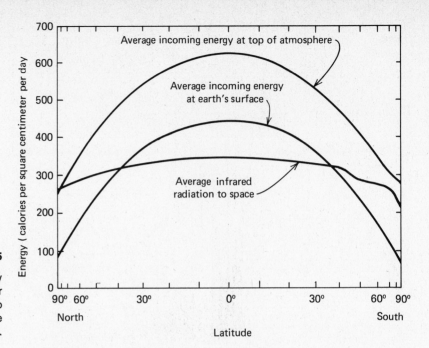

Figure 6-5

About four times as much solar energy reaches the earth's surface at the equator as at the poles, but radiation of energy to space is nearly the same everywhere on the earth.

trace constituents, earth-surface temperatures might drop as low as −20°C, which is the temperature at the top of the cloud layer. Instead, ocean surface temperatures average about 17.5°C and the land about 14°C.

Relatively little heat is transferred to the atmosphere by direct heating *(sensible heat)*. More than 80 percent of the atmospheric heating takes place through release of latent heat during condensation of water vapor, and virtually all of this water vapor comes from the ocean.

The earth's axis of rotation is inclined 23.5° to the plane of the earth's movement around the sun. This causes the amount of radiant energy at the top of the atmosphere to vary throughout the year for any point on earth, creating the change in seasons. Those points where the sun is directly overhead at noon receive the maximum amount of *insolation* (incoming solar radiation). Points north and south receive less, and no incoming radiation at all is received where the sun's rays just graze the earth's surface (see Fig. 6-3(a)).

Averaged over a year, the maximum amount of incoming solar energy is received in low latitudes, so that insolation is unevenly distributed over the earth's surface. The atmosphere, however, radiates energy back to space at a nearly constant rate, as shown in Fig. 6-5. As a result, the earth receives most of its heat between about 40°N and 40°S, whereas there is a net heat loss to space in the high latitudes between 40° and 90° in both hemispheres, but especially pronounced in the hemisphere experiencing winter.

The large input of solar energy into tropical regions causes high ocean-surface temperatures there as seen in Figs. 6-6 and 6-7. Maximum open-ocean temperatures, between 26° and 29°C, occur just north of the equator, at about 5°N (excluding isolated basins). Temperature distribution in the ocean is strongly modified by ocean currents and by the unequal distribution of land masses in the two hemispheres.

Note that the temperature difference between the warmest and coldest month of the year is small—less than 3°C—in the low latitudes (10°N to 10°S) and near the North and South Poles. In the mid-latitudes, around 30° north and south, the annual variation is largest: around 6°C. The average ocean-surface temperature is about 17.5°C, as Fig. 6-8 indicates, whereas average land temperature is about 14.4°C.

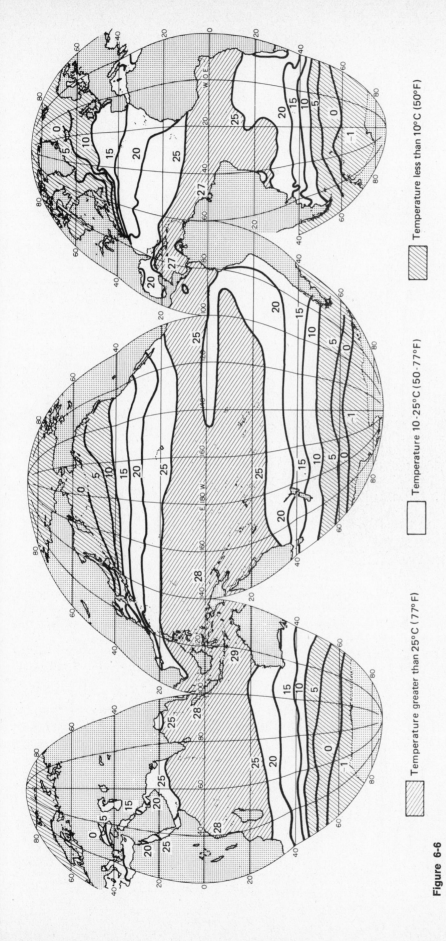

Figure 6-6

Ocean-surface temperatures in February. Note that isotherms (lines connecting points with surface temperature) tend to parallel the equator. (After Sverdrup et al., 1942.)

Temperature less than 10°C (50°F)

Temperature 10-25°C (50-77°F)

Temperature greater than 25°C (77°F)

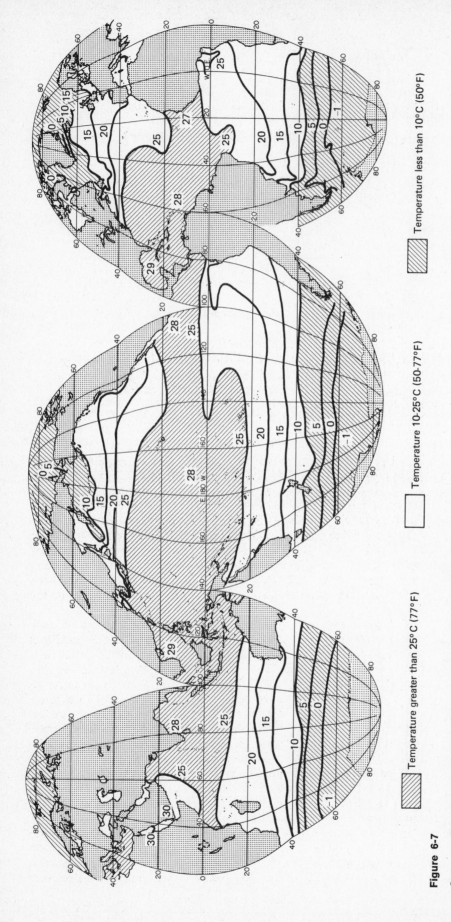

Figure 6-7

Ocean-surface temperatures in August. Below about a hundred meters of the surface, temperatures drop sharply except at very high latitudes. (After Sverdrup et al., 1942.)

Temperature greater than 25°C (77°F)

Temperature 10-25°C (50-77°F)

Temperature less than 10°C (50°F)

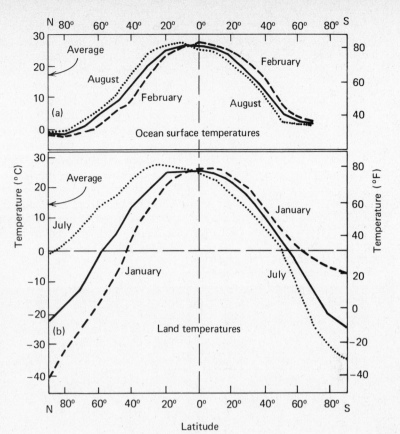

Heat transfer from low to high latitudes takes place in both atmosphere and ocean. In this somewhat oversimplified picture, atmosphere and ocean function as a simple heat engine, transforming energy into motion. Atmospheric motions are experienced as winds, the oceanic movements as currents. Although ocean currents move far more slowly than winds, the large heat capacity of water (compared to air) enables it to carry far more heat by volume than air does. In the Northern Hemisphere, ocean currents are thought to carry about one-third the amount of heat that is carried by the atmosphere. In the Southern Hemisphere, oceanic-heat transport appears to be less important.

Atmospheric circulation is controlled primarily by minute amounts of water vapor, ozone, and carbon dioxide. Although nitrogen constitutes about 78 percent of the atmosphere by volume, it has little direct effect on circulation: carbon dioxide and water vapor together cause the "greenhouse effect" previously mentioned. Absorption of solar radiation (see Fig. 6-5) by ozone (O_3) causes warming of the lower stratosphere, forming a pronounced change in density *(tropopause)* that separates the *stratospheric* (high-altitude) and *tropospheric* (near-earth) circulation systems. In the stratosphere, gases are warmed by absorption of *ultraviolet* radiation; thus temperature of ionized gases increases with increased altitude. In the troposphere, gases are warmed by the earth; consequently temperature decreases with increased altitude. Weather systems, with vigorous vertical convection and formation of clouds, all occur in the troposphere.

On a featureless, water-covered earth, the simplest atmospheric circulation would be a single cell. Warmed moist air would rise in the equatorial regions and move toward the poles in both hemispheres, carrying heat as both latent heat (water vapor) and as sensible heat (warm air). When the air was cooled and water vapor precipitated, it would sink and flow back toward the equator in a surface wind.

Such a simple cell does occur near the equator: the *Hadley cell*, named for the British meteorologist who first postulated its existence. This cell includes the persistent Trade Winds of low latitudes.

The earth is of course not featureless and rotates. Thus the amount of heat that the atmosphere must transport from equator to poles is too great for a simple atmospheric pattern to maintain stability. Instead, the atmospheric circulation breaks down into several cells and exhibits well-developed, wavelike patterns, especially in the mid-latitudes where the westerly winds prevail, as illustrated in Fig. 6-9. These large waves give rise to the series of *fronts* (sharp boundaries between two air masses) that dominate weather patterns through much of the United States and Europe. These waves (or fronts) commonly separate cold, dry polar air masses from warmer, more humid air masses of mid-latitudes.

Figure 6-9

Schematic representation of the planetary circulation of the earth's atmosphere.

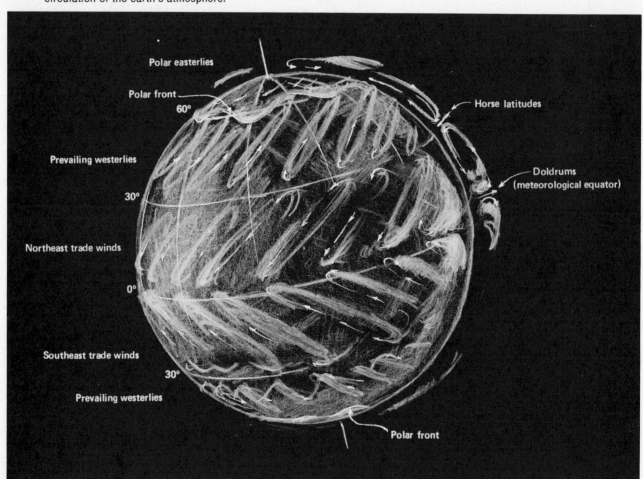

OCEANIC WATER AND HEAT BUDGET

The amount of free liquid water at the earth's surface has remained essentially constant for the past 5000 years, since the retreat of continental glaciers from North America and Europe. The ocean is the major water reservoir; the small amounts in lakes, rivers, and atmosphere are essentially in transit back to their source. Ground water also eventually finds its way back to the ocean, although taking much longer to make the trip.

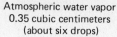

Atmospheric water vapor
0.35 cubic centimeters
(about six drops)

Lakes and rivers
one tablespoon
(15 cubic centimeters)

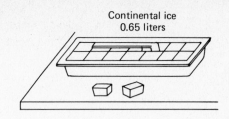

Continental ice
0.65 liters

Sea water
42 liters

Water in sedimentary
rocks and sediments
5 liters

0   30

centimeters

30 cubic centimeters is equivalent to 10^{15} tons

Figure 6-10

Relative quantities of liquid water, ice, and water vapor at the earth's surface, plus water in sediments and sedimentary rocks.

Imagine that for the total amount of free water on the earth's surface, 1.4×10^{24} grams or 1 billion billion metric tons, each 10^{15} tons (1 million billion tons) is equivalent to 30 cubic centimeters (about one ounce). That gives a total amount of water equivalent to about 47 liters (about 12.5 gallons). Most of this (about 42 liters) is contained in the oceans (Fig. 6-10). In addition to the free water, sedimentary rocks and sediments retain an amount of water equivalent to about 15 liters (about 4 gallons).

Another convenient way to visualize the earth's water budget is expressed in Table 6-1. About 97 centimeters is evaporated on the average from the ocean surface each year. Of this amount, 88 centimeters falls as rain on the ocean and its thus almost immediately returned. The remaining 9 centimeters falls on land and eventually makes its way into rivers to return to the ocean.

Yet another way of describing the water budget or the hydrological cycle is to express amounts in thousands of cubic kilometers, as is done in Fig. 6-11. Note that about one third of the land area is desert or has no drainage to the sea.

Evaporation from the ocean surface is most active in subtropical areas, where it is favored by clear skies and dry winds. These are the Trade Wind zones. The sides of ocean basins where dry winds blow from the continents are especially favorable for evaporation.

Neither is precipitation uniform over the entire ocean, as Fig. 6-12 shows. Equatorial and subpolar or polar regions experience heavy precipitation; in the tropics particularly, large amounts of rainfall result from an abundance of warm, moist air rising in the equatorial atmospheric circulation. As rising air cools, its ability to hold water

153

vapor diminishes and precipitation results. Satellite photographs, such as in Fig. 6-13, commonly show a line of clouds in tropical regions because of this circulation; this is known as the intertropical convergence. In the high latitudes, strong atmospheric cooling also favors precipitation.

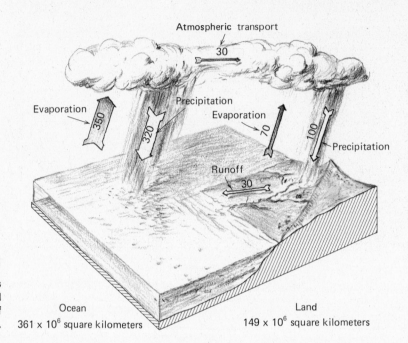

Figure 6-11

Schematic representation of total amounts of liquid water involved in the hydrological cycle each year, in thousands of cubic kilometers.

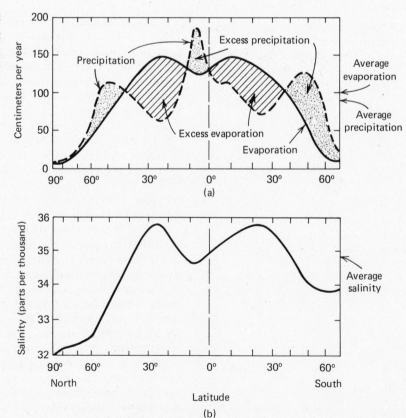

Figure 6-12

Relationship between oceanic evaporation and precipitation at various latitudes (a) and the salinity of surface waters (b). (After Wüst et al., 1954.)

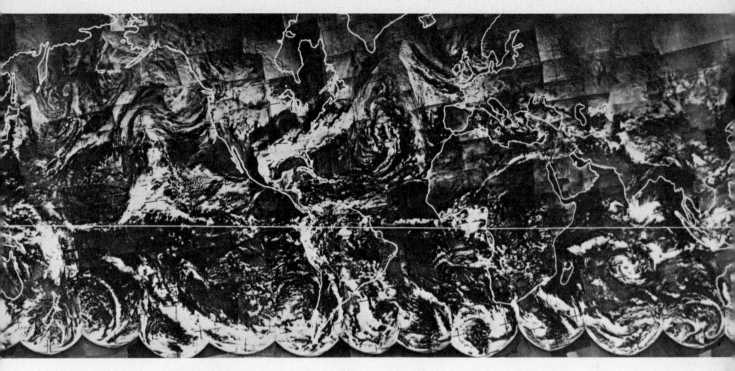

Figure 6-13

Composite photograph (photomosaic) of the earth surface prepared from 450 individual pictures taken by NASA satellite Tiros IX during the 24 hours of February 13, 1965, the first complete picture of the world's weather. Continents are outlined by white lines. Large areas of white are clouds or the permanent ice covers on Antarctica (bottom margin). Note the large spiral weather systems south of Alaska and Greenland. Also note the thin line of clouds along the equator, especially noticeable in the Atlantic. (Photograph courtesy NASA.)

Because of evaporation and precipitation distributions, salinity varies over the ocean surface. Highest salinities occur in open-ocean subtropical areas, where evaporation is pronounced and there are no significant sources of fresh water. Near coasts, rivers discharge fresh

TABLE 6-1
*Water Budget of the Earth ***

AREA	PRECIPITATION		EVAPORATION	
	(10³ km³/yr)	(cm/yr)	(10³ km³/yr)	(cm/yr)
Ocean	320	88	350	97
Continent	100	67	70	47
Entire earth	420	82	420	82

* After Albert Defant, 1961. *Physical Oceanography,* Vol. I, Pergamon Press, Elmsford, N.Y., p. 235.

water back into the ocean, and salinities are lowered. Surface waters of high-latitude oceans are commonly low in salinity owing to relatively heavy precipitation and little evaporation.

Continents (especially those with mountain ranges) and ocean currents affect salinity distributions. The Atlantic Ocean, for example, has a higher salinity than the Pacific. Mountain ranges, especially the Andes of South America, cause winds from the Pacific to give up their moisture before blowing across to the Atlantic. High mountains deflect winds to higher altitudes, where they are cooled and their water vapor condenses to fall as rain on the mountains. On the other hand, the Atlantic's water vapor is carried to the Pacific by winds blowing across the Gulf of Panama.

A water budget can be constructed for a single region and expressed mathematically as:

Evaporation = Precipitation + Runoff + Contribution from currents

$$E = P + R + C$$

where

E = the amount of water lost through evaporation
P = the amount of water gained through precipitation
R = the amount of water brought to the region by rivers
C = the relative amount of fresh water moved by currents.

The last quantity (C) can be positive if water of lower salinity flows into the region or negative if more saline water flows into the region. Unless there is a pronounced salinity change, the amount of water gained by precipitation and river runoff will equal the amount lost through evaporation and carried away by currents.

To summarize, surface-water salinity in the open ocean is controlled primarily by the local balance between evaporation and precipitation. Where evaporation exceeds precipitation, surface waters are more saline than average ocean water (S = 34.7 parts per thousand).

Where precipitation and river runoff from continents exceed evaporation, surface-water salinities are lower than for the average ocean. Low-salinity waters are typical of three regions: the tropics, owing to the large rainfall there; high latitudes, because of relatively high precipitation and limited evaporation; and along continental margins, owing to local rainfall and river runoff. Consequently, high surface-water salinities in the open ocean tend to occur near the centers of the ocean basins, at mid-latitudes (Fig. 6-14).

In coastal ocean areas, highest surface-water salinities occur in partially isolated basins of tropical areas, such as the Red Sea, the Arabian Sea, and the Mediterranean Sea. Here water losses due to evaporation are not replaced by river runoff or precipitation. The restricted communication with adjacent ocean waters prevents ready renewal of the surface waters. Consequently the salinity of surface waters is 40 parts per thousand or more in basins such as the Persian Gulf and the northern Red Sea. These occur at mid-latitudes, around 30°N, where several of the world's great land deserts occur—such as the Sahara in Africa, and Mojave in California. The arid Australian continent also testifies to the low rainfall at comparable latitudes in the Southern Hemisphere.

Heat and water budgets are intimately related. Evaporation of water removes heat from the ocean surface, whereas precipitation contributes large amounts of heat to the atmosphere and indirectly to the ocean.

A *regional heat budget* for a part of the ocean can be expressed mathematically as follows:

Heat gained = Heat lost

$$Q_s + Q_c = Q_e + Q_r + Q_h$$

where

Q_s = energy gained from solar radiation
Q_c = energy gained (or lost) by ocean currents
Q_e = energy lost through evaporation (this may be a positive quantity if more heat is gained from precipitation than is lost through evaporation)
Q_r = energy lost by radiation
Q_h = energy lost by direct heating of the atmosphere (sensible heat)

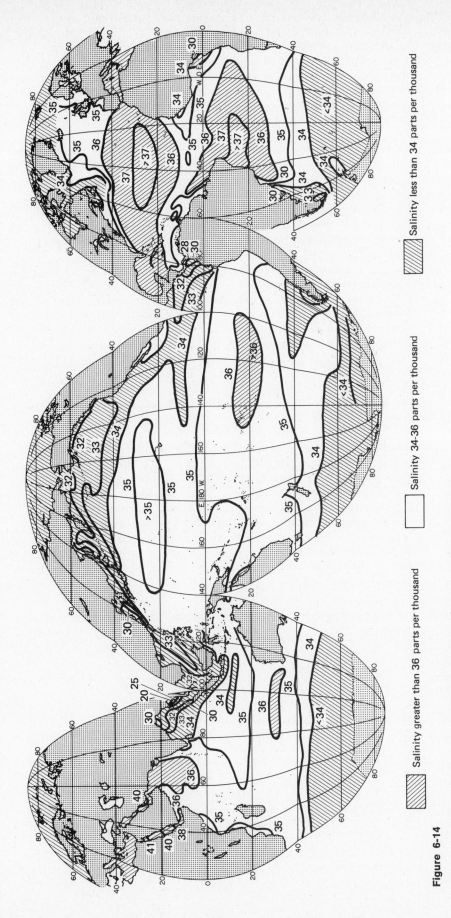

Salinity greater than 36 parts per thousand

Salinity 34-36 parts per thousand

Salinity less than 34 parts per thousand

Figure 6-14

Salinity of ocean-surface water in northern summer. (After Sverdrup et al., 1942.)

Areas of relatively high surface salinity supply heat to the atmosphere in the form of water vapor. This heat is transported by winds to the high latitudes. There, water vapor condenses so that heat is given off, accompanied by precipitation in excess of evaporation. Thus the poleward transport of heat and water is reflected in the relatively low surface salinities at high latitudes, and regional salinity differences are another manifestation of global heat transport.

DENSITY OF SEAWATER

Density of seawater is determined primarily by temperature and salinity. (Pressure is important only in the deep ocean and will be ignored in our present discussion.) Decreases in temperature and increases in salinity cause increases in density. The effect of temperature and salinity can be seen in Fig. 6-15: changing salinity from 19° to 26 parts per thousand (at 30°C) has the same effect as changing the temperature from 31° to 12°C at a constant salinity of 20 parts per thousand.

In the ocean, water masses generally form thin layers arranged according to water density; such waters are said to be *vertically stratified*. In the *stable-density* configuration, the densest water is on the bottom, the least dense on top. If such a stable system is disturbed, it tends to return to its original configuration. A stable situation can also be achieved by supplying energy to a stratified system and mixing the layers so that there is no longer any density stratification. This would be a *neutrally stable system*. If any one of the layers is more dense than the one beneath it, the system is unstable and the denser layer tends to sink to its appropriate density level, thus creating a stable structure.

Since ocean water normally has stable stratification, the most dense water might be expected to occupy the deepest part of a given ocean basin. Where such is not the case, this usually indicates that some physical barrier—say, a shallow sill—is preventing lateral spreading of dense water. For example, dense waters formed in the Arctic Ocean cannot readily flow into the North Atlantic because of submarine ridges stretching from Greenland to Scotland.

In the absence of such barriers, a water mass tends to seek its appropriate density level and to spread laterally. Because of this, lateral changes in water temperature and salinity are much smaller than vertical changes.

Figure 6-15

Variation in seawater density (in g/cm^3) as affected by temperature and salinity. Note the changes in temperature of maximum density and the point of initial freezing caused by changed salinity.

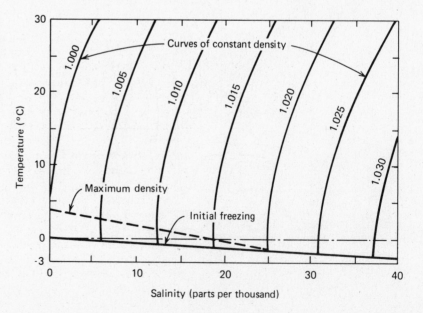

Energy must be supplied if a density-stratified system is to be destroyed. Below the surface of the open ocean, such vigorous mixing is rare. In coastal waters, on the other hand, mixing is common because of tidal currents and movement of waters across rough bottom topography. Near the ocean surface, energy sufficient to cause mixing is supplied by the wind, passage of waves, and surface heating and cooling. In high-latitude areas the density structure is weak enough to permit widespread vertical circulation through the ocean depths.

This pattern is quite different from that of the atmosphere, where strong vertical circulation is a characteristic feature. The ocean is heated and cooled at its upper surface. Heating of the surface inhibits vertical circulation and powers a very inefficient heat engine. In contrast the atmosphere is cooled at the top by radiation to space and heated at the bottom through contact with land and ocean. Consequently, vertical circulation is quite well developed through the lower parts of the atmosphere.

Three general depth zones may be identified in the ocean: the surface, pycnocline, and deep zones, schematically represented in Fig. 6-16. The *surface zone* is in contact with the atmosphere and undergoes substantial seasonal changes because of seasonal variability in local water budgets (evaporation, precipitation) and heat budgets (incoming solar radiation, evaporation). The surface zone contains the least dense water masses in the ocean and is the ultimate source of all marine-food production because there is sufficient light for photosynthesis.

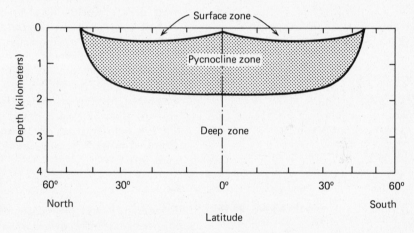

Figure 6-16

Schematic representation of the horizontal layering of ocean waters.

The surface zone averages about 100 meters in thickness and comprises about 2 percent of the ocean's volume. Surface waters are generally quite warm—except at high latitudes, where extensive mixing with deeper layers may prevent development of a warm surface zone, even during summer. Near the surface, waters have nearly neutral stability because of strong mixing, and vertical water movements are quite common. Winds, waves, and strong cooling of surface waters cause mixing in the open ocean.

In the *pycnocline zone,* water density changes appreciably with increasing depth and thus forms a layer of much greater stability than is provided by the overlying surface waters. The formation of a pycnocline zone may be the result of marked changes with depth in either salinity or temperature. Marked temperature changes with depth are common in the open ocean, as Fig. 6-17 indicates, and there the pycnocline coincides with the *thermocline,* or zone of sharp temperature change. Thermoclines are especially prominent in subtropical open-ocean areas, where the surface layers are strongly heated during much of the year.

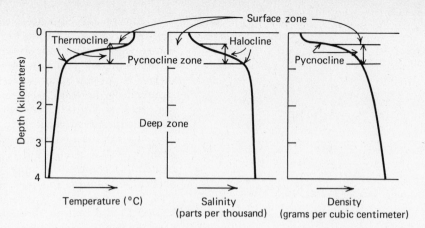

Figure 6-17

Variations with depth in temperature cause the thermocline. Similar variations in salinity cause the halocline. The pycnocline can be caused by either independently or by both occurring together.

At higher latitudes, there is less surface heating by incoming solar radiation. There, the lowered surface salinity caused by abundant precipitation causes a *halocline,* a marked change in salinity with depth. The halocline is well developed over much of the North Pacific Ocean, for example, where it is an important factor in the stability of the pycnocline zone.

The pycnocline acts as the floor to surface processes, because low-density surface waters cannot readily move downward through it. Deep-ocean waters move slowly upward through the pycnocline as part of the worldwide deep-ocean circulation. This is a significant factor in oceanic circulation at mid-latitudes and low latitudes. Seasonal changes in temperature and salinity are quite small in waters below the pycnocline.

The *deep zone* contains about 80 percent of the ocean's waters and lies at depths of about 2 kilometers or more—except at high latitudes, where deep waters are in contact with the surface. The relationship between deep-ocean waters and high-latitude areas accounts for the low temperatures of the subsurface ocean. Dissolved oxygen is also supplied to the deep ocean through these high-latitude exposures of cold water.

WATER MASSES

Figures 6-18 and 6-19 show vertical distribution of temperature and salinity in all three ocean basins. The profiles display temperature and salinity distributions along the western sides of the Atlantic and Indian Oceans and the central portion of the Pacific Ocean. Note both the similarities and the differences between this and the generalized diagram (Fig. 6-16) we previously considered. Note also the complexities of temperature and salinity distribution evident on even such a simplified diagram.

The base of the surface zone corresponds to the 10°C-isotherm contour on the vertical section. The base of the pycnocline can be taken as the 4°C isotherm. Note that there are strong similarities between the distribution of pycnocline and surface zone in the three ocean basins, but that they are by no means identical. Also note that there are differences between hemispheres even within the same ocean basin.

The vertical temperature and salinity sections show that distinctly different strata or water masses exist at intermediate depths. For example, there is a relatively low-salinity water mass at about 1 kilometer extending from the Antarctic region into the South Atlantic Ocean. Through analysis of sections such as these, oceanographers are able to determine where water masses form and how they move below the ocean surface. This technique has been one of the most rewarding means of mapping subsurface currents. Such studies have demonstrated that the dominant movement of near-bottom currents is generally north–south, in contrast to the primarily east–west movements of currents at the surface and intermediate depths.

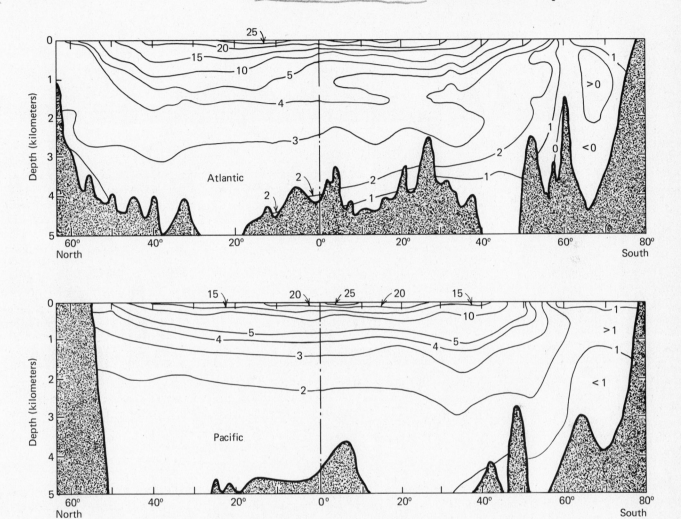

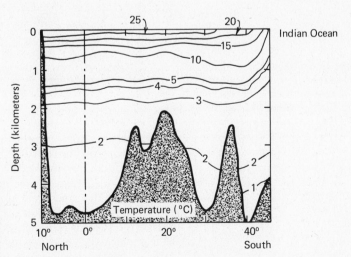

Figure 6-18

Vertical distribution of temperature in the three ocean basins. Vertical exaggeration is approximately a thousandfold. (After Dietrich, 1963.)

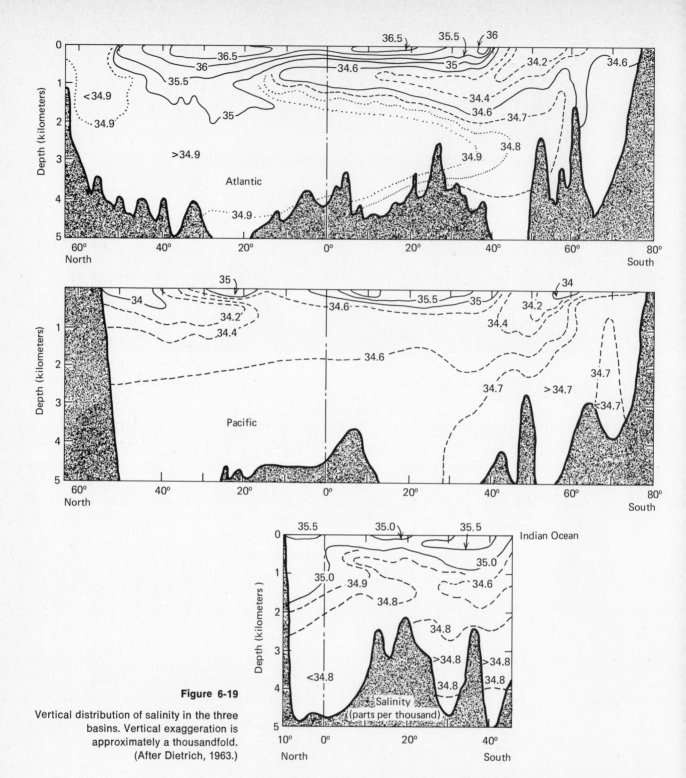

Figure 6-19

Vertical distribution of salinity in the three basins. Vertical exaggeration is approximately a thousandfold. (After Dietrich, 1963.)

Another aspect of the ocean illustrated by the vertical section is that most of the ocean has much lower temperatures, shown in Fig. 6-20, than we might have predicted from examining the ocean surface alone. The average temperature of the entire ocean is 3.5°C; in contrast the average ocean-surface temperature averages about 17.5°C. Thus it is the surface ocean that serves to buffer our climate against short-term changes. Most of the deep ocean is too isolated to have much effect on short-term climatic conditions, although it may influence long-term climatic changes.

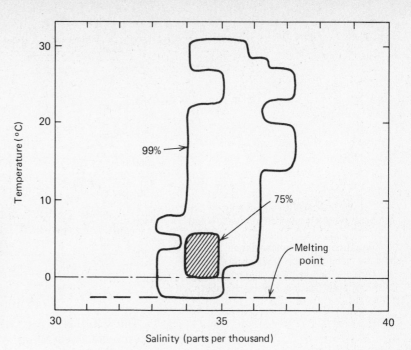

Water masses with characteristic temperatures and salinities are formed in particular ocean areas, as listed in Table 6-2. If a mass of surface water becomes more dense than the water below, it will sink below the surface. From that point, until it returns again to the surface, the temperature and salinity of the waters change primarily by mixing with other water masses. In the deep open ocean, where there is little energy to cause mixing, this process can be very slow. Temperature and salinity can therefore be used in the open ocean as tags to identify individual water masses as they spread laterally between other masses, or move along the ocean floor.

TABLE 6-2

Some Major Water Masses of the Ocean*

WATER MASS	TEMPERATURE (°C)	SALINITY (parts per thousand)
Antarctic Bottom Water	—0.4	34.66
North Atlantic Deep Water	3–4	34.9–35.0
North Atlantic Central Water	4–17	35.1–36.2
Mediterranean Water	6–10	35.3–36.4
North Polar Water	—1–+2	34.9
South Atlantic Central Water	5–16	34.3–35.6
Subantarctic Water	3–9	33.8–34.5
Antarctic Intermediate Water	3–5	34.1–34.6
Indian Equatorial Water	4–16	34.8–35.2
Indian Central Water	6–15	34.5–35.4
Red Sea Water	9	35.5
Pacific Subarctic Water	2–10	33.5–34.4
Western N. Pacific Water	7–16	34.1–34.6
Pacific Equatorial Water	6–16	34.5–35.2
Eastern S. Pacific Water	9–16	34.3–35.1
Antarctic Circumpolar Water	0.5–2.5	34.7–34.8

* After Albert Defant, 1961. *Physical Oceanography,* Vol. I, Pergamon Press, Elmsford, N.Y., p. 217.

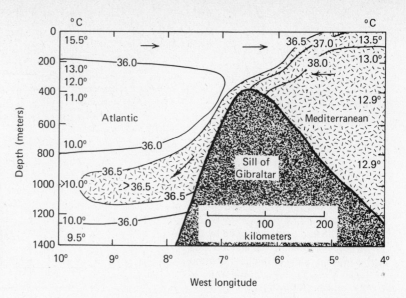

Figure 6-21

Vertical distribution of salinity and temperature in a cross-section (east–west) through the Strait of Gibraltar at the entrance to the Mediterranean. Arrows show the main direction of water movement. (After G. Schott, 1942. *Geographie des Atlantisches Ozeans,* Boysen, Hamburg.)

The so-called *"core" technique* has been used to study movement of well-defined subsurface water bodies as they spread out from a source region. It is a horizontal representation of the most likely paths of water particles. It allows flow across density surfaces since the density of the core water mass usually varies. An example is the warm, saline water (6–10°C, 35.3–36.4 parts per thousand) formed in the Mediterranean Sea which flows out through the Strait of Gibralter near the bottom and spreads laterally at mid-depths in the North Atlantic Ocean, pictured in Fig. 6-21. Its distinctive temperature and salinity permit it to be clearly distinguished from water masses above and below. As the water moves away from its source, it mixes with waters adjacent and gradually loses its distinctive characteristics. Comparison of the values observed at any given point where that mass is identifiable with the original values provides an estimate of the amount of mixing that has occured since the water mass formed.

SEA-ICE FORMATION

Because of low air and water temperatures in polar regions and in certain coastal areas, sea-ice formation is a conspicuous and important ocean phenomenon (shown in Fig. 6-22), illustrating temperature effects at high latitudes. Sea ice exists year-round in the central Arctic and in some Antarctic bays, extending in winter across the entire Arctic and far out to sea around the Antarctic continent. During most winters it can be seen to form in bays as far south as Virginia on the Atlantic coast of the United States. During glacial times, it must have been even more widespread than at present, as shown in Fig. 6-23.

In addition to its local importance, sea-ice formation influences the ocean far beyond the ice-covered area. For instance, Antarctic Bottom Water is thought to form through mixing of extremely cold surface waters and brines released by the freezing and melting of sea ice.

When seawater reaches its temperature of initial freezing, clouds of tiny ice particles form, making the water slightly turbid. The ocean surface becomes dull and no longer reflects the sky. Initially disklike, the ice particles grow and form hexagonal *spicules* 1 to 2 centimeters long. Sea salt is excluded, as it does not fit into the ice-crystal structure.

The water remaining becomes slightly saltier, so that its freezing point is lowered. As freezing continues, convection cells form at the underside of the ice layer. Water comes up from below to replace the

164

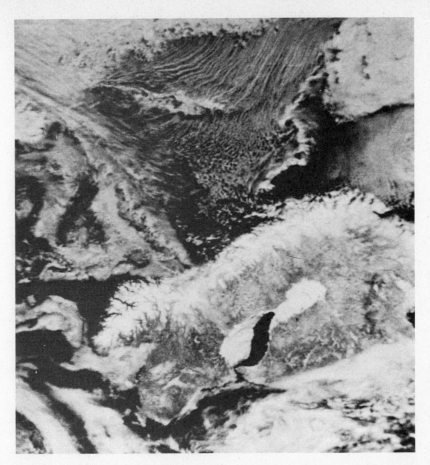

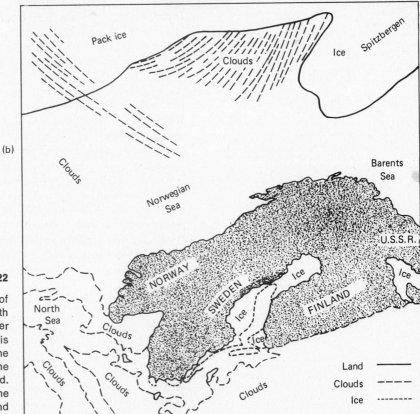

(b)

Figure 6-22

Ice pack in the Arctic Ocean north of Scandinavia, which itself is covered with snow, as seen by the Nimbus-4 weather satellite on April 13, 1970. The pack ice is the irregular white area near the top of the photograph. The Gulf of Bothnia in the Baltic Sea (lower center) is ice-covered. The filmy streaks in the center of the photograph are clouds. (Photograph and explanatory diagram courtesy NASA.)

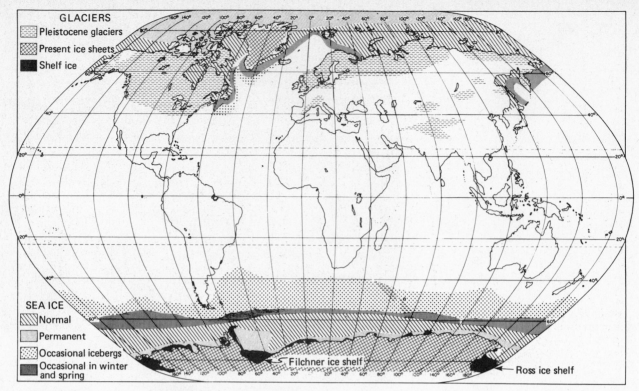

GLACIERS
- Pleistocene glaciers
- Present ice sheets
- Shelf ice

SEA ICE
- Normal
- Permanent
- Occasional icebergs
- Occasional in winter and spring

Filchner ice shelf

Ross ice shelf

Figure 6-23

Distribution of sea ice and probable maximum extent of land ice during glacial Pleistocene times.

increasingly saline brines, which tend to form finger-shaped parcels and stream downward.

Ice crystals form a layer of *slush*, which covers the ocean surface like a blanket. Some crystals grow downward, and cross-connections form between crystals. The result is a thin, plastic layer of ice in which small cells of seawater are trapped. One kilogram of newly formed sea ice contains about 800 grams of ice (salinity 0 parts per thousand) and 200 grams of seawater (salinity 35 parts per thousand). Therefore, the average salinity of newly formed ice is about 7 parts per thousand, depending on the temperature at which it was formed. At temperatures near the freezing point, ice forms slowly so that brines can escape. Relatively little seawater is enclosed in the cells, and the salinity of the ice is low. At lower temperatures, ice forms more rapidly and much more seawater is trapped. In this case, ice salinity is higher, although always less than the seawater from which it formed.

As the weather grows colder and heat is continually removed from the sea surface, new ice is formed. Water in the cells freezes, and their interior walls become thicker. Concentrated brines remain in the ever-smaller cell cavities, and eventually certain salts begin to crystallize out. Sodium sulfate crystals form at − 8.2°C, whereas sodium chloride does not crystallize above − 23°C. Salt crystals and ice crystals may intergrow in a completely solid ice structure if temperatures are low enough.

As sea ice ages, the brine cells tend to migrate. Many reach the surface to escape into the seawater below. Through time the ice retains less and less salt, so that eventually it may be nearly pure fresh water. Pools of water from melting of such nearly pure ice have saved the lives of several polar explorers who ran short of drinking water.

Sheets of newly formed sea ice break into pieces under stress of wind and waves. These are called *pancake ice* because that term so aptly describes their appearance in the early stages of formation, as shown in Fig. 6-24. As pancakes coalesce they form young sea ice, which is usually flexible and retains traces of the pancakes from

Figure 6-24

Wave action on newly formed sea ice forms pancake ice. (Official U.S. Navy photograph.)

which it formed. Currents and winds pile up the ice units so that they form hummocks or pressure ridges. With the freezing of winter snows on the surface, growth occurs from the top as well as from the bottom. This thick, solid mass may break into large *floes* in response to winds and surface currents; these constantly move and shift, freeze together and break loose, buckle up and flatten out as the pack is dragged by currents flowing beneath.

During the course of a freezing season, sea ice may form at the surface to a thickness of 2 meters at high latitudes. When more than 1 year old, it can form fields of coherent ice. Where ice never completely melts away but continues to grow during subsequent seasons, as in the central Arctic Ocean, adjacent units are piled on top and an average thickness of 3 to 4 meters is achieved. During summer the central Arctic ice pack melts to a thickness of about 2 meters.

The behavior of oceanic pack ice should not be confused with the huge masses of land ice that cover Greenland and the Antarctic continent throughout the year. The latter sometimes extends out over bays and sheltered waters as a permanent floating *ice shelf*. Pieces of glacial ice sometimes break off in the ocean, particularly when glaciers flow out of valleys, as in western Greenland. Irregular pieces are known as *icebergs*, as shown in Figs. 6-25 and 6-26; larger, more regular, and usually flat-topped pieces are called *tabular* icebergs, as in Fig. 6-27; these may be tens to hundreds of kilometers long. Icebergs are rarely more than 35 meters above water; 5 times as much is usually below water.

Icebergs are carried by currents out of the polar regions at speeds of about 15 to 20 kilometers per day. In the North Atlantic they are carried southward by the Labrador Current until they reach the vicinity of the New York–Europe shipping lanes. Icebergs from western Greenland travel up to 2500 kilometers, reaching 45°N, which is the vicinity of the Grand Banks.

Currents around Antarctica keep icebergs relatively close to the continent. They are known to move up past the tip of South America to about 40°S in the Atlantic but rarely reach beyond 50°S in the Pacific Ocean. In 1894, a piece of ice was spotted at 26°30'S, only about 350 kilometers from the Tropic of Capricorn, which is commonly taken as the margin of the subtropical ocean. Icebergs typically have about a 4-year life span in the ocean.

Figure 6-25 *(above)*

An iceberg in pack ice off the coast of Labrador in March, 1962. The blocks of ice and snow on top of the berg are as big as houses. (Official U.S. Navy photograph.)

Figure 6-27 *(below)*

A tabular iceberg in Nelville Bay, Greenland. The icebreaker approaching the iceberg is approximately 90 meters long. The iceberg extends 24 meters above the water and is about 1.1 kilometers long and about 0.75 kilometers wide. The iceberg extends about 150 meters below the surface. (Official U.S. Coast Guard photograph.)

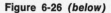

Figure 6-26 *(below)*

An iceberg along the Grand Banks of Newfoundland is tracked by a U.S. Coast Guard airplane for the International Ice Patrol. (Official U.S. Coast Guard photograph.)

The open ocean can be divided into climatic regions. These regions, indicated in Fig. 6-28, generally exhibit characteristic ranges in surface-water characteristics. These regions also have relatively stable boundaries which change little through time. Note that the climate regions extend generally east–west. Both temperature and salinity of surface waters in these regions are related primarily to incoming solar radiation and the relative amounts of evaporation and precipitation.

The Arctic Ocean and the band around Antarctica constitute the major *polar-ocean* areas. Ice occurs at the surface most of the year and therefore keeps surface temperatures at or near the freezing point despite the intense insolation during local summer. In winter there is no sunlight. The relatively low surface salinities, combined with the ice cover, act to inhibit vertical mixing, so that there is little mixing between surface and deeper waters.

Subpolar regions are affected by seasonal migration of sea ice. In winter, pack ice covers the area but it disappears during summer, except for isolated blocks and icebergs. There is enough heating to melt the ice and to raise surface temperatures to about 5°C. Abundant light and the stable layer formed by surface warming combine to provide good conditions for short but intensive phytoplankton (marine algae) blooms. This rich plant crop supports large populations of *zooplankters* (minute marine animals), birds, and whales. Diatom shells (frustules) sink to the bottom forming diatomaceous sediments.

The *temperate* region corresponds to the band of westerlies where there are strong storms, especially during winter, with gales and heavy precipitation. These strong winds cause the West Wind Drift, one of ocean's major currents. Extensive mixing resupplies nutrient elements from deep water to the surface layers, supporting a marine food chain that includes large populations of fish. Most of the world's major fisheries are found in this region.

Subtropical regions coincide with the nearly stationary high-pressure cells of the mid-latitudes. Winds are weak and therefore surface currents tend to be weak also. Clear skies, dry air, and abundant sunshine lead to extensive evaporation, so that surface salinities are

Figure 6-28

Oceanic climatic zones and their relationships to terrestrial climatic zones: (1) tundra, (2) boreal forest, (3) temperate, (4) desert, (5) savannah, (6) steppe, (7) tropical rain forest, (8) icecap. (After D. V. Bogdanov, 1963. "Map of the Natural Zones of the Ocean," *Deep-Sea Research*, 10, 520–23.)

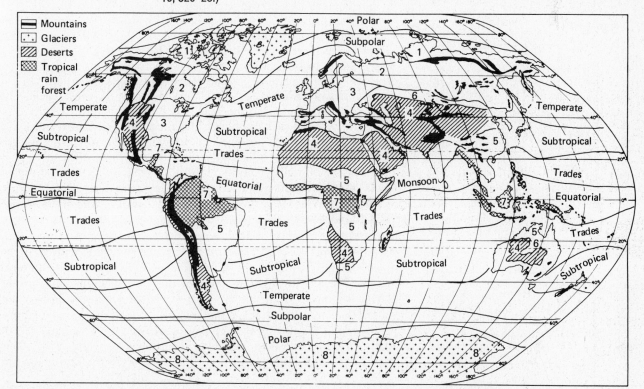

generally high. The pycnocline is well developed and prevents ready exchange with water beneath. Plants growing in surface waters deplete the supply of nutrient elements, and since there is little circulation of deep water to surface layers, plant productivity is limited in this region. Surface waters and floating materials from a vast area tend to converge at the center of ocean basins in subtropical latitudes. For instance, the Sargasso Sea lies in the Atlantic's subtropical current system and derives its name from the floating seaweed that accumulates there.

The *tropical* (trade wind) regions are characterized by persistent winds blowing from the northeast in the Northern Hemisphere and from the southeast in the Southern Hemisphere. These winds cause well-developed equatorial currents. The trade winds also cause moderate seas, in contrast with the relative calm of subtropical latitudes. Waters in tropical oceans originate in subtropical regions and are therefore more saline than average seawater. Moving toward the equator, precipitation increases, causing decreases in surface water salinity.

In *equatorial* regions, surface waters are warm throughout the year. Daily temperature changes equal or exceed annual temperature variations. Warm, moist air generally rises near the equator, causing heavy precipitation and relatively low surface salinities. In the Atlantic and much of the Pacific winds tend to be weak. Sailors named parts of this region the *Doldrums* because their ships were often becalmed there. The Equatorial Countercurrent is located in this belt of weak winds between the strong and persistent trade winds of each hemisphere. In the Indian Ocean, the monsoons complicate the current patterns. Substantial vertical water movement occurs along the equator; this supplies nutrients to the surface layers and causes the area to be quite productive of phytoplankton and other organisms that feed on them.

The ocean moderates climatic conditions on land. Coastal areas typically have a *maritime climate,* where the annual temperature range is generally much narrower than in regions far inland, as shown in Fig. 6-29. Far from the ocean, *continental climates* prevail, with their characteristically wide temperature ranges.

Oceans store large amounts of heat because of water's high heat capacity, permitting heat storage without an accompanying large temperature increase. Furthermore, the ocean (or any large body of water) stores heat through several meters of water due to mixing, and thus retards loss of heat to space by back-radiation at night. In contrast, rocks on land have a low heat capacity, so they become hot during the day. But since this heat is not readily transferred to deeper rocks beneath the surface, most of it is lost at night by back-radiation to space. This may be observed in any desert area, where intense daytime heat is followed by a chilly night.

During summer, continents heat up and warm the atmosphere above them. The relatively less dense continental air mass rises and is replaced by moist air from the ocean. As the oceanic air mass rises, it cools and releases moisture, often as heavy rains. This is the *monsoon* circulation, especially well developed over southeast Asia and India as a result of seasonal atmospheric circulation shifts caused by the warming and cooling of the Asian land mass. A similar air-circulation pattern prevails throughout much of the United States midcontinent during summer. Moist air from the Gulf of Mexico supplies large amounts of moisture during the summer, usually as thundershowers.

Mountain ranges block the flow of air from the ocean and thereby affect climatic conditions over large continental regions. North–south mountain ranges of the western United States block the flow of air from

Figure 6-29

Temperature variation at three locations: San Francisco has a typical maritime climate, with cool summers and relatively warm winters; Denver has a typical continental climate—cold winters, warm summers; Washington, D.C. has a generally continental climate, though tempered by its proximity to the ocean.

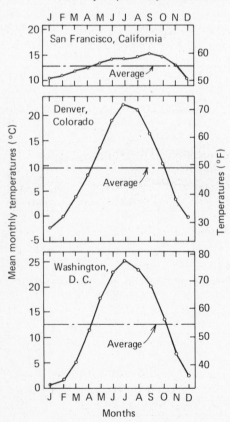

the Pacific, creating a rainshadow over interior valleys so that they receive little precipitation. Air cools as it rises to flow over the mountains. Rainfall is sufficient to support a temperate rainforest on the western slopes of the Coast Range of Washington, whereas deserts or near-desert conditions occur east (downwind) of the mountains. The coastal portion of the western United States thus has a maritime climate, characterized by moderate temperatures, winter and summer, and abundant rainfall. Continental climates with wide temperature ranges are common in the interiors of the continent, away from the ocean's moderating influence (see Fig. 6-29).

On the Eurasian continent, by contrast, the Alps, Himalayas, and associated mountains run generally east–west. These ranges have little effect on the flow of maritime air over Europe and Asia, and the maritime climate extends many hundreds of kilometers inland. These east–west ranges do, however, block north–south air movements and very likely cause the climate of the Mediterranean region to be warmer and drier than it might be in their absence.

SUMMARY OUTLINE

Ocean and atmosphere—closely coupled systems
Approaches used: Budgets and distribution of properties
Light in the ocean
Solar radiation—source of energy for heating ocean and atmosphere and for circulation
Infrared (heat) absorbed at surface
Turbidity—increases absorption and scattering of solar energy; pure water is most transparent to blue-green light
Blue color caused by scattering in particle-free water
Atmospheric circulation and heat budget
Inclination of axis causes seasons
Rotation spreads insolation over surface
Heat received between 40°N and 40°S; lost in high altitudes, especially in winter hemisphere
Heat radiates to space as infrared
Latent heat of water dominates heat transport in atmosphere
Atmospheric circulation typically forms cells and wavelike frontal patterns
Oceanic water and heat budget
Evaporation highest in subtropical areas
Precipitation heaviest in equatorial regions and in high latitudes
Salinity controlled by balance between evaporation and precipitation
High salinity common in central ocean areas
Low salinity in coastal regions caused by river

runoff and by high precipitation and little evaporation in high latitudes
Poleward heat transport reflected in low salinities at high latitudes
Density of seawater
Controlled primarily by temperature and salinity; pressure effect noticeable only at great depths in ocean
Usually stable density structure in the ocean
Vertical stratification inhibits vertical movements of water
Surface, pycnocline, and deep zones
Thermocline—sharp temperature gradient
Halocline—sharp salinity gradient
Pynocline—sharp density gradient
Water masses
Temperature and salinity used to distinguish subsurface water masses
Sea-ice formation
Ice spicules form slush, later plastic layer with enclosed brine cells
Salt crystallizes from brines; brine cells migrate
Pancake ice forms; later coalesces in floes as season advances
Icebergs derived from glaciers on land
Climatic regions
Generally east–west trending
Surface waters respond to local climate, especially winds

SELECTED REFERENCES

ANTHES, RICHARD A., HANS A. PANOFSKY, JOHN J. CAHIR, AND ALBERT RANGO. 1975. *The Atmosphere*. Charles E. Merrill, Columbus, Ohio. 339 pp. Elementary text on atmospheric processes.

PICKARD, G. L. 1975. *Descriptive Physical Oceanography*, 2nd ed. Pergamon Press, New York. 214 pp. Elementary treatment of physical oceanography.

Eighteenth-century navigational instruments: Hanging compass (upper); Log line and glass (lower). The position of a ship can be determined by *celestial navigation*—that is, by observing the sun, moon, or stars. When these are not visible, position can be determined by *dead reckoning:* estimation of a ship's progress by its presumed direction of travel and speed. The *magnetic compass,* indicating north, can be used to determine the ship's heading. Rate of travel is estimated by measuring the apparent speed and making allowances for known currents and the effects of wind and weather.

Measurement of speed was traditionally made using the *log line* and *glass.* The log, or log chip, is a thin wooden triangle weighted with lead on the curved edge so it will float with the point up, providing a point fixed in the water. This is attached to a line, often marked by knots at 15.7-meter intervals for measuring a nautical mile of approximately 1.85 kilometers (6076 feet). A glass designed like a small hourglass would then be used to indicate a 30-second time interval.

To measure the ship's speed, an officer of the watch would heave the log over the stern of the ship on the lee side. The line would run through his hand, and when a marker placed about 20 meters from the log chip appeared, he called to the glass-holder, "Turn." The glass-holder inverted the glass and called, "Done." When the sand had run out, the glass-holder called, "Stop." The officer then observed the point on the line which indicated, by the number of knots which had passed his hand, the velocity of the ship in "knots"—i.e., nautical miles per hour.

Modern taffrail logs, consisting of a spinner towed in the water and a recording instrument attached to the taffrail, continuously display the ship's speed and the distance it has traveled.

7

ocean circulation

C*urrents*—large-scale water movements—occur everywhere in the ocean. The forces causing major ocean currents come primarily from winds and from unequal heating and cooling of ocean waters. Both are the result of unequal heating of the earth's surface. Ocean currents contribute to the heat transport from tropics to poles, thereby partially equalizing earth-surface temperatures.

Modern techniques for current observation involve sophisticated buoys with meters that record the direction and speed of water movements and then relay the data to a ship that services the buoy. Other buoy systems transmit data back to shore stations. Some current-measuring systems use satellites as communication links.

The bulk of current measurements upon which our current maps are based were made in past centuries by navigators who observed how a ship's course was deflected by movements of water through which it had passed. A navigator, having predicted that his ship would be in one location, after steaming for a time found that his final position differed from the prediction. Assuming no errors in navigation and in the absence of winds, the displacement (called the *set*) is caused by currents. Measuring and recording such deflections of ships' courses over many years permitted Matthew Fontaine Maury to compile the observations of navigators in the record files of the U.S. Navy and to draw early charts of ocean currents. Today, we still rely on similar information sources.

Many observations when averaged over many years provide a picture of the average currents. When the effects of variable wind and short-term tidal currents cancel out, only the average current can be detected. Minor variations and even some seasonal variations are difficult or impossible to detect with such data. Detailed knowledge of currents and their variability is obtained by employing more modern methods over large ocean areas.

The movement of floating objects also provides information about surface currents. Wreckage of ships or glass floats used in Japanese fishing nets may move across the Pacific, providing evidence of a current flowing eastward. The wreckage of the ship *Jeanette,* crushed in Arctic ice in 1881 north of the Siberian coast, was discovered near southwestern Greenland three years later. This provided compelling evidence of a current from Siberia toward Greenland and gave an indication of its average speed.

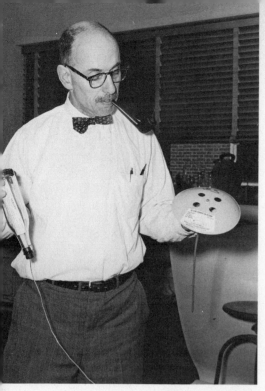

Drift bottles or floating plastic envelopes are often prepared with messages and numbered cards, as shown in Fig. 7-1, to provide similar data, but on a more regular basis. These bottles, released by ships, planes, or offshore installations, carried by surface currents, eventually float ashore to be picked up on the beach or by boats. The finder returns the numbered card and furnishes the data and location of its finding to the organization which released it. This information combined with the release data and location also provides information about surface currents. The data do not, however, indicate the exact path traveled by the bottle, but only its origin, end point, and the time elapsed between.

Direct current observations are made by current meters such as in Fig. 7-2, which work like wind vanes to indicate current direction. Propellers or rotors on the meter measure current speed. There are many ingenious methods for recording these data over long periods so that both short-term fluctuations and long-term average currents can be calculated.

Using meters, current speeds and directions can be determined at the surface and at all depths in the ocean. They can be suspended from anchored ships, but the ship's own motion in the water makes it difficult to interpret the results. Mooring of meters on buoys or fixed platforms is more satisfactory, although expensive in deep-ocean areas.

Figure 7-1 *(above)*

Various types of drifting devices can be used to provide information about movements of waters. These devices are specially weighted so that they float near the bottom and thus provide information about near-bottom currents. Similar devices without weights are used to provide information about surface currents. (Photograph courtesy Woods Hole Oceanographic Institution.)

Figure 7-2 *(right)*

Ekman current meter used to record current speed and direction. The meter is suspended below the surface on a wire hung from the ship. (Photograph courtesy Woods Hole Oceanographic Institution.)

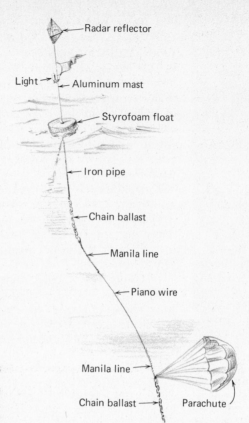

Radar reflector

Light — Aluminum mast

Styrofoam float

Iron pipe

Chain ballast

Manila line

Piano wire

Manila line

Chain ballast — Parachute

Figure 7-3

Parachute drogue rigged to permit tracking movement of subsurface water masses. (Redrawn after Von Arx, 1962.)

Two other ingenious methods are used to study currents be... the surface. Near-surface currents can be studied with *parachute drogues*—parachutes in the water rigged to be moved by currents, as illustrated in Fig. 7-3. The drogue is attached to a surface float that can be tracked by ship. Continual tracking of the float provides a direct indication of current speed and direction at the depth at which the parachute was placed.

The *Swallow float* (invented by the English oceanographer John Swallow) is a cylinder whose density is carefully adjusted so that it floats at a preselected depth, where it moves with the water. In the float is a powerful sound source called a "pinger." A ship at the surface receives the "pings" and by recording their location traces the movements of the float as it moves in the deep current. Just as in the case of the current meter, net currents can be calculated from long-term average movements of these floats.

All of these deep-ocean current-measurement techniques are expensive in terms of both equipment and ship time, and therefore we have limited data on subsurface currents. Lacking a source of subsurface-current information similar to that compiled for hundreds of years by surface ships, most of the data on deep-ocean currents are deduced from changes in ocean-water properties, such as the temperature of deep waters or variations in dissolved-oxygen concentrations.

In a later section we shall see that it is also possible to determine current speed and direction indirectly. This so-called *geostrophic approach* uses precise temperature and salinity determinations to map density variations that can cause currents.

The general surface circulation of the ocean is shown in Fig. 7-4. Circulation patterns are more or less similar in all three major ocean basins, despite their geographic differences. Much of the ocean surface in equatorial regions is dominated by the westerly flow of water in both the North and South Equatorial Currents, set up primarily by the trade winds. These are separate equatorial currents in the Northern and Southern Hemispheres, separated by a narrow current flowing eastward, the Equatorial Countercurrent, in the region of the light and variable Doldrums. Associated with each of these equatorial currents is a current *gyre*—a nearly closed current system. In each case, the current gyre is elongated east–west and lies primarily in the subtropical regions, centered around the 30°N and 30°S latitudes.

In addition to an equatorial current, each of these current gyres includes another east–west current flowing in a direction opposite to the equatorial currents. This is the West Wind Drift, the largest and most important current in the Southern Hemisphere. The North Atlantic Current is the continuation of the Gulf Stream System in the Atlantic and occupies a position similar to the West Wind Drift in the other ocean basins. The Indian Ocean has an equatorial current in the Southern Hemisphere. Near Asia seasonally variable winds caused by the heating and cooling of the land cause a seasonally variable current system—the monsoon currents, indicated in Fig. 7-5. In winter, a North Equatorial Current is present.

In the subpolar and polar regions of all the ocean basins, there are smaller current gyres. These gyres of the high latitudes circulate "opposite" to the subtropical gyres. Because of the position of the continents, subpolar gyres are well developed in the Northern Hemisphere, where their movement is counterclockwise (see Fig. 7-4). (Note the clockwise direction of the North Atlantic Current–Gulf Stream System–Northern Equatorial Current systems which form the North Atlantic subtropical gyre.)

Subpolar gyres can be seen in the Southern Hemisphere as well, but the basic circulation of that region does not favor their development. The Antarctic continent occupies a central position in the Southern Hemisphere circulation, so that the West Wind Drift essentially flows around it rather than being broken up and deflected toward the South Pole by continental land barriers. A few small, clockwise gyres can, however, be seen in the vicinity of Antarctica. (Note the counterclockwise direction of the West Wind Drift–Peru Current–Southern Equatorial Current system.)

Surface currents are changeable and far more complicated than can be shown in Fig. 7-4. Irregularities of the continental coastline cause local eddies in coastal-ocean currents, and seasonal wind shifts cause major changes in currents, especially common in the coastal ocean. The northward-setting Davidson Current of the Washington–Oregon coast, for instance, is well developed in winter, when strong winds from the southwest hold fresh water along the coast and drive the surface waters northward. During summer, when north winds tend to drive water offshore, the Davidson Current disappears until the winds shift back to the southwest. Along the Atlantic coast of the U.S., the extension of the Labrador Current brings cold northern waters as far south as Virginia in winter. During summer, cold waters from the Labrador Current extend no farther south than Cape Cod, Massachusetts.

The most striking of seasonal current changes is the monsoon circulation of the Northern Indian Ocean. This seasonally variable

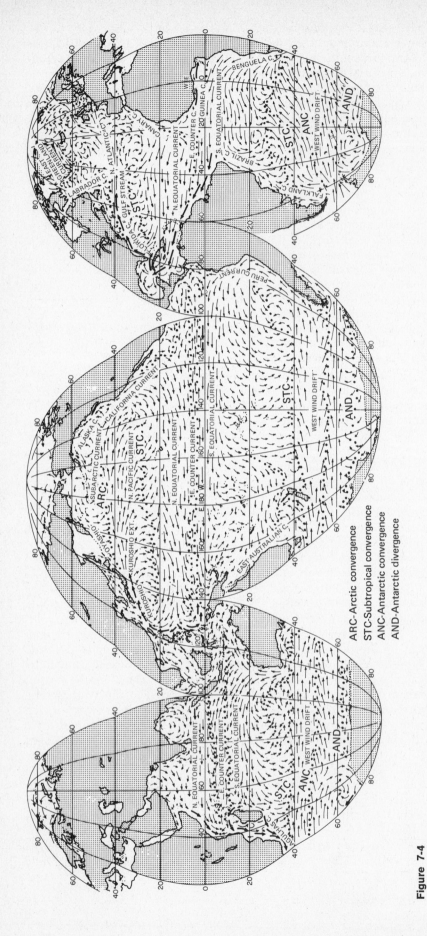

Figure 7-4

Ocean-surface currents in February–March. (After Sverdrup et al., 1942.) The thick arrows indicate especially strong currents.

ARC- Arctic convergence
STC- Subtropical convergence
ANC- Antarctic convergence
AND- Antarctic divergence

current system is intimately associated with the monsoon winds. During summer (May to September), Asia is greatly warmed relative to the adjacent ocean. As the warmed continental air rises, it draws air from the Indian Ocean toward the land, as shown in Fig. 7-5(a). When this occurs, the Southwest Monsoon Current replaces the Northern Equatorial Current, as shown in Fig. 7-5(b), in the Indian Ocean. In winter (November to March), the reverse takes place: air over Asia is much colder than air over the ocean, so that the flow is from the land toward the ocean, as shown in Fig. 7-5(c); the Northern Equatorial Current reappears, as shown in Fig. 7-5(d), and the Indian Ocean circulation resembles that of the other ocean basins.

Figure 7-5

Surface winds (a) and monsoon currents (b) in summer during the Southwest Monsoon.

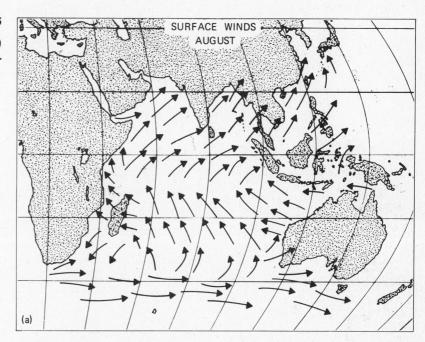

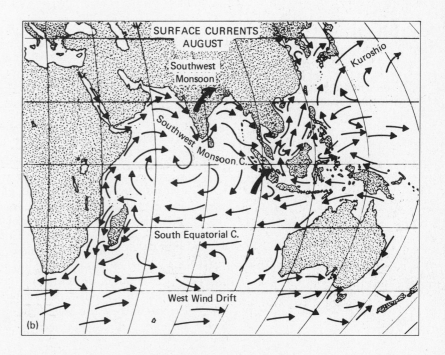

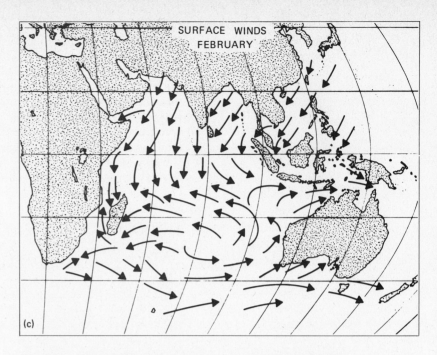

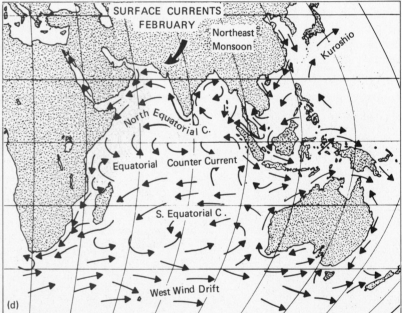

Figure 7-5 (*continued*)

Surface winds (c) and monsoon currents (d) in winter during the Northeast Monsoon.

BOUNDARY CURRENTS Most of the ocean surface is dominated by east–west currents, such as the equatorial currents of the West Wind Drift. These currents involve east–west movements of large volumes of water generally remaining in the same climatic zone. The waters have ample opportunity to adjust to the temperature of the climatic regime in which the current is located. These currents are deflected when they encounter the continents, and the water flows are split into *boundary currents* that flow more or less north–south.

Boundary currents are important to us as land-dwellers. They include the strongest currents in the ocean (the *western boundary currents*); they also play a major role in transporting heat from tropics to polar regions. Furthermore, these currents form boundaries to the

TABLE 7-1

Comparison of Boundary Current Systems in the Northern Hemisphere

TYPE OF CURRENT (example)	GENERAL FEATURES	SPEED	TRANSPORT (millions of cubic meters per second)	SPECIAL FEATURES

Eastern boundary currents
 	
California Current
Canaries Current
	Broad, ≈ 1000 km
	Shallow, ≤ 500 m
		Slow,
		tens
		of kilometers
		per day
			Small,
			typically
			10–15
				Diffuse boundaries
				  separating from
				  coastal currents
				Coastal upwelling
				  common
				Waters derived from mid-
				  latitudes

Western boundary currents
 	
Gulf Stream
Kuroshio
	Narrow, ≤ 100 km
	Deep—substantial
	  transport to
	  depths of 2 km
		Swift,
		hundreds of
		kilometers
		per day
			Large,
			usually
			50 or
			greater
				Sharp boundary
				  with coastal
				  circulation system
				Little or no coastal
				  upwelling; waters
				  tend to be depleted
				  in nutrients, unproductive
				Waters derived from
				  Trade Wind belts

circulation of coastal waters. Where boundary currents are strong, the break between coastal- and open-ocean circulation is sharp, which is the case on the western side of ocean basins, especially in the Northern Hemisphere. Where boundary currents are relatively weak—as on the eastern side of ocean basins—the separation between coastal- and open-ocean circulation tends to be more diffuse.

The characteristics of boundary currents on different sides of the ocean differ substantially, as Table 7-1 indicates. Due to the rotation of the earth on its axis, there is a tendency for current gyres to be displaced toward the west. This results in deep, narrow, fast-moving currents on the western boundaries of ocean basins (Gulf Stream, Kuroshio) in contrast to relatively shallow, broad, and slow currents along the eastern boundaries (Canary Current, California Current). This tendency is well developed and clearly discernible in the Northern Hemisphere but not so obvious in the Southern Hemisphere.

Surface water displaced to the west tends to pile up there, deepening the pycnocline. Phytoplankton grow in this accumulation of surface water and deplete the nutrient supply, which is not replaced by nutrient-rich waters from below the pycnocline. This tends to limit populations of marine organisms on the western side of ocean basins, except where vertical mixing is especially strong. In regions dominated by eastern boundary currents, prevailing winds blow surface waters away from ocean-basin margins. Subsurface waters are drawn upward to replace them, in a process known as *upwelling* which we discuss in Chapter 10. For this reason, productivity of marine plants is highest along eastern basin margins.

The Gulf Stream System is a prime example of a western boundary current. Actually a complex of filamentary flows, eddies, and meanders (see Fig. 7-6), it includes several currents: the Florida Current (see Fig. 7-4), extending from the tip of Florida to Cape Hatteras, North Carolina; the Gulf Stream, extending from Cape Hatteras to the tip of the Grand Banks, Newfoundland; and its eastern extension

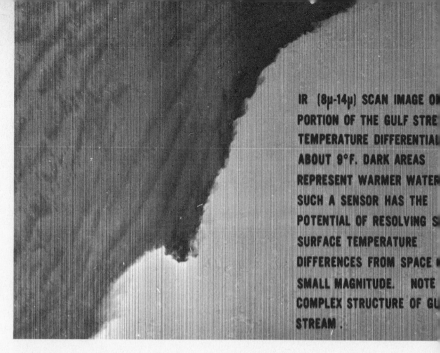

Figure 7-6

Photograph using infrared temperature sensor shows that the edge of the Gulf Stream is marked by a sharp temperature change between colder waters from the coastal ocean (the lighter shade) and the warmer waters of the Gulf Stream itself. Note the complex structure of the small water parcels within the Gulf Stream (left side). An electronic sensor carried by low-flying aircraft detects infrared radiation coming from the surface layer, and an image is produced by electronic equipment. (Photograph courtesy NASA and U.S. Naval Oceanographic Office.)

which forms the North Atlantic Current with its several branches and many eddies.

Water moving in the Gulf Stream System comes from the equatorial currents, especially the Northern Equatorial Current. In the Atlantic substantial amounts of surface waters cross the equator. Part of the Southern Equatorial Current is deflected into the Northern Hemisphere by the eastern projection of South America. These currents feed the Gulf Stream System; details of this process remain to be worked out.

The Gulf Stream System forms a partial barrier between adjacent surface-water masses. Over the continental shelf and slope, the coastal-water masses typically exhibit seasonably variable salinity and temperature, while in contrast, waters in the midst of the Gulf Stream are fairly constant, with temperatures typically about 20°C or higher and salinities of around 36 parts per thousand. There is some movement of surface waters toward the center of the North Atlantic subtropical gyre. This area, known as the *Sargasso Sea,* has indistinct boundaries formed by current systems. Average surface temperatures are 20°C or more and salinities are around 36.6 parts per thousand. The relatively high salinity is due to evaporation in excess of precipitation.

In the core of the Florida Current speeds of 100 to 300 centimeters per second have been measured (corresponding to 2 to 6 knots). Current speeds are highest at the surface, in a band about 50 to 75 kilometers wide. Below the surface, at depths of 1500 to 2000 meters, current speeds are measured at 1 to 10 centimeters per second. This is truly a "deep-draft" current, and therefore does not encroach upon the continental shelf at this point. It flows along the edge of the North American continent until reaching Cape Hatteras, North Carolina where it moves seaward.

After passing Cape Hatteras, the Gulf Stream System tends to form a series of large curves or *meanders,* which form, detach, and reform in a complicated manner. The cause of these meanders is still not known, but may be related to the topography of the ocean bottom over which the current moves; if so, this would be one of the few cases in which ocean-bottom topography appears to influence ocean-surface currents. (In general, surface currents involve only water above the pycnocline—perhaps 100 meters thick—and are therefore unaffected

182

by ocean-bottom topography.) After passing the Grand Banks, off Newfoundland, the flow forms the rather diffuse North Atlantic Current, a relatively shallow surface current.

Separating the Gulf Stream from adjacent, slower-moving water masses are a series of *oceanic fronts*. These are sudden changes in water color, temperature, and salinity. They are often spaced a few kilometers apart, but sometimes nearly overlapping. Even more obvious to an observer aboard ship is the change in color from greenish-gray or bluish coastal waters to the intense cobalt blue of the Gulf Stream waters. Often the front is also marked by choppy waters. Such fronts mark *convergences*—areas toward which surface waters flow from different or opposing directions, tending to sink at the region where they meet.

Eastern boundary currents are much less spectacular. The waters move much more slowly and the boundaries between the boundary current and coastal currents are more diffuse. Eastern boundary currents can flow over the continental margin. Like all boundary currents, they participate in the global heat transport by moving cooler waters toward the equator as in the case of the California Current.

CURRENT MEANDERS AND RINGS

Our knowledge of ocean currents primarily relates to average conditions. As you recall, observations taken decades apart and separated by hundreds of kilometers are combined in order to study currents, especially in open ocean areas. Thus, little is known about short-term processes affecting open ocean currents. The broader views of the ocean surface available from satellites have shown events, previously unknown, as they happen. One example is the development of current meanders and subsequent formation of rings (sometimes called eddies) in the Gulf Stream and its extension, the North Atlantic Current.

A pronounced temperature change—the so-called North Wall—separates the cooler coastal waters from the warmer Gulf Stream waters. On the other side, the Gulf Stream forms the northern boundary of the Sargasso Sea, where the waters are slightly cooler than in the Gulf Stream. Such strong contrasts in surface water temperatures can be detected by infrared sensors on low-flying aircraft or satellites.

Figure 7-7 shows an infrared image of the western Atlantic on March 25, 1973. A storm (in the upper left) has just passed over the area. Cool surface waters are shown as the light gray areas bordering the land, shown in the lower left. The Gulf Stream waters, being much warmer, are the darkest areas in the center and right of photograph.

Note the irregular, or meandering, patern of the Gulf Stream (Fig. 7-8). Such meanders, while present most of the time, are especially well developed after passage of a storm. Gulf Stream meanders have been extensively mapped using several ships operating together or using aircraft with infrared sensors. The meanders move slowly northeastward with the Gulf Stream at speeds of 8 to 25 centimeters per second (7–22 kilometers per day).

If a meander becomes too large, it can form a ring which detaches itself from the Gulf Stream to move with the waters flowing southwestward on either side of the Gulf Stream. A ring is 100 to 300 kilometers (60–200 miles) across, and is bounded by a nearly circular system of swift currents (90 centimeters per second or 78 kilometers per day) that tends to keep the ring together. The ring moves with the waters around it, usually southwestward, at speeds of 5 to 10 kilometers per day. Rings and their associated ring currents extend

Figure 7-8 *(below)*

The axis of the Gulf Stream in early February, 1974, with meanders developed west of Cape Hatteras. Seven rings are shown; four warm (anticyclonic) rings in the slope water north of the Gulf Stream, and three cold (cyclonic) rings in the Sargasso Sea waters south of the Gulf Stream.

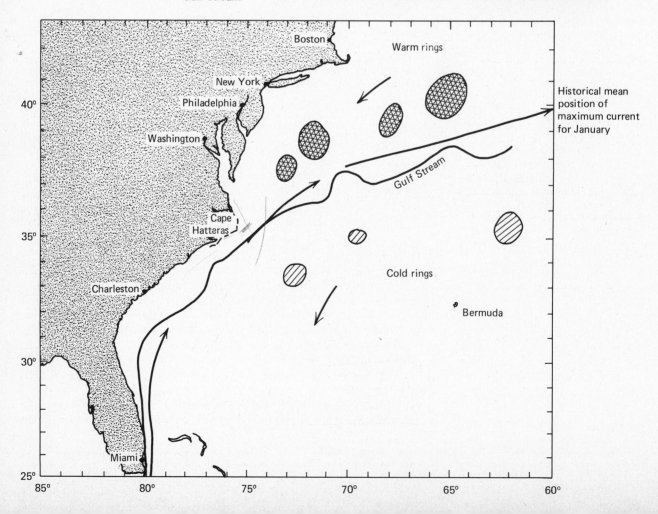

to depths of 2 kilometers so they normally do not go up on the shelf where the waters are only 200 meters deep.

Rings form on both sides of the Gulf Stream. Those forming on the north side (anticyclonic rings) consist of warm water surrounded by colder slope water (shown in Fig. 7-9). These are fairly easily detected by aircraft or satellites. Cold rings (also called cyclonic rings) form on the south side of the Gulf Stream and inject cooler water into the Sargasso Sea. As the rings move southwestward the surface waters warm up. As their temperatures reach those of the surrounding waters, they become more difficult to detect. Consequently we know less about cold rings than about warm ones.

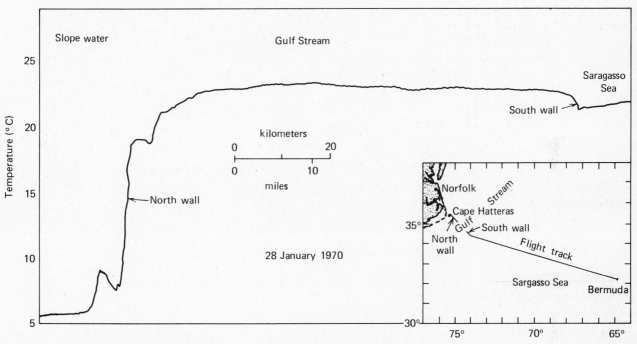

Figure 7-9

Sea surface temperatures recorded by an infrared radiometer on an aircraft crossing the Gulf Stream, between Cape Hatteras and Bermuda on January 28, 1970. Note the sharp temperature change between slope waters and the Gulf Stream (the so-called North Wall) and the much less pronounced South Wall, the boundary between the Gulf Stream and Sargasso Sea. (Data from Naval Oceanographic Office, 1970. *The Gulf Stream*, 5(1), p. 3.)

To see how rings behave in the ocean, we can cite the history of one which probably formed in late August, 1970, north of the Gulf Stream, south of Cape Cod (see Fig. 7-10). When first sighted in early September, it was apparently moving northward. But after encountering the continental slope, the ring was deflected. Subsequent ship and aircraft surveys showed it moving southwestward at speeds between 5 to 10 kilometers per day. By late December it had reached Cape Hatteras where it was deflected into the Gulf Stream by the continental slope. In early January, the ring was resorbed, apparently by a sharp meander of the Gulf Stream. But other rings have been tracked for periods up to three years.

While rings have only recently been discovered, they are significant oceanic phenomena. They transport heat and momentum along the coast. They may also transport tiny floating plants and animals in areas where they would not normally occur; for example, they take slope water organisms into the Sargasso Sea. Similar rings may be formed by other western boundary currents and have comparable biological effects.

These rings also show that the ocean is far more dynamic than our traditional view has shown. While it does not change as rapidly as the atmosphere, some changes occur fairly rapidly. These may have far reaching effects on climate, fisheries, and the removal of pollutants from coastal waters. For instance, Gulf Stream rings moving offshore from the Georgia–Carolina coast cause vertical mixing, injecting large amounts of nutrients into the shelf waters and lowering their temperatures, and replacing volumes of water ten times greater than the local annual river runoff.

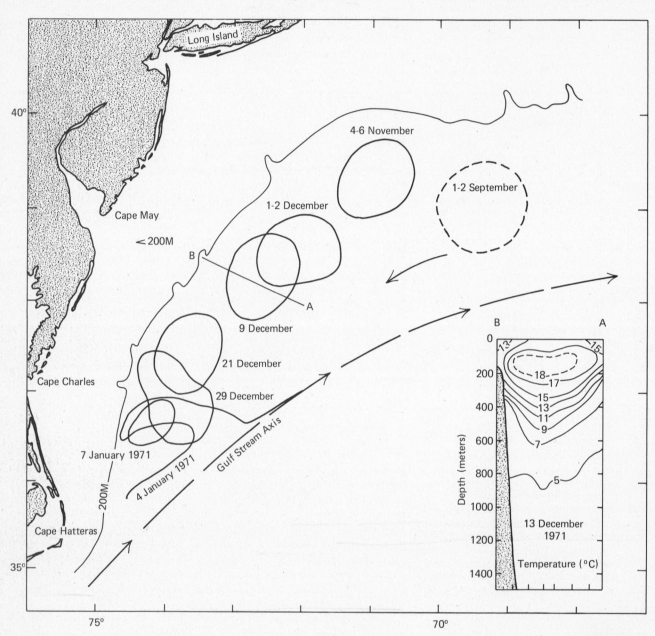

Figure 7-10

Movement of a large ring as shown by the movement of the 15°C isotherm at 200-meter depth that surrounds the mass of warm water. The ring apparently formed in late August 1970 and was resorbed into the Gulf Stream in early January 1971, after moving southwestward at speeds of 5 to 7 centimeters per second for nearly five months. Current speed around the ring reached 90 centimeters per second. (After Naval Oceanographic Office, 1971. *Gulf Stream*, 6(1), Washington, D.C.)

LANGMUIR CIRCULATION

Momentum imparted to surface-ocean waters by winds blowing across the ocean also causes vertical movements in near surface waters. Heat, dissolved gases, and other substances are quickly mixed through surface waters to depths of a few meters to a few tens of meters. Part of this mixing results from wind-generated turbulence—random motion of water parcels of many different sizes.

Winds also cause cellular circulation patterns to be set up; these cells are known as _Langmuir cells_ after their discoverer, Irving Langmuir. In Langmuir cells water moves with screwlike motions, in helical vortices alternately right- and left-handed (see Fig. 7-11). The long

Figure 7-11

(a) Oily surface films and debris form windrows along lines of convergence in the Langmuir circulation set up by a steady wind. (b) Foam lines caused by Langmuir circulation near a floating offshore petroleum drilling platform in the North Sea. (Photograph courtesy Mobil Oil Corporation.)

(a)

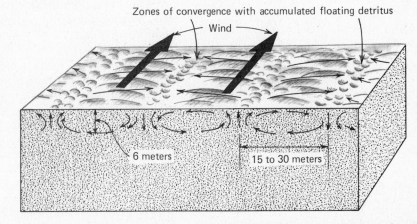

Zones of convergence with accumulated floating detritus

Wind

6 meters

15 to 30 meters

(b)

axes of these cells generally parallel the wind direction. The cells tend to be regularly spaced and are often arranged in staggered parallel rows. Because of the counterrotation of the cells, alternate convergences and divergences are formed at the surface. Floating debris collects in the convergences, such as the conspicuous floating sargassum weed whose distinctive pattern on the water surface can be seen by observers from high-flying aircraft. Between the lines of convergence are lines of divergences where water moves upward and along the surface of each cell toward the next convergence.

Langmuir cells form when wind speeds exceed a few kilometers per hour. The higher the wind speed, the more vigorous the circulation. When evaporation and cooling of surface waters tends to increase water density and favor convective movements, Langmuir cells can form at relatively low wind speeds. Increased stability resulting from surface warming or lowered salinity tends to inhibit cell formation, so that stronger winds are required before it can be set up.

Jet streams of water move downward under the convergences. In large lakes, where this type of circulation has been extensively studied, downwelling streams have been observed to extend 7 meters below the surface. Their speeds have been measured at about 4 centimeters per second. In the adjacent divergences, upwelling waters moved at speeds of about 1.5 centimeters per second.

The vertical dimension of these cells is dependent on wind speed and the vertical density structure of the surface layers. If the mixed layer is deep, there is relatively little hindrance to the vertical extent of the cells. If the mixed layer is shallow, the cells may not be able to penetrate the pycnocline. Such vertical water circulation may partially control the depth of the pycnocline. This is also an important mechanism for transporting heat, momentum, and various substances from the surface to subsurface layers.

FORCES CAUSING CURRENTS

Surface ocean circulation is primarily wind-driven. The prevailing wind systems supply much of the energy that drives surface water movements, as is made clear by comparing the generalized surface winds in February, illustrated in Fig. 7-12(a) with the ocean-surface currents in February and March shown in Fig. 7-4. We have already discussed how seasonal wind changes cause surface-current shifts near Asia in the monsoon areas and along other coastal areas. (In the next section, we discuss how the wind causes water movements and how these are altered by the earth's rotation.)

In addition to the *drift currents* caused by the wind, other currents—called *geostrophic currents*—result from the distribution of water density, which is in turn controlled by water temperature and salinity. For our purposes it is convenient to discuss these two driving forces separately. In the ocean their effects combine to produce the currents we have previously described. Generally currents below the depth to which wind effects penetrate are geostrophic currents.

Both drift and geostrophic currents move water horizontally. Other density-controlled currents are responsible for vertical water movements that supply deep- and bottom-water masses to all ocean basins. Because both heat and salinity are involved, this circulation is referred to as *thermohaline circulation*. This circulation controls the vertical distribution of temperature and salinity in the ocean.

The earth's rotation about its axis, which causes an apparent deflection in the trajectory of artillery shells and rockets, also acts as a modifying influence on winds and ocean currents. Except near the

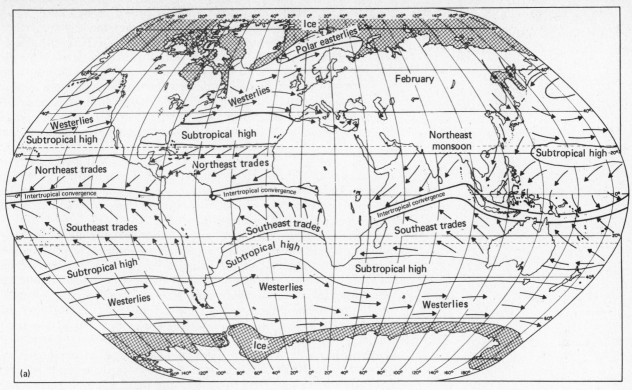

Figure 7-12(a, *above*)

Generalized surface winds over the ocean in February.

Figure 7-12(b, *below*)

Generalized surface winds over the ocean in August.

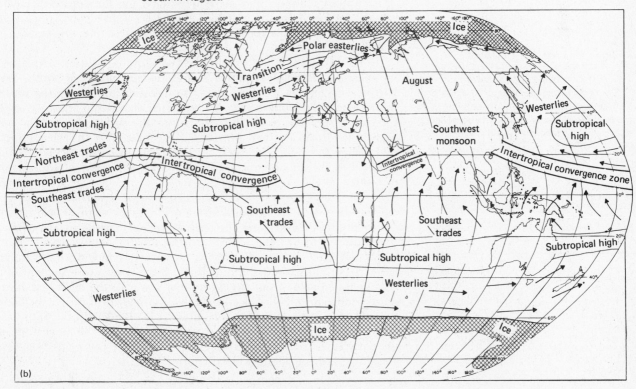

equator, winds and ocean currents follow curved paths instead of the straight ones that a simple theory would predict for their movement if the earth were not rotating. To account for this discrepancy, the *Coriolis effect* (sometimes called the Coriolis force) is included in predictions of long-range wind or water movements. This effect does not set winds or waters in motion; it deflects them after they are in motion.

To understand the Coriolis effect, consider what happens to a rocket fired toward the North Pole from the equator. As it sits poised on the launching pad, the rocket is already moving eastward at a speed of approximately 1670 kilometers per hour by virtue of the earth's rotation (see Fig. 7-13). After launching, the rocket still moves eastward at this speed in addition to the speed imparted to it during the launch. If the rocket were aimed *along* the equator, it would move in a straight line with no deflection—this is, it would not demonstrate the Coriolis effect—just as it would on a stationary earth.

The rocket moving northward from the equator moves over portions of the earth where the surface is moving eastward at progressively slower speeds than that of the equator. At the Pole, the surface has no lateral eastward movement (just as there is no lateral movement at the center of a phonograph record, whereas the outer edge is moving rapidly) relative to the center. At New Orleans, for example, (approximate latitude 30°N), the earth's surface moves eastward with a speed of about 1500 kilometers per hour. But the rocket when it passes overhead still has a net eastward speed of 1700 kilometers per hour, and is therefore moving eastward faster than the earth beneath it. Consequently the rocket seems to veer to the right when an earthbound observer looks along the flight path. An observer on the moon, however, would see that the rocket was traveling a straight line, while the earth turned beneath it. The veering to the right observed by an observer on earth would increase if the launching position were moved northward. This apparent increase is simply a result of more rapid changes in the earth's eastward movement as one moves nearer the pole (see Fig. 7-13).

A similar argument can be developed for the Southern Hemisphere (see Fig. 7-13). The deflection is still eastward but appears, to an observer on earth looking along the path of the rocket, like a deflection to the left.

To this point the discussion of the Coriolis force has explained its effect on particles or water masses moving north or south. But particles moving east or west are also deflected unless moving along the equator. To understand this we must consider a water covered rotating earth. The water is attracted to the surface by gravity but slightly deformed by the centrifugal force caused by the earth's rotation. Thus the water envelope is slightly flattened at the poles and widened at the equator in equilibrium with the speed of the earth's rotation.

The earth rotates toward the east so that a particle that starts moving eastward (say at mid-latitudes) is moving slightly faster than the earth and this increases the centrifugal force acting on it. This in turn causes it to be deflected toward the equator or toward the right.

Conversely a particle moving westward is moving opposite to the earth's rotation and the centrifugal force is therefore slightly less. It is deflected away from the equator or again to the right. In short, the Coriolis effect always acts to deflect particles to the right of their original path in the Northern Hemisphere (to the left in the Southern Hemisphere). It arises from the earth's rotation and the movement of the particle. A particle at rest does not experience this effect nor does a particle moving along the equator.

Figure 7-13

Note the change in speed of the earth's surface moving in an eastward direction going from the equator to either pole. A rocket moving from the equator to the North Pole would apparently be deflected to the right—the Coriolis effect.

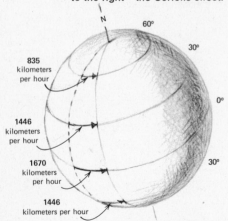

835 kilometers per hour

1446 kilometers per hour

1670 kilometers per hour

1446 kilometers per hour

N

60°

30°

0°

30°

The Coriolis effect (*f*) can be calculated as follows:

$$f = 2(\omega \sin \phi)\,\mu$$

where

ω = angular velocity of the earth, 7.3×10^{-5} sec $^{-1}$, corresponding to the complete rotation of the earth in 1 day

ϕ = geographic latitude; $\sin \phi = 0$ at the equator, $\sin \phi = 1$ at the poles

μ = horizontal velocity of the current

The Coriolis effect acts as a force operating in a direction perpendicular to the current. It causes the current deflections as previously mentioned. It could theoretically cause water parcels to follow spiral paths as they move in the general direction of the current as shown in Fig. 7-14 although these are not recognized in the ocean because other forces, including friction, affect the moving water.

Wind blowing across water drags the surface along so that a thin layer is set in motion. This layer in turn drags on the one beneath, setting it in motion. The process continues downward, involving successively deeper layers. Transfer of momentum between layers is inefficient and energy is lost. As a result, current speed decreases with depth below the surface. In an infinite ocean on a nonrotating earth, the water would always move in the same direction as the wind that set it in motion. A similar effect is seen in "storm surges"—large amounts of water piled up on a coastline by storm winds blowing the water ahead of them. Because the distances and times are relatively short, complications resulting from the earth's rotation are minimal.

Since the earth does rotate about its axis, however, movements of surface waters are deflected to the right of the wind in the Northern Hemisphere, as shown in Fig. 7-15. This tendency was observed by, among others, Fridtjof Nansen, who while studying the drift of polar ice when his ship *Fram* was frozen in polar ice during his 1893–1896 expedition, found that ice moved 20° to 40° to the right of the wind.

Figure 7-14

Some possible paths of particles moving freely over the earth's surface showing the influence of the Coriolis effect.

Figure 7-15

Schematic representation of the Ekman spiral formed by a wind-driven current in deep water. Note the change in direction and decrease in speed with increased depth below the surface.

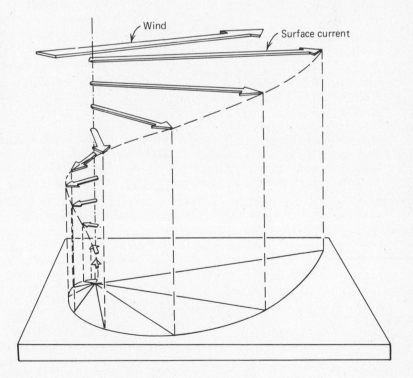

Wind

Surface current

Using Nansen's observations, the young physicist Walfrid Ekman showed mathematically that such effects can be explained using a simple uniform ocean with no boundaries as a model. Each layer, Ekman demonstrated, sets in motion the layer beneath, with the result that the latter moves somewhat more slowly than the former and that its motion is deflected to the right of the layer above. If this effect is represented by arrows (*vectors*) whose direction indicates current direction and whose length indicates speed, the change in current direction and speed with increasing depth forms a spiral, when viewed from above (see Fig. 7-15), now called the *Ekman spiral.* Figure 7-15 shows the Ekman spiral for the Northern Hemisphere. A similar spiral drawn for the Southern Hemisphere would exhibit the opposite sense of deflection, but current speeds would decrease at the same rate.

Wind effects penetrate to some considerable depth below the ocean surface, determined in part by the stability of the water column. The usual limit for wind effects (such as drift currents) is taken to be the depth at which the subsurface current is exactly opposite to the surface current. At that depth, the current speed is about 4 percent of the surface current. Presence of a pycnocline may limit the depth of the drift current. Under strong winds, wind-drift currents have been observed to extend to depths of about 100 meters.

Energy is transferred to this depth by *turbulence* in the surface layer. Turbulence is the disorderly state of motion. Transfer of energy by turbulence allows energy to penetrate at least 100 times deeper into the ocean than could be accounted for by movements of water molecules alone.

Speed of surface currents set up by winds is about 2 percent of the speed of the wind that caused them. For instance, a wind blowing at 10 meters per second would cause a surface current of about 20 centimeters per second.

In shallow waters, wind-generated currents are not deflected as much as predicted theoretically for an infinitely deep ocean. The ocean is not completely uniform. For instance, the pycnocline represents an important barrier inhibiting downward transfer of momentum and materials from the surface into subsurface layers. Furthermore, the wind does not always blow long enough—probably a few days—from a single direction to establish a fully developed Ekman spiral. Under these conditions, the deflection is less than the 45° predicted by the simple case shown in Fig. 7-15. Nansen's original ice observations, for instance, showed that the ideal Ekman spiral is not always encountered in the ocean.

So far we have considered movements of individual layers set in motion by the wind. The net motion of the entire mass of moving water—predicted, using the Ekman spiral, as moving at right angles to the direction of the wind that set it in motion—is called the *Ekman transport,* represented in Fig. 7-16.

In coastal regions, the Ekman transport of surface waters can cause upwelling. Where the wind blows somewhat parallel to a coast, it can cause the surface waters to move offshore. These surface waters are replaced by waters moving upwards, typically from depths of 100 to 200 meters, as seen in Fig. 7-17. Upwelled subsurface waters are characteristically colder and usually contain less dissolved oxygen than the surface waters and can thus be readily recognized. In addition these upwelled waters are usually rich in nutrients (phosphates and nitrates) necessary to support extensive growths of phytoplankton which feed other marine organisms. Consequently upwelling areas are highly productive of fish and other marine organisms.

Ekman transport can also cause surface waters to move toward a coastline. This leads to sinking or *downwelling,* the opposite of up-

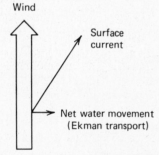

Figure 7-16

Schematic representation of relationship of wind, surface current, and net transport (also called Ekman transport) in a drift current in the Northern Hemisphere.

welling. The surface waters moving toward the coast tend to accumulate there, depressing the pycnocline, and tend to hold river discharge near the coast. Winter conditions along the Washington–Oregon coast causing the Davidson current is one example of downwelling.

Figure 7-17 (right)

Schematic representation of water properties in an area of wind-induced upwelling. Note that the upwelled water comes from a depth of about 200 meters.

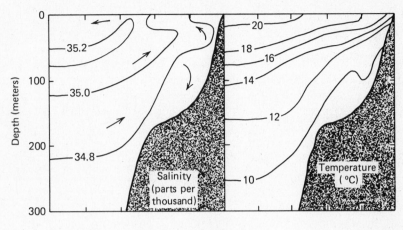

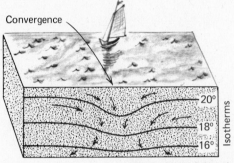

Figure 7-18 (above)

Schematic representation of a convergence in the open ocean. A slight hill of water forms and the surface layer thickens because of water accumulation.

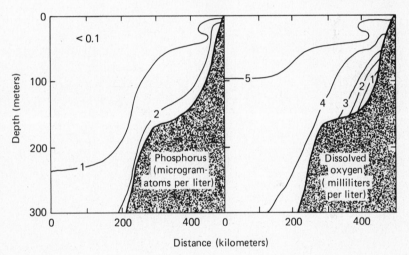

GEOSTROPHIC CURRENTS

Figure 7-19 (below)

Schematic representation of a divergence in the open ocean. Note that the surface layer is thinned.

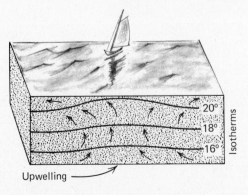

Slight differences in water density set up forces strong enough to cause water movements. Most common are the *geostrophic currents*, in which the flow of water in response to density-related forces is modified by the Coriolis effect.

If you examine the charts of prevailing winds (see Fig. 7-12) and remember that the net transport of surface waters is to the right of the wind in the Northern Hemisphere and to the left in the Southern, you can see that there is a distinct tendency for winds to move surface waters toward the subtropical regions. These regions, as we have already seen, are areas of light winds, and water tends to accumulate there. As Fig. 7-4 shows, these mid-latitude areas are zones of convergence for surface waters. Two things happen in an area of convergence, illustrated in Fig. 7-18: first, the water tends to pile up, forming a hill; and second, there is a depression of the local pycnocline through sinking.

In the reverse case—where surface water is blown away from an area—there occurs a *divergence*, illustrated in Fig. 7-19. A prominent divergence appears around Antarctica (see Fig. 7-4). In a divergence, subsurface waters move upward to replace the water moving away from the region.

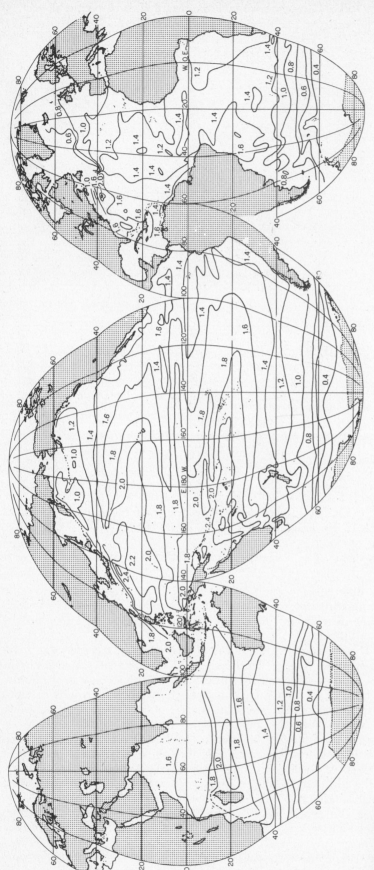

Figure 7-20

Approximate sea-surface elevations, in meters, above an arbitrary level surface. (After H. Stommel, 1964. "Summary Charts of the Mean Dynamic Topography and Current Field at the Surface of the Ocean and Related Functions of the Mean Wind–Stress Studies on Oceanography," *Studies on Oceanography,* ed. Kozo Yoshida, University of Washington Press, Seattle.)

As a result of the above water movements, the sea surface has a subtle topography. The net difference between the hills of water formed in the mid-latitude convergences on the western side of the ocean basins and the divergence surrounding Antarctica is about 2 meters, as Fig. 7-20 indicates. The distances involved are about half the earth's circumference, or at least 20,000 kilometers.

Despite this relatively low relief, water responds to these oceanic hills just as it would on land—by tending to run downhill. Let us consider a simple case of such a hill in the Northern Hemisphere and follow a water parcel along it as illustrated in Fig. 7-21, in order to see how the earth's rotation changes that path. Initially the water parcel moves downslope, just as it does on land. However, the Coriolis effect soon begins to change the path of the water parcel by deflecting it to the right. This continues until the water follows a path that allows it to flow downhill just enough to keep moving but where most of its motion is parallel to the side of the hill. If our hill were contoured to show line of equal sea-surface elevation, the path of the water parcel would nearly parallel these contour lines. On a frictionless ocean, water movement would exactly parallel the side of the hill. This balance—between the gravitational force that pulls the water downhill and the Coriolis effect that deflects it—gives rise to geostrophic currents.

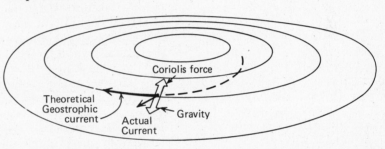

Figure 7-21

Schematic representation of water flow and balance of forces in a geostrophic current in the Northern Hemisphere.

If it were possible to survey the elevation of the ocean surface precisely, oceanographers could map the major features of ocean currents using the topography, making allowance for the slight deviations arising from the frictional effects in the ocean. The currents would appear to be strongest on the steepest hills (where the lines of sea-surface elevation are closest together) and weakest where the slope of the sea surface was most gentle.

Where a western boundary current passes near land, it is sometimes possible to measure such sea-surface slopes. Detailed surveys have shown that the Florida Current (part of the Gulf Stream System between Cuba and the Bahama Banks) has sea-surface slopes of about 19 centimeters over a distance of approximately 200 kilometers. At this location current speeds have been measured at 150 centimeters per second (nearly 3 knots) or more. This rapid current has a much steeper slope than currents in which waters move only a few kilometers per day.

The techniques have not yet been developed to the point where it is feasible to survey the entire ocean surface; therefore, the surface topography must be determined by indirect methods. Usually, a reference depth is chosen where there is some evidence to indicate that no currents are active. This is called the *depth of no horizontal motion* and is designated as the base level for subsequent calculations. Assuming that the weight of a water column of fixed dimensions overlying this chosen surface is equal throughout its area, it is possible to calculate the height of the sea surface and the *dynamic topography*—the topography to which a water parcel responds.

Let us see how this works. Assume that the weight of each water column of fixed base area is equal—that is, that there is an an equal mass of water and salt above each unit of our fixed surface. Now calculate the height of the column of water above any given location for which there are precise measurements of temperature and salinity; knowing the density, one can then calculate relative topography (see Fig. 7-20).

Remember that the mass of water in the column is fixed. Thus a column of less dense water occupies more volume and thus stands higher above the reference surface than a column of more dense water. By this indirect method, it is possible to calculate and prepare maps of dynamic topography for various parts of the ocean. Furthermore, it is possible to prepare such maps for parts of the ocean well below the surface. This is done by selection of the reference surface and the depth interval used in the calculation. The resulting map of dynamic topography is interpreted in the same way that a surveyed map of sea-surface topography would be. The steeper the dynamic topography, the stronger the currents. Using data from various depths, current charts can be prepared for various subsurface circulation systems.

THERMOHALINE CIRCULATION

So far we have discussed only horizontal water movements, but there are also vertical water movements that control temperature and salinity distribution in most of the ocean's depths. The *thermohaline circulation* is responsible for the movement of water from polar regions —specifically, from the North Atlantic and Antarctic—and its circulation throughout all the ocean basins.

Thermohaline circulation is driven by density differences. When the density of waters at the surface equals or exceeds water density at depth, the water column becomes unstable and the more dense water mass sinks, displacing less dense waters beneath. As we know from our previous discussion of density and stability, sinking continues until the water mass reaches its appropriate density level. In a stable water column, the waters beneath will be slightly denser and the waters above will be slightly less dense; at this level, the newly implaced mass of water tends to spread laterally, forming a thin layer. Such processes give rise to the intricately layered structures discovered by precise measurements of temperature and salinity in many ocean areas.

In certain semi-isolated ocean areas, lowered temperature and increased salinity cause the formation of dense water masses. Being denser than all the other waters, they sink to the bottom of our stratified ocean. Low water temperatures are the result of intensive cooling in polar regions, where the surface waters reach the freezing point of seawater. Such a combination of conditions is found in the Atlantic Ocean near Greenland and near Antarctica in the Weddell Sea, as indicated in Fig. 7-22.

The Atlantic is the saltiest of the major ocean waters (see Table 3-4). This salty water is carried into high latitudes by the Gulf Stream. Near Greenland it is intensively cooled to less than 0°C. When surface waters reach a critical density, they may sink suddenly, as a mass (as they do in laboratory experiments), and flow along the bottom of the North Atlantic Basin, especially along the western side.

Submarine ridges between Greenland and Scotland, the region forming the entrance to the Arctic Basin, prevent any bottom waters formed in the Arctic Sea from entering the main part of the Atlantic. The Bering Sill is even more effective in isolating the Arctic from the Pacific Ocean.

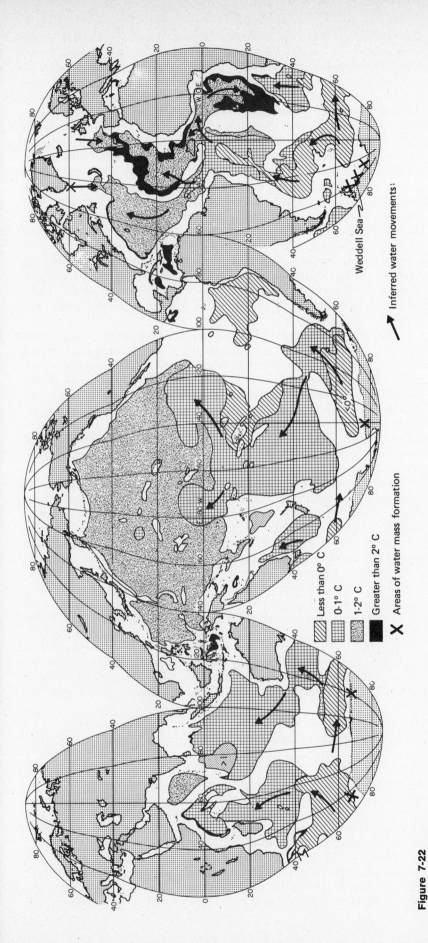

Legend (within figure):

Less than 0° C
0–1° C
1–2° C
Greater than 2° C

X Areas of water mass formation

Weddell Sea

↗ Inferred water movements:

Figure 7-22

Variation in temperature of bottom waters at depths greater than 4 kilometers and the inferred water movements. (After G. Wüst, 1935. "Die Stratosphäre. Deutsche Atlantische Exped. *Meteor*, 1925–27," *Wiss. Erg.* 6(1), 288 pp.)

Large quantities of Antarctic Bottom Water are formed in the Weddell Sea, a partially isolated embayment of the Antarctic continent. There, surface waters are cooled to temperatures of −1.9°C. At this low temperature and a salinity of 34.62 parts per thousand, this water sinks to the bottom of the adjacent deep-ocean basin where it forms the densest water mass in the ocean. In the process of sinking it mixes with other waters and is warmed to −0.9°C. After circulating around Antarctica, and mixing with other water masses, cold dense bottom waters move northward into the deeper parts of all three major ocean basins.

Using temperature as a tracer for this bottom water, we can follow Antarctic Bottom Waters (characteristics given in Table 6-2) as far north as the edges of the Grand Banks (45°N) in the North Atlantic. In the Pacific, mixtures of North Atlantic Deep Water reach the Aleutian Islands (50°N). Deep water movement through ocean basins is controlled by ocean-bottom topography. For example, the Romanche Trench provides a path for Antarctic waters to flow into the deep basins of the eastern portion of the South Atlantic; direct entry of this water mass into the eastern side of the South Atlantic is blocked, however, by the Walvis Ridge.

Near-bottom currents commonly move much slower than surface currents. Speeds of 1 to 2 centimeters per second are typical—except along the western basin margins, where speeds of 10 centimeters per second have been calculated. This is another manifestation of the strong boundary currents along the western side of ocean basins.

Bottom waters are formed in large quantity. To compensate for the production of bottom water there must be a gradual upward movement of waters toward the surface in all ocean basins. This upward movement opposes the downward movement of heat and dissolved gases from the surface.

Various techniques have been employed for estimating deep-water *residence times* (a measure of the time necessary to replace bottom waters by newly formed water masses). Radioactive carbon-14 has been used to determine the time elapsed since the dissolved-carbon compounds in the water were last at the surface. The resultant data suggest residence times ranging from 500 to 1000 years. If we take 1000 years to be a reasonable residence time for deep-ocean waters, this amounts to an upward movement of about 4 meters per year.

Other relatively dense water masses form in high latitudes of the North Atlantic and around Antarctica. There appears to be no dense-water formation in the North Pacific. Most of these high-latitude water masses have relatively low salinity, due to relatively high precipitation and little evaporation in the areas in which they form. Consequently, such surface waters do not achieve the high densities of the water masses formed from Atlantic surface waters.

ATLANTIC OCEAN CIRCULATION— A THREE-DIMENSIONAL VIEW

The distribution and movements of water masses have been worked out using the distribution of temperature and salinity as previously explained. Plotting the distribution of these values as in Fig. 7-23 graphically demonstrates where different water masses form in the ocean. Using the geostrophic approximation discussed above, we can estimate their probable movement rates. Because the Atlantic Ocean is especially well known, we shall use it to illustrate the complexities of circulation throughout an entire ocean basin.

Over most of the ocean, circulation of surface waters is almost completely unconnected with the movement of subsurface waters. Figure 6-19, in which salinity is plotted, shows that the relatively high-salinity surface water (greater than 35 parts per thousand in the

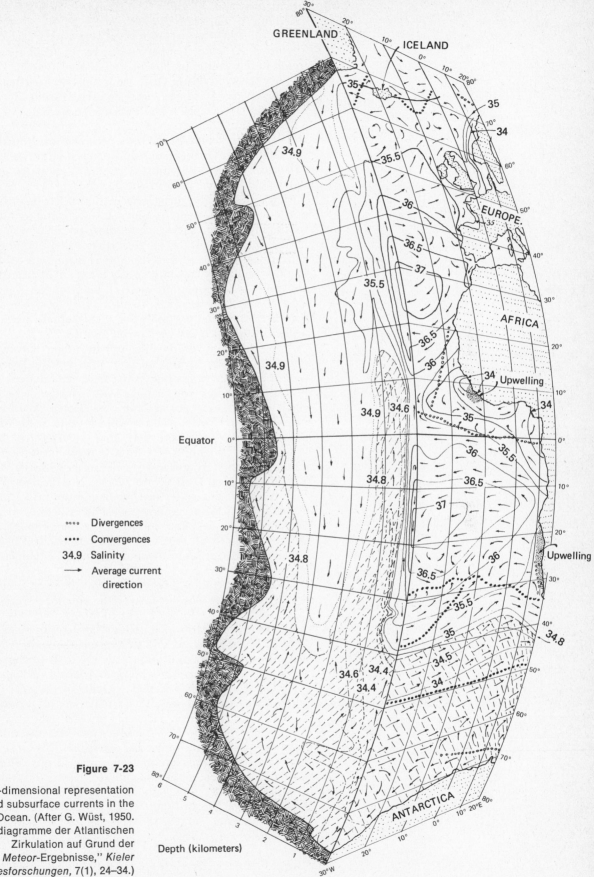

GREENLAND ICELAND EUROPE. AFRICA Upwelling Upwelling ANTARCTICA

Equator

Depth (kilometers)

○○○○ Divergences
•••• Convergences
34.9 Salinity
→ Average current
       direction

Figure 7-23

Schematic, three-dimensional representation
of surface and subsurface currents in the
Atlantic Ocean. (After G. Wüst, 1950.
"Blockdiagramme der Atlantischen
Zirkulation auf Grund der
Meteor-Ergebnisse," *Kieler
Meeresforschungen,* 7(1), 24–34.)

Atlantic) extends down to a few hundred meters at most. There is an exception in the North Atlantic (about 30°N), where the warm saline water from the Mediterranean Sea occurs at a depth of about 1 to 2 kilometers. If water temperature were plotted instead of salinity, we would have a similar picture, as seen in Fig. 6-18 (top).

In the West Wind Drift around Antarctica, there is little separation between surface and subsurface circulation. The small temperature or salinity change with depth does not hinder vertical circulation. This largest of all currents is the prime communication link for surface and subsurface waters of the three ocean basins. Note that the waters in the West Wind Drift are among the lowest-salinity waters found in the Atlantic. (Waters of salinity less than 34.8 parts per thousand are shown by diagonal patterns in Fig. 7-23).

Recall that the surface circulation is dominantly east–west, with north–south movements confined to the boundary currents along the continents. The subsurface circulation, on the other hand, is primarily north–south. As in the surface currents, the strongest near-bottom currents are associated with the western boundaries of the ocean basins. In these boundary currents, speeds of about 10 centimeters per second are common. Throughout most of the deep-water masses, the waters have slow net movements, about 1 centimeter per second, although they may move more rapidly for short periods because of large scale disturbances, much like deep ocean "storms."

We have previously discussed the water masses that flow along the ocean bottom because of their relatively high density. As can be seen in Fig. 7-22, cold bottom water from the Antarctic (called Antarctic Bottom Water) is a conspicuous feature of the deep circulation up to about 45°N. Deep waters from the North Atlantic (called North Atlantic Deep Water) are also conspicuous along the ocean bottom down to the point where they flow out over the Antarctic Bottom Water. North Atlantic Deep Water flows southward at depths between 2 and 4 kilometers, and mixes extensively with the waters circulating in the West Wind Drift before flowing into the deep Pacific basin.

A somewhat smaller intermediate water mass forms in the region of the Antarctic Convergence around 50°S. Because of the low salinities in that region, the water density is not high enough for it to sink to the bottom. Instead, the Antarctic Intermediate Water sinks about 1 kilometer and spreads northward, crosses the equator, and is recognizable to about 20°N.

Two points should be noted about the deep circulation. The first is that cold water flowing toward the equator (which eventually rises to the surface) is a form of heat transport. We have previously mentioned warm-water transport to high latitudes by surface currents. The return of cold water to low latitudes is the counterflow and it takes place in both surface and subsurface waters. It corresponds to the cold-air or cold-water return in a household heating system.

The second point is that currents of the deep ocean transport water across the equator. The circulation of surface waters of the Northern and Southern Hemispheres is almost completely separate. Except for some South Atlantic surface water transported into the Northern Hemisphere where the equatorial current is deflected by the South American continent and some areas in the western Pacific, there is almost no movement of surface waters across the equator. Subsurface currents transport large amounts of water across the equator and balance any surface transport of water, including that by winds.

SUMMARY OUTLINE

Currents—large-scale water movements

Current observations
    Mapping of average current set from displaced ship's course and floating objects
    Direct current observations—current meters, buoys

General surface circulation in the open ocean
    Gyre—nearly closed set of currents, usually elongated east–west
    East–west currents, such as equatorial currents, West Wind Drift
    Boundary currents—eastern and western
    Seasonally variable currents—coastal, monsoon—controlled primarily by wind

Boundary currents
    Western boundary currents—Gulf Stream System—strong, narrow, deep
    Eastern boundary currents—California Current, Alaska Current—broad, shallow, sluggish
    Oceanic fronts—sharp discontinuities in water properties mark convergences

Current meanders and eddies—Gulf Stream
    Meander—wave-like current pattern
    Eddies—nearly closed current systems
        100–300 kilometers across; up to 2 kilometers deep; move 5–10 kilometers per day; last as long as 3 years
        Form on both sides of Gulf Stream
        Move southwestward, counter to Gulf Stream
        Recombine with Gulf Stream

Langmuir circulation—organized set of helical vortices in surface waters
    Transport heat and momentum downward
    Mix surface waters, often to depths below wave-affected zone

Forces causing surface currents
    Prevailing wind systems—primary driving force
        Drift currents—caused directly by wind, primarily in surface layers
        Geostrophic currents—caused indirectly by wind and its effect on density distribution

Coriolis effect—deflecting force arising from earth's rotation; deflects currents to right in Northern Hemisphere, to the left in the Southern

Ekman spiral—systematic decrease in current speed and change in direction with increasing distance below surface
    Surface current flows 45° to right of wind in infinite, homogeneous Northern Hemisphere ocean, to the left in the Southern Hemisphere
    Surface current about 2 percent of speed of the wind causing it
    Net movement of surface layer is perpendicular to wind—known as Ekman transport, to the right in Northern Hemisphere
    Movement of surface waters offshore causing upwelling, vertical movements of subsurface waters
    Sinking or downwelling is caused by surface waters blown landward

Geostrophic currents
    Caused by small density differences
    Involves balance between Coriolis effect and gravitational attraction
    Dynamic topography computed from density distribution (controlled by temperature and salinity)
    Strongest currents where dynamic topography is steepest, weakest where slopes are gentle

Thermohaline circulation
    Vertical water circulation driven by density differences
    Bottom waters formed in North Atlantic and in Weddell Sea (Antarctica)
    Primarily a north–south circulation
    Gradually rises to surface through pycnocline over entire ocean

Atlantic circulation
    Intermediate waters form at high latitudes and flow toward equator
    Antarctic Bottom Waters flow generally northward

SELECTED REFERENCES

STOMMEL, HENRY. 1965. *The Gulf Stream: A Physical and Dynamical Description*, 2nd ed. University of California Press, Berkeley. 248 pp. Describes present understanding of Gulf Stream System; intermediate to technical level.

VON ARX, W. S. 1962. *An Introduction to Physical Oceanography*. Addison-Wesley, Reading, Mass. 422 pp. Physical and geophysical aspects of ocean systems are emphasized; little descriptive material; intermediate level.

Waves at sea, showing chaotic ocean surface. The boundary between air and water becomes progressively more obscured as waves become stronger. (Photograph courtesy of The Portland Cement Association.)

waves

Waves—disturbances of the water surface—can be seen at any ocean beach or along the shore of any large body of water. Seafarers have observed waves for thousands of years. But despite an abundance of observations, understanding of sea waves has come slowly. The ancients knew that waves were somehow generated by wind, but not until the nineteenth century were the first mathematical descriptions of waves developed.

Waves are important to us in many ways. They make and remake ocean beaches each year and in the process entertain millions of surfers and swimmers. Waves from a single storm can kill hundreds of people and can cause millions of dollars in damage to low-lying coastal regions. And waves must be considered in the design and construction of docks, breakwaters, and jetties along the coast, because they are all too often the cause of the failure or even destruction of these structures. Nor can waves be neglected when designing or operating the most modern ships.

IDEAL WAVES

A few minutes' observation of the ocean surface reveals a complex and continually changing pattern—a pattern that never exactly repeats itself no matter how long we watch it. Ocean waves come in many sizes and shapes, ranging from the tiny ripples formed by a light breeze, through enormous storm waves, tens of meters high, to the tides (which are also waves, as we shall see in Chapter 9).

Because of their complexity, ocean waves usually do not lend themselves to accurate description or complete explanation in simple terms. Nevertheless, we commonly work with simplified explanations and descriptions that help us understand wave phenomena; moreover, most advances in the study of waves have come through use of appropriate simplifications. Initially our discussions will involve such simplified or *ideal waves*.

To start, let us consider simple *progressive waves* and their parts as they pass a fixed point—say, a piling. We can make such waves in a laboratory wave tank, or by steadily bobbing the end of a pencil in a basin of water or a still pond surface. Then we see a series of waves, each wave consisting of a *crest*—the highest point of the wave—and a *trough*—the lowest part of the wave. The vertical distance between any crest and the succeeding trough is the *wave height H,* and the horizontal distance between successive crests or successive troughs is the *wavelength L;* these wave constituents are illustrated in Fig. 8-1. The time (usually measured in seconds) that it takes for successive crests or troughs to pass our fixed point is the *wave period T,* which is rather easily measured with a stopwatch. We can express the same information by counting the number of waves that pass our fixed point in a given length of time. This gives the *frequency* $(1/T)$, which is expressed in cycles per second. For individual progressive waves, the speed $(C,$ in meters per second) can be calculated from the simple relationship

$$C = \frac{L}{T}$$

where
L = the wavelength in meters
T = the wave period in seconds

Where the wave height is low, crests and troughs tend to be rounded and may be approximated mathematically by a *sine wave* (a mathematical expression for a smooth, regular oscillation). As wave height increases, sea waves normally have crests more sharply pointed than simple sine waves, and can be approximated by more complicated mathematical curves, such as the sharp-pointed, rounded-trough mathematical curve known as the *trochoid* (see Fig. 8-1).

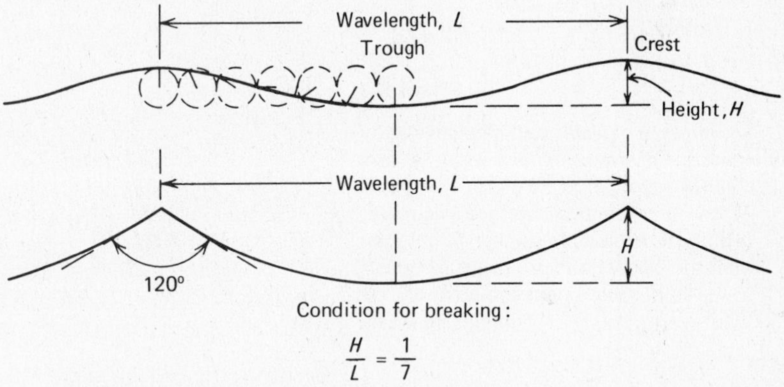

Figure 8-1

Two idealized cases for simple waves, indicating their various parts. Relatively small waves can be described as simple sine waves. Larger waves tend to be more sharp-pointed than a simple sine curve. There are limits on the size to which a wave can grow. Waves commonly break when the angle at the crest is less than 120° or the ratio of wave height to wave length is $H/L = 1/7$.

Wavelength, L
Trough
Crest
Height, H
Wavelength, L
H
120°
Condition for breaking:
$$\frac{H}{L} = \frac{1}{7}$$

Wave steepness (H/L)—the ratio of wave height to wavelength—is a measure of wave stability. It is also the factor that determines whether a small boat will glide smoothly over low waves or pitch through steep, choppy waves. When wave steepness exceeds 1/7, waves become unstable and begin to break (see Fig. 8-1) by raveling of the oversteepened crests, forming spilling breakers. The angle at the crest must be 120° or greater for the wave to remain stable; we examine this phenomenon in detail when we discuss breakers and surf.

So far, we have considered only the movement of the water surface—the crests and troughs moving together that make up a wave train. But what happens to the water itself as the wave form passes? How is the motion of the water related to the motion of the wave form? These questions have been studied using wave tanks with bits of material floating on the surface, or dyed bits of water or oil droplets below the surface.

When small waves move through deep water, individual bits of water move in circular orbits which are vertical and nearly closed. The water moves forward as the crest passes, then vertically, and finally backward as the trough passes; this series of movements is diagrammed in Fig. 8-2. This orbit is retraced as each subsequent wave passes, and after each wave has passed, the water parcel is found nearly in its original position. But there is some slight net movement of the water

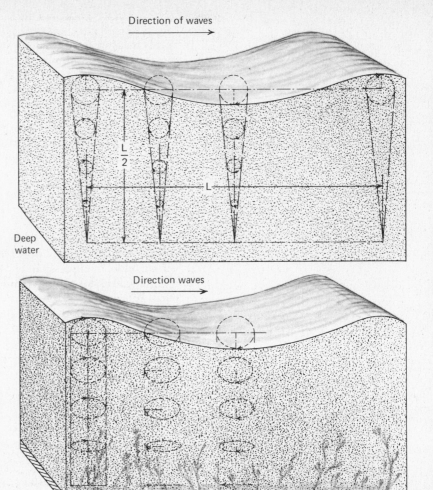

Direction of waves

$\frac{L}{2}$

L

Deep water

Direction waves

Shallow water

Ocean bottom

Figure 8-2

Movements of water particles caused by the passage of waves in deep water (a) and in shallow water (b). Note that the particles tend to move at the surface in circular orbits which become smaller with depth. Near the bottom, the orbits are flattened. Little movement occurs at depths greater than $L/2$.

because the water moves forward slightly faster as the crest passes than it moves backward under the trough. The result is a slight forward displacement of the water in the direction of wave motion and perpendicular to the wave crests, as shown in Fig. 8-3.

Note that if the water moved with the wave form, ships would never have been invented, for no ship ever built could withstand the forces exerted by water movements on such a scale. You may have experienced this yourself. When one floats in the ocean near the beach but beyond the breakers, there is only a gentle rocking motion as waves pass, because there is little net movement of the water. If, however, one tries to stand where waves are breaking, the immediate pounding by the breakers demonstrates large-scale rapid water movements, because in the breakers the water does move with the wave form. Even large ships can be damaged when hit by tons of water from a large, breaking wave.

In deep water, where the water depth is greater than $L/2$, water parcels move in nearly stationary circular orbits; such waves, unaffected by the bottom, are known as *deep-water waves*. The diameter of these orbits at the surface is approximately equal to the wave height. It decreases to one-half the wave height at a depth of $L/9$, and is nearly zero at a depth of $L/2$ (see Fig. 8-2). At depths greater than $L/2$, the water is moved little by wave passage; thus a submarine is essentially undisturbed by waves when it is submerged to depths greater than $L/2$.

In deep water, the *wave speed C,* in meters per second, may be calculated by the following equations:

$$C = \frac{gT}{2\pi} = 1.56T \qquad C = \sqrt{\frac{gL}{2\pi}} = 1.25\sqrt{L}$$

where
T = the wave period in seconds
L = the wavelength in meters
g = the acceleration of gravity, 9.8 meters per second per second

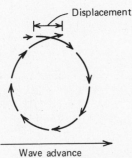

Displacement

Wave advance

Figure 8-3

Orbital motion and displacement of a water particle during the passage of a wave.

Of the two equations, the first is the more useful, because wave period is easily measured, whereas wave length is difficult to determine at sea or on the beach.

Where water depths are less than $L/20$, the motion of the water parcels is strongly affected by the presence of the bottom (see Fig. 8-2); these waves are called *shallow-water waves.* Orbits of water parcels at the surface may be only slightly deformed, usually forming an ellipse (a flattened circle with its long axis parallel to the bottom). Near the bottom, wave action may be felt as the water particles move back and forth; vertical water movements are prevented by the proximity of the bottom. Sometimes we observe movements of water parcels in shallow waters as waves pass over them—for example, where bottom-attached plants are moved with the water as seen in Fig. 8-3. The speed of shallow-water waves, $C,$ in meters per second, can be calculated by

$$C = \sqrt{gd} = 3.1\sqrt{d}$$

where
g = the acceleration of gravity, 9.8 meters per second per second
d = the water depth in meters

So far, we have discussed the behavior of a simple (or ideal) wave, identical with others in a *wave train.* Although we rarely see simple uniform waves at sea, the concept is useful in analyzing the more complicated waves that do occur. Even the most complicated waves may, by appropriate mathematical techniques, be separated into several groups of simple waves which, when added together, reproduce the complexities of the original ocean waves. An observed profile of a sea is shown in Fig. 8-4(a). This group of waves has been analyzed to determine which wave frequencies occur in the wave spectrum; the various components (each a simple sine wave) are shown in Fig. 8-4(b). Combining these waves will result in a wave essentially identical to the one observed.

Now that we have considered individual ocean waves and calculated their speeds we may place them in context: in the ocean, waves usually occur as wave trains or as a system of waves of many wavelengths, each wave moving at a speed corresponding to its own wavelength. Suppose that we produce a wave train in a laboratory wave tank and then carefully observe the result. If we follow a single wave in the resulting wave train, we find that it advances through the group because the individual waves move at a speed double that of the group. The wave approaches the front, gradually losing height as it goes. Finally, it disappears at the front of the wave train, to be followed by other, later-formed waves that have also moved forward from the rear. New waves are continually forming at the back of the wave train, while others are disappearing at the front.

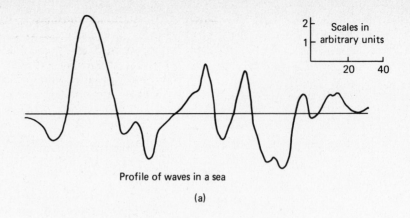

Profile of waves in a sea

(a)

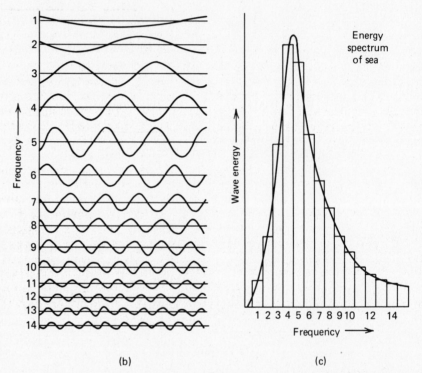

Figure 8-4

An observed profile of waves in a sea (a). Such a complicated wave pattern can be described as consisting of many different sets of sine waves (b), all of them superposed. The lower-frequency waves contain more energy than the higher-frequency ones (c). The energy in a wave is proportional to its height.

(b)   (c)

From this observation we can say that in deep-water waves, wave energy travels at one-half the speed of individual waves. In shallow water, individual waves are slowed down until the individual wave speed equals the *group speed*.

Formation and behavior of waves involve two types of forces: those that initially disturb the water and those that act to restore the equilibrium or still-water condition. The disturbing forces are familiar to us. All of us have made small waves by tossing a pebble into water. If the water surface was initially still, we observe a group of waves changing continuously as they move away from the disturbance. Sudden impulses such as explosions or earthquakes also cause some of the longest waves in the ocean. If the disturbance, such as an explosion, affects only a small area, the waves will move away from that point, much as the waves moved away from our pebble. But if the disturbance affects a large area, as a great earthquake can, the resulting *seismic sea waves* (also called *tsunamis*) may behave as if they had been generated along a line rather than at a point.

Winds are the most common disturbing force acting on the ocean surface, and consequently cause most of the ocean waves (note the large amount of energy associated with wind waves as illustrated in Fig. 8-5). Winds are highly variable, so that wind waves vary greatly in different ocean regions, and with the seasons. (We have more to say about wind waves in the next section.)

The attraction of the sun and moon on ocean water causes the longest waves of all—the *tides*. Since the attractive forces of the sun and the moon act continuously on the ocean water, the tides are not free to move independently as a seismic sea wave does; such waves, where the disturbing force is continuously applied, are known as *forced waves*—in contrast to the *free waves*, which move independently of the disturbance that caused them. An explosion-generated wave is an example of a free wave. Wind waves have characteristics of both free and forced waves.

Once a wave has formed, restoring forces act to damp out the wave and to restore the initial or equilibrium state. Depending on the size of the wave, various forces may be involved. For the smallest waves (wavelength less than 1.7 centimeters, period less than 0.1 second), the dominant restoring force is *surface tension,* in which the water surface tends to act like a drum head, smoothing out the waves. Such waves, called *capillary waves,* are rounded-crested with V-shaped troughs. For waves with periods between 1 second and about 5 minutes, gravity is the dominant restoring force; this range includes most of the waves we see. Because of the influence of gravity, these waves are known as *gravity waves.*

The largest waves—tides and seismic sea waves—involve substantial water movements. In addition to gravity effects, the Coriolis effect is important for waves whose periods exceed 5 minutes.

Waves transmit energy gained from the disturbance that formed them. This energy is in two forms. One-half is potential energy, depending on the position of the water above or below the still-water level; the potential energy advances with the group speed of the individual waves. The rest of the wave energy—known as kinetic energy—is possessed by the water moving as the wave passes. There is a continual transformation of potential energy to kinetic energy and back to potential energy.

Figure 8-5

Schematic representation of the relative amounts of energy in waves of different periods. (After Kinsman, 1965.)

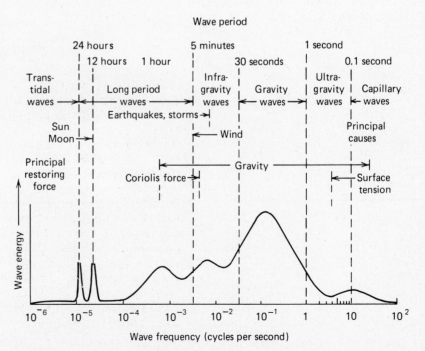

The total energy in a wave is proportional to the square of the wave height; in other words, doubling the wave height increases its energy by a factor of 4. An enormous amount of energy is contained in each wave. For example, a swell with a wave height of 2 meters has energy equivalent to 1200 calories per square meter of ocean surface; a 4-meter swell has 4800 calories per square meter. Nearly all the wave energy is dissipated as heat when the wave strikes a coastline. Sensitive seismographs record the pounding of surf on beaches many hundreds of kilometers away as faint earth tremors.

FORMATION OF SEA AND SWELL

As we have learned from the preceding, waves are disturbances of the ocean surface which pass rapidly across the water, but with little accompanying water movement. Put in another way, waves are the manifestation of energy moving across the ocean surface. Now we shall see how the energy of the wind is supplied to the ocean surface and how wind waves are formed.

Wave formation by the wind is an easily observed phenomenon; in fact wind speed at sea can be estimated from wave conditions as indicated in Table 8-1. Even a gentle breeze results in the immediate formation of ripples or capillary waves, which form more or less regular arcs of long radius, often on top of earlier-formed waves. Ripples are thought by some oceanographers to play an important role in wind-wave formation by providing the surface roughness necessary for the wind to pull or push the water: in short, they provide the "grip" for the wind.

Ripples are short-lived. If the wind dies, they disappear almost immediately; but if the wind continues to blow, ripples grow and are

Figure 8-6
A chaotic sea surface is caused by combined waves of all sizes in an area where waves are formed. (Photograph courtesy Woods Hole Oceanographic Institution.)

TABLE 8-1

Appearance of the Sea at Various Wind Speeds

BEAU-FORT NUM-BER†	WIND SPEED (kilometers per hour)	SEAMAN'S TERM	EFFECTS OBSERVED AT SEA
0	under 1	Calm	Sea like a mirror
1	1–5	Light air	Ripples with appearance of scales; no foam crests
2	6–11	Light breeze	Small wavelets; crests of glassy appearance, not breaking
3	12–19	Gentle breeze	Large wavelets; crests begin to break; scattered whitecaps
4	20–28	Moderate breeze	Small waves, becoming longer; numerous whitecaps
5	29–38	Fresh breeze	Moderate waves, taking longer form; many whitecaps; some spray
6	39–49	Strong breeze	Larger waves forming; whitecaps everywhere; more spray
7	50–61	Moderate gale	Sea heaps up; white foam from breaking waves begins to be blown in streaks
8	62–74	Fresh gale	Moderately high waves of greater length; edges of crests begin to break into spindrift; foam is blown in well-marked streaks
9	75–88	Strong gale	High waves; sea begins to roll; dense streaks of foam; spray may reduce visibility
10	89–102	Whole gale	Very high waves with overhanging crests; sea takes white appearance as foam is blown in very dense streaks; rolling is heavy and visibility reduced
11	103–117	Storm	Exceptionally high waves; sea covered with white-foam patches; visibility still more reduced
12	118–133		
13	134–149		Air filled with foam; sea completely
14	150–166	Hurricane	white with driving spray; visibility
15	167–183		greatly reduced
16	184–201		
17	202–220		

† *Beaufort numbers,* still used to indicate approximate wind speed, were devised in 1806 by the English Admiral Sir Francis Beaufort based on the amount of sail a fully rigged warship of his day could carry in a wind of a given strength. Modified from U.S. Naval Oceanographic Office, 1958. *American Practical Navigator* (Bowditch), Rev. ed., H.O. Publ. No. 9, Washington, D.C., p. 1069.

gradually transformed into larger waves, usually short and choppy ones. These latter waves continue to grow so long as they continue to receive more energy than is lost through processes such as wave breaking. Energy is gained through the pushing-and-dragging effect of the wind. The amount of energy gained by the waves depends on factors such as sea roughness, the specific wave form, and the relative speed of the wind and waves. Choppy, newly formed *seas* provide a much better grip for the wind than smooth-crested *swells*.

The largest wind waves (shown in Fig. 8-6) are formed by storms —often a series of storms—at sea. The size of the waves formed depends on the amount of energy supplied by the wind. The relevant factors

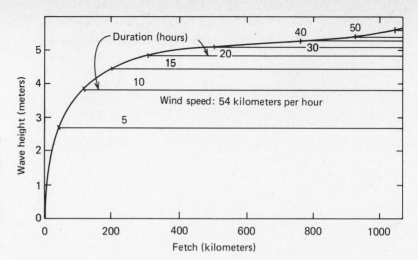

Figure 8-7

Growth of wave height under a constant wind of increasing duration acting over different length and fetch. Note the rapid change in wave height during the first 10 hours as compared with the change between 40 and 50 hours. (After Sverdrup et al., 1942.)

operating here are wind speed; the length of time that the wind blows in a constant direction; and the *fetch*—the maximum distance the wind flows in a constant direction. The process is illustrated in Fig. 8-7. Usually there will be waves present on the ocean at the time the new waves begin to form. Either the older waves will be destroyed by the storm, or newly formed waves will be generated on top of the old ones. There is continuous interaction between waves. Wave crests coincide, forming momentarily new and higher waves. Seconds later, the wave crests may no longer coincide, but instead cancel each other; then the wave crests disappear.

As the winds continue to blow, waves grow in size, as illustrated in Fig. 8-7, until they reach a maximum size—defined as the point at which the energy supplied by the wind is equaled by the energy lost by breaking waves, called *whitecaps*. When this condition is reached, we refer to it as a *fully developed sea*.

Given wind speed, duration, and fetch, it is possible to predict the size of the waves generated by a given storm. Waves of many different sizes and periods are present in a fully developed sea, but waves with a relatively limited range of periods will predominate for a steady wind with a fixed speed (see Fig. 8-8). Such predictions are complicated by the fact that winds almost never blow at a constant speed; winds are just as likely to be gusty at sea as on land, and just as likely to change direction as not.

Figure 8-8

In a fully developed sea, most of the wave energy occurs in a relatively restricted range of wave periods (or wave frequencies). Note that changes in wind speeds cause marked changes in wave energy and wave period. (After G. Neumann and W. J. Pierson, 1966. *Principles of Physical Oceanography*, Prentice-Hall, Englewood Cliffs, N.J.)

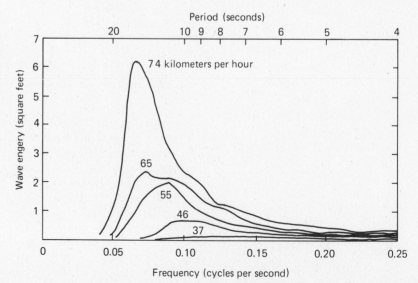

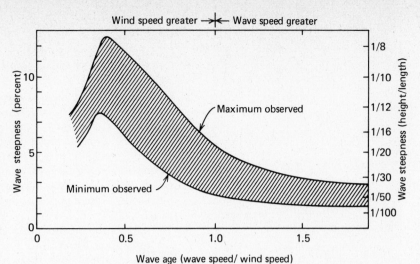

Figure 8-9

Variation in wave steepness with wave age (wave speed/wind speed). The upper and lower curves are probable maximum and minimum values of wave steepness for a given wave age. (After H. U. Sverdrup and W. H. Munk, 1947. *Wind, Sea and Swell: Theory or Relations for Forecasting.* U.S. Naval Oceanographic Office H.O. Publ. 601, Washington, D.C.)

Initially, the waves in a sea are steep, chaotic, and sharp-crested, often reaching the theoretical limit of stability ($H/L = \frac{1}{7}$), when the waves either break or have their crests blown off by the wind; the factors affecting wave steepness are illustrated in Fig. 8-9. As waves continue to develop, their speed approaches, then equals, and finally exceeds the wind speed; as this happens, wave steepness decreases. As waves travel out of the generating area, or if the wind dies, the sharp-crested, mountainous, and unpredictable sea is gradually transformed into smoother, long-crested, longer-period waves—*swell*. These waves can travel far because they lose little energy due to viscous forces.

Let us examine some of the processes that cause waves to change from sea to swell, as they move through calmer ocean areas. One of these processes is the spreading of waves due to variations in the direction of the winds that formed them. Unless destroyed or influenced in some way by ocean boundaries, waves continue to travel for long distances in the direction that the wind was blowing when they formed. Since winds are rarely constant for long in terms of either speed or direction, waves formed in a storm will move away from the area and "fan out" as they move (angular dispersion). As this happens, the wave energy is spread over a larger area, causing a reduction in wave height.

Figure 8-10

Waves moving away from the generating area change from sharp-created chaotic seas to longer, smoother crested swell.

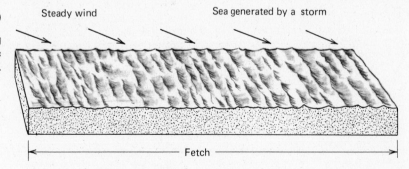

At the same time that the waves are fanning out, they are also separating by wavelength, a process known as *dispersion*. Remember that the speed of deep-water waves C, in meters per second, can be calculated by

$$C = \frac{L}{T} = \sqrt{\frac{gL}{2\pi}} = 1.25 \sqrt{L}$$

where

$L =$ the wavelength in meters
$T =$ the wave period in seconds

Thus longer waves tend to travel faster than shorter waves. As a result, complex waves of varying wavelengths formed in the generating area are sorted as they move away from the storm area, the long waves preceding the shorter waves. Consequently, the first waves to reach a particular coast from a distant large storm at sea will be the waves having the longest periods. Island-dwellers, sensitive to the normal wave period on their coasts, may be warned of approaching hurricanes by the arrival of such abnormally long waves, which travel faster than the storm.

Swells can travel great distances, crossing entire oceans before they encounter a coastline. For example, storms in the North Atlantic form waves which end up as surf on the coast of Morocco, about 3000 kilometers from their point of origin. On extremely calm summer days, very-long-period swell strikes the southern coast of England after traveling about 10,000 kilometers (roughly one-quarter of the way around the earth) from storm areas in the South Atlantic Ocean. Similarly, in the Pacific, waves from Antarctic storms have been detected on the Alaskan coast, more than 10,000 kilometers away.

WAVE HEIGHT

Despite an abundance of data, observations of wave height often leave much to be desired. An observer on a moving ship with no fixed reference points for use in making estimates does not always provide the most reliable information. Still, more than 40,000 observations made from sailing ships, as classified in Table 8-2, indicate that about one-half of the waves in the ocean are 2 meters or less in height. Only

TABLE 8-2

Relative Frequency, in Percent, of Wave Heights in Various Ocean Areas *

OCEAN	WAVE HEIGHT (meters)					
	<1	1–1.5	1.5–2	2–4	4–6.1	>6.1
North Atlantic (Newfoundland to England)	20	20	20	15	10	15
North Pacific (Latitude of Oregon and south of Alaska Peninsula)	25	20	20	15	10	10
South Pacific (West Wind belt latitude of Southern Chile)	5	20	20	20	15	15
Southern Indian Ocean (Madagascar and Northern Australia)	35	25	20	15	5	5
Whole ocean	20	25	20	15	10	10

* After H. B. Bigelow, and W. T. Edmondson, 1947. *Wind Waves at Sea, Breakers and Surf,* U.S. Naval Oceanographic Office H.O. Publ. No. 602, Washington, D.C., 177 pp.

about 10 to 15 percent of the ocean waves exceed 6 meters in height, even in such notoriously stormy areas as the North Atlantic or in the Roaring Forties of the southern oceans.

How *does* one report wave height? When one looks out over a stretch of water, there are waves present covering quite a range of heights. It turns out, however, that the scene is not totally random and can be described statistically. Detailed studies of ocean waves show a nearly constant relationship between waves of various heights. One useful index is the one based on the height of the *significant waves*— the average of the highest one-third of the waves present—as shown in Table 8-3. Setting the height of the significant waves at 1.00, we find that the most frequent waves are about one-half as high, and the average waves are about 0.61. The highest 10 percent will be about 1.29 times higher than the significant waves. Thus, given the wave height for part of the wave spectrum, we can predict the other parts of the spectrum.

TABLE 8-3
*Wave-Height Characteristics **

WAVES	RELATIVE HEIGHT
Most frequent waves	0.50
Average waves	0.61
Significant (highest one-third)	1.00
Highest 10 percent	1.29

* U.S. Naval Oceanographic Office, 1958. p. 730.

The largest waves occur in the open ocean, where they are formed by strong winds blowing over large bodies of water. Such waves occur most frequently at stormy latitudes, where the storms tend to come in groups traveling in the same direction, with only short periods separating them. Thus the waves of one storm often have no chance to decay or travel out of the area before the next storm arrives to add still more energy to the waves, causing them to grow still larger. Typhoons and hurricanes (as the one in Fig. 8-11) do not form exceptionally large waves, because their winds, although very strong, do not blow long enough from one direction.

Figure 8-11

Hurricane driving waves against the North Bayshore retaining wall at Biscayne Bay, Miami, Florida, Sept. 21, 1964. (Photograph courtesy NOAA.)

There are reliable reports of waves 13 to 15 meters high in the North and South Atlantic and the southern Indian Ocean. It appears that these ocean regions rarely produce waves much higher, for several reasons. First, winds rarely blow from one direction long enough to produce waves that are significantly higher, and when the wind changes direction, waves produced under previous wind systems are destroyed or greatly modified. Also, the stormy areas of all oceans experience equally severe storms at one time or another. Thus the major difference between ocean areas is the maximum fetch on which the wind can act. In the North Atlantic the maximum effective fetch is about 1000 kilometers. With such a fetch, a wind blowing about 70 kilometers per hour can produce waves about 11 meters high; with an unlimited fetch, the same wind could produce waves about 15 meters high.

As might be expected from consideration of ocean-basin dimensions, the Pacific holds the records for giant waves. The largest deepwater wave for which we have reliable data was measured in the North Pacific on February 7, 1933. The Navy tanker *U.S.S. Ramapo* encountered a prolonged weather disturbance which had an unobstructed fetch of many thousands of kilometers. The ship, steaming in the direction of wave travel, was relatively stable, and the ship's officers were able to measure wave height as shown in Fig. 8-12, which showed that one wave was at least 34 meters high. The wave period was clocked at 14.8 seconds and the wave speed at 102 kilometers per hour somewhat faster than the theoretically predicted wave speed.

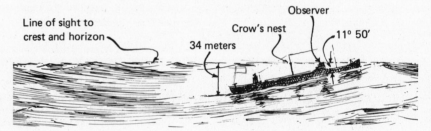

Figure 8-12

Measurement of a wave 34 meters high by the *U.S.S. Ramapo* in the Pacific Ocean, February 7, 1933. This is the largest wave ever measured in a reliable manner.

An impressive amount of energy is dissipated by breaking waves in the surf. A single wave 1.2 meters high, with a 10-second period, striking the entire West Coast of the United States, is estimated to release 50 million horsepower. (For comparison, Hoover Dam produces about 2 million horsepower per year.) Most of this energy is released as heat, but it is not detectable, because, water has a high heat capacity, and, perhaps more important, there is extensive mixing in the *surf zone,* where waves break forming surf, so that the heat is mixed through a large volume of water and carried away from the beach.

WAVES IN SHALLOW WATER

Some wind waves are destroyed when they encounter opposing winds, others interact and cancel each other, but most end up as breakers when they encounter the bottom at a coastline. Except for the very longest waves—such as seismic sea waves or the tides—most waves move through the deep ocean without experiencing any effects of the bottom. As waves approach the coast, they are increasingly affected by the presence of the bottom, changing gradually from deep-water to shallow-water waves. The wavelength and speed continually decrease, while the wave period remains constant. At the same time, the wave height first decreases slightly, then increases rapidly as water depths decrease to one-tenth the wavelength, and the wave crests crowd closer together. (This series of events is illustrated in Fig. 8-13).

Figure 8-13

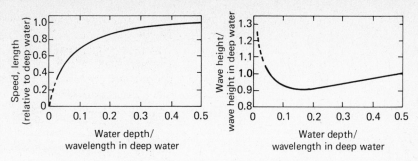

Waves change as they enter shallow water. Speed and wave length decrease as the water becomes more shallow; wave height first decreases, then increases.

Furthermore, the direction of wave approach changes upon entering shallower water, so that we commonly see the breakers nearly parallel to the coastline when they reach the beach, although they may have approached the coast from many different directions. This process, known as *wave refraction,* occurs because the part of the wave still in deeper water moves faster than the part that has entered the shallower water. This rotates the crest to be parallel to the bottom depth contours in the shallow water, as shown in Fig. 8-14.

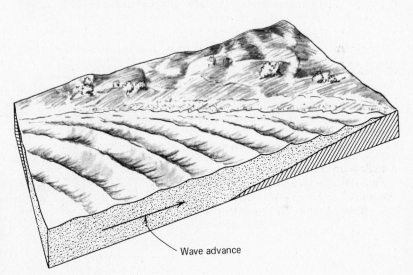

Wave advance

Figure 8-14

Refraction of a uniform wave train advancing at an angle to a straight coastline over a gently sloping, uniform bottom. Note the bend in the crests as the waves approach the beach. Such waves would cause a longshore current moving to the right near the beach.

In the simple case just discussed, the ocean bottom was sloping uniformly away from the beach. Obviously, this is not always the case, and ocean bottom irregularities cause pronounced wave refraction. For example, submarine ridges and canyons cause wave refraction such that the wave energy is concentrated on the headlands and spread out over the bays, as shown in Fig. 8-15. This results in more rapid erosion of the headlands; the eroded material is usually deposited in adjacent bays, eventually creating a simpler, less rugged coastline. An example is the refraction of waves by the Hudson Submarine Canyon off the New York–New Jersey shore, where waves of certain periods are focused on the southern shore of Long Island (illustrated in Fig. 8-16); much less energy reaches parts of the New Jersey coast.

As waves encounter shallow water, wave height increases and wavelength decreases. Consequently, wave steepness (H/L) increases, the wave becomes unstable when the wave height is about 0.8 the water depth, and it forms a breaker. The belt of nearly continuous breaking waves along the shore or over a submerged bank or bar is known as *surf.* These breaking waves are distinctly different from the breaking of oversteepened waves in deeper water, where the tops are blown off by the wind.

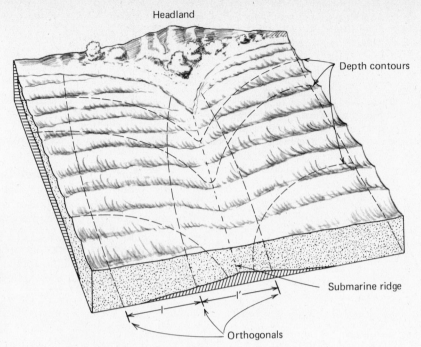

Headland

Depth contours

Submarine ridge

Orthogonals

Figure 8-15

Wave refraction causes energy initially equal at *l* and *l′* to be concentrated on the headland and to be spread out over the adjacent bay.

Figure 8-16

Submarine topography, especially Hudson Channel (a), causes complicated wave-refraction patterns near the entrance to New York Harbor (b). (After W. J. Pierson, G. Neumann, and R. W. James, 1955. *Practical Methods for Observing and Forecasting Ocean Waves by Means of Wave Spectra and Statistics,* U.S. Naval Oceanographic Office H.O. Publ. 603, Washington, D.C.)

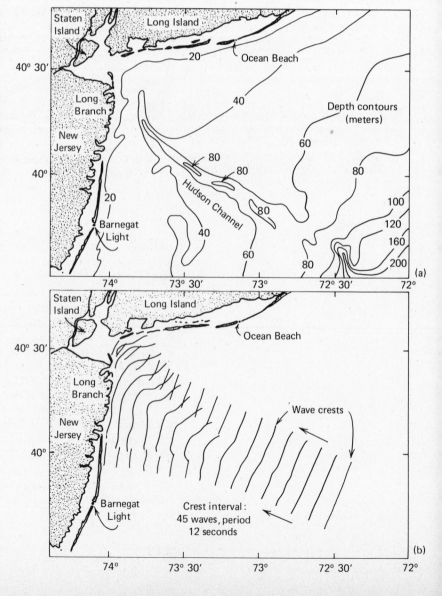

Staten Island

Long Island

20   Ocean Beach

40° 30′

40

Depth contours (meters)

Long Branch

60

New Jersey

80   80   80

40°

80

20

100

Barnegat Light

40

120

Hudson Channel

160

60

80   200

74°   73° 30′   73°   72° 30′   72°   (a)

Staten Island

Long Island

Ocean Beach

40° 30′

Long Branch

Wave crests

New Jersey

40°

Barnegat Light

Crest interval: 45 waves, period 12 seconds

74°   73° 30′   73°   72° 30′   72°   (b)

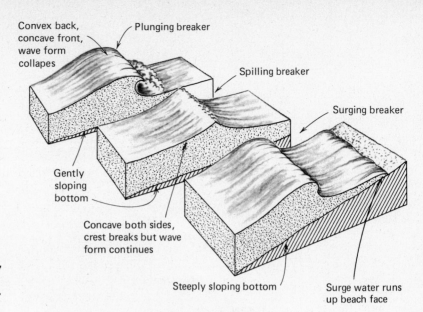

Convex back,
concave front,
wave form
collapes — Plunging breaker

Spilling breaker

Surging breaker

Gently
sloping
bottom

Concave both sides,
crest breaks but wave
form continues

Steeply sloping bottom

Surge water runs
up beach face

Figure 8-17

Types of breakers.

Several types of breakers are shown in Fig. 8-17 and classified in Table 8-4. The *spilling* type and the *plunging* type behave differently and form under different circumstances. The spilling breaker (see Fig. 8-18) is easily visualized as an oversteepened wave where the unstable top spills over the front of the wave as it travels toward the beach. In a spilling breaker, the wave form advances, but wave height (i.e. wave energy) is gradually lost.

TABLE 8-4

*Types of Breakers and Beach Characteristics Associated with Each ***

BREAKER TYPE	DESCRIPTION	RELATIVE BEACH SLOPE	RATIO OF WATER DEPTH TO WAVE HEIGHT
Spilling	Turbulent water and bubbles spill down front of wave; most common type	Flat	1.2
Plunging	Crest curls over large air pocket; smooth splash-up usually follows	Moderately steep	0.9
Collapsing	Breaking occurs over lower half of wave; minimal air pocket and usually no splash-up; bubbles and foam present	Steep	0.8
Surging	Wave slides up and down beach with little or no bubble production	Steep	Near 0

* After Cyril J. Galvin, 1968. "Breaker Type Classification on Three Laboratory Beaches," *Journal of Geophysical Research*, 73 (12), 3655.

Figure 8-18

Spilling breakers on a rocky beach, Stradbroke Island, Queensland, Australia. (Photograph courtesy Australian Tourist Commission.)

Plunging breakers are more spectacular (see Fig. 8-19). The wave crest typically curls over, forming a large air pocket. When the wave breaks, there is usually a large splash of water and foam thrown into the air. These waves are excellent for surfing.

Figure 8-19

A plunging breaker on the south shore of Long Island. The spray is blown seaward by strong winds.

Plunging breakers tend to form from long, gentle swells ($H/L = 0.005$) over a gently sloping bottom with rocky irregularities. On even steeper bottoms, such waves may break over the lower half of the wave with little upward splash. This is known as a *collapsing* type of breaker, shown in Fig. 8-20.

Figure 8-20

Changes in the water surface through time (0.06-second intervals) as a breaker advances toward the beach. (After C. J. Galvin, 1968. "Breaker Type Classification on Three Laboratory Beaches," *Journal of Geophysical Research*, 73, 3651–3659.)

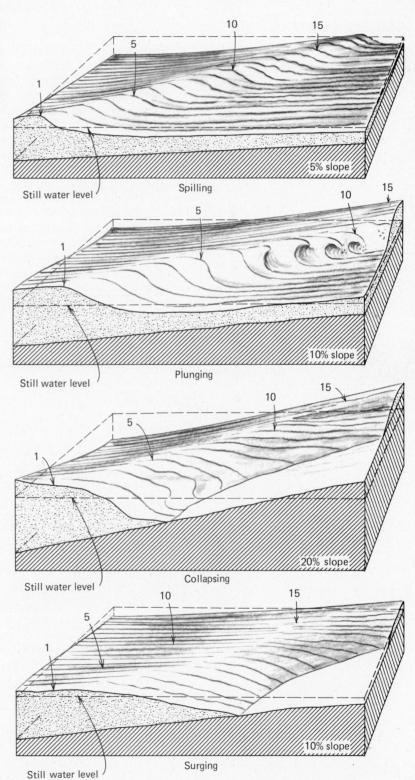

Usually surf consists of a mixture of various types of breakers, the result of different type waves coming into the beach and the complex and uneven bottom topography offshore, which is usually changing because of tidal action.

If a wave strikes a barrier, such as a vertical wall, it may be reflected, its energy being transferred to another wave, which travels in a different direction. *Wave reflection* may be seen when small waves in a bathtub are reflected from the sides. In other cases, a wave striking a steeply dipping barrier may form a surging breaker or a turbulent wall of water, which then moves up the barrier as the wave advances and runs back down when the wave retreats.

Surf height depends on the height and steepness of the waves offshore, and to a certain extent on the offshore bottom topography. Thus the breakers and surf may be only a few centimeters high on a lake or a protected ocean beach, or many meters high on an open beach. Reports of spectacular surf come from lighthouses built in exposed positions. For example, Minot's lighthouse, 30 meters tall, on a ledge on the south side of Massachusetts Bay, is often engulfed by spray from breakers, and the glass in the lighthouse at Tillamook Rock, Oregon, 49 meters above the sea, has often been struck by surf.

We have no record observations of the waves that cause such surf, but breakers about 14 meters high twice damaged a breakwater at Wick Bay, Scotland, moving blocks weighing as much as 2600 tons. Breakers about 20 meters high have been reported at the entrance to San Francisco Bay, and at the Columbia River estuary on the Pacific coast when onshore gales were blowing. Waves and breakers at a river or harbor are likely to be especially high when the incoming waves encounter a current setting in the opposite direction. Ships may have to wait for days before finding a time when incoming waves and tidal currents are right, permitting them to enter the harbor with safety.

Large *tsunamis, or seismic sea waves* having very long periods, are apparently caused by sudden movements of the ocean bottom.

Figure 8-21

Diagonally hatched areas are affected by seismic sea waves.

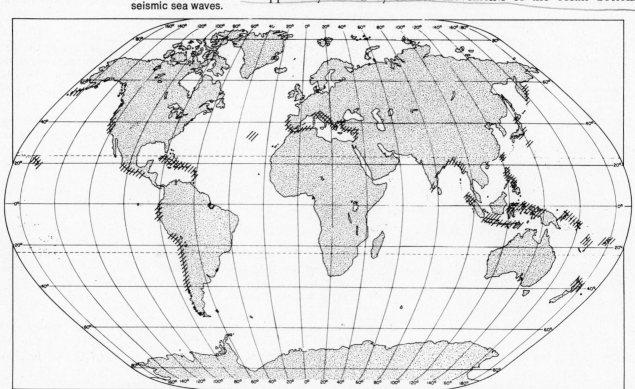

These behave like shallow-water waves even when passing through the deep ocean. For example, an earthquake in the Aleutian Islands on April 1, 1946 caused a tsunami with a 15-minute period and a wavelength of 150 kilometers. Even being in the Pacific Ocean, where the average depth is about 4300 meters, the wave speed was controlled by the bottom ($L/d = 150/4.3$); yet it still traveled at a speed of about 800 kilometers per hour. In deep water such wave crests were estimated to be about a half-meter high, which would be undetectable to ships, especially considering the extremely long wavelength.

Like wind waves, tsunamis eventually encounter the coast (Fig. 8-21), often with catastrophic results due to their great speed and height. As the waves from the Aleutian earthquake hit the Hawaiian Islands, they were driven ashore in a few places as a rapidly moving wall of water up to 6 meters high. These waves also formed enormous breakers that towered up to 16 meters above sea level, where the water was funneled in a valley. More than 150 people were killed in Hawaii, and property damage was extensive. Apparently the wave was highest in the Aleutian Islands, where a reinforced-concrete lighthouse and radio tower 33 meters above sea level were destroyed at Scotch Cap, Alaska. Historical records show that Japan has been hit by about 150 tsunamis.

Because of their frequent occurrence in areas bordering the Pacific (see Fig. 8-21), a warning net operates to detect large earthquakes likely to cause tsunamis and to warn areas likely to be hit. As a result, the tsunami of 1957 killed no one in Hawaii, even though water levels were locally higher than in 1946.

INTERNAL AND STANDING WAVES

So far, we have spoken only about progressive surface waves. But there are other types of waves in the ocean which are not as easily observed, and consequently not as well known. *Internal waves,* as in Fig. 8-22, occur within the ocean rather than at the ocean surface, although in principle they are similar to surface waves. Surface waves occur at an interface between air and water, and internal waves are found at an interface between water layers of different densities—for example, the pycnocline. Since the pycnocline is associated either with the halocline or the thermocline, internal waves can be indirectly observed by studying changes in temperature or salinity at a given depth and fixed location.

Internal waves, it is thought, move as smoothly undulating shallow-water waves. Neglecting the earth's rotation, the speed C of an internal wave in meters per second is given by

$$C = \sqrt{g\left(\frac{\rho - \rho'}{\rho}\right)\left(\frac{dd'}{d + d'}\right)}$$

where

ρ and $\rho' =$ the densities of the lower and upper layers, respectively
$d =$ the depth in meters below the interface
$d' =$ the height of the free surface above the interface
$g =$ the acceleration of gravity

Surfaces where internal waves form involve only small density differences between two water layers rather than the much larger density difference between air and water. As a result, internal wave heights can be much greater than surface waves and they generally move much slower.

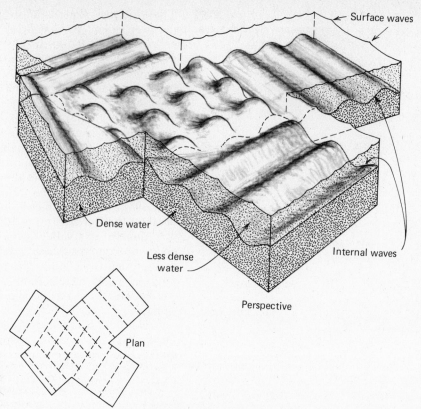

Dense water

Less dense water

Internal waves

Perspective

Plan

Figure 8-22

Simple internal waves at interface between water layers of different densities and their interaction to form more complex patterns.

Standing (or _stationary_) _waves_ are yet another type of wave phenomenon in the ocean; standing waves are also important in lakes and play an important role in tidal phenomena. Standing waves can be generated experimentally by first tilting a round-bottomed dish of water and then setting it on a table: the water's surface will appear to tilt first one way and then the other. This type of movement is distinctly different from the progressive waves we have been discussing

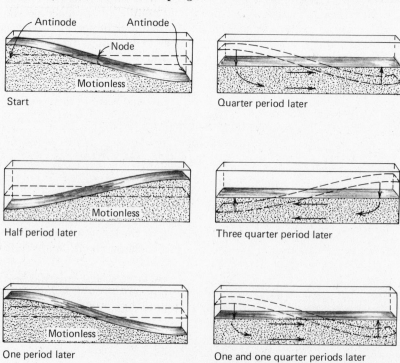

Antinode   Antinode
Node
Motionless

Start

Quarter period later

Motionless

Half period later

Three quarter period later

Figure 8-23

Simple standing wave, with one node, shown at quarter-period intervals.

Motionless

One period later

One and one quarter periods later

wherein the wave moves across the body of water—as if we had dropped a pebble in the dish of water.

One may notice that in the standing wave, part of the water surface does not move vertically but acts as a sort of hinge about which the rest of the water surface tilts. This stationary line or point is known as the *node*, and the parts of the water surface having the greatest vertical movement—known as the *antinodes*—are situated at the walls of the container. More complicated stationary waves may have more than one node and several antinodes in addition to those at the boundaries of the container.

In a stationary wave (as shown in Fig. 8-23) the maximum horizontal water movement occurs at the nodes when the water surface is exactly horizontal. When the water surface is tilted most, there is no water motion. In contrast to the continual orbital motion of the water in progressive waves, in a stationary wave the water flows for a distinct period, stops, and then reverses the flow direction. Also, the wave form alternately appears and disappears, and the water does not move in the circular or nearly circular and continuous orbits associated with progressive waves. Standing waves, also known as *seiches* (pronounced "saysh"), are characteristic of steep-sided basins and are well known in many lakes and in the tidal phenomena of nearly closed basins, such as the Red Sea.

$$T = \left(\frac{1}{n}\right)\left(\frac{2L}{\sqrt{gd}}\right)$$

where
n = the number of nodes
L = the length, in meters, of the basin measured in the direction of wave motion
d = the depth, in meters, of the basin
g = the acceleration of gravity

Like progressive waves, standing waves can be modified by their surroundings. They are reflected by vertical boundaries and partially absorbed or obliterated by gently sloping bottoms. Standing waves are refracted by moving into depths substantially less than one-half their wave length. Standing waves in large basins such as the Great Lakes are complicated by the influence of the Coriolis effect, and the resulting wave, instead of simply sloshing back and forth, has a rotary motion.

STORM SURGES

In a storm surge, strong winds pile up water along a coast, causing sea level to rise. When a northwest gale blows across the North Sea, for example, with a fetch of 900 kilometers from Scotland to the Netherlands, sea level can rise more than 3 meters. On January 31–February 1, 1953, a storm surge combined with high tides and strong waves broke through dikes and dunes, causing disastrous flooding of the low-lying Dutch coast (Fig. 8-24).

We can predict the approximate height of a storm surge by making calculations that include wind speed and direction, fetch, water depth, and the shape of the ocean basin. Other factors, such as currents, astronomical tides, and seiches set up by the storm may further complicate the calculation, making accurate prediction difficult.

A second type of storm surge resembles a large wave which moves with the storm or hurricane that caused it. First comes a gradual change in water level, the *forerunner,* a few hours ahead of the storm's arrival. This is apparently caused by the regional wind

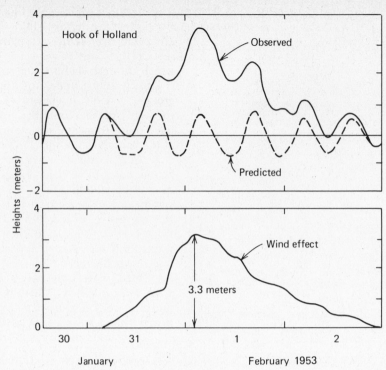

Figure 8-24

Storm surge of 1953 in the North Sea.
Sea-level changes resulting from the surge
were estimated by subtracting the predicted
level from the observed sea levels.
(After Groen, 1967.)

system and may cause sea level to fall slightly along a wide stretch of coastline.

When the hurricane center passes, it causes a sharp rise in water level called the *surge*. This usually lasts about 2½ to 5 hours; rises in sea level of 3 to 4 meters have been observed—usually localized, but slightly offset from the center of the storm. Combined with extremely high waves generated by the storm, hurricane surges can be extremely destructive.

Following the storm, sea level continues to rise and fall as the resurgences or oscillations set up by the storm pass. These are more-or-less free, wavelike motions of the water surface, and have been termed the *wake* of the storm, like the wake left by the passage of a ship through the water. These resurgences can be quite dangerous, particularly as they are often not expected after the storm itself has subsided.

Tropical storms and hurricanes frequently cause storm surges along the Gulf Coast of the United States. In 1900, Galveston, Texas was destroyed and about 2000 people were killed by a storm surge resulting from a hurricane. In 1969, the second strongest recorded storm to hit the Gulf Coast caused millions of dollars of damage; even with advance warning, it killed several hundred people. A disastrous storm surge occurred in 1876, on the Bay of Bengal, when 100,000 people were killed. In 1970 a storm surge hit the same area, killing an estimated 500,000 persons.

SUMMARY OUTLINE

Ideal waves—disturbances of the water surface

Crest—highest part; trough—lowest part of wave form

Wave height—vertical distance from bottom of wave trough to top of wave crest

Wavelength—horizontal distance between successive wave crests (or troughs)

Wave period—in progressive waves, time for successive wave crests or troughs to pass a fixed point; in standing waves, time for water surface to assume initial position

Sine wave—smooth-crested, smooth-troughed ideal wave form used to describe low ocean waves (a mathematical expression)

Trochoid—sharp-crested, flattened trough; ideal wave form useful to describe larger waves

Wave steepness—ratio of wave height to wavelength

Deep-water wave—wave unaffected by ocean bottom, water deeper than $L/2$

Shallow-water wave—wave affected by ocean bottom, in water less than $L/20$ deep; wave speed $C = \sqrt{gd}$

Seismic sea waves—formed by sudden movement of ocean bottom due to earthquake or sediment slump

Forced waves—formed by disturbance continuously applied

Free waves—move independently of disturbance

Capillary waves—ripples, round-crested, V-troughed; wave length less than 1.7 cm; wave period 0.1 sec; surface tension is dominant restoring force

Gravity waves—most common waves; gravity is dominant restoring force; period 1 sec to 5 min

Formation of sea and swell

Sea—waves under influence of the wind that formed them; short, sharp crests, chaotic, unpredictable

Swell—waves outside generating area; long smooth crests

Energy for waves usually supplied by wind

Ripples form first, grow into larger waves, and eventually into fully developed seas

Swells form by gradual transformations

Angular dispersion occurs because of difference in direction

Waves separate according to wave length, longer waves travel faster than short ones

Wave height

About 90 percent of waves less than 6 meters high

Highest waves: Pacific—34 meters; other oceans 15 meters

Waves in shallow water

Refraction—change in wave advance direction upon entering shallow water; energy focused on headlands

Reflection—energy reflected, transferred to newly formed wave moving in direction opposite to original wave

Breakers—unstable waves losing energy; classified as plunging, spilling, collapsing, or surging

Surf—band of breakers parallel to coast

Internal and standing waves

Internal waves—occur at boundaries between water layers

Standing waves (seiches)—occur in nearly closed basins; entire water surface tilts

Nodes—water level constant; maximum currents

Antinode—maximum vertical water movements and changes in water surface level

Storm surges—strong winds move waters toward coast

Creates standing wave in some basins—e.g., North Sea

Surge—preceded by forerunner and followed by pronounced oscillations of water level

SELECTED REFERENCES

Bascom, Willard. 1964. *Waves and Beaches: The Dynamics of the Ocean Surface.* Doubleday Anchor Books, Garden City, N.Y. 267 pp. Elementary.

Kinsman, Blair. 1965. *Wind Waves: Their Generation and and Propagation on the Ocean Surface.* Prentice-Hall, Englewood Cliffs, N.J. 676 pp. Advanced mathematical treatment; good descriptions.

Russell, R. C. H., and D. M. Macmillan. 1954. *Waves and Tides.* Hutchinson, *London.* 348 pp. Elementary.

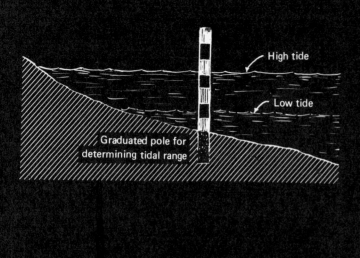

High tide

Low tide

Graduated pole for
determining tidal range

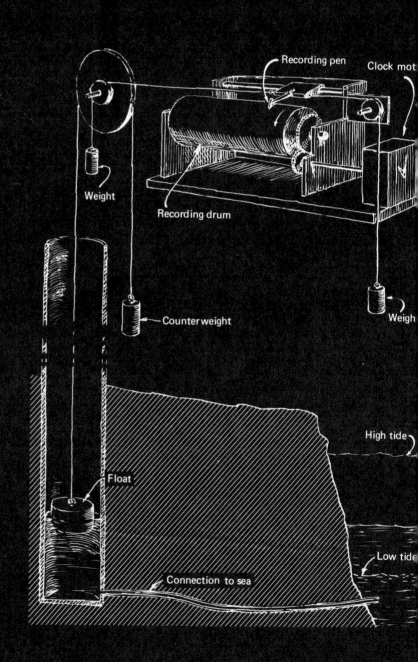

Recording pen

Clock mot

Weight

Recording drum

Counterweight

Weigh

Float

High tide

Low tide

Connection to sea

tides
and
tidal
currents

Tides are the pulse of the ocean. Their effects are felt most keenly in coastal areas, where periodic rise and fall of the ocean surface, with alternate submersion and exposure of the intertidal zone, profoundly influence plant and animal life. Nor is man oblivious to the tide. Extremely low tides are occasions for digging clams; extremely high tides combined with storms have often caused flooding of low-lying areas and damage to coastal installations. Tidal currents, accompanying the rise and fall of the tide, are by far the strongest currents in the coastal ocean; even large, modern ships prefer to depart a port with an outgoing tide and to enter on an incoming tide, just as sailors did in antiquity.

Like waves, tidal phenomena are easily observed and have been studied since Roman times, at least. Pliny the Elder (A.D. 23–79) correctly attributed tides to the effect of the sun and the moon. The small tides of the Mediterranean Sea could easily be ignored, and most ancient writers did so. Two famous men of action got into trouble by ignoring the tides when their military conquests took them into other ocean areas: Julius Caesar lost part of his fleet and sustained damage to most of his ships during a night of high tide on the coast of England; and Alexander the Great got into similar difficulties at the mouth of the Indus River in the Northern Indian Ocean.

Tides, however, were certainly not strange to those who lived around the North Atlantic. The Venerable Bede (A.D. 673–735) wrote of the extensive tidal observations made by medieval British priests and exhibited a clear understanding of the relationship of tidal phenomena to the moon. He pointed out that separate tidal predictions must be made for each area and that 19 years of observations (corresponding to a complete lunar cycle) were required in order to compile accurate tide tables.

Among the oldest records of oceanographic observations is the "Flod at London Brigge," a set of tide tables for the London Bridge dating from the late twelfth or early thirteenth century. Probably other seafaring people also had some means of predicting tides, but any such records have been lost, so that we can now only speculate about them. Tables of tides and tidal currents represent one of our most successful efforts at predicting physical processes in the ocean.

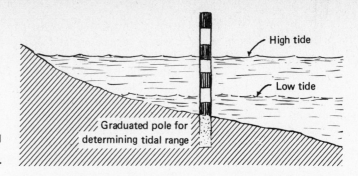

Figure 9-1

Graduated pole for determining tidal range.

High tide

Low tide

Graduated pole for determining tidal range

TIDAL CURVES Compared with most oceanographic observations, measurements of the tide can be quite simple, requiring only a pole marked at appropriate intervals and firmly attached to a piling or stuck in the bottom; such a pole is shown in Fig. 9-1. At timed intervals we can record the height of the still water surface on the pole—for instance, every hour. When we plot the height of the water surface at each interval, the result is a *tidal curve*.

More elaborate installations are needed for the continuous tidal observations made in most major ports. A simplified diagram of such a *tide station* is shown in Fig. 9-2. An enclosed basin is constructed

Figure 9-2

Simplified diagram of automatic tide gauge.

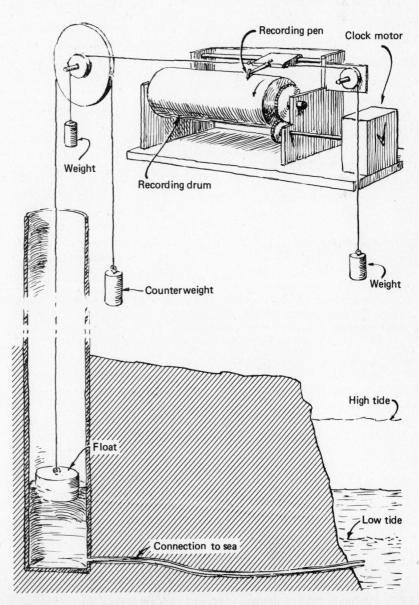

Recording pen

Clock motor

Weight

Recording drum

Counterweight

Weight

High tide

Float

Low tide

Connection to sea

with a narrow connection to the ocean so that the water level in the basin is always equal to the undisturbed sea level outside but is not itself disturbed by normal waves. This avoids some of the uncertainty in tidal observations caused by waves or other disturbances of the water surface. A float on the water surface in the basin is connected to a marker, often a pencil, which draws the tidal curve on a clock-driven, paper-covered drum, indicating the changing sea level. The mechanism operates continuously and requires a minimum of maintenance; normally such a gauge is checked once a day.

Examination of tidal curves from many ports shows that each one has a different tide. Tides are grouped into three types, based on the number of highs and lows per day, the relationship between the heights of successive highs or lows, and the time between corresponding high (or low) stands of sea level. Most tidal curves show two high tides and two low tides per *tidal day*—about 24 hours, 50 minutes. This period corresponds to the time between successive passes of the moon over any point on the earth. The time, either 12 hours and 25 minutes or 24 hours, 50 minutes, between high (or low) tides is known as the *tidal period*.

A few ocean areas, such as parts of the Gulf of Mexico, have only one high tide and one low tide each day (see Fig. 9-3); these are called *daily tides* or *diurnal tides*. Most North Atlantic ports have two high and two low tides that are approximately equal; these are *semidaily* or *semidiurnal* tides. They are relatively easy to predict because high tides tend to occur at a regular time after the moon has crossed the meridian for that port. Tidal predictions for ports with semidaily tides have been made for many centuries, based primarily on an understanding of the lunar cycle.

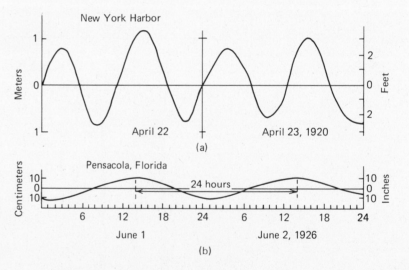

Figure 9-3

Typical tide curves for (a) semidaily tide, New York Harbor, April, 1920 and (b) daily tide, Pensacola, Florida. (After H. A. Marmer, 1930. *The Sea*, D. Appleton, New York.)

Tidal curves from U.S. Pacific ports also show two high tides and two low tides per tidal day, but the highs are usually quite different in height, and the low tides differ as well. These *mixed tides* are shown in Fig. 9-4. The higher of the two high tides is called *higher high water* (abbreviated HHW); the other is called *lower high water* (LHW). There is a similar nomenclature for the low tides—*lower low water* (LLW) and *higher low water* (HLW).

Mixed tides are not as easy to predict as semidaily tides because the timing of the high- and low-tide *stands* (when there is no appreciable change in the height of the tide) does not bear a simple relationship to the passage of the moon over the meridian of the station. More sophisticated means of predicting the tides must be used.

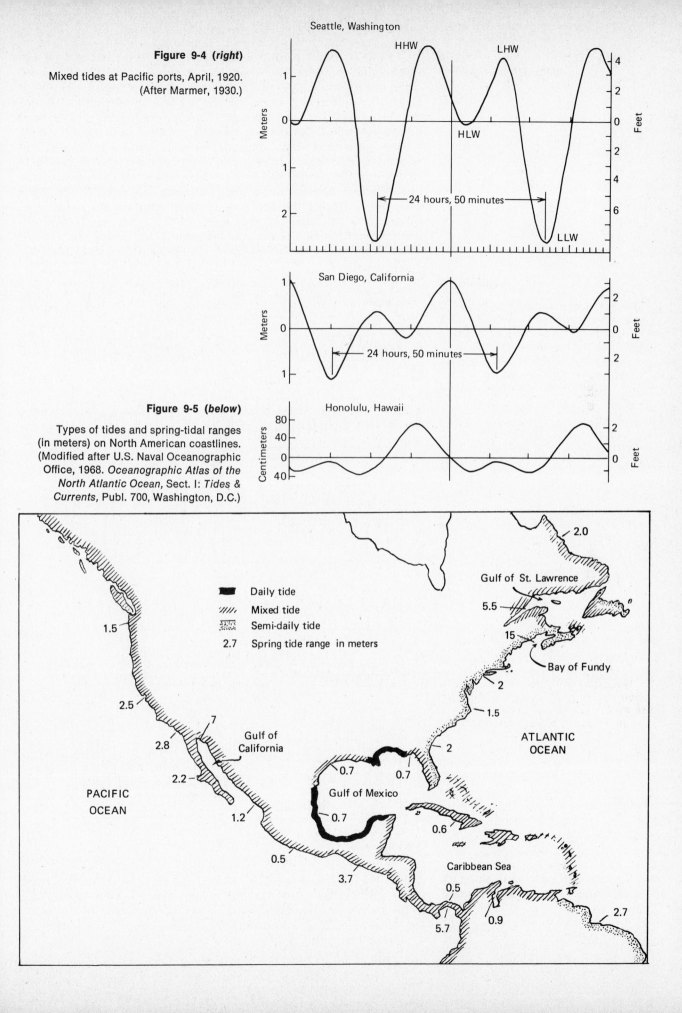

Seattle, Washington

HHW   LHW

HLW

24 hours, 50 minutes

LLW

Figure 9-4 (right)

Mixed tides at Pacific ports, April, 1920.
(After Marmer, 1930.)

Meters   Feet

San Diego, California

24 hours, 50 minutes

Meters   Feet

Figure 9-5 (below)

Types of tides and spring-tidal ranges
(in meters) on North American coastlines.
(Modified after U.S. Naval Oceanographic
Office, 1968. *Oceanographic Atlas of the
North Atlantic Ocean,* Sect. I: *Tides &
Currents,* Publ. 700, Washington, D.C.)

Honolulu, Hawaii

Centimeters   Feet

PACIFIC
OCEAN

ATLANTIC
OCEAN

2.0

Gulf of St. Lawrence

5.5

15

Bay of Fundy

2

1.5

2

Daily tide
Mixed tide
Semi-daily tide
2.7   Spring tide range in meters

1.5

2.5

7

Gulf of
California

2.8

2.2

1.2

0.5

3.7

0.7   0.7

Gulf of Mexico

0.7

0.6

Caribbean Sea

0.5

0.9

5.7

2.7

From a record of only a few days' length, we can classify the type of tide characteristic of any harbor (see Fig. 9-5). Typical tidal curves are shown in Fig. 9-6. For example, we can measure the _tidal range_ (the difference between the highest and lowest tide levels) and the _daily inequality_ (the difference between the heights of successive high or low tides). But the tide also changes from week to week. With a record of several weeks' duration, we see a pattern in the changes of tidal range. _Spring tides_ occur near the times of full and new moons, and the spring-tidal range is larger than the _mean-tidal range_ (the difference between mean high and mean low tides) or the _mean daily range_. During the first and the third quarters, the tidal range is least; these are called the _neap tides_. As Fig. 9-6 shows, there is substantial variation in the tides at the same place during the month. Other, less striking variations occur over periods of several years.

Figure 9-6

Tidal variations at certain ports during the course of a month. Note the variations in the timing of the spring and neap tides relative to the new and full moon. (After U.S. Naval Oceanographic Office, 1958. _American Practical Navigator,_ rev. ed., (Bowditch), H.O. Publ. No. 9, Washington, D.C.)

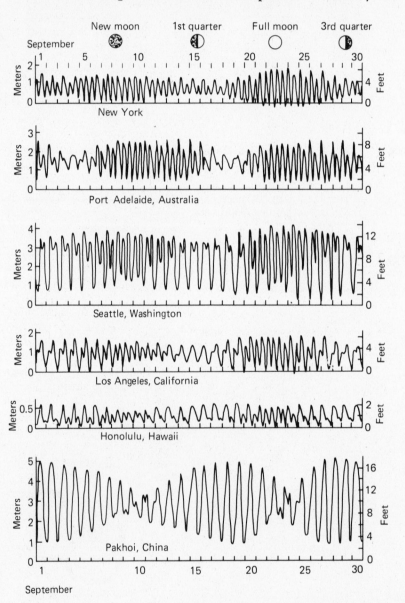

TIDE-GENERATING FORCES AND THE EQUILIBRIUM TIDE

Although it had long been known that the tides were closely related to the sun and moon, it remained for Sir Isaac Newton (1642–1727) to lay the foundation for understanding the mechanics of the tides. He began by making several simplifying assumptions; his equilibrium

Earth

Tide
producing force

r

M

Gravitational
attraction

Common center of mass

Moon

59r

61r

Figure 9-7

Schematic representation of the Earth–moon
system revolving around their common
center *M*. Note that the ocean nearest
the moon is about 59 earth radii (*r*) from the
moon compared to 61 earth radii for the
side farthest from the moon. At the center
of the earth, gravitational attraction of the
moon is balanced by the centrifugal forces.

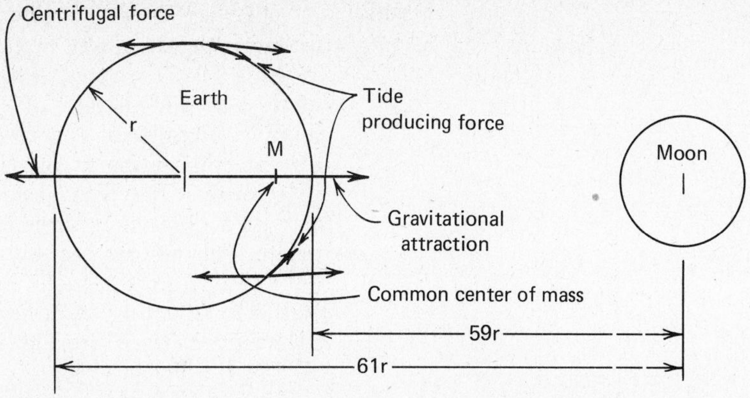

theory of the tides assumes a static ocean completely covering a non-rotating earth, with no continents. Let us consider the effect of the moon on such a simplified ocean.

Gravitational attraction pulls the earth and moon toward each other, while centrifugal forces, acting in the opposite direction, keep them apart, as illustrated in Fig. 9-7. In all respects, the earth and moon act like twin planets revolving about a common center, which in turn moves around the sun. If the earth and moon were the same size, the center of revolution of the system would be located midway between them. The moon, however, is only about 1/82 the mass of the earth; consequently the center of revolution of the earth–moon system is located nearer the earth, about 4700 kilometers from the earth's center. This is analogous to an adult and a small child on a see-saw: the adult must sit closer to the pivot to achieve a balance.

There are, however, small unbalanced forces in the system. Consider the gravitational attraction of the moon on the earth's surface. Remember that the force of gravity is inversely proportional to the square of the distance separating the two objects. In other words, doubling the separation reduces the force of gravity to one-fourth of its former strength. A parcel of water located on the earth's surface is only 59 earth radii away at a point nearest the moon, but 61 earth radii away when it is on the opposite side of the earth (see Fig. 9-7). Hence the moon's gravitational attraction is greatest on the side of the earth nearest the moon, and least on the opposite side of the earth. The centrifugal forces, however, are equal over the earth's surface. On the side nearest the moon, the attraction of the moon exceeds the centrifugal force, so that the water is attracted toward the moon. On the opposite side of the earth, the centrifugal forces overbalance the attraction of the moon so that there, too, a force acts on the water effectively dragging it away from the earth. These unbalanced forces on the earth's surface, shown in Fig. 9-8, are the tide-generating forces associated with the moon.

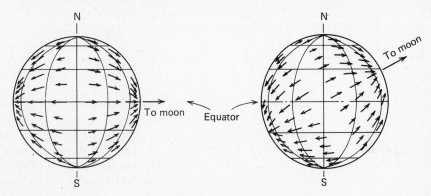

Figure 9-8

Horizontal component of the tide-producing
forces acting on the ocean surface when
the moon is in the plane of the earth's
equator (a) and when moon is above the
equatorial plane (b). Note that the
tide-producing forces shift their orientation
as the moon's position changes. (After
U.S. Naval Oceanographic Office, 1958.)

To see how the tides are created, let us look at these forces in more detail. Such forces can be represented by vectors (shown in Fig. 9-9 as arrows) which point in the direction in which the forces act. The length of the arrow (the vector) corresponds to the relative strength of the force. Each vector can be resolved into two components, acting at right angles to each other. One component acts in a vertical direction, perpendicular to the earth's surface; the other acts in a horizontal direction, parallel to the earth's surface at that point. The vectors point toward the moon on one side of the earth (and away from the moon on the opposite side). The relative strength of the force acting in either a horizontal or vertical sense varies over the earth's surface (see Fig. 9-9).

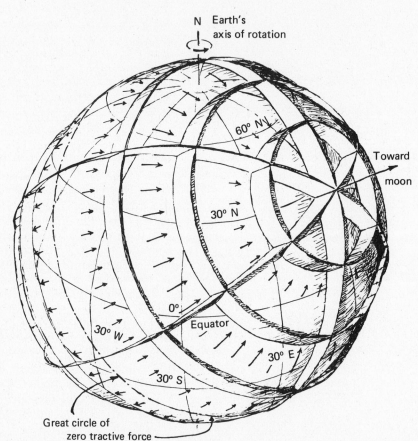

Figure 9-9

Equilibrium tide (shown by the "fences") when the moon is above the plane of the earth's equator. The height of the tidal bulge is shown by the height of the radial "fences." Note variations in the tide on any parallel of latitude as the earth rotates. (After W. S. von Arx, 1962. *An Introduction to Physical Oceanography*, Addison-Wesley, Reading, Mass.)

Directly beneath the moon and on the opposite side of the earth, tide-generating forces act solely in a vertical direction. But these vertical components have little effect on tides, since they are counteracted by gravity, which is about 9 million times stronger. Horizontal components of the tide-generating force are also rather weak, but they are comparable in strength to other forces acting on the ocean's surface and thus are not overwhelmed.

Newton pointed out that these horizontal or tractive forces cause the *equilibrium tide,* which consists of a tidal bulge on the side of the earth nearest the moon and on the side opposite the moon. As a result of these horizontal forces, the water covering the earth is deformed slightly to form an egg-shaped water envelope on our imaginary, nonrotating, water-covered earth. The solid earth itself also responds to these tide-generating forces and deforms slightly, but much less than the ocean waters.

Now if we permit the earth to rotate beneath its deformed, watery covering, we can see how the equilibrium-tide theory explains successive high and low tides. If the moon is in the plane of the earth's equator, the tidal bulges of the equilibrium tide will also be centered on the equator. A tide gauge at any point of the equator would register high tide when that point is directly under the moon. After the earth rotates 90°, the tide gauge would register low tide, which is located midway between the two tidal bulges. After rotating another 90°, the tide gauge is directly opposite the moon and registers high tide.

This simplified model explains semidaily tides with two equal high tides and two equal low tides per tidal day. Remember that the earth rotates beneath its slightly deformed water cover. The more-or-less egg-shaped deformed water surface, corresponding to the equilibrium tide, remains fixed in space, its location determined by the location of the moon in our simplified case.

The moon, however, does not maintain a fixed position relative to the earth. It moves from a position 28.5° north of the equator to 28.5° south of the equator. When the moon changes its position, so does the orientation of the tide-generating forces and the position of the equilibrium tide.

Let us imagine a purely theoretical case of tide-generating forces on a water-covered earth, in order to demonstrate the effect of the earth's rotation on the height of the tides. We shall identify points on the earth's surface by their real names, but the conditions described do not necessarily apply on the real earth at those points, because the example illustrates conditions as they would be without continental barriers to water movements.

If the moon were, for instance, over the site of Miami, there would be a tidal bulge there and also one near the west coast of Australia. These bulges would be particularly well-developed by virtue of being in line with the tide-generating forces. After the earth rotated 180°, these points would again experience a high tide, but a lower one because Miami and Australia would no longer be in line with the moon. Tide-generating forces would now be greatest off northern Peru and in the Bay of Bengal–South China Sea area. (Use of a globe will be helpful in demonstrating this point.) This example shows that the equilibrium theory of the tides not only explains the semidaily tides but the daily inequalities as well.

A similar analysis could be made for the sun–earth system to determine the sun's influence on the tides. There are, however, several important differences. First, the greater distance of the sun from the earth (approximately 23,000 earth radii) is only partially compensated by its greater mass (330,000 earth masses), so that the solar tide-generating forces are only about 47 percent as powerful as those of the moon. Second, solar tides have a period of about 12 hours rather than the 12 hour-25 minute period of the lunar tide. Finally, the sun's position relative to the earth's equator also changes, from 23.5° north to 23.5° south of the equator, but it requires a full year to make the complete cycle, in contrast to the monthly changes of the moon's position from 28.5° north to 28.5° south.

The equilibrium theory also explains variations in tidal range between the spring and neap tides. Spring tides occur every two weeks, usually within a few days of the new and full moons (see Fig. 9-6). Considering their positions in space (shown in Fig. 9-10), we see that during the time of the full and new moons, the solar and lunar tide-generating forces act together, causing large tidal bulges and consequently greater tidal ranges. During the first and third quarters of the moon, tide-generating forces partially counteract each other by

Figure 9-10

Relative positions of sun, moon, and earth during spring and neap tides.

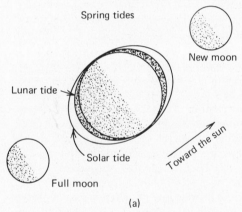

Spring tides

New moon

Lunar tide

Solar tide

Toward the sun

Full moon

(a)

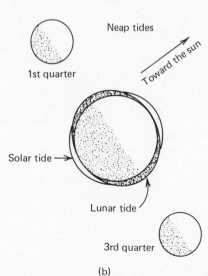

Neap tides

1st quarter

Toward the sun

Solar tide

Lunar tide

3rd quarter

(b)

acting in different directions, resulting in the lowest tidal range—the neap tides.

There are other, more subtle variations in tides associated with the variable distance of the moon and sun from the earth. When either the sun or moon is closer to the earth, the equilibrium theory predicts that tides will be higher (or lower) than when they are farther from the earth. All these complicated relationships are exactly duplicated every 19 years; all possible positions of the sun and moon relative to the earth occur during that period of time.

Despite the many simplifying assumptions involved in the equilibrium theory, it accounts for many of the features of the tides, and its simplicity helps us to understand such aspects as the daily inequalities. It fails, however, to predict the timing of the tides. According to the equilibrium theory, high tide should occur when the moon crosses the meridian of a given port. This is clearly not the case. If it were true, prediction of the tides could have been done for centuries, whereas in fact it is only in comparatively recent times that tide-predicting computers have been designed to make the necessary complicated calculations.

Newton himself was aware that his equilibrium theory was not a complete explanation of tidal phenomena. The theory assumed an idealized ocean and dealt only with the tide-generating forces, without taking into consideration the response of the oceans to these forces.

DYNAMICAL THEORY OF THE TIDE

Since Newton's time, mathematicians and physicists have investigated the tides by considering the response of the oceans to the tide-generating forces. This is known as the *dynamical approach,* since it involves a dynamic rather than a static ocean. This approach, like the equilibrium theory, also requires some assumptions in order to deal with the complicated interactions. Among these is the neglect of vertical forces, which is not too serious because they are quite small compared with the earth's gravitational attraction. Perhaps the most important difference between the dynamical treatment and the equilibrium theory is that the former considers the effect of the shape and depth of the basins on the tides. Furthermore, it recognizes that ocean tides do not occur in a static ocean but involve substantial water movements, and thus must take the Coriolis effect into consideration.

The dynamical theory treats tides as another type of wave phenomenon. As we have seen, tides have many characteristics in common with both progressive and standing waves. Furthermore, treating tides as long-period waves permits the use of powerful mathematical techniques.

Tides can be resolved mathematically into several components and each treated separately. The tide-generating forces may be resolved into *tidal constituents,* the most important being those due to the moon and the sun; some important constituents are shown in Table 9-1. Because of the changing positions of these bodies relative to the earth, as many as 62 tidal constituents are used to make tidal predictions, although the four principal ones usually account for about 70 percent of the tidal range.

The response of a bay or harbor—or an entire ocean—to each of the tidal constituents can be considered separately as a *partial tide.* The tide for any location thus consists of the combination of these partial tides (illustrated in Fig. 9-11), just as the complicated waves in a sea can be reconstructed by combining several simple wave trains.

Study of tidal curves indicates how the partial tides must be combined for a given port. Then tidal predictions are made by com-

Figure 9-11

A daily and a semidaily tide when combined produce a mixed tide. (After H. A. Marmer, 1926. *The Tide,* Appleton-Century-Crofts, New York.)

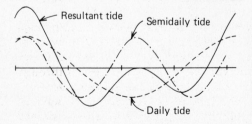

Resultant tide

Semidaily tide

Daily tide

bining partial tides in the appropriate way, normally using some type of computer. Some of the earliest tide-predicting machines were actually simple computers. Development of large, high-speed computers permits more sophisticated tidal models to be studied in greater detail with fewer simplifying assumptions than was previously possible.

Comparison of a tidal record with the profile of a simple wave clearly shows that tides can be considered to be long-period waves. Neglecting the earth's curvature, the two water bulges are the crests and the intervening low areas are the troughs of a simple wave. Because of their immense size relative to the ocean basins, such waves should behave as shallow-water waves. Their wavelength, one-half of the earth's circumference, is about 20,000 kilometers, while the ocean basins have an average depth of only about 4 kilometers, so

TABLE 9-1
Some Important Constituents of Tide-Generating Forces *

CONSTITUENT	SYMBOL	PERIOD (hours)	RELATIVE AMPLITUDE	DESCRIPTION
Semidaily tides				
	M_2	12.4	100.00	Main lunar constituent
	S_2	12.0	46.6	Main solar constituent
	N_2	12.7	19.2	Lunar constituent due to changing distance between earth and moon
Daily tides				
	K_1	23.9	58.4	Soli-lunar constituent
	O_1	25.8	41.5	Main lunar constituent
	P_1	24.1	19.3	Main solar constituent
Long-period tides (spring and neap tides)				
	M_1	327.9	17.2	Main fortnightly constituent

* After A. Defant, 1958. *Ebb and Flow*. University of Michigan Press, Ann Arbor.

that $d/L = 4/20,000$. This is much smaller than the limit of 1/20 for shallow-water waves. If tides were free waves, they would move at a speed of $\sqrt{gd}$ or $\sqrt{9.8 \times 4000}$ meters per second, or about 200 meters per second (720 kilometers per hour). But in order to "keep up with the moon," the tides need to move around the earth in 24 hours, 50 minutes, which means that they would have to move 1600 kilometers per hour at the equator.

In order for the tide waves to move fast enough at the equator to keep up with the moon, the ocean would have to be about 22 kilometers deep. Since it is much shallower, the tidal bulges move as forced waves, whose speed is determined by the movements of the moon. The position of the tidal bulges relative to the moon is determined by a balance between the attraction of the sun and moon and frictional effects of the ocean bottom.

Because the world ocean is cut by the north–south-trending continents into various basins, it is not possible for the tide to move east–west across the earth as a forced wave except around Antarctica, where the absence of continents permits the tide to sweep unimpeded through the ocean. Let us see how the tide behaves in the Atlantic Ocean.

After the tide enters the South Atlantic, it moves northward as a free wave. In the South Atlantic, its course is relatively simple, although doubtlessly influenced by the irregular ocean boundaries and probably by the irregular bottom topography as well. As it progresses northward, the wave from the Antarctic tide also interacts with the independent tide of the Atlantic.

Even if the southern end of the Atlantic Ocean were closed off, it would still have tides. The tide in a basin such as the Atlantic behaves to some degree like a standing wave, as well as having some characteristics of a progressive wave. Every basin has its own natural period for standing waves. If that natural period is approximately 12 hours, the standing-wave component of the tide will be well developed. If the period of the basin is substantially greater than or less than 12 hours, the resemblance of the tide to a standing wave is lessened.

Even if a basin lacks a standing-wave tide, it often has a tide that is distinctly different from that in the adjacent ocean. Tides in the North Atlantic are altered by reflection of the tide wave from

Figure 9-12
Amphidromic motion of a standing wave tide in a Northern Hemisphere embayment. (After von Arx, 1962.)

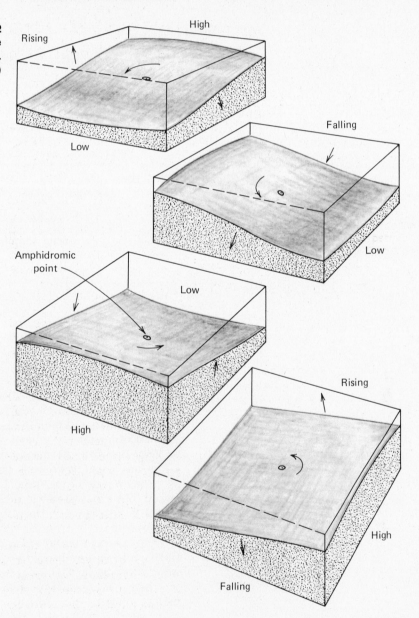

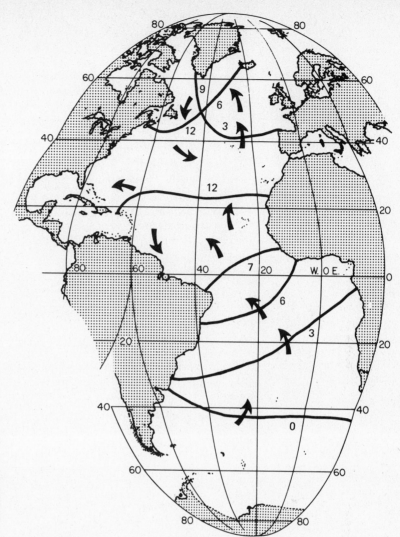

Figure 9-13

Locations of the high water of the principal
lunar partial tide—the major tidal component
caused by the moon. The numbers indicate
the number of hours since the moon
crossed the Greenwich meridian.
(After Defant, 1958.)

the complicated coastline and a standing wave is set up, but this is
not the simple standing wave discussed previously. Since substantial
water movements are involved, it is modified by the Coriolis effect.
The resulting standing wave is a swirling motion such as might result
if we swirled water in a round-bottomed cup. Let us see why such a
wave forms and how it behaves.

We shall choose a channel in the Northern Hemisphere and
set the water moving northward associated with a standing wave.
Because of the Coriolis effect, it will be deflected toward the right,
causing a tilted water surface, as shown in Fig. 9-12. When the water
flows back again, its surface will be tilted in the opposite direction.
The tilted wave surface rotates around the basin, once for each wave
period. Near the center of such a system, called an *amphidrome system,* is a point where the water level does not change, the *amphidromic
point.*

Figure 9-13 shows the movement of the high water associated
with the principal lunar partial tide in the Atlantic. There we see
a large amphidrome system in the North Atlantic. Other, smaller
amphidrome systems not shown on the map occur in the English
Channel and in the North Sea. Points located near an amphidromic

(a)

(b)

(c)

point for one partial tide will not be affected by that partial tide, but will likely be affected by another. For example, an area near the amphidromic point for a semidaily tide may well have a daily-type tide.

Because of their dimensions and consequently their natural periods, each ocean basin responds more readily to certain constituents of the tide-generating forces than to others. The Gulf of Mexico appears to have a natural period of about 24 hours; consequently, it responds more to the daily tidal constituents than to the semidaily constituents, and much of the Gulf has a daily tide (see Fig. 9-5). The Atlantic Ocean, on the other hand, responds more readily to the semidaily constituent of the tide-generating forces, and thus tends to have a semidaily type of tide. Because of their size and complex shapes, the Caribbean Sea and the Pacific and Indian Oceans respond to both the daily and semidaily tidal forces and thus have mixed types of tides.

Each marginal sea, or bay, is affected by the tide in much the same way as the Atlantic is affected by the wave from the Antarctic. A tide wave advancing through the bay is reflected by the coast. If the basin has a natural resonance of the appropriate period, a standing wave may be excited as well. The tide in a bay, harbor, or sea is greatly influenced not only by the magnitude of the ocean tide at its mouth, but also by the natural period of its basin and by the cross-section of the opening through which the tide wave must pass. For example, the small tidal range (less than 0.6 meters) of the Mediterranean Sea can be explained as the result of the small opening through the narrow Strait of Gibraltar into the Atlantic which inhibits exchange of water during each tidal cycle. The small tidal range in several marginal seas of the Pacific can be similarly explained.

Where the natural period of a basin is near the tidal period, it is possible to set up large standing waves, giving rise to exceptional tidal ranges. One famous example is the Bay of Fundy, whose natural period is apparently about 12 hours. Spring tides in the inner part of the bay have ranges of 15 meters (see Fig. 9-5), due to an especially favorable situation. A standing wave is combined with a narrowing valley, which funnels the tide and increases its range. The extreme tidal range and the resulting tidal bore are illustrated in Fig. 9-14. Large tidal ranges (about 13 meters) also occur on the Normandy coast of France and at the head of the Gulf of California (about 7 meters).

The most successful techniques for studying and predicting these complicated interactions involve harmonic techniques, in which the response of the basin to each of the various tide-generating forces is more-or-less isolated and then combined with all the other responses to other forces, to construct a prediction of the tide for a given location at a given time. With even more sophisticated computers it should eventually be possible to make such predictions based solely on physical principles and using the known characteristics of the oceans and of the particular area for which a prediction is being calculated.

At present, tidal predictions still involve the recombination of information gained from many years of observations for each port.

Figure 9-14 (facing page)

(a) High tide and (b) low tide at Market Slip, Saint John, New Brunswick, on the Bay of Fundy. At the northern end of the bay, on the Petitcodiac River, the profile of the incoming tide surges up the river as a steep wave, known as a tidal bore (c), photographed here at Moncton, New Brunswick. (Photographs courtesy New Brunswick Travel Bureau.)

TIDAL CURRENTS   *Tidal currents* are horizontal water movements associated with the rise and fall of the sea surface. The relationship between these two types of water movement is not necessarily a direct one. Not every seacoast has tidal currents, and a few areas have tidal currents but no tides.

First, let us look at the currents we would expect to find in a tide consisting only of a simple progressive wave. As we saw in an earlier section, tides must be considered as shallow-water waves because of their extreme length and the relative shallowness of the world ocean. In such a tide, the crest of the wave is high tide and the trough is low tide. Orbital motions of water caused by the tides are ellipses, greatly flattened circles with their long axes parallel to the ocean bottom. In other words, most of the water motions associated with the tides consist of horizontal motions, with little vertical motion involved.

Near coasts, we find the familiar reversing tidal currents in which water periodically flows in one direction for a while, then reverses to flow in the opposite direction. Such tidal currents can be compared to water movement associated with the passage of a progressive wave as illustrated in Fig. 9-15. As the wave crest moves toward the coast, the water also moves toward the coast; this corresponds to the *flood current*. As the wave trough moves toward the coast, the water moves away from the coast, corresponding to the *ebb current*. Each time the current changes directions, a period of no current, known as *slack water*, intervenes.

In fact, tides and tidal currents are rarely that simple. Progressive waves are reflected by the shore, so that the tide observed in most coastal areas consists of several progressive waves moving in different directions, as well as the standing-wave component that we discussed in previous sections. Other complicating factors include friction effects on the waves and nontidal currents such as river currents. As a result, no universally applicable generalization can be made concerning high and low tides and their relation to the times of slack water or maximum currents. Just like the tides, tidal-current predictions depend on the analysis of observations of tidal currents taken over a long period of time.

In contrast to the simple observations of the tides, tidal current studies are difficult and expensive, since they involve measurements of currents. For example, a study of the tidal currents in Long Island Sound in the 1960's required 3 years to make the measurements and involved 160 separate buoy stations, each with one to three current meters operating for periods of 4 to 15 or 29 days, depending on the station location. Because of the expense involved, our knowledge of tidal currents is usually restricted to coastal areas with much shipborne traffic (for example, the North Sea) or harbors and bays.

Figure 9-15

Relationship of tide and tidal currents in an idealized tide consisting of a simple progressive wave. Note that slack water occurs at high and at low tides.

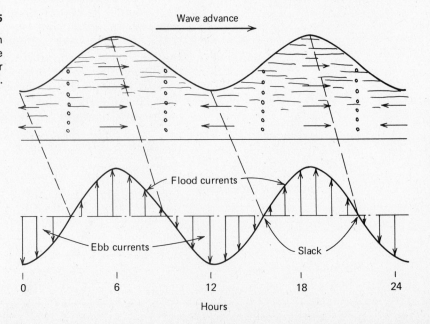

Tidal currents through a tidal period in New York Harbor are illustrated in Fig. 9-16. Beginning with slack water before high tide, the flood current increases until it reaches a maximum and then decreases again until it is slack water about an hour after high tide. After this slack the current ebbs as the tide falls. The current again reaches a maximum, then decreases until about 2 hours after low tide, when it is slack water again. Elsewhere in New York Harbor, slightly different tidal-current patterns may be observed. For example, tidal currents near shore usually turn sooner than in the middle of the channel, because of friction along the channel walls. Tugboat captains often take advantage of these nearshore tidal currents to avoid bucking strong mid-channel currents.

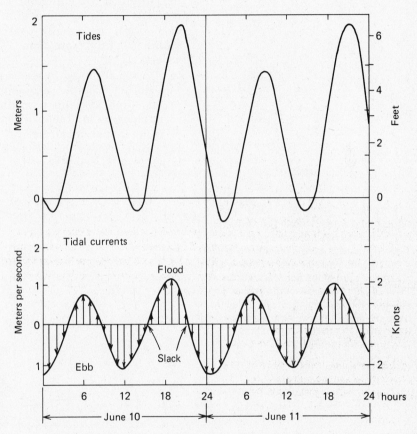

Figure 9-16

Tides and tidal currents at the Narrows, New York Harbor.

Tidal currents are substantially altered by winds or river runoff. For example, river runoff tends to prolong and strengthen ebb currents because more water must move out of the harbor on the ebb than comes in on the flood. Also, different tides will have different tidal currents. In areas with large daily inequalities between successive high (or low) tides, there may be days with continuous ebb (or flood) currents which change in strength during the day.

The strength of a tidal current depends on the volume of water that must flow through an opening and the size of the opening. Thus it is not possible to predict the strength of the tidal current given only the tidal range. For example, the large tidal range in the Gulf of Maine is accompanied by weak tidal currents. Conversely, Nantucket Sound has strong tidal currents but a small tidal range. At a given location, however, the strength of the tidal currents is generally proportional to the relative tidal range for that day. A spring tide, for instance, is usually accompanied by stronger tidal currents than is a neap tide.

Tidal currents are generally the strongest currents in coastal regions. In the English Channel and the Strait of Dover, the tidal currents locally have speeds of 2.5 to 3 meters per second (about 9 to 10 kilometers per hour). Tidal currents of 1.2 meters per second (2.4 knots) occur at the Narrows, the entrance to New York Harbor. At Admiralty Inlet, the entrance to Puget Sound, tidal currents have velocities of 2.4 meters per second (4.7 knots).

In the open ocean, tidal currents are not restricted by the coastline, and they exhibit rotary patterns quite different from the reversing currents observed in coastal areas. Open-ocean tidal currents continually change direction. Instead of slack-water periods with no current, such as we find in coastal areas, there are periods in which the current is at a minimum. In the Northern Hemisphere the current usually changes direction in a clockwise sense. Tidal currents in the open ocean are usually weaker, typically about 30 centimeters per second (slightly more than 1 kilometer per hour), than in coastal waters.

A log anchored by an elastic tether near Nantucket Shoals Lightship during a simple semidaily tide would move in a clockwise sense in an elliptical path, returning to its initial point after 12 hours, 25 minutes, as indicated in Fig. 9-17. This again assumes the absence of wind or nontidal currents. If currents or winds are present, the log moves around the elliptical path at the same time that the whole water body is being moved.

Figure 9-17

Average tidal currents during one tidal period at Nantucket Shoals Lightship off the Massachusetts coast, The arrows indicate the direction and the speed (shown by the arrow length) of the tidal currents for each hour during a tidal period. The outer ellipse indicates the movements of a log (tied by an elastic line) in the water in the absence of wind or other currents. (After Marmer, 1926.)

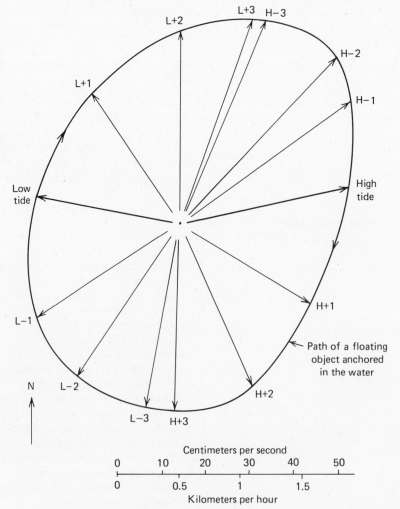

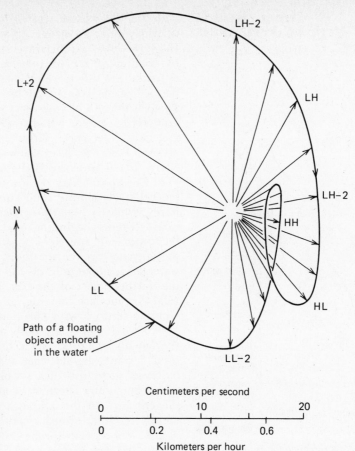

Centimeters per second

0	10	20
0	0.2 0.4 0.6	

Kilometers per hour

Figure 9-18

Average tidal currents during one tidal day at San Francisco Lightship. The radiating arrows indicate the direction and speed (shown by arrow length) of the tidal currents for each hour of a tidal day. The outer ellipse indicates the movements of an object in the water (tied by an elastic line) in the absence of wind or nontidal currents. (After Marmer, 1926.)

In areas with mixed tides, where there is a substantial inequality, the current ellipse is more complicated, as shown in Fig. 9-18, because there are two different ellipses. A log placed in such a current pattern moves through two elliptical paths and returns to its original position after 24 hours, 50 minutes, again assuming no wind or nontidal currents.

Since rotary tidal currents constantly change their direction and speed, we use vectors to indicate direction and current strength at hourly intervals (see Fig. 9-17). The pattern formed by the vectors is sometimes called a *current rose*. To understand the current rose, let us imagine that we are on a ship anchored at the center of the vectors in Fig. 9-17. We pump some dye overboard at a time corresponding to high tide at that location on a windless day with no other currents. The dye moves generally eastward in the direction shown by the arrow. One hour later (H + 1), we pump more dye overboard. It now moves generally southeastward. If we continue this each hour, we shall find that the dye patches have radiated out from our ship like spokes on a wheel. Each of these dye patches indicates the direction of the tidal current at the time of release, with the length of the patch being proportional to the current strength. After a tidal period (12 hours, 25 minutes) the pattern repeats itself. Similar results would be obtained in an offshore area with mixed tides.

A log floating in the water near our ship would also be moved by tidal currents. Its movements, however, would follow a pattern resulting from a combination of all the instantaneous currents in that region, including local wind-driven currents and major geostrophic currents.

Tides in the coastal ocean affect estuaries and their tributaries. The tide may enter an estuary as a progressive wave and travel upstream. In many estuaries, the wave is gradually damped (reduced in height) by friction of the channel and opposing river flow. But in some estuaries, the tide wave reaches the end of the estuary and is reflected back. The net result is a standing-wave type of tide, where the tide range at the end of an estuary exceeds that close to its entrance. In the Hudson River, for instance (Fig. 9-19), the tidal range is about 1.1 meters at New York Harbor, decreasing to slightly less than 1 meter at West Point. At Troy, New York (the uppermost extremity of tidal effect), the tidal range is slightly greater than 1.1 meters.

Chesapeake Bay on the mid-Atlantic coast of the United States has relatively simple tidal currents associated with a tide which behaves primarily like a progressive wave. We shall follow the changing tidal currents beginning with the time when waters are slack at the entrance (Cape Henry), as shown in Fig. 9-20(a). From just inside the entrance, in the southern part of the bay, up to a second slack-water area, about midway up the bay, tidal currents are ebbing. In the northern part of the bay, beyond the second area of slack water, tidal currents are flooding, and we can see that areas of slack water separate sections with ebb and flood currents.

Two hours later, both areas of slack water have moved into the bay about 120 kilometers, as shown in Fig. 9-20(b). By this time, tidal currents are flooding at the bay entrance and the pattern we observed previously has simply been displaced northward.

Four hours later, the southernmost slack-water area has reached the entrance to the Potomac River, and flood-tidal currents at the mouth have diminished somewhat in strength, as shown in Fig. 9-20(c). A final look at Chesapeake Bay 6 hours after slack water before flood, shows that the area at the bay mouth is again experiencing slack water, but this time it is slack water before ebb. The arrows in the initial picture are now essentially reversed, flood currents being substituted for ebb currents.

Because of its narrowness, there is little tidal flow across the axis of Chesapeake Bay. The reversing tidal currents are therefore relatively uncomplicated, flowing generally along the axis of the bay. In a wider channel, the current pattern would likely be complicated by currents flowing across the channel.

Tidal currents in Long Island Sound do not exhibit the features of a progressive wave as in Chesapeake Bay. Instead, the tidal currents are nearly the same over most of the sound at any given time. For instance, slack water occurs nearly simultaneously over the entire sound, and flood or ebb currents do likewise, although current strength varies between areas as shown in Fig. 9-21. The area near the western end of the sound is an exception. An area of slack water separates the small area of opposing tidal currents which have advanced into Long Island Sound from the tidal system in New York Harbor, which connects with the sound through several narrow channels.

The situation where simultaneous slack water is followed by ebb or flood currents throughout the water body is typical of a standing-wave type of tide. This type of tide in Long Island Sound results in part from the natural period of the sound, which is apparently about 6 hours.

For our final example, let us consider tidal currents in the North Sea, illustrated in Fig. 9-22, an area whose tidal currents are relatively well known and rather complicated. This basin is wide enough to permit currents across as well as along the basin axis.

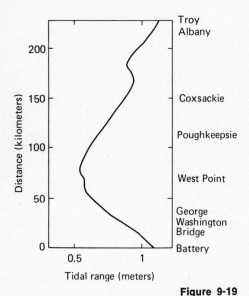

Figure 9-19

Changes in tidal range on the Hudson River.

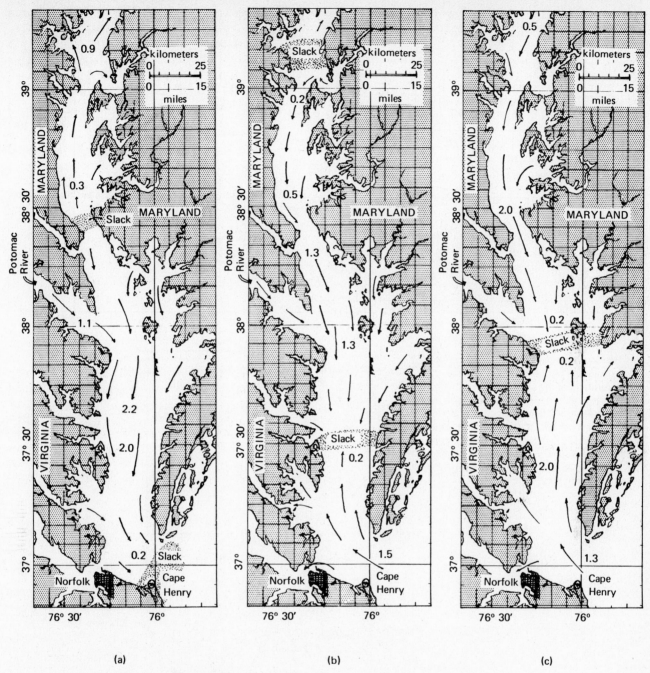

Figure 9-20

Tidal currents in Chesapeake Bay: (a) slack water before flood begins at Cape Henry; (b) 2 hours after flood begins at Cape Henry; and (c) 4 hours after flood begins at Cape Henry. Arrows indicate current direction and the numbers give maximum current velocities in kilometers per hour during spring tides. (After U.S. Naval Oceanographic Office, 1968. Publ. 700, Sect. I.)

The tide enters the North Sea through the Strait of Dover at the south end and from the large opening at the north. In the English Channel–Strait of Dover, the tide behaves essentially as a progressive wave, but in the North Sea the tide is an amphidromic system. The area of slack water nearest England corresponds to the center of the amphidromic system. Some of the currents parallel the coastline (for example, the coast of Belgium and the Netherlands), as we have discussed in the simple reversing currents; other tidal currents are directed primarily in toward the coast (The Wash, on the east coast of England) or directly off the coast (North Germany–Denmark). (Compare this with the simple rotary tide discussed earlier, in which there is no slack water, only tidal currents whose strength and direction change constantly.)

249

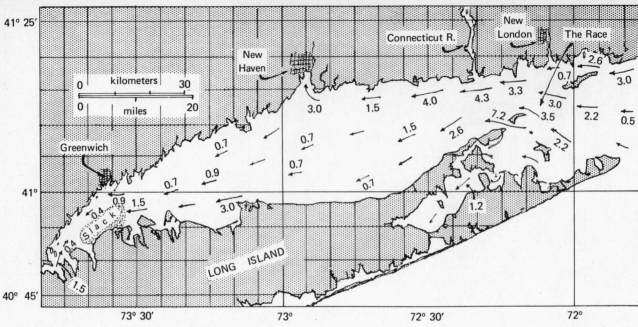

Figure 9-21 (above)

Tidal currents in Long Island Sound after the slack before flood at the Race, the eastern end of the sound. Arrows indicate current direction; the numbers give the current speed in kilometers per hour during spring tides. (After U.S. Coast and Geodetic Survey, 1958. *Tidal Current Charts: Long Island Sound and Block Island Sound,* 4th ed., Serial 574, Washington, D.C.)

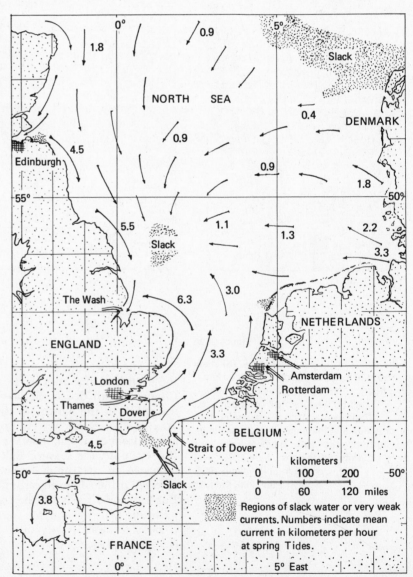

Figure 9-22 (right)

Tidal currents in the North Sea. Arrows indicate current direction; numbers give the current speed in kilometers per hour during spring tides. (After U.S. Naval Oceanographic Office, 1968. Publ. 700, Sect. I.)

Each of these three areas discussed above exhibits different aspects of tidal phenomena and illustrates some aspects of the tidal currents associated with them. Collectively, they show some of the complexities resulting from interactions between tide-generating forces and ocean basins.

Tidal currents are affected by many nonperiodic processes in the coastal ocean. In large estuaries, river flow modifies tidal currents, altering their timing so that current prediction is more difficult than predictions of tidal height. Both tides and tidal currents are affected by winds. Times and heights of tides in bays and coastal areas are commonly changed by storms. Hence tide tables often fail to predict the tides as they may be observed under a variety of unusual conditions.

Reversing tidal currents can control the timing of river water discharge (Fig. 9-23). As sea level rises with a flooding tide, denser seawater flows under the low-salinity river water. For a time, surface and subsurface currents are in opposite directions; flow is seaward at the surface and landward near the bottom. Because of the rising sea

Figure 9-23

Idealized representation of a tide rip. On a rising tide (A), seawater flows into the estuary under the outward-flowing river water, effectively preventing seaward river flow. Both the subsurface seawater and the less-dense river water above it move upstream, toward the head of the estuary. At slack high water (B), river water flows seaward above the subsurface layer. On the ebb tide (C), river water and seawater flow downstream, so that most river water enters the ocean on the ebbing tide. (After J. P. Tully and A. J. Dodimead, 1957. "Properties of the Water in the Strait of Georgia, British Columbia, and Influencing Factors," *Journal of the Fisheries Research Board of Canada*, 14(3), 241–319.)

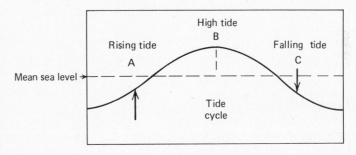

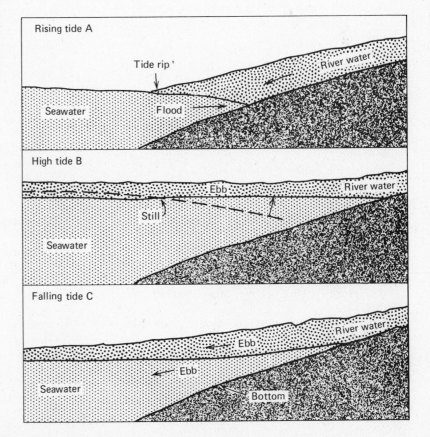

level, flow in the upper layer diminishes until it finally equals the speed of the subsurface current flowing in the opposite direction.

A convergence forms (known as a *tide rip*) marking the point where the opposing flows are equal. Note (in Fig. 9-23) that the boundary is inclined. Little or no river water flows past the convergence, since the landward flow is exactly equal to the seaward flow in the surface layer. Rips are easily recognized by the floating debris that collects there.

As the tide continues to rise, surface flow is eventually reversed, and at this time both surface and subsurface waters flow upstream. The denser, subsurface seawater mass and the rip move up the estuary. For a time, both fresh water and seawater accumulate in the estuary, and there is no river discharge to the coastal ocean.

Approaching high tide, but before high water is attained, the flood current diminishes to a point where it is weaker than the river flow. At this time, subsurface water is still flowing landward, although the rip is being pushed seaward at the surface by river water. Finally, there is no longer any opposing flow at the surface, and seaward flow accelerates as the accumulated water is discharged from the estuary. In the coastal ocean, the momentum of the river flow is quickly dissipated and the newly discharged waters are moved by currents and winds away from the mouth of the estuary. Tidally-timed discharge of river water masses into the coastal ocean is discussed further in the next chapter.

SUMMARY OUTLINE

Tides—periodic rise and fall of the ocean surface

Tidal curves—record of changing sea level
    Tidal day—time between successive transits of the moon over a given point, 24 hours 50 minutes
    Tidal period—time between successive high (or low) tides
    Types of tides:
        Semidaily tide—2 nearly equal high and 2 nearly equal low tides per tidal day
        Daily tide—1 high and 1 low per tidal day
        Mixed tide—2 unequal high tides and 2 unequal low tides per tidal day
    Tidal range—vertical distance between high and low tide
    Spring tide—greatest tidal range
    Neap tide—least tidal range

Tide-generating forces and the equilibrium tide
    Equilibrium tide would apply to static ocean completely covering a smooth earth; only the tide-generating forces are considered
    Unbalanced forces (gravitational attraction of the moon and centrifugal force from earth–moon revolution) acting on the earth's surface cause the tides
    Vertical and horizontal components, of which the horizontal is the stronger
    Equilibrium theory explains daily and fortnightly inequalities
    Variable position of the sun, moon, and earth causes spring and neap tides

Dynamical theory of the tide
    Includes ocean's response to tide-generating forces
    Tidal constituents may be considered separately, as responsible for partial tides
    Partial tides combined for tidal predictions
    Tides as long-period waves, in shallow water; length is one-half earth's circumference
    Natural period of each basin favors a standing wave; forms amphidrome system affected by Coriolis effect.
    Factors controlling tides in nearly enclosed basins:
        Range of tide at ocean entrance
        Natural period of basin
        Size of ocean entrance

Tidal currents—horizontal motions with little vertical component
    Comparable to water movements associated with progressive waves
    Reversing tidal currents near coasts
        Flood current—water moves landward
        Ebb current—water moves seaward
        Slack water—little or no current
    Complex relationship between tides and tidal currents
    Open-ocean rotary currents flow continuously, changing direction
        Particles move in elliptical paths
        Current rose—formed by vectors indicating direction and speed

Tidal effects in coastal areas
    Chesapeake Bay tidal currents behave like a progressive wave
    Long Island Sound has a standing-wave type of tide
    North Sea has an amphidromic system set up by tidal currents
    Tidal currents may be modified by river flow, winds, storms

River discharge controlled by tides
    Minimum discharge on rising tide, flow reversed on flood tide
    Maximum river discharge on falling tide, ebb-tidal current
    Tide rip forms at point where water masses meet

SELECTED REFERENCES

DARWIN, G. H. 1962. *The Tides and Kindred Phenomena in the Solar System*. W. H. Freeman, San Francisco. 378 pp. Originally published in 1898; classic, exhaustive treatment of the subject; elementary to intermediate.

DEFANT, ALBERT. 1958. *Ebb and Flow*. University of Michigan Press, Ann Arbor. 121 pp. Descriptive and mathematical; elementary to intermediate.

RUSSELL, R. C. H., AND D. H. MACMILLAN. 1954. *Waves and Tides*. Hutchinson, London. 348 pp. Elementary.

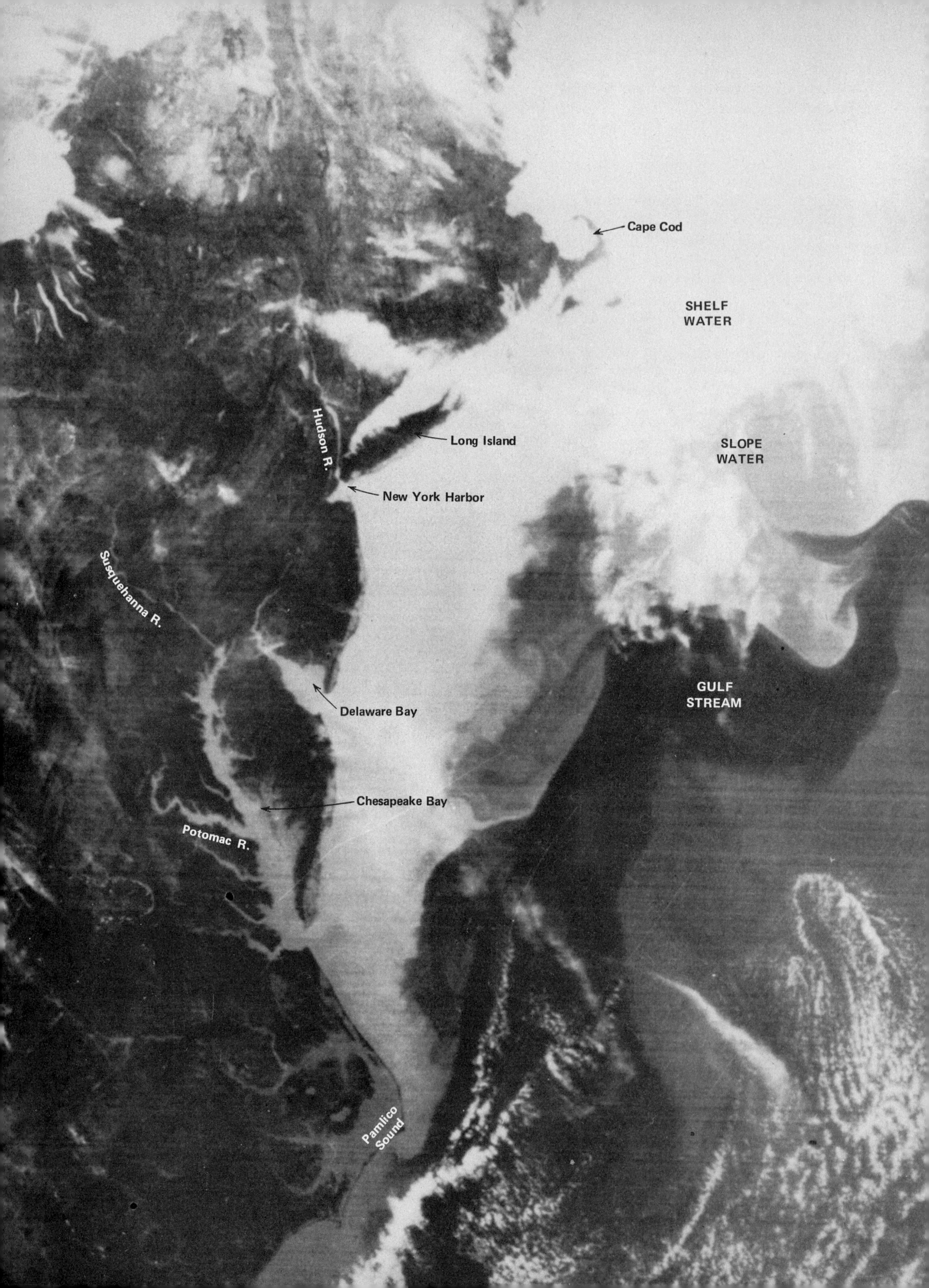

Cape Cod

SHELF
WATER

SLOPE
WATER

Hudson R.

Long Island

New York Harbor

Susquehanna R.

GULF
STREAM

Delaware Bay

Chesapeake Bay

Potomac R.

Pamlico
Sound

Satellite view of the coastal ocean off eastern North America. Note that a band of cool (light-colored) water separates warmer Gulf Stream waters from the coast in this infrared photograph. (Photograph courtesy National Environmental Satellite Service.)

estuaries and coastal processes

In the open ocean where depths typically range from 4 to 6 kilometers, surface processes are bounded only by the air–water interface and the current-deflecting continents. The coastal ocean, however, is only about 70 meters deep on the average. Its properties are usually confined to continental shelf waters, but there are exceptions; for instance where shelves are narrow on the United States West Coast, boundary currents define the limits of the coastal ocean. Its properties thus may extend seaward of the continental shelf break, with a portion of the continental slope serving as a floor.

TABLE 10-1

Environmental Aspects of U.S. Coastal-Oceanic Regions *

COASTAL REGION	SHORELINE	CONTINENTAL SHELF	BOUNDARY CURRENT
Gulf of Maine	Rocky, embayed by drowned river valleys; fjords	Rocky, irregular, 250–400 km wide	Labrador
Middle Atlantic	Smooth coast, embayed by drowned river valleys and lagoons	Smoothly sloping, 90–150 km wide	Labrador
South Atlantic	Smooth, low-lying, marshy; many lagoons	Smoothly sloping, 50–100 km wide	Gulf Stream
Caribbean	Irregular; mangrove swamps, coral reefs	East side 5–15 km wide; west side 300 km wide	Gulf Stream
Gulf of Mexico	Smooth, low-lying; barrier islands, marshes, lagoons	Smoothly sloping, 100–240 km wide	Seasonally variable
Pacific Southwest	Mountainous, few embayments	Irregular, rugged, average 15 km wide	California
Pacific Northwest	Mountainous, few embayments, some fjords	Irregular (Oregon) to smoothly sloping (Washington), 15–60 km wide	California
Southeast Alaska	Mountainous, many embayments and fjords	Rocky and irregular in places, 50–250 km wide	Alaska

* Data from U.S. Dept. of Interior, 1970. National Estuarine Pollution Study.

TABLE 10-2

Oceanographic Aspects of U.S. Coastal-Oceanic Regions *

COASTAL REGION	TEMPERATURE average (range) (°C)	SALINITY average (range) ($^0/_{00}$)	TIDE †	RIVER DISCHARGE Water (km³/yr)	Sediment (10⁶ tons/yr)
Gulf of Maine	8 (18–0)	30.3 (31–29)	SD	64	?
Middle Atlantic	12 (23–2)	31.5 (31.8–31.3)	SD	95	15.3
South Atlantic	24 (30–10)	35.5 (36.0–32.5)	SD	137	58.1
Caribbean	27 (30–22)	35.7 (36.3–34.9)	SD	10	?
Gulf of Mexico	23 (30–12)	32.3 (37.0–30.3)	D	712	360
Pacific Southwest	15 (20–13)	33.6 (33.8–33.3)	M	21	24
Pacific Northwest	10 (13–7)	30.8 (32.5–28.5)	M	113	127
Southeast Alaska	5(13 to −1)	31.9 (32.1–31.2)	M	?	?

* Data from U.S. Dept. of Interior, 1970. National Estuarine Pollution Study.

† SD—semidiurnal tide

  M—mixed tide

  D—diurnal tide

The shallow bottom and nearby shorelines affect oceanic processes so markedly that coastal areas act as separate oceanographic entities (Table 10-1). Like marginal seas (Chapter 3), coastal ocean regions are partially isolated by shoreline features. In the U.S., for instance, Cape Cod separates the Gulf of Maine region from the Mid-Atlantic region, and Cape Hatteras separates the Mid-Atlantic and South Atlantic regions (Fig. 10-1). Characteristic oceanographic properties generally prevail in each region (Table 10-2).

In this chapter we reexamine wind and temperature effects, evaporation, and dilution, this time in the context of the coastal ocean.

Figure 10-1

Major North American coastal-oceanic regions and estuarine systems.

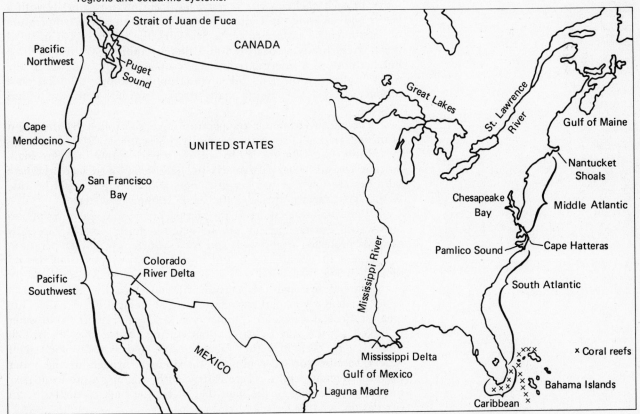

Large temperature and salinity changes occur in shallow, partially isolated coastal waters, a degree of variability well beyond anything observed in the open ocean. This is in part due to the proximity of the land.

Winds blowing over the continent toward the ocean, for instance, affect both water temperatures and salinities to a marked degree. On the U.S. Atlantic coast, winds coming from the continent are much hotter than the ocean in summer and much colder in winter, and they exert substantial influence on coastal water temperatures.

Winds from the continent are also dry, having lost much of their moisture through precipitation; therefore they cause extensive evaporation when they blow across coastal waters. In contrast, winds blowing across the ocean toward the coast of a continent, as in the northwestern United States during winter months, have less effect on the coastal ocean.

Salinity extremes occur in coastal waters, usually in isolated embayments where mixing with a larger body of water is severely restricted. In the western Mediterranean and northern Red Sea, for example, evaporation is high and surface water salinities exceed the average for their latitude by more than four parts per thousand (see Fig. 6-14). As we know, there is evaporation (Fig. 6-12) from all ocean surfaces. But along coasts, excess fresh water is discharged from continents to be mixed back into the ocean. For this reason, the lowest ocean salinities also occur in coastal waters, near major rivers such as the Amazon or Congo. Isohalines tend to parallel ocean boundaries, reflecting the fact that the greatest net evaporation occurs near centers of oceans.

The ocean's highest and lowest surface temperatures occur in coastal waters. Surface-water temperatures exceed 40°C in the Persian Gulf and in the Red Sea during the summer, whereas the surface temperature of the open ocean rarely (if ever) exceeds 30°C. Again, water in such enclosed seas is restricted from mixing laterally with a larger body of water, in contrast to the open ocean where temperature is distributed by surface currents. Secondly, the limited fetch in small ocean areas limits wave action and thereby inhibits vertical mixing. Furthermore, the intertidal zone heats up markedly and warms the water brought in on flood tides. Water ebbing from a tidal flat on a summer's day is commonly many degrees warmer than the waters offshore.

Lowest surface-water temperatures are controlled by the point of initial freezing of seawater; generally this is about −2°C, depending on salinity (see Chapter 5). In winter, sea ice forms in many high-latitude coastal areas, particularly in shallow bays and lagoons where salinities are low due to river discharge, and cooling is rapid because of the large surface area and the small volume of water.

Time scales are short in the coastal ocean, and warming of surface waters during the day can be detected fairly readily, especially in areas protected from winds and waves. Surface water temperatures are highest in midafternoon and lowest at dawn; variations are indicated in Fig. 10-2.

Marked seasonal temperature differences in the coastal ocean occur particularly at mid-latitudes, where air temperature changes are greatest. Coastal waters are commonly mixed by storms, and cold water is mixed all the way to the bottom during winter (Fig. 10-3). Offshore waters, even in winter, never get as cold as continental shelf waters because surface cooling is distributed throughout the water column, to great depths. There is frequently a temperature front near the edge of the edge of the shelf that marks the limit of coastal ocean circulation.

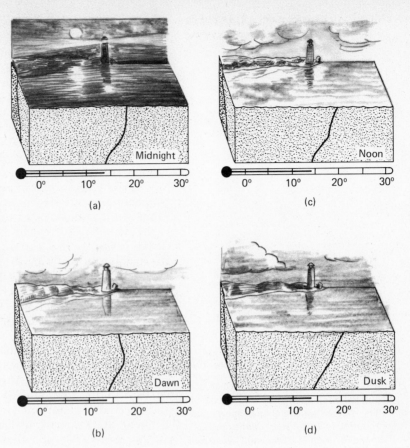

In spring, when surface waters are warmed, the depth of the warmed layer depends on mixing. As the summer progresses, the warmed, mixed layer deepens and a pronounced thermocline develops. During autumn, surface layers cool by radiation, evaporation, and continued mixing with deeper waters, caused primarily by storms. By the time winter begins, coastal waters are thoroughly mixed, with virtually no temperature or salinity differences between surface and bottom waters.

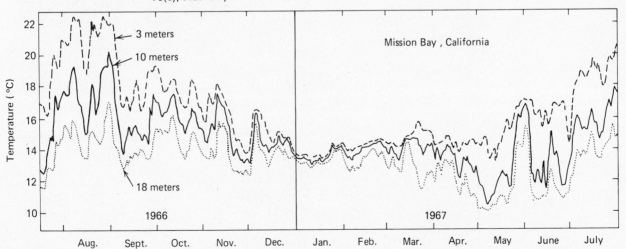

Coastal currents are horizontal water movements that generally parallel the shoreline. They are primarily geostrophic in origin, driven by winds (especially storms) and by river discharge. Such currents may form, disappear, or change flow direction within a matter of hours or days. During periods of light winds or reduced river flow they often become so weak as to be barely discernable.

Coastal currents are strongest when runoff is large and winds are strong. Along the West Coast of the United States, for example, surface waters are blown shoreward and held there by continuing winds, depressing the pycnocline. A sloping sea surface results, creating geostrophic currents which parallel the coast and are quite variable.

Coastal geostrophic circulation is bounded to seaward by the boundary currents that move water around the major ocean basins. Along United States coasts, currents often set in the opposite direction to the boundary current offshore. When winds diminish, or river discharge is low, coastal currents weaken or disappear. Such conditions are common in late summer, when coastal waters tend to form large, sluggish eddies, rather than distinct coastal currents. With the onset of winter storms, these eddies disappear, the surface layers are mixed, and the coastal circulation is reestablished within a day or two.

Water discharged by rivers or derived from coastal precipitation eventually moves across coastal currents to mix into the open ocean. Since currents tend to parallel the coast, transfer of fresh water across the system can be very slow. We say that the *residence time* is long. For example, within the Middle Atlantic Bight between Nantucket Shoals off Cape Cod, and Cape Hatteras, North Carolina, the fresh water necessary to account for the lowered water salinity is equivalent to about $2\frac{1}{2}$ years of discharge from the region's rivers. (By comparison, in the Bay of Fundy it takes about 3 months for the fresh water to be replaced, and about 3 to 4 months in Delaware Bay).

Figure 10-4

Schematic representation of surface and subsurface circulation in an area of coastal upwelling. Nearly horizontal lines show equal density surfaces. (After H. U. Sverdrup, 1938. "On the Process of Upwelling," *Journal of Marine Research*, 1(2), 155–64; and T. J. Hart and R. J. Currie, 1960. "The Benguela Current," *Discovery Reports*, 31, 123–298.)

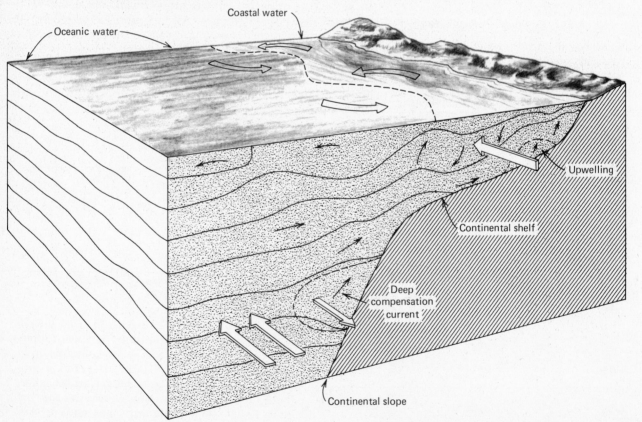

Coastal water

Oceanic water

Upwelling

Continental shelf

Deep compensation current

Continental slope

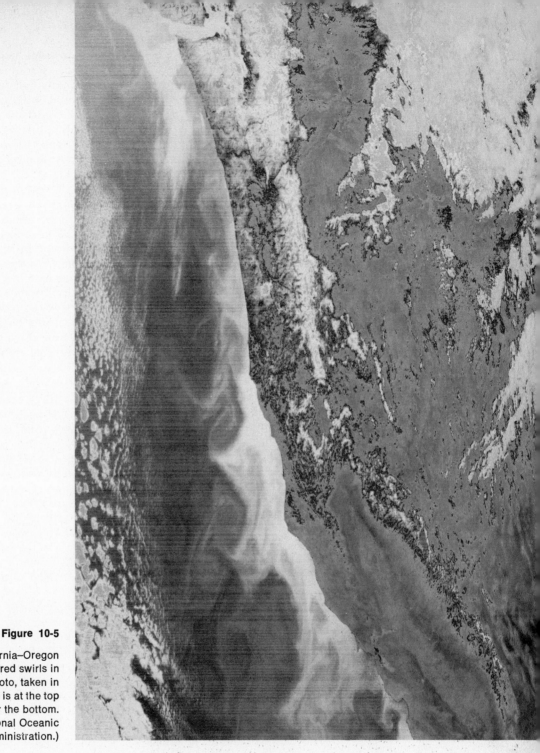

Figure 10-5

Upwelled water off the California–Oregon coast shows up as light-colored swirls in this infrared satellite photo, taken in September, 1974. Puget Sound is at the top and San Francisco Bay is near the bottom. (Photograph courtesy National Oceanic and Atmospheric Administration.)

When winds move surface waters away from the coast, subsurface waters move upward to replace them, as seen in Fig. 10-4. This process, called *upwelling,* is common along the western sides of the continents, in the eastern boundary-current areas. Coastal upwelling occurs along the California–Oregon coast, the western coast of South America, western Africa, northwestern Africa (Morocco), and the western coast of India.

During upwelling, deeper, much colder waters are brought to the surface, usually from depths of 200 meters or less. They form cold surface or near-surface water masses near the coast and cause significant cooling of adjacent land areas (Fig. 10-5).

This phenomenon is particularly important for the biological productivity of coastal regions. Phosphate and nitrate ions, depleted in sunlit surface waters by phytoplankton growth, move vertically during upwelling at a rate of about 2 meters per day. This resupplies food for active plant growth, which in turn feeds small animals and fish. Areas of pronounced upwelling have photosynthetic-carbon production rates from 2 to 10 times that of average open ocean waters. In fact, a map of upwelling regions nicely outlines the most biologically productive ocean areas (see Fig. 12-23) and the world's richest fishing grounds.

The cooling effects of upwelling are most noticeable along arid coasts in subtropical areas, such as Southern California. There, little rainfall or river runoff are present to form a stable, low-salinity surface layer with a pycnocline that would prevent denser waters from coming to the surface. But near small rivers, as along the Oregon coast, upwelling can occur when winds move the low-salinity surface layer seaward and the cold, upwelled waters underneath are exposed. Along the Washington State coast there is upward movement of subsurface water, but an extensive low-salinity layer formed by the discharge of many large rivers prevents it from reaching the surface. For this reason upwelling is not readily discernible along the Washington coast.

ORIGIN OF ESTUARIES, FJORDS, AND LAGOONS

We have already learned that deep-ocean basins were formed in the last 200 million years. Coastal oceans and estuaries, on the other hand, formed during the past 20,000 years as sea level rose from the continental shelf break to its present level following the recent Ice Age. The sea has been at its present level for only about 3000 years, and coastal ocean features are still adjusting to recent changes of sea level.

We can observe the progress of coastal evolution by comparing the recently glaciated and therefore relatively young Canadian and New England coasts with regions farther south. On far northeast and northwest U.S. coasts, glaciers removed the soil cover and carved deep *fjords* through which rivers now run to the ocean. The ancient continental margin was partly destroyed, leaving a narrow, irregular continental shelf. These regions have remained virtually unaltered for thousands of years.

In the Mid-Atlantic region, where glaciers had less effect on the topography, there are *coastal-plain estuaries* where the drowned mouths of ancient rivers are now below sea level. Southern coastlines tend to have a wide coastal plain, from which an abundance of sediment is discharged to estuaries and the nearshore ocean. Shallow, lagoon-type, bar-built estuaries and deltas and smooth, heavily sedimented coastlines are the rule, both above sea level and on the continental shelf. In the Far West, where tectonic processes have been active within recent times, coastal plain estuaries or shallow bays (*tectonic estuaries*) are cut off from the ocean by young mountain ranges. Some characteristics of these four estuarine types are summarized in Table 10-3.

From the Strait of Juan de Fuca northward along the Pacific coast and along most of the Canadian coastline, the shoreline has been greatly modified by glacial action. This has created narrow, deep, steep-sided fjords and fjordlike estuaries like those in Fig. 10-6, perhaps the most spectacular of all estuarine systems. Fjords occur in other high-latitude coastal areas such as Norway and Tierra del Fuego in South America. Glaciers moving down mountain valleys carved gorges hundreds of meters deep, and often deposited boulders and clay at the mouths of the valleys. This glacial debris forms an underwater sill at the entrance to many fjords, restricting movement of bottom waters.

TABLE 10-3

Characteristics of Some North American Estuarine Systems

ESTUARINE SYSTEM	ESTUARY AREA (square kilometers)	ESTUARY VOLUME (cubic kilometers)	MEAN WATER DEPTH * (meters)	ANNUAL FRESH WATER DIS- CHARGE (cubic kilometers per year)	LAND AREA DRAINED (thousands of square kilometers)	MAJOR RIVERS
DROWNED RIVER VALLEYS						
Chesapeake Bay System (Maryland–Virginia)	11,000	67	6.1	65	110	Susquehanna
Potomac River	1,280	7.3	5.7	12	36	Potomac
James River	650	2.3	3.5	10	26	James
Raritan Bay (New York–New Jersey)	230	1.1	4.5	1.8	3.4	Raritan
New York Harbor	159	1.2	7.5	19.4	34.7	Hudson
Long Island Sound (New York–Connecticut)	3,180	62	19.4	21	40.7	Connecticut Housatonic
LAGOONS						
Pamlico–Albemarle Sound (North Carolina)	6,630	23.9	3.6	7.8	51	Neuse Pamlico
Laguna Madre (Texas)	158	1.1	0.9	—0.85‡	nd†	
FJORDLIKE						
Strait of Juan de Fuca (Washington–British Columbia)	4,370	490	112	nd	nd	
Puget Sound (Washington)	2,640	185	70	36.5	37.6	Skagit Snohomish
Strait of Georgia (British Columbia)	6,900	1,025	156	145	270	Fraser
TECTONIC						
San Francisco Bay (California)	1,190	6.2	5	40	161	Sacramento San Joaquin

* Mean depth = volume/area.
† nd = no data.
‡ Evaporation exceeds river runoff plus rainfall.

Figure 10-6

The rugged topography of a small inlet on Canon Fjord, Ellesmere Island, in the Canadian Arctic was cut by a glacier of which a remnant (the large, white mass) can be seen at the upper left. The icebergs in the water come from such glaciers. Note the small floes of sea ice in the foreground. (Photograph courtesy National Film Board of Canada.)

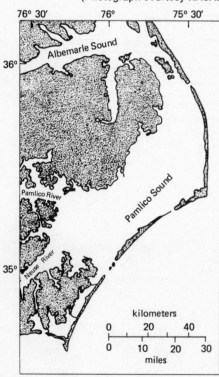

Modern coastal plain estuaries assumed their present size and
shape about 3000 years ago, when rapid melting of glaciers ceased and
the sea reached its present level. About 30,000 years earlier these rivers
entered the ocean through ancient estuaries that now lie tens of kilometers offshore and 100 or more meters below present sea level. Coastal
plain estuaries are typically broader than any part of the river upstream, and they gradually deepen seaward except where mud and
sand have been deposited. Some, such as the Hudson or Columbia
Rivers, have associated submarine canyons extending across the former
coastal plain, now the submerged continental shelf.

A *lagoon* is a broad, shallow estuarine system. Flow of water between estuary and coastal ocean is restricted by a barrier beach offshore, generally paralleling the shoreline. Lagoons are invariably
shallow, because barrier beaches form only in waters a few meters
deep, where wave action suspends sediment and transports it along the
coast. The barrier beaches enclosing a lagoon are interrupted by narrow
inlets through which tidal currents move water in and out of the
lagoon (Fig. 10-7).

On the United States Gulf Coast and on the Atlantic coast south
of New England, long stretches of barrier beach separate lagoons from
the coastal ocean. Often a single stretch of beach will isolate several
lagoons and estuaries. All parts of such a system communicate with
the coastal ocean through the same set of narrow inlets.

In New England and the Pacific Northwest, where the coastline
has been scoured by glaciers, lack of sediment moving along the coast
prevents formation of extensive bars or spits. On much of the Pacific
coast, the continental shelf is too narrow and steep to permit barrier
beaches to form.

When all inlets to a lagoon are closed by storms or changes in
wave or sediment conditions, the lagoon is transformed into a lake.

264

Fresh water flowing into such a lake from local rivers would raise its level until it overflowed its boundaries, if it were not for the fact that sandy beaches are permeable to water. Seepage through barrier beaches keeps lake levels fairly constant.

Estuaries and lagoons make up 80 to 90 percent of the Atlantic and Gulf coasts but only 10 to 20 percent of the Pacific coast. Mountain building around the Pacific has left little low-lying coastal plain, and few rivers have cut through the mountains to reach the sea. Furthermore, the deserts behind mountain ranges in this region yield little water; many areas do not discharge any water to the ocean. The large tectonic estuarine systems of San Francisco Bay and the Strait of Juan de Fuca (Fig. 10-1) formed when sections of the continent containing former river valleys sank below sea level because of active mountain building.

ESTUARINE CIRCULATION

Estuaries are semienclosed arms of the ocean, where fresh water from the land is mixed with seawater. Their characteristic circulation patterns are governed by the amount and rate of fresh water entering the estuary, size and shape of the basin, and the effects of winds and tides. In an ideal estuary, low-salinity water from the river flows in at the head of the bay and spreads out over the seawater beyond because it is less dense (Fig. 10-8). There is a more or less horizontal pycnocline zone with a marked density discontinuity separating the two layers.

Figure 10-8

Schematic representation of a simple salt-wedge estuary showing the two-layered structure with a landward flow in the salty subsurface layer and a seaward flow in the less saline surface layer. (Redrawn from D. W. Pritchard, 1955. "Estuarine Circulation Patterns," *American Society of Civil Engineers Proceedings,* 81(717), 11 pp.)

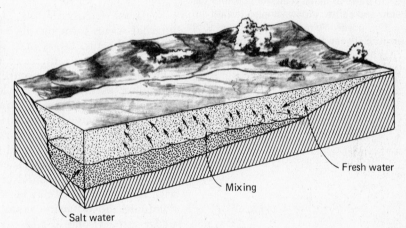

Fresh water

Mixing

Salt water

In fact, however, this oversimplified picture is modified by several factors. First, friction between seaward-moving fresh water and the underlying seawater causes currents as shown in Fig. 10-9. Water is entrained (dragged up from below) and mixed with the surface layer.

Figure 10-9

Variation in salinity and current speed with depth in a simple salt-wedge estuary.

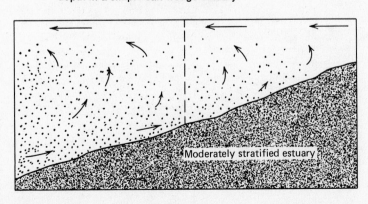

Moderately stratified estuary

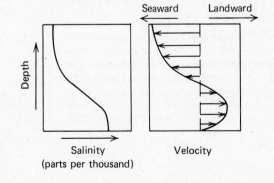

Depth

Salinity (parts per thousand)

Seaward   Landward

Velocity

This newly mixed water cannot re-enter the lower seawater layer because its reduced salinity makes it less dense. Consequently, ocean water flows along the bottom, into the estuary, to replace that drawn seaward by the surface flow. Each volume of fresh water ultimately mixes with several volumes of seawater, so that surface salinities increase in a seaward direction. Landward flow along the bottom is much greater in volume than the flow of river water into the estuary. The subsurface salt water in such an estuary often forms a wedge with its thin end pointed upstream; hence in its simplest form, this is called a *salt-wedge estuary.*

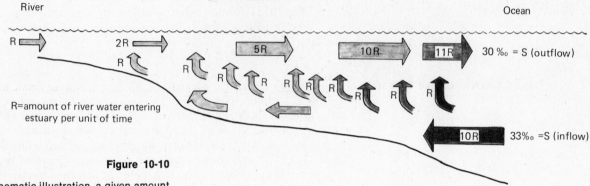

Figure 10-10

In this schematic illustration, a given amount of river water, designated as *R,* draws 10 times as much seawater upward from below and carries it downstream, toward the ocean. To replace the outflow, 10 *R* amount of seawater moves into the estuary. Thus the effect of estuarine circulation is to draw subsurface, high-salinity water into the surface layer. Note that this is essentially an upwelling process.

If the frictional drag between the two layers is great enough, internal waves form on the pycnocline. Such waves at the boundary between the two layers may actually break, like the surf seen at the beach. This greatly increases the rate of mixing between the layers. The more mixing that occurs in the estuary, the greater is the landward flow in the subsurface layer (see Fig. 10-10).

Simplified salt-wedge stratification such as we have described is only rarely observed in estuaries. Certain systems closely approximate it, however, especially where river flow is large and tidal range is low. Salt-wedge estuaries are found where a large river discharges through a relatively narrow channel, such as the lower passes of the Mississippi River or the Columbia River during flood stages. Forces associated with large river flows may be so much stronger than the tides that they dominate the estuarine circulation. This forms a nearly ideal salt-wedge estuary, one in which relatively little salt water is mixed into the upper layer. Nearly fresh water is discharged into the coastal ocean, where it mixes with ocean water over the continental shelf. The Amazon is a particularly striking example of such a river.

Figure 10-11

Schematic representation of water movements in a moderately stratified estuary. There is a net landward movement in the subsurface layers and a seaward flow in the surface layers, although the difference between the layers is not as pronounced as in the salt-wedge estuary. (After Pritchard, 1955.)

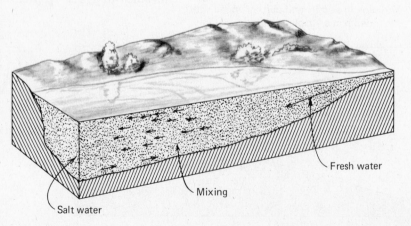

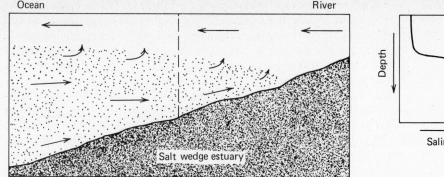

Figure 10-12

Variation in salinity and current speed
in a moderately stratified estuary.

As river flow decreases in an estuary, the importance of tidal effects increases. A single estuary may act like a salt-wedge estuary during floods, but during low-flow periods of the year it is usually influenced much more by tides and tidal currents, departing substantially from the pattern of a simple salt-wedge stratification. It then becomes a *moderately stratified estuary*.

Tidal currents cause mixing between layers, so that the waters become only moderately stratified, as Fig. 10-11 illustrates. Tidal flow causes *turbulence* (random movements) throughout the water column, and this in turn increases mixing so that more salt water is transferred from the subsurface to the surface layer. Some fresh water from the surface also mixes downward. Consequently, salinity decreases in a landward direction in both the surface and subsurface layers. As in the salt-wedge estuary, however, there is a net landward flow in the subsurface layer, replacing salt water lost from the system, and a net seaward flow in the surface layer removing both fresh water and salt water from the estuary, as Fig. 10-12 indicates.

Consider the volume of flow in these two layers. Remember that river water has almost no salt as it enters the head of the estuary. Assume that the seawater moving in along the bottom has a salinity of 35 parts per thousand. A mixture of equal volumes of fresh and salt water gives a salinity of 17.5 parts per thousand; mixing 2 volumes of salt water with 1 volume of river water results in a salinity of about 23.4 parts per thousand as shown by Fig. 10-13. By the time surface-

Figure 10-13

Relationship between salinity of water
masses formed in an estuary and the relative
amounts of seawater (salinity 35 parts per
thousand) and fresh water (salinity 0 parts
per thousand) involved in the final mixture.

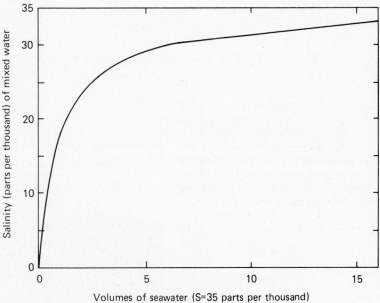

water salinity has reached 34 parts per thousand, 1 volume of river water has mixed with 39 volumes of seawater. In other words, at this point, landward flow in the subsurface layer is 39 times the volume of river water discharged at the head of the estuary. The volume of surface water moving seaward in the surface layer will be 40 times the original river discharge.

Because of their greater depth, fjord systems have more complicated circulation systems. Many fjords have a *sill,* or submerged ridge at the entrance deposited by the glacier that cut the fjord. This sill cuts off most of the deeper water from communication with the adjacent ocean, as can be seen in Fig. 10-14. Alberni Inlet on Vancouver Island, British Columbia, for example, has a sill which comes to within 40 meters of the surface. The basin behind the sill is 330 meters deep, so that nearly seven-eighths of its volume is partially isolated from communication with the ocean.

Fresh water flowing into the fjord forms a low-salinity surface layer which moves seaward, setting up a typical estuarine circulation. This layer, however, often extends no more than a few tens of meters below the surface and usually involves only the waters above the top of the sill.

The deeper waters of a fjord are almost completely isolated from the surface circulation, but are strongly affected by conditions at the fjord mouth, outside the estuary. Since river flow, tides, and winds have little effect on this deep water, its circulation is controlled primarily by density differences. If the water at or slightly above the sill outside the estuary becomes more dense, it can flow into the fjord and displace the deeper water there, which then moves out slowly. Strong winds can affect the deep waters by setting up seiches (standing waves), which cause the waters to "slosh" back and forth.

Tides dominate the circulation in most estuaries. Any current seen in an estuary is most likely of tidal origin. To observe estuarine circulation, it is necessary to make current measurements extending over many tidal cycles. During this period, water flows upstream on the flood current a distance nearly equal to its downstream travel on the ebb current. When averaged, tidal currents cancel one another, leaving a nontidal estuarine circulation. A current meter placed in the surface layers would indicate a net flow seaward, and one in the deeper layers a net flow landward.

Where tidal effects are relatively large, the estuary tends to be less stratified. Conversely, increased river flow causes a greater degree of stratification. In an estuary with small river flow but large tides and tidal currents, the waters may be mixed almost completely from top to bottom.

Figure 10-14

Schematic representation of circulation in a simple fjordlike estuary. Note the water mass formed by mixing surface and subsurface waters while flowing over the sill. (After M. Waldichuk, 1957. "Physical Oceanography of the Strait of Georgia, British Columbia," *Journal of the Fisheries Research Board of Canada,* 14(3), 321–86.)

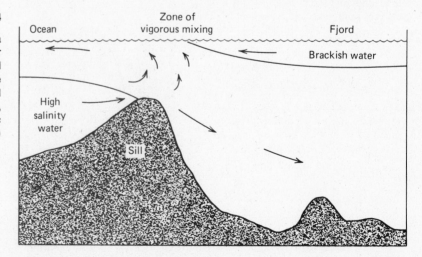

Figure 10-15

Small estuaries along the coast of South Carolina discharge turbid waters into the Atlantic. Note the small size of these sediment-filled systems, in contrast to the much larger Chesapeake Bay or New York Harbor, which remain unfilled by sediment. (Photograph courtesy NASA.)

The rise and fall of tides control the timing of river-water discharge into the coastal ocean. Most fresh water is discharged on the ebb tide and enters the ocean as a series of pulses. River discharge tends to form cloudlike parcels of low-salinity water, which, when turbid with suspended sediment, are easily recognizable from the air (Fig. 10-15). Each tends to overrun the previous one, forming a complex of irregular fronts (Fig. 10-16). As the cloudlike masses move away from the river mouth, waves and tidal currents cause mixing over the continental shelf. Boundaries are obscured, and eventually a large lens or *plume* of low-salinity water forms. Such a plume can be traced for many tens or hundreds of kilometers off the mouths of major rivers (Fig. 10-17), especially where there are no other large rivers discharging in the area.

Figure 10-16

Fresh water from the Quinault River discharges into the northeast Pacific Ocean at Taholah, Washington. Electronic sensors detect differences in water temperature which outline the cloudlike water masses of low-salinity colder water mixing with coastal-ocean waters, which appear dark because they are warmer. (Photograph courtesy NASA.)

Figure 10-17

Muddy water discharged by the Irrawaddy and nearby rivers moves through the Gulf of Martaban on the eastern side of the Bay of Bengal (Indian Ocean). This photograph was taken in November, 1966, from an altitude of 295 kilometers. (Photograph courtesy NASA.)

Having examined the basic processes acting in estuaries and coastal ocean regions, let us now consider some examples.

CHESAPEAKE BAY Chesapeake Bay is the classic example of a coastal-plain estuarine system. It formed from the drowned valleys of the Susquehanna River and its tributaries, including the Potomac, the James, the York, and the Rappahannock. In addition, dozens of smaller tributaries drain into arms, inlets, and bays on all sides, giving Chesapeake Bay the complicated, branching form seen in Fig. 10-18. The southernmost arm of the system, the James River, was probably a separate river flowing into the ocean through its own estuary before the area was flooded in the last few thousand years.

The seaward boundary of Chesapeake Bay is a line between Cape Charles and Cape Henry. The upper boundary is at the "fall line" on the Susquehanna River. Here, waterfalls or rapids in the river mark the boundary between soft, easily eroded coastal plain sediments and sedimentary rocks, and hard rock underlying the rolling hills of the Piedmont. The fall line marks the limit of navigation for ocean vessels, and the landward extent of tidal influence.

The total volume of fresh water discharged each year by rivers flowing into Chesapeake Bay approximately equals the total volume of the estuary. The Susquehanna River supplies 49 percent of the annual fresh-water discharge, the Potomac River contributes 18 percent, and the James River about 15 percent. Nearly all fresh water is discharged into the bay along its western margin, where rivers drain the slopes of the Appalachian Mountains. Very little fresh water comes from the small rivers and swampy areas along the eastern shore of the bay.

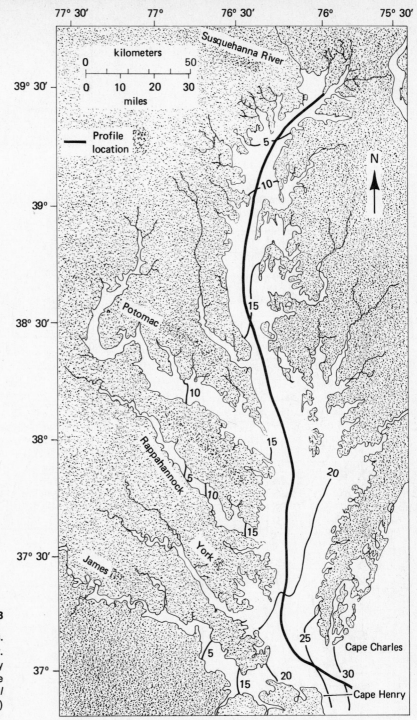

Figure 10-18

Chesapeake Bay and its tributary estuaries. Contours show surface salinity in summer. (After D. W. Pritchard, 1952. "Salinity Distribution and Circulation in the Chesapeake Bay Estuarine System," *Journal of Marine Research,* 11(2), 106–123.)

Tidal currents in Chesapeake Bay are mainly the reversing type common to coastal waters. Depending upon location, maximum current speeds may be as low as 25 centimeters per second or as high as 50 centimeters. As in all estuarine systems, net flow at the surface is seaward, resulting in ebb currents that begin sooner and persist longer at the surface. Inflow at the bottom is reflected in the early initiation of the flood tides and their greater persistence.

Salinities are lowest during spring (February through April) when high river flow prevails throughout the region. In summer and fall, low river discharge results in maximum salinities. Because of the large

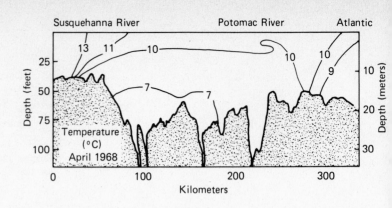

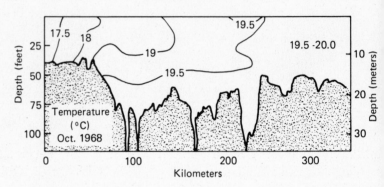

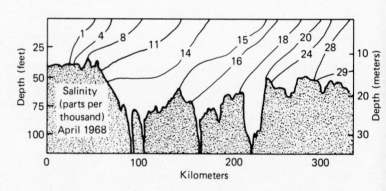

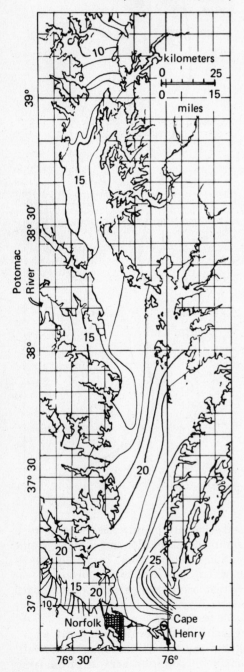

Figure 10-19 (below)

Typical distribution of surface salinity in Chesapeake Bay in winter. (After Pritchard, 1952.)

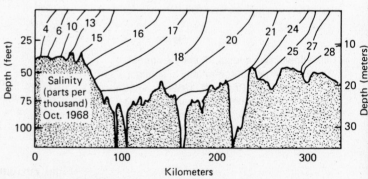

Figure 10-20 (above)

Temperature and salinity distribution in Chesapeake Bay, winter and summer. Profile location is given in Fig. 10-18; vertical scale is highly exaggerated. Note that salinity varies with depth, surface waters being less saline. Temperature in the relatively shallow bay is controlled primarily by surface weather conditions. Steam-powered generating plants use waters for cooling; when returned to the Bay they may warm local areas by as much as 10°C. (Data from Chesapeake Bay Institute, The Johns Hopkins University.)

discharge of fresh water along the western shore and the deflecting effects of the Coriolis effect, surface-water salinity is generally greater on the eastern side of Chesapeake Bay than along the western side (see Fig. 10-19). Estuarine currents contribute to this effect by moving surface water parallel to the eastern shore. In any given area, there are daily variations which are related to the tidal current and to local river discharge.

Salinity variation with depth is typical of estuarine systems (see Fig. 10-10). Near-bottom salinities are highest at the mouth and decrease toward the head of the bay due to mixing, at all seasons. Salinities at the surface also increase seaward. Temperature does not show as clearcut a distribution as does salinity (Fig. 10-20). The system is relatively shallow, so that water temperatures are controlled primarily by local weather conditions. In winter, water temperature generally increases toward the mouth of the bay, as a result of warmer oceanic inflow (Fig. 10-21). Lower-salinity waters at the bay's head often freeze during January and February, with ice forming first in shallow, protected bays and inlets, then gradually extending toward open water. In especially cold winters, ice covers large areas from shore to shore in the upper estuary.

Figure 10-21

Schematic representation of changes in temperature relationships between surface and bottom waters of Chesapeake Bay. (After R. C. Seitz, 1971. *Temperature and Salinity Distribution in Vertical Sections along the Longitudinal Axis and Across the Entrance of the Chesapeake Bay.* Chesapeake Bay Institute, The Johns Hopkins University, Baltimore, Md., p. 25.)

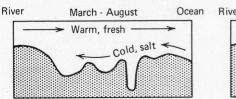 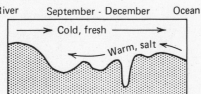 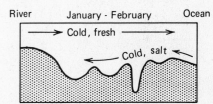

Strong winds have a pronounced effect on the bay. In addition to the waves they generate, storm winds can also change sea level locally, sometimes causing flooding of low-lying coastal areas. In general, winds only modify currents in Chesapeake Bay; currents set up by river discharge and tides are usually dominant influences.

LAGUNA MADRE

In Chesapeake Bay, local river runoff and precipitation exceed evaporation; consequently, salinities are lower than those found in the open ocean. One of the few *hypersaline* (above-average salinity) *lagoons* in the United States is Laguna Madre, a series of long, narrow lagoons on the coastal plain of the Texas Gulf Coast, between the Rio Grande and Corpus Christi (Fig. 10-22). A sandy barrier island (Padre Island) parallels the coast for approximately 160 kilometers and separates the lagoon from the open ocean. Circulation in this lagoon is quite different from the typical estuarine system.

The Laguna Madre barrier island complex was formed when sea level approached its present position, and it is subject to continuing natural changes. At each end of Padre Island, narrow inlets connect Laguna Madre with the ocean. Before these were jettied, they were frequently closed and reopened by storms and longshore currents.

A hurricane in 1919 formed large shallow areas, which now divide the lagoon into two smaller basins. The deeper basin, Baffin Bay at the northern end, averages about 1.6 meters. Much of Laguna Madre is extremely shallow (average depth 0.9 meter); many areas are covered by only a few centimeters of water at high tide.

Since the inlets at each end are quite narrow, there is relatively little tidal interchange with the open ocean. Also, this section of the Gulf of Mexico has a tidal range of less than 0.5 meter. Consequently, the tidal range in Laguna Madre is only a few centimeters. Wind

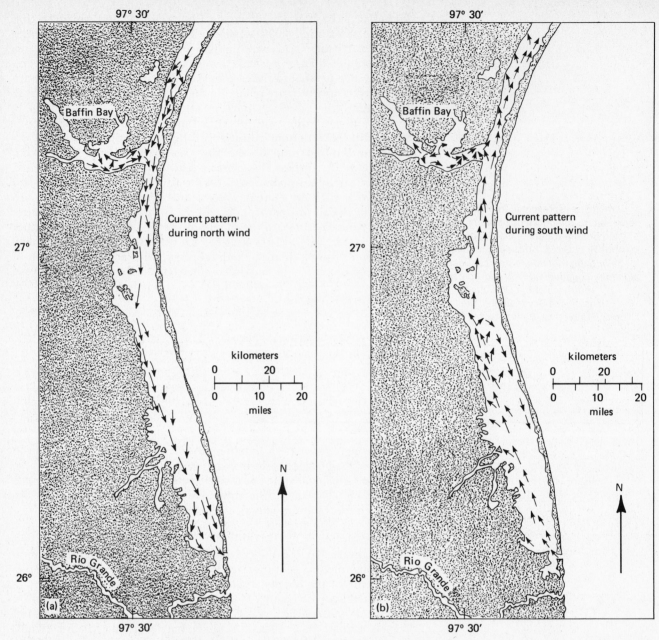

Figure 10-22

Postulated current patterns in Laguna Madre during north (a) and south (b) winds. (After G. A. Rusnak, 1960. "Sediments of Laguna Madre, Texas," in F. P. Shepard, et al. (eds.), *Recent Sediments, Northwest Gulf of Mexico*, American Association of Petroleum Geologists, Tulsa, Okla., pp. 82–108.)

effects are usually more important than the tide. Wind tides produce changes in water level, predicted to exceed 1 meter, based on observed wind velocities.

Winds also control circulation in Laguna Madre. Winds from the north cause currents to set southward, with flow out of the southern inlet. From the south, they cause essentially the opposite current direction, as indicated in Fig. 10-22. Note that the shallow areas in the center of Laguna Madre greatly influence currents set up by winds from the south. Back-and-forth seichlike water movements are sometimes set up in the lagoon by winds blowing steadily from a single direction. These act to stir up sediments, so that lagoon waters are often turbid.

Laguna Madre is located in a semiarid region where evaporation usually exceeds rainfall. Runoff comes from a few small streams or irrigation ditches. Infrequent but intense local storms may bring more than 10 centimeters of rain in a few hours, providing large amounts of fresh water. Between storms, salinity in the lagoon is usually substantially higher than in the adjacent ocean (Fig. 10-23), and may exceed 80 parts per thousand. Such erratic freshwater discharge results in extreme salinity variations. The highest salinities recorded are between 110 and 120 parts per thousand. Surface-water salinities as low as 2 parts per thousand have been recorded immediately following a cloudburst, but within a relatively short time salinity is increased by evaporation and mixing.

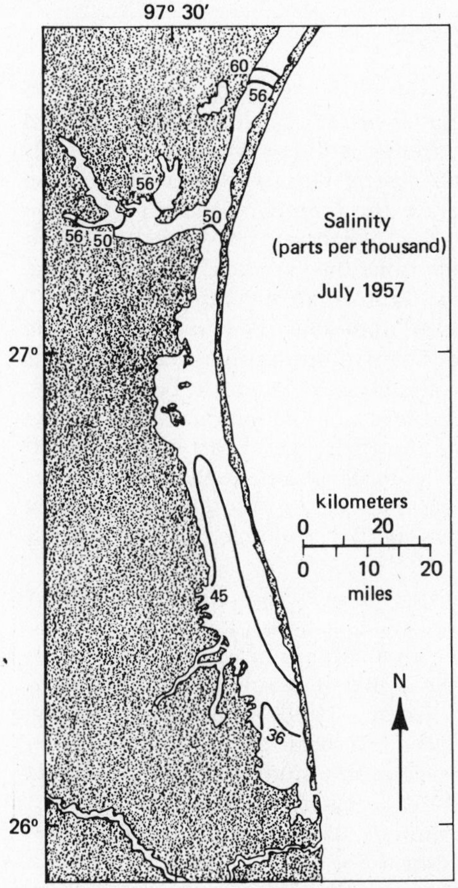

Surface-water temperatures are also extremely variable. In winter, cold air comes down from the north, chilling surface waters. Because of the extreme cooling and lack of escape channels for fish in shallow bays, there are mass mortalities among temperature-sensitive fish during unusually cold weather. Substantial fish mortalities ("fish kills") can also occur in summer at times when water temperatures and salinities become intolerable to some species. Dredging of ship channels and inlets has apparently reduced fish mortality by providing escape channels and shelter from the highly variable water conditions.

Vigorous wind mixing permits relatively little stratification within this shallow bay. Near inlets, mixing is caused by tidal currents. Excess evaporation sets up a circulation system opposite to the estuarine pattern, since the lowest-salinity water comes from the adjacent ocean and thus occurs at the surface. As it flows into the lagoon, ocean water evaporates and sinks with the increased density. Salty, denser subsurface waters flow out into the ocean.

Dredging of channels and construction of tidal inlets, jetties and other engineering works have modified water circulation and biology in Laguna Madre. The estuary was substantially altered in 1949 when dredging was completed on the Intra-Coastal Canal. This channel increased water exchange between the northern and southern portions of the lagoon, and prevented the near-drying-up of one or the other of these basins, as had happened before construction of the canal. A bridge and solid-fill causeway at the northern end of the lagoon has greatly restricted passage of water from one side to the other.

Laguna Madre is highly productive of fish. In the early 1940's, slightly more than half of the total fish production for Texas was taken there. In part, this is a consequence of the deterioration of such nearby areas as Galveston Bay, once an excellent fishing ground. There is also some evidence that the high salinity of the waters causes certain fishes to grow larger.

Bay waters receive wastes discharged by urbanized and industrialized areas, as well as runoff from agricultural land, which brings silt, pesticides, and excess fertilizer into the water. Declines in Laguna Madre fish production during the 1960's were ascribed to the cumulative effects of these waste discharges, especially pesticides.

Figure 10-23

Surface-water salinity in Laguna Madre, July, 1957. (After Rusnak, 1960.)

NORTHEAST PACIFIC COASTAL OCEAN

High rainfall and river runoff make the entire North Pacific Ocean an area of net dilution, with a margin of especially low-salinity water bordering its coasts (see Fig. 6-14). These conditions cause a basically estuarine circulation north of 40°N, with net flow of surface water toward the center of the basin.

Flowing nearly due east from Asia, the mid-latitude N. Pacific Current (Fig. 7-4) brings relatively cool water of 33 to 34 parts per thousand salinity (see Fig. 6-14) toward the North American continent. There it is deflected to form the eastern boundary currents—Alaska Current setting northward and California Current setting southward parallel to the continental margin. The point of divergence shifts seasonally; it is about 50°N in summer and about 43°N in winter.

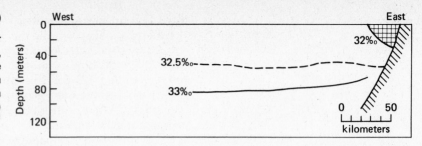

West   East

Figure 10-24 (right)

Distribution of salinity with depth near the mouth of the Columbia River, January–February, 1962. Note that the low-salinity waters form a wedge rather than a layer. Location of the profile is shown in Fig. 10-27. (After Duxbury et al., 1960.)

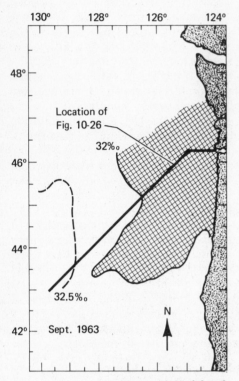

Figure 10-25 (above)

Distribution of surface salinity in September, 1963. Note the large area covered by surface water with salinity less than 32 parts per thousand, derived primarily from the Columbia River. (After T. F. Budinger, L. K. Coachman, and A. C. Barnes, 1964. *Columbia River Effluent in the Northeast Pacific Ocean, 1961, 1962; Selected Aspects of Physical Oceanography,* University of Washington, Dept. of Oceanography Tech. Rept. no. 99 (ref. M63-18), 78 pp.)

Since current speeds are low and flows rather diffuse, there is no sharp boundary separating the coastal-current system from the eastern boundary current, in contrast to the situation prevailing in the Gulf Stream system along the Atlantic coast. We must, therefore, choose some criterion by which to separate the boundary current from the coastal circulation. Along the Pacific Northwest coast of the United States, the 32.5 parts-per-thousand isohaline is taken to separate the low-salinity coastal waters from the California Current offshore.

During the rainy winter season, numerous rivers discharge to the coastal ocean. An almost continuous band of low-salinity water parallels the coast from central Oregon to the Strait of Juan de Fuca (see Fig. 10-1). In summer and autumn, however, the Columbia River is the largest source of fresh water along the entire west coast of the United States. Relatively little mixing occurs in its estuary so that these waters are still rather fresh as they enter the ocean. The effluent forms a distinctive water mass, the Columbia River plume, having salinity less than 32.5 parts per thousand.

Within 50 kilometers of the river mouth, the plume is mixed by tidal currents and waves; farther offshore, currents and regional winds move it parallel to the coast much as a column of smoke is moved in the air by gentle winds. During the winter it is held near the coast by southwesterly winds (Fig. 10-24), forming a band of northward-moving, low-salinity water derived from all the coastal rivers—the *Davidson Current.* This coastal current is recognizable from southern California to the Strait of Juan de Fuca at the U.S.–Canadian border.

From July until October, Columbia River discharge forms a low-salinity plume extending southwestward (Fig. 10-25) as a layer about 30 meters thick (Fig. 10-26). In October, 1961, for instance, it covered an area of 200,000 square kilometers, extending 800 kilometers southward. Following major summer floods, the plume may be even larger. (See Fig. 6-14, where it appears as a southward dip of the 33 parts-per-thousand contour in the east Pacific.) It persists until the first winter storm, usually in October, after which surface waters are rapidly mixed and the plume can be recognized only near the river mouth (Fig. 10-27).

Over the continental shelf, the landward flow of subhalocline waters (part of the estuarine-like circulation) produces near-bottom currents; these currents are studied using *seabed drifters*—plastic devices (usually brightly colored for ease in spotting) designed to float near the ocean bottom. Drifters are dropped from ships or low-flying aircraft, in groups held together by a "collar" made of rock salt, which dissolves after its weight has dragged the group of drifters to the

Figure 10-26 (right)

Distribution of salinity with depth near the Columbia River mouth, September, 1963. Location of profile is shown in Fig. 10-25. Waters with salinities less than 32 parts per thousand form a relatively thin layer (less than 30 meters thick).

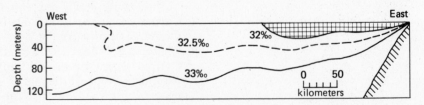

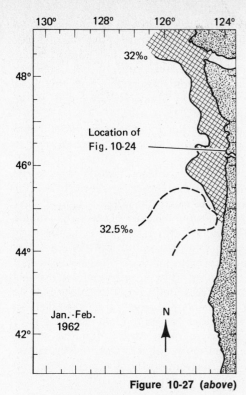

Location of
Fig. 10-24

Figure 10-27 (*above*)

Distribution of surface salinity in the Pacific
Northwest, January–February, 1962. Note
that low-salinity surface waters form a
narrow band along the coast. (After A. C.
Duxbury, B. A. Morse, and N. McGary, 1966.
*The Columbia River Effluent and its
Distribution at Sea, 1961–1963,* University
of Washington, Dept. of Oceanography
Tech. Rept. no. 156 (ref. M66-31), 105 pp.)

bottom. Position and time of release of each group is noted, as well as
the serial number of each drifter.

Drifters are recovered by beachcombers or by fishermen. For a
small reward, the finder sends the date and location of recovery with
the serial number of the drifter, to the laboratory that released it. The
apparent direction and distance traveled by the drifter, and its apparent speed, are calculated by comparing release and recovery points,
and elapsed time.

Recovery of seabed drifters released on the Washington–Oregon
coast continental shelf indicates a net northerly movement of near-
bottom waters at depths exceeding 50 meters. In waters less than 50
meters deep, near-bottom currents move toward the coast. This appears
to be a result of near-bottom water movements and wind-induced
upwelling in the summer (Fig. 10-28).

Estuarine circulation causes net landward movement of near-
bottom waters toward the Columbia River within about 10 kilometers
of its mouth. A similar situation prevails within 70 kilometers of the
Strait of Juan de Fuca. Less water is discharged from that system, but
the landward subhalocline flow is from 5 to 20 times greater due to
extensive mixing in the large, silled Strait of Juan de Fuca–Strait of
Georgia system. More seawater is carried out in the surface layers
for each volume of fresh water entering the estuary, and therefore more
seawater must come in to replace it. Strong currents are set up by the
flow of large amounts of water through the strait, and the associated
near-bottom currents affect a large part of the continental shelf.

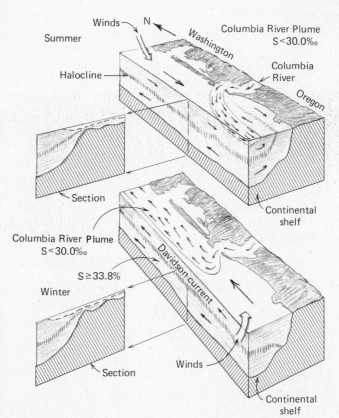

Figure 10-28 (*right*)

Block diagrams showing generalized surface
and subsurface circulation in the Pacific
Northwest region, near the mouth of the
Columbia River in winter and summer.
Compare the surface and subsurface
currents for the two seasons. Also note the
effects of upwelling along the Oregon coast
in summer and the downwelling in winter.

In the Columbia River system, the relatively large amounts of fresh water discharged through the narrow, deep river mouth preclude much mixing in the estuary itself. Most of the initial mixing occurs over a relatively large area outside the estuary, setting up an ill-defined estuarine circulation. Near-bottom currents associated with this weak circulation are thus observed within only about 10 kilometers of the river mouth.

Despite low population densities along the Washington–Oregon coast, the harbors and the coastal ocean itself have been subject to contamination. Waste chemicals and paper mill wastes in various estuaries cause locally severe pollution problems. Indeed, marine life in some of the harbors has been nearly destroyed, for short periods, by large accidental releases of wastes into estuarine waters.

This particular coastal area is unique because of the large volume of radioactive wastes that it has received. From the 1940's to the early 1970's, plutonium-producing reactors on the Columbia River about 600 kilometers upstream from the river mouth released large volumes of slightly radioactive waters to the river. These radionuclides escaped primarily through corrosion in the pipes used to cool the reactors. The warm and radioactive cooling waters were discharged to the river. In 1957, these waste discharges accounted for more than 95 percent of all radioactivity released to the environment by the non-Communist world, excluding fallout from nuclear weapons testing in the atmosphere. Modern reactors, however, recirculate the cooling waters and use their heat for electrical-power generation.

SUMMARY OUTLINE

Coastal ocean—average 70 meters deep, generally above continental shelf
    Coastal oceanic regions bounded by shoreline features

Temperature and salinity
    Winds from continent evaporate, heat, cool
    Extremes of temperature, salinity
        Highest temperatures, salinities in enclosed basins
        Lowest salinities near rivers
        Rapid cooling, sea ice, in shallow areas
        Large seasonal temperature changes

Coastal currents and upwelling
    Geostrophic currents set up by winds, river discharge
    Boundary currents are outer margin
    Upwelling on western sides of continents, especially caused by winds, currents
    Brings deep water to surface; important for productivity

Origin of four types of estuaries
    Fjords—northern coasts altered by glaciers
    Coastal plain estuaries—drowned river mouths
    Lagoons—shallow; sedimentation forms barrier islands, with inlets
    Tectonic estuaries—mountain building regions; estuaries separated from continent

Estuarine circulation—fresh water mixes with seawater
    Salt-wedge estuary
        Less saline upper layer from river; denser deep layer from ocean
        Seawater entrained by fresh water flowing out
        Deep, denser water circulates toward surface; a type of upwelling
    Moderately stratified estuary—greater volume of seawater circulated
        Tides, tidal currents cause mixing
        Relatively less river flow, less well-defined pycnocline
    Fjords—sill at mouth, marked density stratification
    Tides usually dominate estuarine circulation
    River discharge enters coastal ocean with tidal pulses, forms low-salinity plume

Chesapeake Bay—coastal plain estuary, many tributary rivers
    Volume of estuary discharged annually
    Reversing tidal currents
    Salinity varies with depth
    Temperature distribution controlled by local weather because of shallowness

Laguna Madre—a hypersaline lagoon
    Shallow, small tidal range
    Winds control currents

Net evaporation except after storms; little density stratification

Reverse of estuarine circulation—more saline water flows toward ocean

Northeast Pacific Coastal Ocean—area of net dilution
Estuarine-type circulation due to heavy river discharge
Boundary currents—Alaska setting northward; California setting southward
    Point of divergence shifts seasonally
    Boundary with coastal currents diffuse
Low-salinity Columbia River plume

Forms northward-setting Davidson current—winter
Moves southwestward—summer
Generally northward near-bottom currents studied using seabed drifters
Net landward bottom-water flow near coast caused by estuarine-like circulation
Strait of Juan de Fuca—estuarine circulation; mixing in estuary
Columbia River—heavy river flow; ill-defined estuarine circulation outside estuary; wastes discharged by industries

SELECTED REFERENCES

KETCHUM, B. H. 1972. *The Water's Edge: Critical Problems of the Coastal Zone*. MIT Press, Cambridge, Mass. 393 pp. Studies of management of coastal zone resources and man's impact on coastal ocean.

LAUFF, G. H., ed. 1967. *Estuaries*. American Association for the Advancement of Science, Washington, D.C. Publ. 83. 757 pp. Collection of technical papers on many aspects of estuaries and coastal-ocean processes, emphasizing biology.

PICKARD, G. L. 1975. *Descriptive Physical Oceanography*, 2nd ed. Pergamon Press, New York. 214 pp. Elementary treatment of general oceanography, including coastal processes.

shorelines and shoreline processes

T he *shoreline*—the line where land, air, and sea meet—is the most dynamic part of the coastal zone. Here the continents respond to the same forces that affect the coastal ocean: tides, winds, waves, changing sea level, and man.

The shore, as typically shown in Fig. 11-1, extends from the lowest tide level to the highest point on land reached by wave-transported sand. It is affected by the aforementioned forces each day.

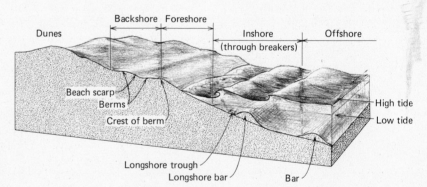

Figure 11-1

Profile of a typical beach and adjacent coastal zone.

The broad zone (known as the *coast* or *coastal zone*), directly landward from the shore is also strongly affected by the processes that shape the shore. Usually these forces act on the coastal zone only under extreme conditions, such as storms or exceptionally high tides. Violent storms or exceptional tides can be expected to strike most coasts every 10 to 100 years. For example, southern New England has been hit by destructive hurricanes in 1635, 1815, 1938, and 1954. The intervals involved are short in the billion-year history of a continent, and storms of this intensity have a major effect on the coastal zone.

Processes affecting shores act on time scales ranging from a few seconds to tens of thousands of years. Among the more obvious forces are the waves which break, rush up the beach face, and then retreat in a matter of seconds. Along open coasts there are few days without some wave action to move materials on the beach face, dissipating the energy originally imparted by winds at sea.

The daily rise and fall of sea level under the influence of the tides submerges or exposes parts of the beach every 6 or 12 hours. To a great extent this controls the type and vertical extent of the beach, as well as the types of plants and animals that live there. Tidal currents are important in inlets and coastal embayments, but have relatively little influence on open, straight coasts.

Storm surges—relatively sudden large changes in sea level resulting from strong winds blowing across a shallow and partially enclosed body of water—also cause large changes in the shoreline and the coastal zone landward because of the flooding of normally protected areas. Although major storm surges may come only every few 10 or 100 years, the changes they cause often remain visible for decades. For instance, some inlets on the south shore of Long Island broke through barrier islands during hurricanes in the 1930's, changing previously isolated and brackish water ponds and lakes into lagoons and estuaries which are now strongly affected by the tides. The inlets are now kept open by man-made jetties and continued dredging, in order to prevent accumulations of sediment from filling them in and forming solid barrier beach once more.

Other processes controlling the shoreline act so slowly that their effects are difficult to detect over a human lifetime, or even in historical records. Such changes occur over periods ranging from 100 to 10,000 years; even areas with the longest historical records have little reliable information about coasts and their features that predate the classical historians, who wrote about 2000 years ago. To detect and interpret long-term changes it is necessary to rely on the record preserved, often indistinctly, in elevated beaches such as those occurring on the west coast of the United States, an area which is generally rising because of continuing mountain building, or in Scandinavia, where the continent is still rising because of the unloading of the crust caused by melting of the continental ice sheets over the past 20,000 years. In parts of the Mediterranean area, some ancient Roman port cities are now submerged as a consequence of the sinking of the crust, also under the influence of large-scale crustal movements; others have been elevated.

Perhaps the most profound influence on coasts as we now see them was the repeated advance and retreat of the continental glaciers and resulting sea-level changes. When sea level stood at its lowest, about 20,000 years ago, the shoreline stood near the present continental-shelf break. As the glaciers waned and water was released, the shoreline moved back across the continental shelf until it reached its present level about 3000 years ago. If the remaining ice sheets (Antarctica, Greenland) melt in the future, the sea surface may eventually stand as high as 50 meters above its present level. The shoreline we now see will be submerged, and a complex of beach ridges and lagoon or estuarine deposits will mark its former location.

Thus the present shoreline is the result of processes acting on different time scales. The beach itself has the shortest "memory," recording distinctly only what happened when the most recent waves rushed up and then retreated, or the level of the last high tide; perhaps a recent exceptionally high tide will have left its mark. The beach as a whole usually records the effects of storms and seasonal changes. In general, it is difficult to detect records of events that occurred more than a few years ago on a beach; even the extensive changes resulting from a storm or storm surge in winter are usually quickly obliterated during the following summer unless humans intrude by dredging inlets or building seawalls, groins, or jetties in order to inhibit erosion and reduce sand movements.

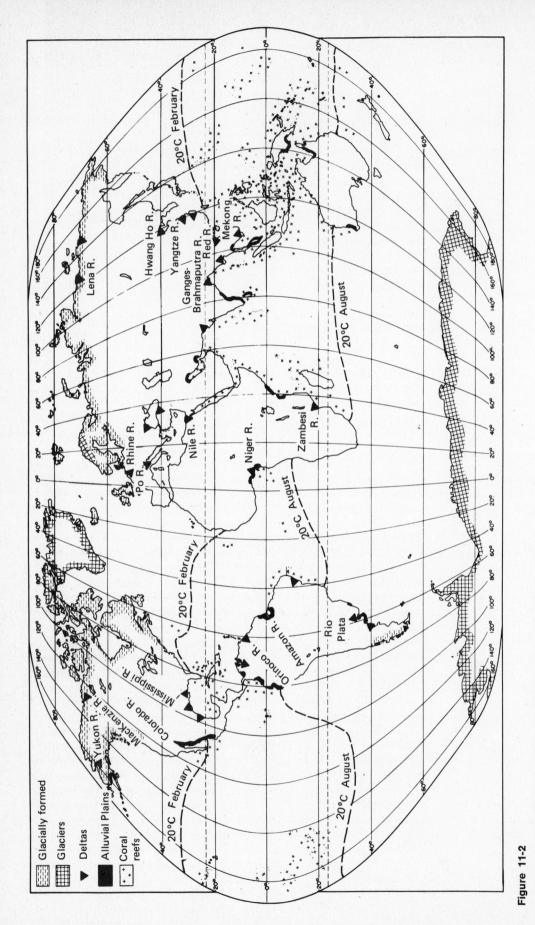

Figure 11-2

Coastlines of the world. (After J. T. McGill, 1958. "Map of Coastal Landforms of the World," *Geographical Review*, 48, 402.)

COASTLINES Geomorphologists—those who study land forms—have classified coastlines according to certain common features and have attempted to determine the dominant forces that have created them. Of the coastlines of the world, shown in Fig. 11-2, most exhibit signs of the recent rise of sea level, modified by wave action and by deposition of sand and gravel to form beaches, or mud to form deltas and marshes.

One major type of coastline, called a *primary coast,* is formed primarily by terrestrial forces; the ocean has not been at its present level long enough to reshape the land features along its margin. An example of such a coastline is seen in the drowned river valleys of the United States' east coast, where the former valley of the Delaware River is now Delaware Bay, and the former valley of the Susquehanna River and its lower tributaries form Chesapeake Bay. The original shapes of these valleys are virtually intact, although largely underwater and somewhat modified by recent sediment deposition. In glacially carved regions, often mountainous, movements of the great ice sheets created deep valleys with U-shaped cross-sections. Now filled by seawater, the U-shaped bottoms are concealed, but the valleys cutting back into mountains form typical Norwegian coastal fjords. Fjords also occur in Canada, Alaska, and southern Chile.

Not all processes acting on the coastline have had the effect of cutting away land surface; other forces were active in leaving behind deposits which now dominate the coastline. Long Island, in New York State, is part of the terminal moraines formed in two periods of Pleistocene glaciation. The moraine is a deposit of sand and gravel orginally scoured from the continental surface, carried toward lower latitudes, and left behind at the terminus of the glacier when it melted and retreated. The delta at the mouth of a sediment-laden river is another example of coastline modification caused by deposition of material eroded from the land.

Spectacular but rather rare coastline features are formed by volcanoes. Good examples are found in the main Hawaiian islands where, as shown in Fig. 11-3, volcanic cones and lavas come directly

Figure 11-3

The coastline at Waikiki Beach on the island of Oahu in the Hawaiian Islands. Diamond Head in the background is a remnant of an extinct volcano. The original volcanic features of this island have been altered by erosion and by man as demonstrated in the large amount of dredging visible in the yacht basins in the foreground. (Photograph courtesy Hawaii Visitors Bureau.)

down to sea level. In other areas, the volcanoes have been worn down to sea level, and in still other areas the sea has broken into the top of the volcano, producing a round bay, usually steep-sided and deep, making an excellent harbor. The harbor at Pago Pago (American Samoa) is thought to have such an origin.

A *secondary coast* is shaped by marine processes or by marine organisms. Such coastlines are often cut in rocks or sediments soft enough so that even the limited time of the present sea-level stand has been sufficient to cut back the coast. Where bluffs of unconsolidated sand and gravel rise above the shore, they are often cut back by waves to provide a wavestraightened coastline, as illustrated in Fig. 11-4. The materials derived from the erosion of these bluffs are deposited near shore, or form small beaches or *baymouth bars* across bays and inlets near the bluffs. Sand moves along the coast by the action of longshore currents, forming *barrier islands* and *spits* that separate such inlets and bays from the open ocean.

Figure 11-4

The coastline on the northern shore of Long Island has been modified by wave erosion and sediment transport. Sand and gravel eroded from bluffs at the points (sharp bends in the coastline) move along the coast and smooth original irregularities. In the left center, a small inlet connects a salt marsh (known as Flax Pond) to Long Island Sound. Sand and gravel beaches separate the marsh from the sound. On the right side of the photograph, a relatively large, shallow harbor has been created by a long beach which separates submerged former valleys from the sound. (Photograph courtesy National Ocean Survey.)

DELTAS The sediment load carried downstream by most rivers and streams is either deposited in the estuary or at the river mouth. For most streams, the only sign of their sediment load near the river mouth is a small shifting sand bar which is moved by the tides. The bulk of this sediment is eventually resuspended by wave action and moved along the coast by longshore or tidal currents, to be carried onto nearby beaches or deposited on the continental shelf.

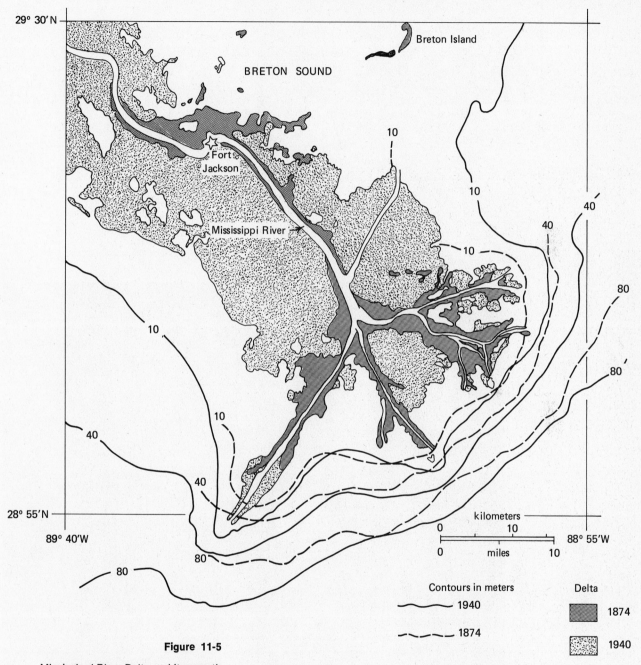

Figure 11-5

Mississippi River Delta and its growth between 1874 and 1940. Note the characteristic "bird-foot" shape resulting from each distributary having built seaward, with flooding between the distributaries.

Some rivers, however, carry far more sediment than can be dispersed along the adjacent coast. In this case, sediment usually settles at the river mouth, forming a delta. This is a fairly common occurrence along coastlines; well-known examples include the Mississippi (as shown in Fig. 11-5), Rio Grande, Fraser, Nile, Tigris–Euphrates, and Rhine deltas. Delta lands are flat, well-watered, and often exceedingly

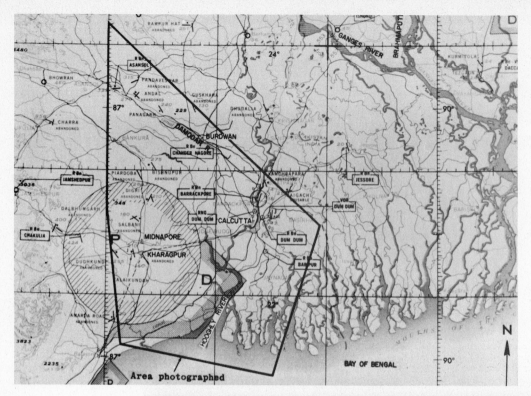

Area photographed

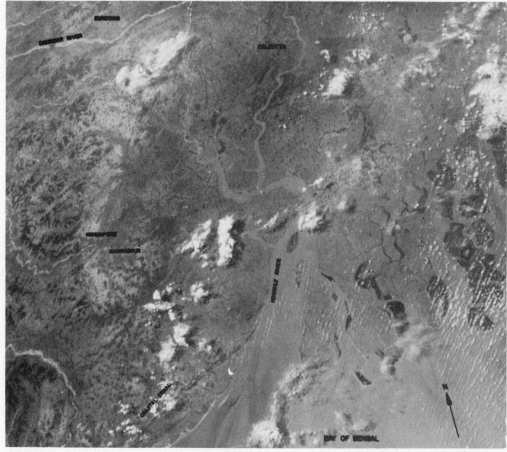

Figure 11-6

A portion of the delta at the mouth of the Ganges–Brahmaputra Rivers as photographed from a satellite. (Photograph courtesy NASA.)

fertile because sediment deposition during flood periods provides a continuous supply of rich soil. Many primitive agricultural societies have developed and flourished on river deltas, the Egyptian culture on the Nile being a familiar example. Lower Mesopotamia, site of the early Sumerian civilization, was an ancient delta, and all over the world deltas are still valued as farmlands. (Their low elevation above sea level makes them vulnerable to flooding by storm surges.)

The relative capacity of the coastal ocean to transport sediment is determined by the tidal range and associated strength of tidal currents, the wave energy dissipated on the coast, and the strength of the longshore currents. The strong tidal currents set up by a large tidal range cause scouring of recently deposited sediment. Waves act to resuspend sediment in the surf zone; they also cause *longshore currents,* which move sediment along the coast in a direction determined by prevailing wave approach.

The Columbia River on the west coast of the United States, discharging between 10 and 30 million tons of sediment per year into the ocean, lacks a delta due to its relatively large tidal range and strong waves at its mouth. On the other hand, the nearby Fraser River, with approximately the same sediment load, has formed a delta in the protected waters of the Strait of Georgia (British Columbia) because strong wave action is inhibited there by the limited fetch for wave generation. The Mississippi River has formed the largest delta in the United States as a consequence of its large sediment load (about 300 million tons of sediment per year) and the low tidal range in the the Gulf of Mexico.

Some river systems supply so much sediment that a delta forms in spite of a large tidal range and extensive wave action. The region where the Ganges–Brahmaputra Rivers in India discharge at the head of the Bay of Bengal, shown in Fig. 11-6, is an example. These two rivers, which reach the ocean through the same delta complex, carry about 700 million tons of sediment each year; a large delta has formed as a result of its deposition.

The first step in delta formation, illustrated in Fig. 11-7, is for a stream to fill its estuary with sediment. The Mississippi, for instance, probably filled its estuary soon after sea level reached its present position. Rivers with large estuaries and relatively moderate sediment loads, such as the Amazon River, are still filling their estuaries. Until the estuary is completely filled, little or no sediment escapes to be deposited at the river mouth or to form a delta.

As a delta forms, the river builds channels across it called *distributaries,* as shown in Fig. 11-8, through which the water flows on its way to the ocean. There are often a series of these channels, complexly interconnected; they extend across the delta as long, radiating, and often branching fingers. An active distributary continually builds its mouth further seaward until the distance to the sea is so great that the river can no longer maintain flow through that channel. At this time the river shifts course, often during flood, so that the flow proceeds through a different set of distributaries to reach the ocean, and the whole process begins again. The Mississippi River Delta reveals a series of abandoned distributaries, each with its own subdelta, forming a complex *lobate delta* (see Fig. 11-5). Distributary abandonment is not always sudden, but may occur gradually as one channel becomes too shallow to carry a large amount of water and another gradually receives more of the flow so that it becomes enlarged.

Many deltas occur in areas where the land is subsiding. For example, the Mississippi Delta region as a whole is subsiding at a rate of about 1 to 4 centimeters per year. Furthermore, the sediment beneath the delta continually compacts and expels water from the sediment.

Figure 11-7 (*right*)

Stages in the development and filling of an estuary by sediment deposits, converting a river valley to a salt marsh.

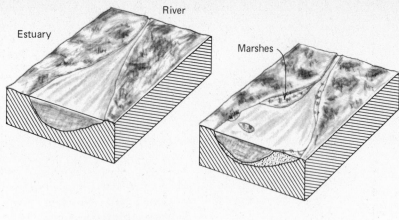

Figure 11-8 (*below*)

A portion of the modern delta of the Mississippi River is pictured here looking south from the "Head of Passes," where one distributary separates into three. The left-hand pass is a natural artery; the other two are kept open for navigation by dredging, and by construction of jetties to direct the river flow—thus increasing its velocity so that less sediment is deposited in the channels. (Photograph courtesy U.S. Army Corps of Engineers, New Orleans District.)

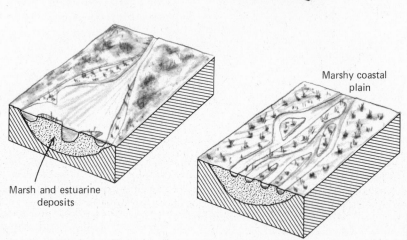

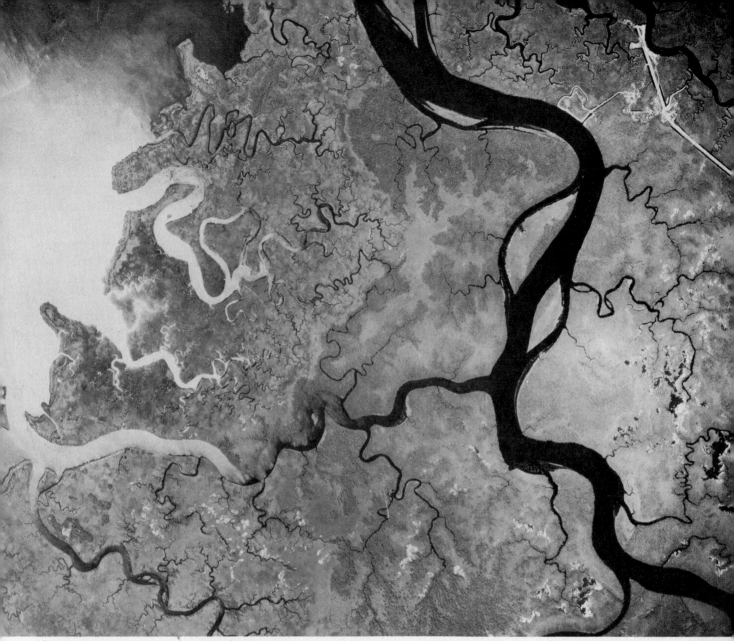

Figure 11-9

Distributaries of the Suwannee River, Florida, showing the tidal marsh and cypress swamps lying between them. Note the meandering tidal creeks in the marsh. A road and dredged channel are conspicuous in the upper-right portion of the photograph. (Official U.S. Geological Survey photograph.)

Consequently, the delta surface is continually subsiding, except in those areas currently receiving sediment.

When a distributary is active, adjacent areas are usually occupied by ponds or lakes with many marshes, as shown in Fig. 11-9. Areas between distributaries receive less sediment, and the deposits there tend to be finer-grained and contain bulky plant debris (peat). These deposits compact (and subside) more than the coarse-grained material along the distributary banks. The banks, therefore, subside less than the adjacent marsh and eventually stand higher, forming natural levees on which towns and roads are built.

As the marsh subsides below sea level, it is transformed into a shallow bay. Seawater in these bays is diluted by ground water discharged through the delta, so that bay waters are usually brackish. These areas, even though their appearance to an outsider may be unprepossessing, are highly productive of marine life. About 90 percent of commercially important coastal fish and game fish are dependent on these bays or similar marsh areas during some critical stage of their life cycle.

291

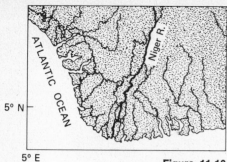

Figure 11-10

Delta of the Niger River on the western coast of Africa.

Deltas are rarely formed by rivers draining areas that were covered by ice during the last glacial period. Lakes, gouged out by ice, are common in these areas, and act as traps to prevent the escape of river-borne sediment to the sea. An example is the St. Lawrence River, which drains an area where the Great Lakes act as effective sediment traps. After lakes in the river's drainage basin and the estuary at the river's mouth are filled with sediment, the stream may then be able to transport enough sediment to the ocean to build a delta.

The shape of a delta is controlled in a complex way by the balance between erosion and sediment transport along the coast, on the one hand, and the rate of sediment supply on the other. Where coastal processes predominate, the margin of the delta tends to be rounded, as in the Niger Delta, as shown in Fig. 11-10. Where the sediment supply from the river exceeds the capacity of the coastal ocean to transport it, the delta shape is primarily controlled by river processes rather than coastal processes. The Mississippi Delta's *"birdfoot"* shape results from each distributary having built seaward, with flooding between the distributaries, as previously discussed.

BEACHES

Figure 11-11

Trunk Bay, a sheltered cove on St. John, Virgin Islands, a beach between rocky headlands. (Photograph courtesy U.S. Virgin Islands.)

Beaches are probably the most familiar shoreline features. Sand beaches, barrier islands, and bays border the Atlantic coast of the United States from Long Island, New York, to Key West, Florida; in the Gulf Coast region, barrier islands and lagoons are also a characteristic feature. On mountainous coasts, beaches are usually less extensive, being generally restricted to low-lying areas between rocky headlands (as illustrated in Fig. 11-11).

In a very real sense, a *beach* is a sediment deposit in motion. At any time the motion of grains in the surf may be obvious, while the rest of the beach appears quite stable. But a visit to a beach following a major storm will show substantial changes, demonstrating that large segments of the beach move fairly frequently. Higher or more protected parts may move only during exceptionally powerful storms; nonetheless, the whole beach does move, and has justly been called a "river of sand."

To most of us, the word "beach" brings to mind a sand beach composed of grains with diameters between 0.062 and 2 millimeters. On mid-latitude coasts, beach sands are usually derived from chemical alteration and mechanical breakdown of silicate rocks. The mineral quartz is a common constituent of such beaches. In fact, some beach sands are so pure that they are mined to obtain nearly pure quartz sand for making glass. In tropical and subtropical areas where silicate rocks are rare or absent, beach sands come from broken carbonate shells and skeletons of marine organisms. Such beaches are often white or slightly pink—as in the Bermuda Islands, where the broken shells of a red formaminifer *(Homotrema rubrum)* color the sands. In the Hawaiian Islands, all of the famous black-sand beaches are formed from recently erupted lava (volcanic glass) that make up the islands.

Not all beaches are composed of sand. Where wave and current action is especially vigorous, sand may be washed away faster than it is being brought in, leaving behind gravel or even *cobbles*. Still other beaches consist of mixtures of gravel and sand where wave action is not strong enough to completely remove the sand. Sand–gravel beaches occur on the north shore of Long Island, in New England, and on many Pacific coast beaches.

In general, beaches are accumulations of materials that are locally abundant and which are not immediately removed by waves, tidal currents, or winds. In areas like Long Island, beach sands and gravels are derived from erosion of glacial deposits, originally containing unsorted gravels, sands, and clays. Only the gravels and sands remain on the beaches; silt and clay-sized particles are usually washed out of beach areas by even weak waves or tidal currents. Fine-grained sediments tend to accumulate in areas with little wave action or tidal currents, either on the continental shelf at depths below about 30 meters or in lagoons, bays, or tidal marshes.

Most beaches consist of fine sand and tend to slope gently, with a hard-packed *foreshore* (the zone between high and low water). As the average grain size of the material forming the beach increases, the slope of the beach also increases. Beaches composed of fine sand ($\frac{1}{8}$ to $\frac{1}{4}$ millimeter) have average beachface slopes of about 3 degrees. Pebble (4 to 64 millimeter) beaches typically slope about 15 degrees and cobble (64 to 256 millimeter) beaches typically slope about 24 degrees. Even on a single beach, one can see the slope of the land increase as the grain diameter increases. Changes are easily observed as you go from areas with weak currents, where fine sands can accumulate, to areas where stronger currents remove fine sands, leaving gravels and a more steeply sloping beach. Fine-sand beaches are usually hard-packed, permitting easy walking and even the driving of cars on them. Coarse-sand or gravel beaches are much looser, because the larger grains do not pack as compactly as fine sands.

Beaches typically form near a major sediment source—at the base of a cliff or near a river mouth. Sediment is moved onto beaches by waves and currents, replacing materials either moved out into deeper water or transported along the coast. On the west coast, the Columbia River is a conspicuous example; near its mouth are some

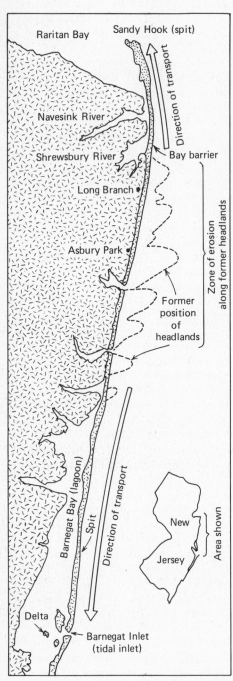

Figure 11-12

Several types of barrier beaches occur
along the New Jersey coast. The beaches
are formed by sand moving north and
sand from an area of headlands in the
vicinity of Long Branch and Asbury Park.
(After D. L. Leet and S. Judson, 1971.
Physical Geology, 4th ed., Prentice-Hall,
Englewood Cliffs, N.J., 704 pp.)

of the largest beaches and sand dunes on the Washington–Oregon
coast. In most other coastal areas, including much of southern California, beaches tend to be small and located near the rivers which
supply their sand.

On the North Atlantic coast of the United States, virtually no
river-borne sediment escapes the many large estuaries to enter the
ocean. Hence many Atlantic coast beaches are formed either from
erosion of nearby cliffs or from sands deposited offshore during a
time of lower sea level. In many cases a beach forms where transport
of sand along the coast is interrupted by an obstruction such as a
headland downstream.

Not all beaches are located on coastal plains. If the land along
a coast is low-lying and slopes gently toward the ocean, sand deposits
will probably form parallel to the coast and a short distance offshore.
If submerged these are called *longshore bars;* but where large enough
they form *barrier islands* which typically have a shallow lagoon or bay
between the island and the mainland. Such *barrier beaches,* formed
by the onshore movements of sands and the longshore movements of
currents, are probably the most common type of beach occurring along
the low-lying coastlines of the world (Fig. 11-12). When connected
to the mainland, usually at some *headland* (a point of land that
juts out from the coast), they are known as *barrier spits.* At small
indentations a barrier of sediment may build completely across the
mouth of a bay, in which case the barrier is known as a *baymouth
bar.* Along rocky coasts, offshore rocks and stacks are sometimes connected to the mainland by beach complexes called *tombolos.* A wide
variety of beaches can form within even a single bay.

Two theories have been advanced to explain why barrier beaches
form where they do. One is that they develop where there is a base
of ancient, submerged sediment. The other is that sand moving past
headlands tends to be deposited rather than being moved further
downcurrent, forming a series of spits that bridge the mouths of bays
in series along a coast, as shown in Fig. 11-13. In any case, barrier
bars result from a dynamic balance between longshore currents, wave
attack, sediment supply, and bottom topography. They tend to be
long and straight, broken at intervals by inlets through which water
is borne in and out of the bay or lagoon by tidal currents. The force
of these currents keeps the inlets open; thus they act in opposition to
the longshore currents, which tend to deposit sediment across the
channel.

The barrier beach, like the beach on a coastal plain, is normally
in a state of being washed away and replaced at a fairly constant rate.
It is extremely sensitive to changes in the force with which waves
break on it, or changes in the amount of sand carried toward it by
those waves. In a few hours, a major storm can move large quantities
of sand to form new inlets or to close old ones.

Coming toward the beach from the ocean, an echo sounder on
a small boat would reveal one or more submerged low sand ridges
on the ocean bottom, called *longshore bars,* parallel to the shore, and
generally situated in a few meters of water (see Fig. 11-1). At extreme
low tides, the tops of these bars are often exposed. Separating the longshore bar from the beach proper is the *longshore trough,* which
usually remains filled with water at low tide. On coasts with a restricted tidal range, there may be several such bars and troughs developed. Coasts with large tidal ranges usually have only one set of
bars, at or near the low-tide mark. Bars can usually be identified, even
when submerged, by the fact that waves break on them; on many
coasts, several sets of bars can be spotted from the lines of breakers
offshore.

294

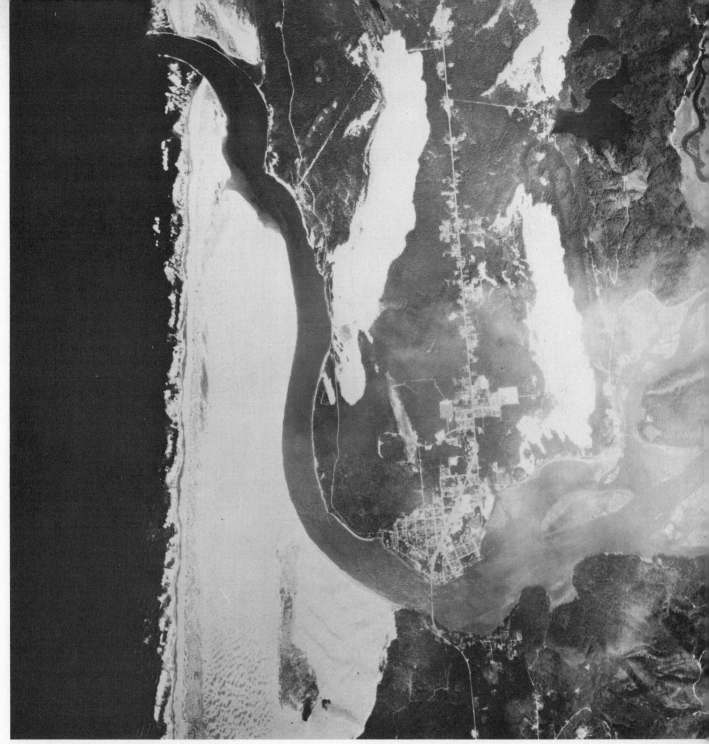

Figure 11-13

A baymouth bar formed by longshore currents moving sediment northward along the Oregon coast at the mouth of the Siuslaw River. The light-colored spit is an area of active dune migration with no vegetation. Darker areas on the spit are depressions between dunes; some have small lakes in them. Dark-colored areas landward of the spit are heavily vegetated, and sand there is completely stabilized and not readily moved by wind. (Official U.S. Geological Survey photograph.)

Leaving the offshore portion of the beach and coming onto the beach proper, there is a sandy area dipping gently seaward, the *low-tide terrace*. This part of the beach is completely bare at low tide and submerged at high tide. A small scarp or vertical face often occurs at or near the upper limit of the low-tide terrace. This is commonly left by a recent cycle of more intensive wave action which has caused erosion into the beach profile formed during a preceding cycle. The seaward-dipping portion of the beach, collectively known as the *fore-shore,* leads up to the *berm crest* or *berm,* the highest part of the beach. There may be several berms present on a beach at any time. Berm crests usually form during storms, and represent the effective upper

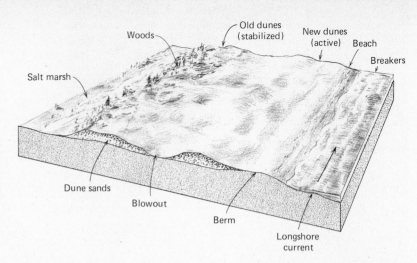

Old dunes (stabilized)
New dunes (active)
Woods
Beach
Breakers
Salt marsh
Dune sands
Blowout
Berm
Longshore current

Figure 11-14

Typical sand beach with active dunes, stabilized dunes and salt marshes. Such a beach–dune complex is found on the south shore of Long Island and at many points along the U.S. Atlantic coast. (Redrawn from F. D. Larsen, 1969. "Eolian Sand Transport on Plum Island, Massachusetts," *Coastal Environments of Northeastern Massachusetts and New Hampshire,* University of Massachusetts Coastal Research Group, Amherst, Mass., pp. 356–67.)

Figure 11-15

Beach grasses stabilize the dunes at Lighthouse Point near Gainesville, Florida. (Photograph courtesy Florida News Bureau.)

limit of wave action during that storm. Usually the highest berm on a given beach is formed during winter storms and is often referred to as the "winter berm."

The berm slopes gently downward toward the base of the cliffs or dunes behind the beach. These sands are usually not moved often, as indicated by the large trees that grow there (and often by the accumulations of beer cans and other human artifacts). Where wind action is especially strong, the sands may be winnowed, removing the lighter particles and leaving behind the heavier mineral grains. On some beaches, the heavy minerals remaining give the surface a deep red or purple color, especially noticeable near the base of sand dunes.

Where a beach is not backed by cliffs, dunes are often formed from beach sands blown by the prevailing onshore winds, as illustrated in Fig. 11-14. Where dunes are not actively gaining or losing sediment, they may be colonized by various salt-tolerant plants or trees, to become stabilized through time; this process is shown in Fig. 11-15.

Dunes protect the low-lying sand behind them; for example, in the Netherlands the dunes form a vital part of the defense against flooding from the North Sea. Frequently, access to the dunes is limited or prohibited in beach park areas to preserve the slow-growing plants which protect dunes against active erosion by storm winds.

On the Atlantic coast of the United States, extensive areas of dunes occur near Cape Canaveral, Florida, and near Provincetown, Massachusetts. On the West Coast, dunes are associated with the beaches near river mouths (see Fig. 11-13) and also occur on isolated beaches along the rest of the coast. Since they are controlled by the winds, dunes can form obliquely to the coast. Several intersecting sets of dunes may occur in the same region.

BEACH PROCESSES   Waves dominate beach processes. Currents and turbulence generated by waves stir up sediment, and the currents associated with the waves and tidal currents transport sediment parallel to the coast. Transport usually takes place between the upper limit of wave advance on the beach and depths of about 15 meters. Large amounts of sand are transported in suspension; relatively little is transported along the bottom.

Beaches undergo definite seasonal cycles. During seasons of low, long-period swell, sand is moved back onto the beach, usually causing increases in height and width. Longshore bars migrate shoreward, filling in the troughs, and a new berm forms, usually at a level lower than the one preceding.

During periods with high, choppy waves (usually in winter), beaches are commonly cut back. The beach foreshore becomes more gently sloping and a beach scarp may occur as erosion proceeds. Strong longshore currents caused by the waves develop deep channels. Bars develop because of the offshore movement of sand from areas seaward of the breakers. Most of the sand removed from the beach is deposited nearby in the offshore zone to be moved back onto the beach during the next period of smaller waves.

Waves commonly approach the beach obliquely—rarely at right angles to the beach. Even though they are refracted upon entering shallow water so that crests are more nearly parallel to the coast, the process is rarely complete, and most waves approach the shore obliquely (see Fig. 11-17). Wave energy acting parallel to the coast causes longshore currents to move generally in the same direction as the waves when they approached the coast. The current is strongest in the band between the surf zone and the beach. The strongest currents are predicted to occur when the waves approach the shore from a 45° angle. This rarely happens; crests of most waves usually deviate less than 20° from being parallel to the beach when they strike the shoreline.

Along some trade wind coasts, the amount of sand moved by longshore currents exceeds 4 million cubic meters per year. On the south shore of Long Island, these currents move about 500,000 tons of sediment westward every year. A comparable quantity of sand moves along New Jersey beaches.

Each wave hitting the beach causes an uprush of a relatively thin sheet of water or *swash* onto the beach face, as Fig. 11-16 illustrates. The water rises until all the energy of the oncoming wave is dissipated or until the water moved by the wave percolates downward into the sand. Any water remaining on the surface runs back down the slope of the beach face.

Since waves rarely strike the beach head-on (see Fig. 11-17), but usually strike it at some angle, the swash rises obliquely across the

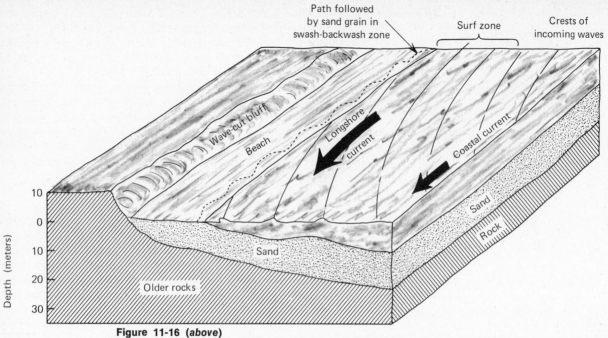

Figure 11-16 (above)

Incoming waves are refracted and change
directions as they enter shallow water.
Striking the beach at an angle, they cause a
longshore current in the surf zone that
moves sand on the beach and along the
bottom. Coastal currents are outside
the breaker zone.

Figure 11-17 (right)

Waves strike the beach north of Oceanside,
California; their approach at an angle moves
sediment along the beach parallel to the
shoreline and also causes longshore currents.
(Photograph from R. L. Wiegel, 1964.
Oceanographical Engineering, Prentice-Hall,
Englewood Cliffs, N.J., p. 372.)

beach face. When the water with its entrained sediment runs back, it goes directly down the slope of the beach. Sand moving along the beach as a result of wave effects is called *littoral drift*. Its direction can change during a single day or over a season; this is a small-scale phenomenon associated with waves hitting the beach. Longshore currents are large-scale phenomena which also result from waves striking the coast.

Speeds of wave-induced sediment movement can be surprisingly high—up to 25 meters per hour and up to 1 kilometer per day. A more typical rate would be 5 to 10 meters per day on the average. Direction of net movement along the beach is determined by the direction of the strongest and longest acting waves. On most beaches this is the direction from which storm winds come.

Within an individual wave, there is substantial movement of water as it breaks. Water at the surface and along the bottom moves toward the beach, carrying with it materials floating or dragged along the bottom. This accounts for the fact that a beach acts as a convergence zone, collecting all sorts of debris as well as sediment. Return flow of the water occurs at mid-depth within the water column as illustrated in Fig. 11-18. In general, this flow occurs through several meters of water and is not terribly strong.

Rip currents are another manifestation of the movement of water toward the beach in the surf and its return flow. After moving toward the beach, water tends to flow parallel to the beach for short distances until it enters a highly localized stream of return flow through the breaker zone—the rip current. In rip currents, the most rapid flow is relatively narrow, and has speeds of up to 1 meter per second until it reaches a distance of perhaps 300 meters from the coast. In this section of the current, it often forms a channel deeper than the surrounding ocean and is therefore recognizable from above by the different color of the water and its high sediment content.

Seaward of the breaker zone, the current becomes more diffuse and spreads out, forming a "head" to the current. The water is caught up at this point in the general flow toward the beach. Rip currents and their associated water movements form a cell-like, nearshore circulation system within the breaker zone.

As an example of the processes that control beach development and sediment movement, let us consider the New Jersey coast (see Fig. 11-12). This 200-kilometer stretch of coast consists of a series of barrier islands separated from the mainland by bays, lagoons, and tidal marshes. Along this coast, beaches are interrupted by three rocky headlands and ten major inlets leading to lagoons or bays. Tides here are about 1.5 meters; northeasterly storms are common, and there are occasional hurricanes.

At the northern end of the coast, near Sandy Hook, the net littoral drift is northerly because nearby Long Island shelters this stretch from waves approaching the coast from the north and northeast. Over a period of about 100 years; sand accumulation on Sandy Hook amounted to about 400,000 cubic meters per year. Near Cape May, at the southern end of this stretch of coast, the net littoral drift is directed southerly and amounts to about 150,000 cubic meters per year. The total amount of sand in transit is substantially greater than the net littoral drift. For example, the total littoral drift at Cape May is estimated to be about 900,000 cubic meters per year, equivalent to about 1 million tons per year.

Between the northern and southern portions of the New Jersey beaches, there is a nodal point—located about 60 kilometers south of Sandy Hook—where the net littoral drift is zero. Sand accumulating at Sandy Hook comes from erosion of this section of headland, as there are no major rivers in the region. Between 1838 and 1953, this section of coast was cut back 150 meters at a rate of about 1.5 meters per year. The materials eroded were about two-thirds sand, which was added to local beaches; about one-third was silt and clay-sized material, which moved seaward and was lost from the coastal area. Extensive construction of seawalls and groins in this section effectively reduced the littoral drift by only about 12 percent.

Inlets and lagoons behind barrier islands act as traps for sediment moving along the coast. Sand is stirred up and put in suspension by waves and then moved into the inlets or lagoons by flood-tide currents. When these slacken, or when sediment encounters the dense vegetation of the tidal flats, it settles out. Without resuspension by waves, the ebb-tide currents fail to move this sediment back out of the inlet. Thus each of the New Jersey inlets accumulates about 200,000 cubic meters—equivalent to 250,000 tons—per year, filling in the navigation channels and necessitating dredging.

MINOR BEACH FEATURES

As a wave breaks and runs up on the beach as a thin film of water (the swash), it carries with it various floating debris, sea foam, and small shells (Fig. 11-18). When the wave reaches its highest point on the beach, the water percolates into the permeable sand, leaving a line of foam known as a *swash mark*. If the next wave runs higher up the beach, it incorporates the preceding swash mark and moves it higher. Where seaweed, shells, driftwood, straw from marshes, or man-made trash are thrown up by storm waves, the line of debris may be quite substantial. If the tide is falling, swash marks are left successively lower, and an entire series may be preserved on the low-tide terrace until the next high tide erases them. Several of the minor beach features are illustrated in Fig. 11-19.

As sea level drops during the tidal cycle, it leaves the sands or gravels of the beach saturated with seawater, which gradually drains out. This dewatering of the beach usually manifests itself in the wetness of the lower part of the low-tide terrace. Where larger volumes of water are discharged, a series of small *rills* (miniature channels) may be cut by the water as it runs down the beach.

Rills cutting deeply enough into the beach, or a trench dug perpendicular to the trend of the beach, will usually reveal a series of buried layers known as *laminations*. These layers are generally

Figure 11-18

Schematic representation of breaking waves and water movement at and below the water surface. Water depth at the outer boundary is approximately half the wave length for storm waves. (Redrawn from R. L. Miller, and J. M. Zeigler, 1964. "A Study of Sediment Distribution in the Zone of Shoaling Waves over Complicated Bottom Topography," in R. L. Miller, (ed.), *Papers in Marine Geology*, Shepard Commemorative Volume, Macmillan, New York, pp. 133–53.)

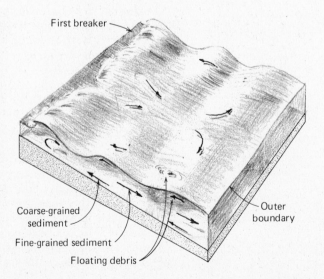

First breaker

Coarse-grained sediment

Fine-grained sediment

Floating debris

Outer boundary

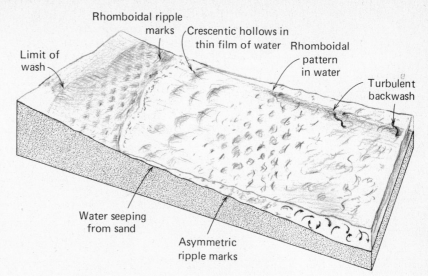

Figure 11-19

Examples of some small-scale features commonly observed on sand beaches. (After A. Guilcher, 1958. *Coastal and Submarine Morphology*, John Wiley, New York, 274 pp.)

formed by different types of materials deposited as thin sheets. For example, during periods of little wave activity, mica flakes may be deposited instead of being removed, as they usually are during strong wave activity. When the beach is dry, as during periods of strong wind, some quartz grains are often blown away (as previously described), leaving behind a layer of dark-colored, heavier minerals as in Fig. 11-20. In general, these layers parallel the present beach surface, dipping seaward on the foreshore. There may also be disturbances of older beach surfaces recorded; for instance, air can be entrapped in the sand, which then migrates to form a dome structure which later collapses, disrupting the layering of the beach. Burrowing of organisms also frequently disrupts the regular layering. On public beaches, it is sometimes possible to find buried automobile tracks that were left on a hard-packed surface and later covered by sand. The depth to which beach-cleaning machines have sifted the sand can often be clearly seen.

Figure 11-20

A scarp cut by waves. The boardwalk has been undermined, and snow fences have been placed in an effort to trap wind-blown sand. The dark patches and streaks at lower left are heavy mineral grains left behind by the waves. Some lamination is visible along the scarp. (Photograph courtesy Fire Island National Seashore, National Park Service.)

Figure 11-21

Beach cusps at El Segundo, California, photographed from the air. (From Wiegel, 1964. p. 34.)

Some beaches have well-developed *beach cusps*: rather uniformly spaced tapering ridges, with rounded embayments between them, as illustrated in Fig. 11-21. The regular spacing ranges from less than 1 meter to several tens of meters, and seems to be related to wave height: higher waves are associated with wider spacing of cusps. These cusps seem to form and reform rather quickly, especially on fine-sand beaches. Their origin has long puzzled those who study beaches, but there seems to be general agreement that they are related to the to-and-fro motions of the water running up the beach. When waves strike a beach obliquely, causing longshore currents, the cusps do not form and are usually obliterated. Apparently, the rounded embayments between the points are channels through which water brought onto the beach by waves returns seaward.

Ripple marks on the sand are another common feature. There are generally two kinds, both formed by water movements. Oscillation ripple marks tend to be symmetrically sloping on both sides, with pointed crests. They form roughly perpendicular to the back-and-forth motion of the water. The other kind of ripple mark is formed by currents; these are asymmetrical, with a gently sloping side facing upcurrent, the sharp side of the ripple mark facing downcurrent. This type is formed by unidirectional currents, although both types are commonly found together.

SALT MARSHES

Salt marshes are low-lying portions of coastline which are submerged by high tides but protected from direct wave attack. Their surfaces are nearly flat and generally overgrown by a salt-tolerant vegetation, adapted to periodic submergence.

Figure 11-22

Small salt marshes in nearly filled bays on the north shore of Long Island (Stony Brook Harbor and West Meadow Creek). Both marsh areas have been modified by dredging. (Photograph courtesy National Ocean Survey.)

Marshes form where sediment is available from rivers or from resuspension in waves or strong tidal currents. In some areas, sediment depositing in marshes comes from local sources such as erosion of nearby headlands; this is the source of sediment for many New England marshes, as illustrated in Fig. 11-22. Size and shape are de-

termined by the general outline of the depression in which the marsh forms; vertical extent is controlled by the tidal range. The upper limit of a marsh is generally controlled by the spring tides in an area, which is the highest level to which ocean water can periodically transport sediment.

Most of the marsh area consists of nearly flat-topped banks consisting of sand or mixtures of sand and silt. The tops of the banks, known as *tidal flats,* are commonly exposed at low tide and submerged at high tide. Where a marsh is well protected, the flats are usually covered by dense growths of marsh grass. Where the marsh is big enough and open enough for wind waves to be set up across its surface, tidal flats may be completely barren of plants.

Cutting through the tidal flats are many channels, through which seawater enters and drains from the marsh in response to the tidal cycle. The largest and deepest of these channels contain water even at low tide, and plants do not usually grow in them. The bottom material is generally shifting sand or gravel because strong tidal currents resuspend the finer-grained materials, to be deposited on the flats. These large channels connect with smaller branching and meandering channels which cut back into the marsh and are often exposed at low tide.

Sediment is transported into the marsh by strong currents in the tidal channels. As rising water moves out into the smaller channels and then onto the tidal flats, current velocities decrease. At some point, the current velocity is too low to keep sediment in suspension, so that particles settle out and are eventually deposited. When the tide goes out, current velocity is inadequate to resuspend the sediment grains, which therefore remain where they settled out of the water.

Figure 11-23

Mangroves growing out into the water at Ten Thousand Islands, Florida. (Photograph courtesy Florida News Bureau.)

Plants on tidal flats also tend to retain sediment, and marshes thus act as effective traps for fine-grained sediment moving in the coastal ocean. Many marshes mark the locations of former small, shallow estuaries.

Some of the largest and best-studied marshes are those lying behind the Friesian Islands on the Dutch North Sea coast and the adjacent North German coast along the edge of the Rhine Delta. In the Netherlands much of the marshland that formerly bordered that delta has been reclaimed by building dikes to exclude ocean waters, and then draining the salt water to make fields suitable for agriculture.

Salt marshes, or *wetlands,* are important biological features of the coastline. In temperate climates, salt-tolerant plants and grasses produce an abundance of food for large and small animals. Much plant debris escapes to be deposited with the sediment accumulating on the nearby ocean bottom.

In tropical climates, grasses play a less conspicuous role, and *mangroves* such as those in Fig. 11-23 dominate marshes along the estuarine border. These large, treelike plants have extensive root systems. The roots form dense thickets at the water level which provide shelter for both marine and land animals—a zone truly intermediate between land and water. Many forms of life are specially adapted to survive in this environment; the mangrove oyster, for instance, attaches itself to roots and branches that are exposed at low tide, presenting the spectacle of oysters growing on trees.

Mangrove roots trap sediment and organic matter, and eventually the swamp is filled in to be replaced by a low-lying tropical forest. Over a 30- to 40-year period, 1500 acres of new land were created in Biscayne Bay and Florida Bay by colonization of shallow-water areas by mangrove seedlings.

MODIFICATION OF COASTAL REGIONS

Coastal areas have been substantially modified by man's efforts to adapt them to his own needs—especially near urban centers, where population pressures are greatest. Improvement of shipping facilities in estuaries and harbors, filling-in of marshes for industries, airports, and housing, prevention of beach erosion, and use of coastal areas for waste disposal probably account for the bulk of coastal modifications in the United States. Because of their importance to shipping, estuaries have perhaps been more extensively modified than any other coastal feature.

The dredging of deeper (and usually straighter) channels for safer passage of ships is perhaps the oldest form of coastal alteration commonly undertaken in this country. Most estuaries have bars—shallow sand deposits—at their mouths, whose greatest depth in channels would be around 3 to 5 meters below sea level. These shallow channels are kept open by river flow and tidal currents, but they are far too shallow to accommodate modern vessels; furthermore, they tend to shift location frequently. Dredging them to depths of up to 15 meters has improved navigation in large harbors, but it has also introduced a host of other problems to the estuary.

Deeper channels act as sediment traps for sands moving along the coast as well as for river-borne sediment, so that a continuous program of dredging is required to keep the channels open. Alteration of normal circulation patterns affects salinity distributions; for instance, enlargement of inlets may permit larger amounts of seawater to enter, and water may flow through man-made channels toward remote areas that would normally be relatively fresh. Fish and bottom life are affected by such changes.

While less altered than most harbors, beaches have often been subjected to conflicting uses and have been greatly altered in the process. In many areas, beaches are exploited directly as sources of sand for construction and land-fill operations, making them unsuitable for recreational use. In other cases, changes in the local regime of sand transport has caused erosion of beaches. When a harbor or inlet, for instance, is dredged for improvement of navigation facilities, the resulting deep channels or basins may act as traps for sediment moving down the coast. This cuts off the sand supply to beaches downcurrent.

Efforts to prevent accumulation of sediment in inlets have met with limited success. Seawalls have been built to stem erosion, and groins to inhibit sediment movement along beaches. Other efforts might include construction of jetties to keep sand out of the inlets, perhaps combined with pumping of sand past the inlet. Sand may also be dredged from bays and used to replenish beaches. These approaches are expensive, and some are undesirable because of side-effects—such as destruction of large and valuable marsh areas by dredging.

Once lost, a beach is not easily restored. Several approaches have been used to protect or rebuild beaches. One of the most direct approaches has been to construct *groins* (low stone walls) such as those in Fig. 11-24, built at regular intervals along a beach to retard sand movement. On the side from which sand is moving, the beach is widened; but the beach downcurrent is usually narrowed unless particular care is taken in construction and placing of the groins. In some instances where groins have been improperly placed, erosion has actually increased. On the New Jersey coast, where groins have been used extensively, it is estimated that they have only reduced the rate of sand movement by about 12 percent.

The other approach to beach stabilization is *replenishment,* a procedure whereby large volumes of sand are brought in and put on the beach. In some instances, the sand is dredged from shallow, protected bays behind the beach. The effects are noticeable but transitory, because the sand moves downcurrent as part of the regional movement of sediment along the beach. Often, a single storm will remove the newly added sand.

Salt-marsh areas were commonly bypassed in the early development of coastal areas in the United States. Unsuited for most agricultural purposes and difficult to build on, they were often used initially for pasture, then as waste-disposal sites and only later for building purposes. About 20 percent of Manhattan Island in New York City is built on "reclaimed" marsh and shallow harbor areas. Continued population growth has often resulted in accelerated use of salt-marsh areas as they come to represent the only available open land. La-Guardia and Kennedy Airports in New York City are built on filled marsh and estuarine areas, as are San Francisco's International, Boston's Logan, and Washington, D.C.'s National Airport; Fig. 11-25 illustrates how this was done for Newark International Airport. Once dredged and filled, salt marshes are attractive sites for industrial development because of their convenient proximity to water transport. Housing developments have also taken their toll of salt-marsh areas.

In addition, coastal marshes are commonly dredged and filled to form numerous narrow peninsulas and boat channels, offering access to waterways for small-boat operators. In the process the marshes are altered or destroyed, and circulation within any remaining wetlands is greatly modified. Frequently, the dredged basins and channels provide poor circulation, and the resultant accumulations of wastes—including untreated sewage from housing developments

Figure 11-24

Groins have been constructed along the ocean beach (left side of the picture) at Sandy Hook, N.J., in an attempt to halt the flow of sand along the beach.

and boats—have caused severe local problems including odors, and fish kills.

Even where marshes are not developed for housing or other purposes, they are often modified by extensive ditching operations as part of mosquito-control measures. Ditches are laid out and dug on regular patterns to improve drainage in the interior of the marsh and thereby eliminate mosquito-breeding areas. Part of the original marsh surface is destroyed and circulation locally altered in the process. Erosion may be accelerated owing to improper location of drainage channels.

Through the combination of all these alterations, it was estimated in 1970 that about 23 percent of estuarine systems (including salt marshes) in the United States had been severely modified, and about 50 percent moderately altered. In 1968 the United States government alone spent more than $870 million on coastal-channel and harbor improvement projects. Many millions more were spent by state and local governments and by private concerns.

In the past, planning and engineering of large-scale coastal modification projects depended primarily on experience accumulated over many years of building (and sometimes rebuilding) coastal installations. During the Middle Ages, for instance, Dutch engineers were already constructing dikes to reclaim shallow areas of the North Sea for agriculture and industry.

Figure 11-25

Former marshland on the western side of Newark Bay, N.J., has been developed by dredging channels and filling low areas to build terminals for handling containerized freight (center of photograph) and Newark International Airport (left side). Relatively undisturbed marsh is visible in the background. (Photograph courtesy Port Authority of New York and New Jersey.)

Modern coastal engineering depends increasingly on various types of models to predict the effect of structures on the coastal ocean. Where processes are well understood and computation facilities adequate, it is possible to construct mathematical models—expressions describing in precise mathematical terms the nature and effects of physical processes. Using computers, it is then possible to predict the effects of projected structures. Designs can be altered if necessary to achieve the desired objective at minimum cost, or to meet other criteria.

Although the use of mathematical models has become increasingly common since the development of computers, engineers have successfully calculated the effects of tides and winds on coastal structures for the past several decades. The great Dutch Afsluitdijk ("enclosure dike"), for instance, has kept seawater out of the former Zuider Zee estuary since 1932. Extensive studies of tides and current forces and of the probable strength and frequency of storms preceded its construction. Much valuable agricultural land has been reclaimed from shallow coastal areas behind the dike by first pumping out the salt water, then permitting rainfall to flush away the salt remaining in the soil (see also Chapter 15).

Many problems cannot be handled by mathematical techniques. Either the processes are too poorly understood to be described mathematically or too complicated to be solved by existing computers. For these problems, *hydraulic models* such as the one shown in Fig. 11-26 have proven extremely useful. The model simulates an area on a greatly reduced scale so that physical processes can be modeled and studied relatively inexpensively. Furthermore, the model permits the collection, in a few days, of data that would require perhaps years to collect in the field.

Hydraulic models are carefully constructed to duplicate the original feature. Then careful observations are made in the field, and the model is adjusted to reproduce physical processes accurately. Hydraulic-model studies of estuarine systems have been especially useful in studying such diverse phenomena as the effects of breakwaters on waves in harbors, movement of wastes in estuaries, and sediment movement in harbors.

Figure 11-26

Hydraulic model of Galveston Bay showing the inlet in the barrier beach. The model was constructed on a scale of 1:100 vertically and 1:3000 horizontally. Tides and tidal currents are reproduced in the model by a tide generator located in the Gulf of Mexico portion of the model, on the right-hand side. The strips in the model on the bay side are necessary to adjust flow in the model to make it reproduce conditions observed in Galveston Bay. (Photograph courtesy U.S. Army Corps of Engineers, Waterways Experiment Station.)

Shore—extends from low-tide to high-tide levels

Coast—broad zone extending landward from shore

Processes affecting shores are effective over long as well as relatively short times

Storm surges—relatively sudden sea level changes, usually caused by storm winds

Coastlines

> Primary coast—formed by terrestrial forces: drowned river valleys, volcanoes
> Secondary coast—shaped primarily by marine processes or organisms; barrier beaches, coral reefs

Deltas—large deposits of river-borne sediment at the river mouth

> Modified or prevented by tidal currents or waves
> Steps in delta formation: filling of estuary, formation of distributaries, shifting of distributaries
> Shape controlled by river processes and coastal processes

Beaches—deposits of loose sedimentary material moved by waves

> Usually sand (sometimes gravel) derived from local sources, such as rivers or bluffs; nature of material depends on coastal processes
> Barrier beaches—islands or spits built by action of waves and currents
> Beach coastlines—generally smooth outlines of pre-existing topography
> Offshore—sand bar and trough
> Onshore—low-tide terrace, beach face, berm crest (high point, formed by storms)
> Dunes—commonly behind large sand beaches

Beach processes

> High, choppy waves erode beaches, usually in winter
> Long-period, low swell builds beaches, usually in summer

> Waves approaching beach obliquely cause longshore currents and beach drift
> Water movement within breaking wave is toward land at surface and along bottom, seaward at mid-depths
> Rip currents—return flow of water moved toward beach by breakers

Minor beach features

> Swash—thin film of water from wave moving up beach carries debris
> Rills—minute channels formed by water draining out of beach
> Laminations—layers of colored minerals left behind during periods of strong wind erosion
> Beach cusps—uniformly spaced tapering ridges with rounded embayments between them
> Ripple marks—formed by current movements

Salt marshes—low-lying area covered by vegetation, usually submerged only at highest tides

> Tidal flats—covered by high tide, bare at low tide; may or may not have vegetation cover
> Tidal creek channels—flooded at all times, avenues for drainage
> Sediment moves into marsh during flood current, not readily eroded by ebb current
> Vegetation also traps sediment—e.g., mangrove swamps

Modification of coastal regions

> Estuarine modification
> Includes dredged navigation channels, stabilization of harbor entrances with jetties, waste disposal
>> Alters estuarine circulation patterns and sediment flow
> Salt-marsh modification
>> Used for waste disposal, filled for construction of houses and airports
>> Navigation-channel dredging and mosquito-control drainage alters circulation

SELECTED REFERENCES

BASCOM, WILLARD. 1964. *Waves and Beaches: The Dynamics of the Ocean Surface.* Doubleday Anchor Books, Garden City, N.Y. 267 pp. Elementary, well written.

BIRD, E. C. F. 1969. *Coasts.* M.I.T. Press, Cambridge, Mass. 246 pp. Wave, current, and wind effects on the structure of coastline; elementary.

JOHNSON, D. W. 1965. *Shore Processes and Shoreline Development.* Hafner, New York. 584 pp. A classic study, first published in 1919.

KOMAR, PAUL D. 1976. *Beach Processes and Sedimentation.* Prentice-Hall, Englewood Cliffs, N.J. 429 pp.

STEERS, J. A. 1969. *Coasts and Beaches.* Oliver & Boyd, Edinburgh. 136 pp. Elementary discussion of beaches and beach processes; examples primarily from Great Britain.

VAN VEEN, JOHAN. 1962. *Dredge, Drain, Reclaim: The Art of a Nation.* Fifth Ed. Martinus Nijhoff, The Hague. 200 pp. Elementary discussion of Dutch reclamation work.

ZENKOVICH, V. P. 1967. *Processes of Coastal Development.* Interscience, New York. 738 pp. Thorough treatment of shore processes, with emphasis on Russia and recent Russian research.

12

biological oceanographic processes

T he main difference between land and ocean as environments for life is that on land an organism is surrounded by air, and in the ocean by water. Some implications of this difference will be examined in this study of interactions between marine organisms and the environments in which they live.

THE MARINE ENVIRONMENT If we compare conditions in our environment with those in the ocean, we note that on land we live at the bottom of our "ocean" rather than all through it. Some spores and seeds, as well as bacteria and viruses, do indeed remain for relatively long times in the atmosphere and some spiders can float attached to a fine thread. Birds and flying insects often use updrafts to support themselves for a while, but even they alight from time to time.

But the inhabited layer of atmosphere is relatively shallow. The tallest trees extend only about 70 meters into the air and few animals penetrate into the soil more than a meter. Most terrestrial organisms live in a zone about 20 meters thick.

The sea, on the other hand, is three-dimensional. Many marine organisms spend at least part of their lives attached to a solid surface, but most float or swim in the seemingly boundless, relatively homogeneous ocean. Life exists at all depths—from the sunlit surface layer to the dark ocean bottom. The ocean averages about 4 kilometers deep, and covers about 70 percent of the earth's surface, a living space about three hundred times larger than that inhabited on land.

The density of seawater, like that of living protoplasm, is around 1.025 grams per cubic centimeter, whereas the density of air is much less—around 0.0012. This means that the effective weight of animals and plants in the sea is only a negligible fraction of what it would be in the air. Support requirements for organs and appendages of the body are quite different for these two environments. The heavy internal skeletons of land animals, developed to hold us upright against the pull of gravity and to facilitate motion and leverage, are unnecessary in the ocean. Neither are the rigid cellulose structures characteristic of large land plants.

One problem unique to the ocean is sinking, which has no counterpart on land. Most marine organisms are adapted for survival in a particular layer of the ocean, having specific light and temperature requirements. An organism less dense than seawater tends to float toward the surface. But most marine plants and animals, being slightly denser than seawater, tend to sink unless they can swim upward or are equipped with some means of maintaining buoyancy.

Marine environments are far less variable than the land. Seawater temperatures change slowly, as we know, owing to seawater's large heat capacity. Although surface water temperatures vary widely with latitude (Fig. 12-1), daily and seasonal fluctuations are much smaller than in land areas at the same latitudes (see Chapter 6). Thus marine animals are rarely subjected to large or rapid temperature changes, as many land forms are.

Figure 12-1

Plants and animals live at all temperatures in the world ocean. Note that temperature variations are greatest in temperate waters, and least at high latitudes. (After J. W. Hedgpeth, ed., 1957. *Treatise on Marine Ecology and Paleoecology,* The Geological Society of America, New York, p. 364.)

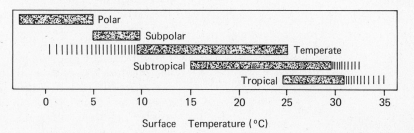

Most marine organisms are cold-blooded, meaning that their internal temperatures are essentially uniform with the environment. Within the temperature tolerance range for a species, such processes as growth and development, decomposition, photosynthesis, oxygen consumption, and metabolic rate accelerate with increased water temperature. As a rule, cold-blooded organisms can more easily tolerate temperature changes toward low extremes than toward the high end of their tolerance range. Pronounced cooling is likely to bring about a period of quiescence, in which organisms require little or no food and less oxygen than normal. Extreme heating may cause death.

Most marine organisms in the open ocean have body fluids whose salinity is not very different from the seawater in which they live. Thus they need not develop mechanisms to protect themselves from losing water, as land forms do, nor are they in danger of taking in too much or too little salt. But salinity is often variable in intertidal, estuarine, and coastal environments. Some fish and invertebrates have ways of excreting water, others of excreting salt, and yet others of protecting their internal salinity with outer shells or scales. In estuaries, where there is marked salinity change going from fresh water to seawater, the different kinds of plants and animals are distributed according to their tolerance for a fresh, brackish, or high-salinity environment (Fig. 12-2). Note that the fewest species live in brackish water, where salinity is most variable.

Figure 12-2

Distribution of plants and animals according to salinity tolerance in the Tees River estuary, on the northeast coast of England. (After W. A. Alexander et al., 1935. *Survey of the River Tees,* Water Pollution Research Technical Paper No. 5, Her Majesty's Stationery Office, London, p. 66.)

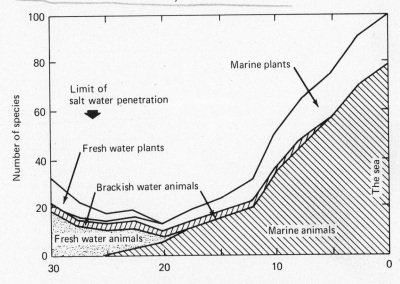

Another important aspect of the ocean as an environment for life is that it is so nearly homogeneous, in contrast to land environments where plants absorb nutrients and water from the soil, and gases from the air. Seawater carries nutrient salts, oxygen, and carbon dioxide dissolved throughout the ocean, though not in uniform concentration. Marine plants need no roots to absorb nutrients and water from soils, nor leaves to exchange gases with the atmosphere and catch the sun's light. In a typical marine plant these functions are carried out within a single cell which absorbs nutrients, light, and water from the surrounding ocean.

In describing marine life styles, three principal categories are recognized: drifting, swimming, and attached. The drifters or *plankton* are organisms that have limited ability to propel themselves against the currents. Included here are *bacteria, phytoplankton* (one-celled plants), and *zooplankton* (animal plankton), many of which are microscopic. The strong swimmers, or *nekton,* include most adult fishes, as well as squid and whales. *Benthos,* or bottom-dwelling organisms, include large plants which occur only in shallow waters, and bottom-dwelling animals at all depths.

Marine communities are classified according to habitat as *intertidal, neritic* (coastal ocean), *pelagic* (free-swimming), *benthic* (bottom-dwelling), and *abyssal* (deep-ocean) (Fig. 12-3).

Figure 12-3

Classification of marine environments.

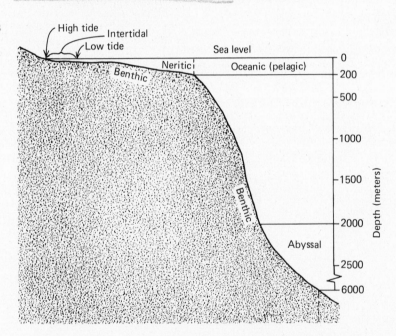

Perhaps the most important distinction, in terms of marine ecology, is between the *photic,* or sunlit surface zone, and the *aphotic,* essentially dark waters below. In the photic zone, phytoplankton produce the carbohydrates and amino acids that support virtually all marine life.

The depth of the photic zone depends on water turbidity (see Chapter 6) as well as the intensity of sunlight. In the clearest open ocean about 50 percent of light penetrating the surface remains at 18 meters depth, with about 1 percent extending to 100 meters. But near coasts, for example off Cape Cod or in the English Channel, phytoplankton grow so abundantly in the nutrient-rich waters that they block the sunlight. As a result, only about 50 percent of the light penetrates to 3.5 meters and 10 percent to 8 meters. At 17 meters depth, only 1 percent of insolation may remain.

Plants and animals in the ocean, as on land, depend on one another to provide the conditions which support life. Mutually interdependent members of a community exchange matter and energy with other organisms and with the nonliving environment. In a self-sustaining community, or *ecosystem*, *autotrophic* organisms (plants) produce food from inorganic compounds, using energy from the sun. *Heterotrophic* organisms (animals) consume plant food, and *decomposers* (bacteria) break down uneaten organic matter into its original inorganic components (Fig. 12-4). Dynamic patterns of growth and interaction may cause the amount of a particular element to vary from time to time, but the basic components remain in balance so that the ecosystem as a whole maintains a steady state.

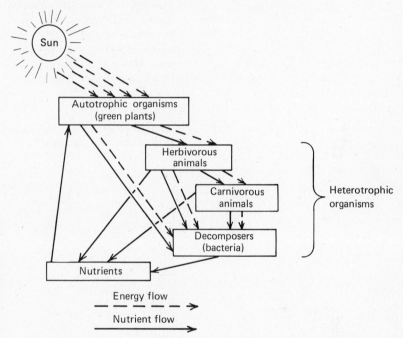

Figure 12-4

Flow of matter and energy through the marine food web. Nutrient salts are returned to solution in seawater to be recycled. Energy, originally derived from the sun, and stored by plants in the form of carbohydates, is dissipated as work and as heat.

Population size is controlled by *limiting factors*. This means that population growth is limited by the availability of food, living space, or some other resource. Factors that limit phytoplankton production have the most effect on marine ecosystems, since most life in the ocean depends on food production by these microscopic, floating plants. Phytoplankton growth is limited by three major factors: availability of *light, nutrients,* and *trace organic materials* in the water.

During winter, lack of sunlight limits plant growth over much of the world ocean, specifically in mid- and high-latitude waters. When sunlight becomes stronger in early spring, phytoplankton begin to *bloom* locally, that is, to reproduce rapidly. Under favorable conditions, they may become so abundant that plants growing near the surface cut off light from those below. If this happens, light again becomes a limiting factor, since growth of plants is then limited to the topmost surface waters. In clear water, photosynthesis may take place to depths of 100 meters, or more.

Scarcity of nitrates or phosphates in near-surface waters limits plant growth over much of the world ocean. This is particularly true in the centers of ocean basins, and in deep, tropical waters, for two reasons. First, in very deep waters, dying plants and animals tend to sink below the photic zone before decomposition releases their nitrates and phosphates into the water. The rate of nutrient return from deep to surface waters in these areas is very slow, on the order of hundreds to thousands of years. Secondly, in the tropics where light levels are uniform throughout the year, and there are no seasons, phytoplankton use available nutrients throughout the year, so that they never build up in the water. Outside the tropics, by contrast, nutrient levels rise during the winter when photosynthesis of plants is light-limited.

Besides phosphorus and nitrogen, many other elements are essential to marine plants, and some are typically present in quantities that limit the growth of a particular species of organism. Lack of silicon, for example, can limit growth of *diatoms,* a common high-latitude phytoplankter, or of *silicoflagellates* (these organisms are described in Chapter 13).

The last column, (S/P) in Table 12-1, is the ratio of supply (S) to demand (P) for each element, a figure which varies widely for the different elements. S/P is small for nitrogen, phosphorus, and silicon, normally available in seawater as NO_3^- (nitrate), PO_4^{3-} (phosphate) and SiO_3 (silica). Distribution of these nutrients is affected by uptake and release by organisms; thus they behave as nonconservative elements in seawater (see Chapter 5).

TABLE 12-1

*Abundance of Biologically Important Elements
in Plankton and Seawater**

ELEMENT	PLANKTON† (P) (grams per 100 grams of dry weight)	SEAWATER (S) (grams per cubic meter, 35 parts per thousand)	RATIO OF ABUNDANCE IN SEAWATER RELATIVE TO PLANKTON (S/P)
O	44.0	8.57×10^5	19,600
C	22.5	28	1.25
Si	20	3	0.15
K	5‡	380	72
H	4.6	1.08×10^5	23,500
N	3.8	0.5	0.13
Ca	0.8	400	500
Na	0.6	1.05×10^4	17,500
S	0.6	900	1,500
Cl	0.5‡	1.9×10^4	38,000
P	0.4	0.7	0.17
Mg	0.32	1.3×10^3	4,100
Fe	0.035	10^{-2}	0.29
Zn	0.026	10^{-2}	0.39
B	0.01‡	5	500
I	0.003‡	6×10^{-2}	20
Cu	0.002	3×10^{-3}	1.5
Mn	0.00075	2×10^{-3}	2.7
Co	0.00005	3×10^{-4}	6
V	0.00005	2×10^{-3}	40

* After H. J. M. Bowen, 1966. *Trace Elements in Biochemistry,* Academic Press, London.
† Mainly diatoms.
‡ Concentrations poorly known.

Plankton exhibit remarkably constant atomic ratios of carbon, nitrogen, and phosphorus (see Table 12-2). The latter two elements are incorporated in animal protein and skeletal materials, and a small amount is lost through undecomposed shells and skeletons contributing to biogenous sediments. Most nutrients, however, are returned to seawater during decomposition of organic matter by bacteria.

Table 12-2

Atomic Ratios of Carbon, Nitrogen, and Phosphorus in the Elementary Composition of Plankton *

TYPE	CARBON	NITROGEN	PHOSPHORUS
Zooplankton	103	16.5	1
Phytoplankton	108	15.5	1
Average	106	16	1

* After A. C. Redfield, B. H. Ketchum, and F. A. Richards, 1963. "The Influence of Organisms on the Composition of Sea-water," in M. N. Hill (ed.), *The Sea*, Interscience, New York, Vol. II, pp. 26–77.

A third factor limiting phytoplankton growth may be the presence (or absence) of *organic materials in trace amounts*. Substances such as vitamins, antibiotics, toxins (poisons), hormones, and trace nutrients are frequently secreted into the water by bacteria, plants, or animals. One or more of these substances may be necessary in some cases, in other cases detrimental, to the success of another plant or animal species. It is important to remember that in marine ecosystems no process is carried on in isolation; every organism is directly or indirectly affected by the activities of other members of the community.

PHOSPHORUS AND NITROGEN CYCLES

Nitrate and phosphate cycles have been extensively studied, particularly in coastal environments. Concentrations of these nutrients vary with water mass, season, and depth. Figure 12-5 shows that after nutrients are incorporated in organic matter, they tend to sink into deep water. Since much decomposition occurs below the photic zone, organic nitrates, phosphates, and silicates (S_iO_4) are released there. In tropical and subtropical surface layers, as well as in temperate environments following a phytoplankton bloom, inorganic nitrate and phosphate concentrations may be undetectable by chemical analyses.

Figure 12-5

The photic zone in all oceans is depleted in phosphate and nitrate, due to uptake by phytoplankton. Most of these data were taken between 30°N and 30°S, where there is little seasonal variation in productivity. (After Sverdrup et al., 1942.)

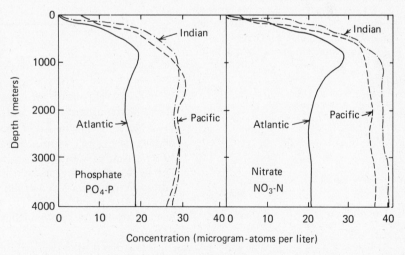

In shallow coastal waters nutrients recycle fairly quickly. Winter storms cause mixing throughout the water column, so that when light increases in early spring, a bloom can take place. Seasonal variations in phosphorus content of subpolar coastal waters are shown in Fig. 12-6. During the winter, when phytoplankton populations are small, dissolved inorganic phosphorus concentrations are as high as 0.002 milligrams of atomic phosphorus per liter, or more. The first spring bloom,

317

in about March, causes the total soluble phosphorus level to drop sharply, while the particulate phosphorus (in plants and later in animals) rises correspondingly. When nutrient levels drop off the organisms die and release nutrients, or, in some cases, a zooplankton bloom results from the abundance of plant food, and the animals release nutrients during metabolism. In any case, there is often a second bloom in April or early May, again causing a drop in soluble phosphorus and a rise in particulate organic matter.

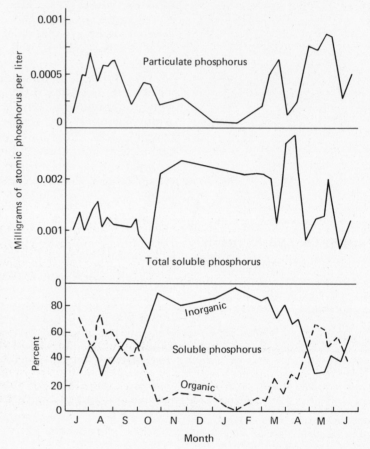

Figure 12-6

Seasonal variations in three forms of phosphorus in a coastal embayment, Departure Bay, British Columbia. Particles containing phosphorus, e.g. plants and animals, are abundant in surface waters during spring and fall blooms. Inorganic soluble phosphorus content of surface waters is highest between blooms, that is during the winter months. The curve for organic soluble phosphorus shows that it is present when plants and animals are in the water, thus it is close to the curve for particulate phosphorus. (After Parsons and Takahashi, 1973. *Biological Oceanographic Processes,* Pergamon Press, Oxford, p. 37.)

Zooplankton are an important source of dissolved phosphorus in many environments. For instance, the shrimplike copepod, *Calanus,* in a North Atlantic study, retained 17.2 percent of its dietary phosphorus for growth, eliminated 23.0 percent in solid wastes, and excreted 59.8 percent as soluble phosphorus. Another study showed that during ten weeks of a phytoplankton–zooplankton bloom in the North Sea, inorganic phosphate remained at or above 0.0006 milligrams of atomic phosphorus per liter. But in the absence of zooplankton, phosphate levels dropped to 0.0001 milligrams of atomic phosphorus per liter after two weeks.

Dissolved and particulate organic phosphorus released by phytoplankton and animals is utilized by bacteria and by certain heterotrophic phytoplankton. Since many bacteria are consumed by one-celled animals, the organic phosphorus→bacteria→microzooplankton chain (Fig. 12-7) provides a pathway by which organic phosphorus can reenter food webs directly without being changed to inorganic phosphate.

Phosphates dissolve readily from organic compounds, aided by bacterial and phytoplankton enzymes that are often abundant in seawater. But nitrogen cycles more slowly, because energy is required to "fix" it in organic form, to reduce it, and finally to oxidize it back to nitrogen. Zooplankton facilitate this process by releasing soluble organic nitrogen compounds such as urea and ammonia, which can be taken up by phytoplankton.

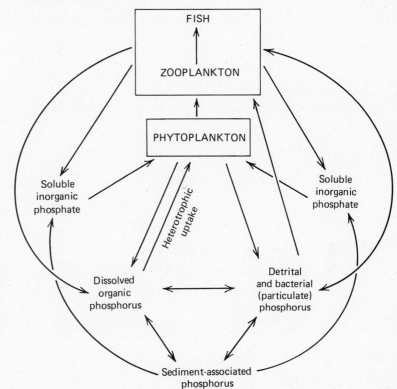

Figure 12-7

Major pathways in the phosphorus cycle of the sea. Fish and zooplankton release soluble organic and inorganic phosphates and phytoplankton release dissolved and particulate organic phosphorus. Plants take up soluble inorganic phosphates, and some are able to utilize dissolved organic phosphorus. Some organic phosphorus enters sediments and is eventually released, after decomposition, as inorganic phosphate. (Modified from Parsons and Takahashi, 1973.)

Under conditions of nitrate or phosphate depletion, marine plants can sometimes survive with greatly lowered internal concentrations of these nutrients, at least for short periods. Phytoplankton can extract phosphate from seawater very rapidly, and "hoard" much greater amounts than they actually need for growth. For this reason, they may continue to grow for several generations after the supply in the water has been depleted.

Nitrate uptake and release in temperate environments generally follows a pattern similar to that for phosphorus, except that nitrates may virtually disappear from the surface waters during summer. But in tropical and subtropical oceans, scarcity of nitrates probably limits productivity at all times.

Bacteria play an important part in the biogeochemical cycle for nitrogen (Fig. 12-8) because they oxidize ammonia to nitrites and subsequently to nitrates. Reduction of nitrates and nitrites also occurs, as Fig. 12-8 shows, because these reactions are reversible. Some nitrate is lost by formation of nitrogen gas through denitrification, but certain algae and bacteria also "fix" dissolved nitrogen, that is, assimilate it directly from seawater, especially in tropical regions.

Figure 12-9 summarizes seasonal variations in concentrations of limiting nutrients in the English Channel. In spring and late summer intense diatom blooms deplete the nutrients; the curve for silicon shows this especially well. During winter, when lack of sunlight and mixing of surface waters below the photic zone prevent much plant growth, nutrient levels are at a maximum

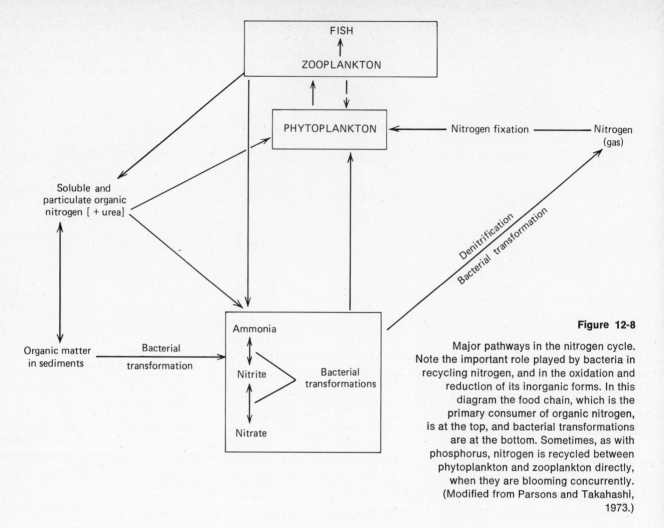

FISH

ZOOPLANKTON

PHYTOPLANKTON

Nitrogen fixation → Nitrogen (gas)

Soluble and particulate organic nitrogen [+ urea]

Organic matter in sediments

Bacterial transformation

Ammonia

Nitrite

Nitrate

Bacterial transformations

Denitrification
Bacterial transformation

Figure 12-8

Major pathways in the nitrogen cycle. Note the important role played by bacteria in recycling nitrogen, and in the oxidation and reduction of its inorganic forms. In this diagram the food chain, which is the primary consumer of organic nitrogen, is at the top, and bacterial transformations are at the bottom. Sometimes, as with phosphorus, nitrogen is recycled between phytoplankton and zooplankton directly, when they are blooming concurrently. (Modified from Parsons and Takahashi, 1973.)

Figure 12-9

Seasonal variation in nitrate, silicate, and phosphate in the English Channel. The curve for silicate reflects the fact that, in the Channel, diatoms (yellow-green algae with siliceous shells) bloom profusely in spring and again in late summer. Their shells are voided by animals, rather than being incorporated into animal tissue. Channel waters are so shallow and well-mixed that there is little difference in nutrient concentration throughout the water column at any time. (After J. E. G. Raymont, 1963, *Plankton and Productivity in the Ocean*, Macmillan, New York, 625 pp.)

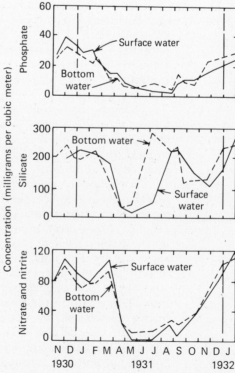

ORGANIC GROWTH FACTORS   Scarcity of dissolved organic compounds, such as *vitamins,* can also limit plant growth. Organisms that secrete these substances pre-condition waters for other organisms that require them; thus the production of such "micronutrients" often determines succession of species.

One or more vitamins, specifically B_{12}, thiamine, and biotin, are required by many phytoplankton species. Bacteria are the main producers of B_{12}, but other bacteria require it, and some phytoplankton secrete it. Complex relationships arise between producers and consumers of vitamins. For instance, the diatom *Nitzschia punctada* requires B_{12} but synthesizes biotin and thiamine. A related species, *N. closterium,* requires both B_{12} and thiamine; *N. putrida* does not require any. A dinoflagellate plankter, *Gymnodinium brevis,* requires all three. In general, about half or less of the diatom, green algae, and blue-green algae species are known to have vitamin requirements, whereas among other groups of phytoplankton the vitamin-requirers dominate.

In the open ocean, vitamin B_{12} concentrations are commonly higher below 50 meters than in surface waters. In some areas, lack of it may limit plant growth. Coastal waters normally contain an adequate supply, though this may be temporarily depleted by a heavy diatom bloom. B_{12} is absorbed in large quantities by detritus and is thus ingested by filter-feeding animals.

Soluble organic *chelating agents* (e.g., amino acids, nucleotides, and hydroxy acids) can be growth factors. They form compounds with certain metals that otherwise cannot be utilized by phytoplankton. At the normal pH of seawater, in the absence of chelating agents, iron, manganese, and probably other trace elements are insoluble, and hence unavailable to plants. Different organisms require different concentrations of trace metals and chelating agents and in general, coastal waters are less likely to be deficient in these substances. Iron, for example, may be growth-limiting where organic chelators are lacking in tropical and subtropical oceanic waters.

These growth factors can have other effects. For instance, the characteristic form of certain marine plants depends upon bacterial by-products. The sea-lettuce, *Ulva,* loses its normal broad-leafed shape and grows into long, tubular filaments in the absence of certain bacteria. The microorganisms in turn seem to be stimulated by the seaweed.

TROPHIC LEVEL
AND BIOMASS   Phytoplankton, the main *primary producers* of the ocean; they are eaten by the smallest marine animals, the herbivorous zooplankton. These *primary consumers* are in turn eaten by *secondary consumers,* either carnivorous zooplankton or fish. The position of an individual or group of individuals in a *food web*—who eats it and whom it eats—(Fig. 12-10) defines its *trophic level.* Plants are the first trophic level, herbivores the second, primary carnivores the third, secondary carnivores the fourth, and so on. At each trophic level, dissolved and suspended organic matter (*organic detritus*) is released from living organisms into the water. (This should not be confused with the release of inorganic nutrient salts during animal metabolism.) Organic detritus is released during excretion of animals and in various other metabolic processes, and it is an important source of food for many kinds of small heterotrophic organisms.

Some plant and animal matter sinks below the surface, where it supports life at all depths, so that energy produced at the surface is transported through food webs to all parts of the marine environment. When organisms die they are eaten by scavengers or decomposed by bacteria, which are themselves a food source for some marine animals. Ultimately, nearly all the chemical energy stored during photosynthesis (see p. 323) is used up in respiration. Energy, unlike matter, is a one-way flow, as illustrated in Fig. 12-4.

Figure 12-10 shows food being transferred from one trophic level to the next. But most of the food energy available to a plant or animal cell is used by the cell itself in growth, reproduction, and other functions. Animals use more energy than plants because they are active in seeking and consuming prey. To calculate how much energy remains at each trophic level, available to the next level of consumers, *ecological efficiency* (E) is defined as follows:

$$E = \frac{\text{amount of energy extracted from a trophic level}}{\text{amount of energy supplied to a trophic level}}$$

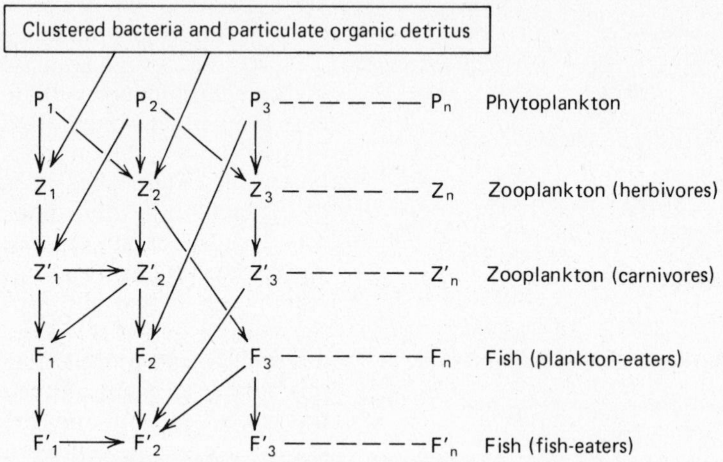

Figure 12-10

A theoretical pelagic food web. Many species (P_1, P_2, P_3, etc.) of organisms at each of 5 trophic levels are shown. A nearly infinite number of food relationships is possible, though at any given time and place abundance of one organism may dominate a particular trophic level. Note that an organism (e.g. F_2') may feed on at least three trophic levels. (Adapted from Parsons and Takahashi, 1973, p. 123.)

As a general rule we assume E to be on the order of 10 to 20 percent in marine ecosystems. An ecological efficiency of 10 percent means that it takes 500 grams of phytoplankton to support a population of herbivorous zooplankton that weighs 50 grams. This zooplankton population provides a fish with enough food to increase its body weight by 5 grams.

The value of E is determined by a variety of factors, such as the amount of detritus recycled in a given food chain, and the amount of work an animal must do to locate and capture its prey. In general, energy transfers in oceanic food webs are of lower ecological efficiency than those in coastal ocean food webs.

To quantify the amount of organic material in an ecosystem we use the concept of *biomass*. Biomass is the weight of plants or animals in a given population, expressed as weight of organic carbon per volume of seawater. For example, we might find that the biomass of primary producers at a given location is 880 milligrams of carbon per cubic meter of seawater (mgC/m^3). A cubic meter of water taken from the same location might contain 90 milligrams of carbon per cubic meter of herbivorous zooplankton, measured by filtering the water through a fine sieve and picking out zooplankton known to consume only plant food. The same water might also contain 10 milligrams of carbon per cubic meter of carnivorous zooplankton.

Biomass can also be expressed as total weight of organisms in the water under a specified area (e.g. a square meter) of seawater, or grams per square meter. Bottom-dwelling populations are usually measured in terms of weight per square meter of the bottom. Sometimes plant biomass is given as plant pigment (e.g. chlorophyll) units per volume of water.

Marine plants, like most of their terrestrial counterparts, are auto-trophs. The presence of *chlorophyll* or other light-absorbing pigment in autotrophic organisms enables them to make their own food, typically *carbohydrates,* complex organic substances in whose bonds chemical energy is stored. In this process, known as *photosynthesis,* carbon dioxide combines with water and 120 kilocalories of radiant energy are absorbed per mole of carbohydrate formed. The end product is a sugar containing chemical energy with oxygen being released as a by-product. Thus

$$6CO_2 \; + \; 6H_2O \; + \; 120 \, kcal \xrightarrow{\text{chlorophyll}} C_6H_{12}O_6 \; + \; O_2$$

carbon   water   energy   carbohydrate   oxygen
dioxide

This simplified statement shows that oxygen is given up and energy is stored, but it does not describe the sequence of one hundred or more chemical steps involved. Energy exchanges during each step are mediated by specific enzymes.

The process is reversible during plant or animal *respiration* when oxygen is taken up and chemical bonds are broken. Carbohydrate is *oxidized* and energy released. All *heterotrophic* organisms, that is, animals and most bacteria, obtain energy from breaking down and rearranging chemical-energy-containing materials originally synthesized by autotrophs.

It is important to remember that plants utilize part of the energy they produce. During respiration, plants and animals take up oxygen and expend energy on growth, reproduction, and work. The net production of food by a green plant depends on its synthetic processes exceeding the amount of energy used in respiration.

The first stages in photosynthesis involve absorption of solar energy by chlorophyll. These stages take place in sunlight only and are called the "light reaction." Subsequent parts of the synthetic process are carried on irrespective of light conditions, and respiration takes place in light and dark.

There are two basic steps in the "light reaction." The first involves synthesis of a substance which will later accept oxygen (this is defined as a *reduced* substance). Hydrogen is added to a complex organic molecule, *nicotinamide adenine dinucleotide phosphate* (NADP), yielding NADPH:

$$NADP + H \longrightarrow NADPH$$

The added hydrogen is now available to create other reduced substances, meaning that NADPH has *reducing potential.*

In the second step, energy-containing molecules are created. Energy from the sun is conserved in a chemical form by means of a cyclical electron-transport chain, whereby ATP (adenosine triphosphate) is formed from ADP (adenosine diphosphate) and phosphate (PO_4^{3-}).

$$ADP \; + \; P_i \longrightarrow ATP$$

adenosine   inorganic   adenosine
diphosphate   phosphate   triphosphate

This process is called *phosphorylation.*

The chemical energy contained in ATP and the reducing power of NADPH now remove oxygen from CO_2, forming carbohydrates as well as synthesizing proteins and fats. Collectively, this part of the process is referred to as the "dark reaction."

Within all cells, plant and animal alike, energy is transferred as shown in Fig. 12-11. A certain amount of energy is conserved in the internal recycling of ATP and ADP, but some is permanently lost, as shown by the two black arrows which represent energy leaving the system.

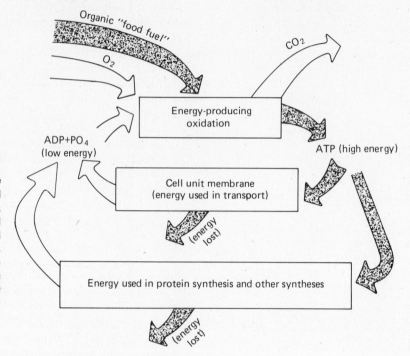

Figure 12-11

The cyclic transfer of energy within single cells. Conservation of energy is achieved by the ADP–ATP recycling process. Remember that in plants, the organic "food fuel" is derived from compounds formed by photosynthesis within the cell itself. In animals, food must come from consumption of a plant or another animal. (Adapted from W. D. Russell-Hunter, 1970. *Aquatic Productivity,* Macmillan, New York, p. 15.)

MEASURING PRIMARY PRODUCTIVITY

The net amount of food produced in a given area during a given time defines the *productivity* of the area's plant population. Productivity is thus a rate phenomenon, whereas *standing crop,* or biomass, is a measure of the amount, by weight, of plant (or animal) material in a given area at any time. For example, productivity of a population may be high even though its biomass is quite small. This is the case when grazing by zooplankton sharply depletes a rapidly reproducing, highly productive phytoplankton crop as shown in Fig. 12-12.

Figure 12-12

Standing crop represents only a small part of total plant production. Data based on a study in the English Channel, showing that, of 85,000 plant pigment units per cubic meter produced (according to predictions from PO_3^{4-} uptake), only 2500 units per cubic meter, or about 3 percent, were measured as a standing crop. Difference was assumed to be due to grazing. (After Raymont, 1963.)

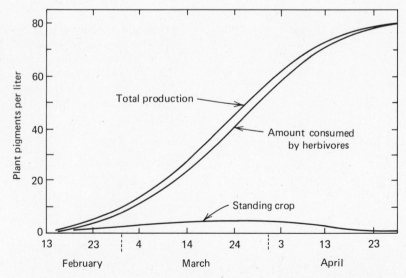

Productivity may be estimated by approximating the average standing crop of a population and multiplying that figure by its rate of generation (doubling time) over short intervals. If the average standing crop of zooplankton is 500 milligrams of carbon per square meter and the population doubles once a month, then its productivity is 6000 milligrams or 6 grams of carbon per square meter per year (assuming constant production throughout the year).

Productivity is hard to measure accurately in the ocean. One way to quantify *primary production* (plant productivity) is to introduce radioactive $^{14}CO_2$ into a sample of seawater and monitor its rate of uptake by phytoplankton. This measures *gross productivity*, which is defined as the mass of carbon fixed per unit area per unit time, usually in grams or kilograms of carbon per square meter per year. Net productivity is calculated from gross productivity by correcting for losses due to respiration, which are usually 10–50 percent.

Another method for measuring productivity is to determine the amount of oxygen produced by the plants in seawater, because oxygen is given off by plants in direct proportion to the amount of organic carbon synthesized. In this "oxygen-bottle" technique, sealed bottles of seawater containing phytoplankton are lowered to selected depths below the surface. After a known period, the oxygen content of the water in the bottles is measured and is compared with the amount present when the bottles were sealed.

Near the top of the photic zone, oxygen content increases markedly. The rate of oxygen increase diminishes with depth until a point of no net increase is reached. This is known as the *compensation depth*, meaning that at that point the oxygen produced by photosynthesis is exactly equal to the amount utilized in respiration. At greater depths, respiration exceeds photosynthesis, and no plant production occurs. Finally, no oxygen is produced at all and the plants eventually die. The concept of compensation depth, then, is used in reference to metabolism of an individual plant sample at a specific depth, which is determined by extent of light penetration.

Marine and land plants contribute about equal amounts annually to global food production. The productivity of phytoplankton is several orders of magnitude greater than that of land plants, but the biomass of plankton in the world ocean appears to be several orders of magnitude less than that of terrestrial producers, such as trees, grasses, and food crops.

FACTORS IN PLANT PRODUCTIVITY

The most important single factor in the productivity of an ocean area is usually the rate at which nutrients are recycled to surface waters. For this reason, most continental shelves are regions of high productivity; they are shallow, well-mixed areas where an estuarine circulation tends to retain nutrients in the coastal ocean. Photosynthesis at mid- and high latitudes is frequently inhibited during local winter, however, due to lack of sunlight. Plants can often grow in the top few centimeters of water, but they are constantly carried below the sunlit zone due to mixing by winter storms in the absence of a well-stabilized surface zone. Thus there is no net plant production.

In the spring, increased sunlight permits photoplankton to grow and reproduce rapidly, provided the water is well supplied with nutrients. There can be no productivity increase, however, as long as plants are carried below the photic zone causing net respiration in excess of production.

The concept of *critical depth* expresses the relationship between a stable surface layer and an increase in plant productivity. That is the depth above which there is enough light to support the growth of phytoplankton. The critical depth depends on the penetration of sunlight below the surface; it is at a maximum in early summer and at a minimum in early winter at mid-latitudes. Until a thermocline is formed by warming at the surface (or until sunlight penetrates to the bottom of the water column), plants are transported below the critical depth, and no effective production can occur.

As light increases with the coming of spring (as early as January in Long Island Sound), phytoplankton begin to bloom. At this time the compensation depth is very near the surface because sunlight is still relatively weak. The critical depth (always below the compensation depth because it represents a lower limit for the growing population as a whole) may occur at some point within the mixed surface layer. If that is the case, plant growth is retarded because part of the population is continually being mixed below the critical depth. As sunlight becomes stronger, the critical depth may extend to the bottom in shallow areas. In deeper waters, a thermocline may form at or above the critical depth. In either case, the plant population expands rapidly as soon as the critical depth exceeds the depth of the mixed surface layer. This has been well documented in the North Sea as shown in Fig. 12-13.

Figure 12-13

Observations at a weather station (66°N, 2°E) in the North Sea show that plankton do not appear in large numbers until the critical depth exceeds depth of the mixed layer. (After R. S. Wimpenny, 1966. *The Plankton of the Sea,* American Elsevier, New York.)

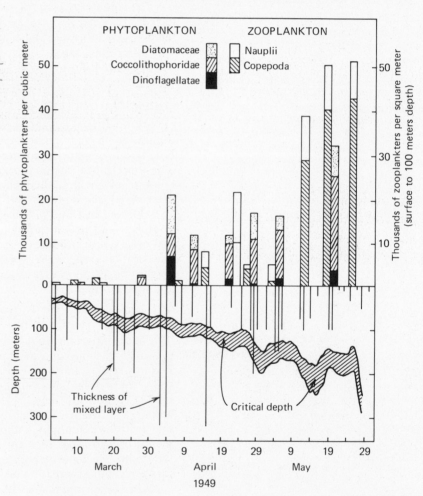

Differences between the biological properties of coastal and offshore waters are illustrated by a sequence of blooms that occurs off the New England coast. In shallow, nearshore waters the spring diatom increase begins sometime between late January and early March. Further offshore, on Georges Banks, diatoms do not bloom until late March or April, and this occurs up to a month later in continental slope waters and the open Gulf of Maine. Flowerings near the coast are smaller than offshore, and there may be additional blooms in late summer or fall, which does not usually happen in deeper water. Another difference is that the number of individuals in nearshore areas never falls as low as in continental slope and Gulf of Maine waters. In both areas, heavy grazing by zooplankton terminates the blooms within a few weeks.

The different oceanographic properties of these waters explains the difference in their productivity. Near the coast, river discharge lowers surface salinities and forms a stable surface layer during much of the year. Phytoplankton production begins to exceed consumption soon after there is an increase in sunlight (Fig. 12-14). Since some photosynthesis goes on all winter, nutrients never accumulate to the extent that they do offshore and the spring flowering is less dramatic. There is some nutrient enrichment of near-bottom waters during the course of the summer as animals feed, release organic detritus into the water, and die. Detritus and decomposition products sink to the bottom but are recycled from time to time through the shallow waters, bringing nutrients to the surface and causing late summer blooms.

Farther out on the continental shelf, in deeper waters, the spring bloom does not occur until a thermocline develops. Then, as organic matter decomposes it sinks far below the photic zone. Nutrients released are not recycled to the surface until winter storms break down the thermocline and a new annual cycle begins. Thus one spectacular bloom is the rule rather than the two that occur closer to shore.

Red tides are particularly striking examples of plankton blooms. They happen when dinoflagellates become so abundant that they discolor the water, often giving it a "tomato-soup" color extending for hundreds of yards or even many kilometers. In some instances red tides are accompanied by cases of paralytic shellfish poisoning in which neurotoxins that have been produced by the dinoflagellates become concentrated in shellfish. This often has no discernible effect on the organisms themselves, but when they are eaten by humans the toxins cause symptoms ranging from tingling in the hands and feet, to paralysis and even death. Shellfish beds are usually closed to harvesting during red tides.

Red tides may result from a redistribution of organisms due to several coincident circumstances. First is stratification of the water column caused by heavy rains or river discharge. Then in the presence of strong sunlight dinoflagellates reproduce rapidly and swim toward the light. They become concentrated from a large volume of water into the topmost surface layer.

Figure 12-14

The seasonal cycle of phytoplankton off Woods Hole, Mass. (Redrawn from G. A. Riley, 1969. "Seasonal Fluctuations of the Phytoplankton," in D. J. Reish, ed. *Biology of the Oceans,* Dickenson Publishing Co., Belmont, Calif., p. 72.)

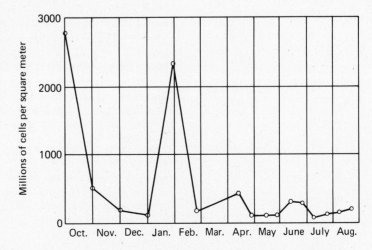

In addition to the problem of paralytic shellfish poisoning, decomposition of many organisms at high temperatures uses up the dissolved oxygen in bottom waters, killing bottom-dwelling fish and other organisms.

A cycle of red tide outbreaks has been noted in New England coastal waters with occurrences in May and June and again sometimes in late August and early September where nutrient levels are high in a stable surface layer. In Southern California, red tide is restricted by the longshore current to a narrow band of waters along the coast, or to local embayments.

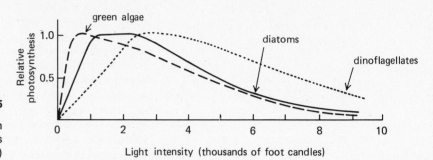

Figure 12-15

Relative photosynthesis–light curves in some marine phytoplankton. (After Parsons and Takahashi, 1973. p. 73.)

There is frequently a direct relationship between increased insolation and increased productivity. But not all plants respond to light level in the same way. Figure 12-15 shows how three types of marine phytoplankton respond to increased light intensities. In the study from which these data were taken, light saturation occurred between 500 and 750 foot-candles for green algae, between 2500 and 3000 foot-candles for dinoflagellates, and at intermediate light levels for diatoms. Above these intensities, photosynthesis decreased. "Sun-type" species which flower in the summer, or at shallow water levels, commonly photosynthesize most efficiently at high light intensities. During the winter and in deep waters, however, "shade-type" communities flourish because they experience maximum photosynthesis at lower light intensities. They can evidently utilize weak sunlight more efficiently than the "sun-type" species can.

NONLIVING ORGANIC MATTER

A large fraction of the ocean's organic content is nonliving. Most of it is dissolved, or else present as very small, nonliving particles (Table 12-3). Similar data appear graphically in Fig. 12-16 though here no

TABLE 12-3

Relative Abundance of Various Forms of Organic Matter in Seawater

FORM OF ORGANIC MATTER	RELATIVE ABUNDANCE (percent)
Dissolved organic matter ($<10^{-6}$ mm)	80
Colloidal organic matter (10^{-6}–10^{-3} mm)	18
Nonliving particulate organic matter ($>10^{-3}$ mm)	1.8
Phytoplankton	0.2
Zooplankton	0.02
Fish	0.0002

Data from J. H. Sharp, 1973(b). "Size Classes of Organic Carbon in Seawater," *Limnology and Oceanography*, 18, 441–447; and Raymont, 1963.

distinction is made between living and nonliving matter. Note the tiny fraction of particulate matter (greater than 10^{-3} millimeter), the larger proportion of *colloidal* sized particles (10^{-3} to 10^{-6} millimeter) and the very large percentage of dissolved organic matter (particle size less than 10^{-6} millimeter). Open ocean waters contain about 0.4 to about 2 milligrams of carbon per liter (mgC/1) of dissolved material, and particulate organic matter ranges from about 0.01 milligrams of carbon per liter in deep water to about 0.1 to 0.5 milligrams carbon per liter near the surface. This enormous reservoir of nonliving organic material reenters food webs largely through bacterial activity.

Sources of dissolved organic compounds are:

1. Decomposition of dead plants and animals. Initially, up to 50 percent of an organism's total mass may dissolve in seawater as cell membranes are decomposed by bacterial enzymes. Subsequent bacterial action releases still more amino acids, peptides (a type of protein derivative), carbohydrates, and fatty acids into the water.

2. Secretion by plants. Sugars, amino acids, fats, organic phosphates, vitamins, enzymes, toxins, and a variety of other organic solutes are released by living phytoplankton. Less than 10 percent of the carbon assimilated during photosynthesis is usually released into the water as dissolved organic compounds during active plant growth and reproduction. But under conditions of stress, such as in very high light intensities, 50 percent or more of the photoassimilated carbon may be released. This is also likely to occur near the end of a bloom, or when a population is for some reason inhibited from reproducing. Amounts released by healthy populations generally vary inversely with the availability of nutrients. A greater proportion of organic matter (12–27 percent) is secreted on a daily basis in nutrient-poor waters of the open ocean than in more fertile coastal waters (6–12 percent).

3. Excretion and exudation by animals. Animals excrete some dissolved organic waste products during normal metabolic activity. Certain animals also release dissolved materials while moulting, or sloughing off one shell to grow another.

The principal sources of particulate organic detritus are:

1. Dead organisms, and parts of them.

2. Solid wastes excreted by animals.

3. Dissolved and colloidal organic matter which adheres to solid particles or surfaces in seawater. This will be discussed in more detail below.

Organic carbon concentrations in the North Atlantic Ocean are highest near the surface because there is more life there than in deeper waters (Fig. 12-17). For the same reason, shallow and nearshore waters contain proportionately more particulate organic carbon than the open ocean does (Table 12-4). In open-ocean surface waters, 10 to 50 percent of the organic matter may be alive, hence particulate. But in the thousands of meters of dark, cold water below the photic zone, more than 90 percent of organic matter is nonliving. Thus the particulate fraction in open ocean water from surface to depth is only about 1 percent of total organic carbon and usually less than that in deep waters.

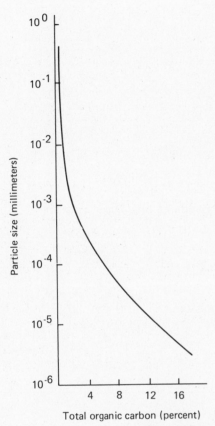

Figure 12-16

Generalized size distribution of organic carbon in seawater. Only about 2 percent is in particles larger than 1 micron (10^{-3} millimeter). Only about 18–20 percent is in particles larger than 10^{-6} millimeter. (After Sharp, 1973(b).)

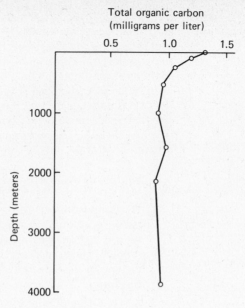

Total organic carbon
(milligrams per liter)

Figure 12-17

Depth curves of total organic carbon from
two studies in the western North Atlantic
Ocean. (Data from J. H. Sharp, 1973(a).
"Total Organic Carbon in Seawater,"
Marine Chemistry, 1, 211–29.)

TABLE 12-4

*Particulate Organic Carbon as a Percent of Total Organic Carbon
for Various Marine Environments*

AREA	PERCENT
North Central Pacific Ocean (subtropical)	0.7
Central Western North Atlantic Ocean (incl. Gulf Stream, Caribbean)	1.5
Strait of Georgia, North Pacific (nearshore)	13
Chukchi Sea, Arctic Ocean (nearshore, shallow)	24

Data from Sharp, 1973(a).

BACTERIA IN FOOD WEBS

Bacteria, being less than 2 microns in diameter (<0.002 millimeter), have negligible biomass. But their role in marine food webs cannot be evaluated on the basis of weight, as it is for plants and animals, because decomposition, rather than energy transfer, is their most important function.

For example, a total of $2.5–3.0 \times 10^{10}$ metric tons of CO_2 is 'fixed" annually by photosynthesis. An equivalent amount is therefore released through respiration during the same period, with remineralization of organic compounds to inorganic nutrients. Bacterial activity may account for roughly 90 percent of this turnover. Populations of bacteria turn over quickly, so that any one time the bacterial biomass may represent only about one-tenth of annual production. Populations can double in a few hours to a few days, but predation and death usually keep their number constant.

There are usually a few hundred to several thousand living bacteria per milliliter (cubic centimeter) of coastal water, though as few as 10 and as many as 1 million have been counted in different water masses. In the open ocean there are generally about 10 per milliliter,

Figure 12-18

Alvin, the submersible research vessel in which the effect of retarded deep-sea decomposition was documented. (Photograph courtesy Office of Naval Research.)

but below 100 meters depth there may only be a few per liter. Some species live free in seawater; others must be attached to a solid surface in order to grow and reproduce.

At surface temperatures and pressures bacteria have a high oxygen consumption and metabolic rate, but at the low temperatures and high pressures of the deep ocean they seem to be less active. This was inadvertently demonstrated in 1968 when the Woods Hole Oceanographic Institution's submersible research vessel, *Alvin* (Fig. 12-18), sank with the hatch open in 1540 meters of water and remained on the bottom for ten months. When recovered, the crew's lunch of apples, bologna sandwiches, and bouillon in thermos bottles was found in a plastic box, wrapped but soaked with seawater. The food smelled and tasted fresh when it was first opened, but after four weeks of storage at 3°C in sterile seawater it became putrid. Additional experiments showed that organic material, wrapped to prevent water or air from circulating past it, decomposes 10 to 100 times more slowly at deep ocean pressures than at the surface, even at very low temperatures. Studies of digestive tract fauna in deep-ocean animals, however, indicate that bacteria play the same role in decomposing undigested material as they do at sea level.

Bacteria are an important source of concentrated protein for some filter-feeding animals, and for organisms that scrape accumulated detritus from underwater surfaces. Individual microbes cannot usually be captured directly from the water because of their small size, but colonies of bacteria become locally concentrated on suspended particles and other submerged surfaces in the ocean (Fig. 12-19(a)). Bacteria collect on the surface-active or oily organic residues that adhere to sediment particles, piers, and ships' hulls, forming a greasy film. Dissolved and colloidal organic compounds also have a tendency to clump together in seawater, aggregating into amorphous, suspended masses called "marine snow." Any such organic mass attracts bacteria, and any solid object in the water soon becomes coated with a film of organic matter containing bacterial microcolonies and their metabolic prod-

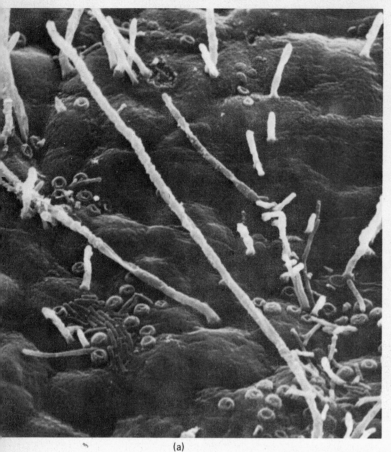

(a)

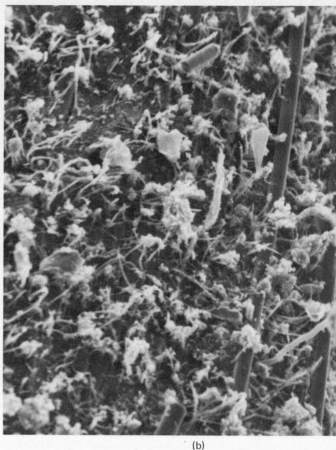

(b)

Figure 12-19

Bacteria colonize submerged surfaces, creating a slimy matrix that supports many benthic organisms. (a) Rod-shaped, ring-shaped, and filamentous bacteria, attached to the red seaweed *Rhodymenia palmata.* The longest filament here is about 0.05 millimeters (50 microns) in length. (b) A "garden" of diatoms and bacteria grows on fiberglass rods under water. An area about 0.37 millimeters square is shown in this photograph. (Photographs courtesy of Dr. J. M. Sieburth and University Park Press, Baltimore, Md.) (c) The tunicate *Molgula,* a common fouling organism in Chesapeake Bay, feeds on organic detritus from the side of an aquarium. (Photograph courtesy Michael J. Reber.)

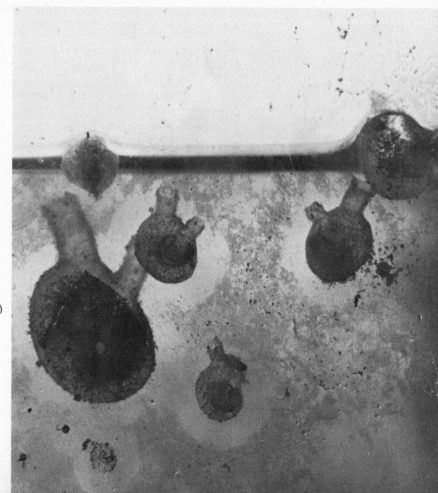

(c)

ucts. Many kinds of benthic organisms colonize surfaces where bacterial slime has accumulated, in a process known as "fouling" (Fig. 12-19(b)). Some organisms may simply be seeking a firm surface to attach to; others live by scraping off and consuming the film with its aggregated organic components (Fig. 12-19(c)).

Figure 12-20 summarizes major pathways by which materials are recycled in marine ecosystems. Organic matter derived from byproducts of plant and animal metabolism is shown cycling to the right of the living components, products of decomposition to the left. Note that bacteria decompose organic material, and release carbon dioxide and inorganic nutrients; these reenter the water to support photosynthesis. Some inorganic nutrients are also returned directly to the environment by animal excretions. The term "heterotrophic uptake" refers to direct assimilation of dissolved substances from seawater by bacteria and certain kinds of heterotrophic phytoplankton. For example, some bacteria quickly take up and assimilate dissolved organic matter that has been exuded by photosynthesizing plants, such as amino acids and glucose.

Processes involving bacteria in sediments and in the abyssal zone are discussed in Chapter 14.

Figure 12-20

Pathways of transfer and nutrient remineralization from dissolved and suspended organic matter in the marine environment. Compare with Fig. 12-7. Note that both plants and animals contribute dissolved and suspended organic matter to the vast reserves in seawater; some of it reenters the food web via filter-feeding animals and heterotrophic plants and the rest is decomposed by bacteria. (Modified from Parsons and Takahashi, 1973. p. 98.)

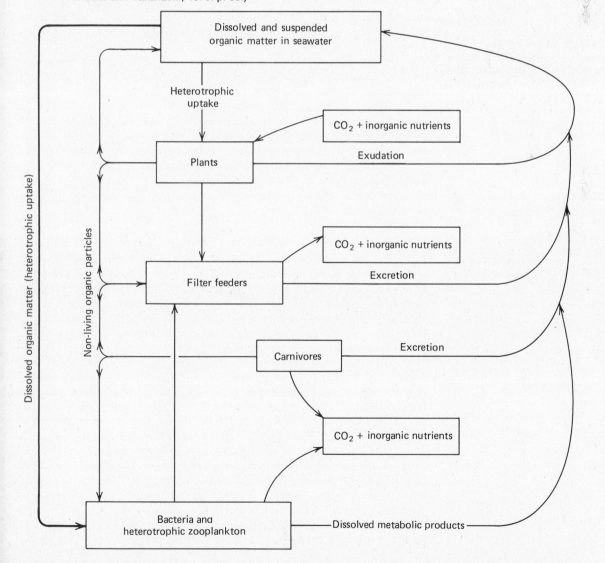

Bacteria utilize oxygen in decomposition of organic matter. For this reason, curves showing distribution of oxygen in the ocean (Fig. 12-21) are usually a mirror image of nutrient distributions, at least in the top 1000 meters (see Fig. 12-5). Note in Fig. 12-21 that in the North Atlantic, where freshly oxygenated deep waters are moving toward the bottom, oxygen levels are high throughout the water column. But throughout most of the ocean, although surface layers are saturated with oxygen, its concentration diminishes with depth for a few to several hundred meters due to respiration and decomposition of animals and bacteria in near-surface waters.

Oxygen demand at great depths, however, is limited. Low temperatures reduce the metabolic rate of marine organisms, and high pressures seem to reduce bacterial activity, so that the total rate of oxygen consumption is lowered. Scarcity of food keeps deep-water populations sparse, and decomposition of detrital material is minimal since most of it has been consumed near the surface. Consequently, deep-ocean waters contain much of the oxygen with which they were saturated when they sank below the surface.

It takes hundreds to thousands of years for deep waters to reach the surface at mid- and low latitudes. The relatively oxygen-rich profile of the Atlantic Ocean (see Fig. 12-21) reflects the fact that surface waters are introduced at both poles, whereas the northern boundaries of the Pacific and Indian Oceans are blocked by land barriers. South of Greenland, where the North Atlantic Deep Water sinks throughout the year, there is little variation in dissolved-oxygen profile throughout the water column.

As deep water rises in mid- and low latitudes, its dissolved oxygen content is gradually reduced due to utilization by marine organisms. At these latitudes, in warm and temperate waters, there is a zone, varying in depth from 150 to 1000 meters, where dissolved oxygen concentrations are at a minimum (see Fig. 12-21). Little or no dissolved oxygen from the surface is mixed downward to this layer, and little oxygen remains in the water that has risen from great depths.

In areas of restricted circulation such as some fjords and deep-ocean basins, deep waters may be devoid of dissolved oxygen, and in summer, some relatively shallow estuaries may also have no oxygen in near-bottom waters.

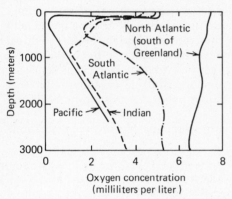

Figure 12-21

Characteristic oxygen profiles of major ocean basins. Positions: 12°N, 137°W; 0°N, 80°E; 9°S, 5.5°W; 55°N, 45°W. (After Dietrich, 1963.)

What controls productivity in the ocean? Ocean waters contain vast amounts of nutrients, and ample sunlight for plant growth occurs over most of the earth. Furthermore it has been shown that productivity of the richest ocean areas (shallow-water estuarine and coral reef ecosystems) compares favorably with intensively cultivated agricultural land; gross productivity in these areas can be as high as 20 grams of carbon per square meter per day (Fig. 12-22).

But there are important differences between production of organic matter on land and in the ocean. Land plants can grow for years before soils are depleted of nutrients, whereas phytoplankton can only sustain maximal production for a few days before nutrients in the waters must be replenished. Highly productive ocean water contains only about 5 parts usable nitrogen per ten million parts of seawater, enough to produce perhaps 5 grams of dry organic matter. Over most of the ocean, nitrogen concentrations average only about one-fourth as much. But in fertile soil about 30,000 times as much nitrogen is available for plant growth, on the order of 5 parts per thousand.

One key to productivity of ocean waters is the rate at which nutrients can be renewed. Another is the quantity of light absorbed by plants. Primary productivity in polar regions is light-limited through-

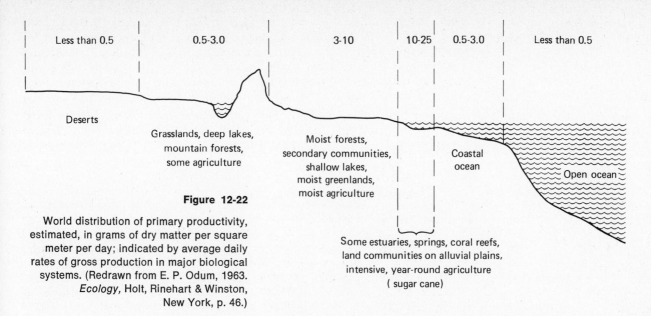

Less than 0.5 | 0.5-3.0 | 3-10 | 10-25 | 0.5-3.0 | Less than 0.5

Deserts

Grasslands, deep lakes,
mountain forests,
some agriculture

Moist forests,
secondary communities,
shallow lakes,
moist greenlands,
moist agriculture

Coastal
ocean

Open ocean

Some estuaries, springs, coral reefs,
land communities on alluvial plains,
intensive, year-round agriculture
(sugar cane)

Figure 12-22

World distribution of primary productivity,
estimated, in grams of dry matter per square
meter per day; indicated by average daily
rates of gross production in major biological
systems. (Redrawn from E. P. Odum, 1963.
Ecology, Holt, Rinehart & Winston,
New York, p. 46.)

Figure 12-23

Distribution of primary productivity in
surface waters of the world ocean. (Redrawn
from O. J. Koblentz-Mishke, V. V.
Volkovinsky and J. G. Kabanova, 1970.
"Plankton Primary Production of the World
Ocean," in *Symposium on Scientific
Exploration of the South Pacific,* National
Academy of Sciences, Washington, D.C.,
pp. 183–93.)

out the year, because the growing season is short and a heavy ice cover
absorbs much of the available insolation. Lack of stability in the surface zone, where plankton are carried below the critical depth, can
limit production in temperate latitudes. But in general there is ample
light in surface waters, and productivity is controlled by the rate at
which nutrients are recycled to the photic zone.

Figure 12-23 shows relative productivity of surface waters in the
world ocean. Note that the northernmost Arctic is thought to be an
area of low productivity due to its limited insolation. Note also that
the centers of ocean basins have low productivity because nutrients
sink out of the photic zone and are not rapidly renewed. It has been

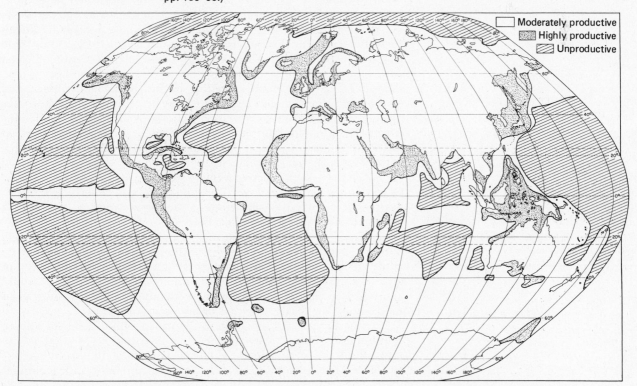

Moderately productive
Highly productive
Unproductive

estimated that subsurface waters move upward through the open ocean thermocline (at about 1000 meters depth) at rates between 0.5 and 1.6 centimeters per day. Nitrogen concentrations at this depth are usually about 20 milligrams atomic nitrogen per liter, so that 3 to 9 grams of carbon per square meter per year might be a normal productivity if it were not for the fact that zooplankters excrete nutrients in the surface layer. In fact, primary productivity in central ocean basins averages around 50 grams of carbon per square meter per year (Table 12-5).

TABLE 12-5

Primary Productivity of Ocean Areas *

PROVINCE	PERCENT-AGE OF OCEAN	AREA (millions of square kilometers)	MEAN PRODUC-TIVITY (grams of dry carbon/ m²/yr)	TOTAL PRODUC-TIVITY (10⁹ tons carbon per year)
Open ocean	90	326	50	16.3
Coastal zone	9.9	36	100	3.6
Upwelling areas	0.1	3.6	300	0.1
Total	100.0		—	20.0

* After John H. Ryther, 1969. "Photosynthesis and Fish Production in the Sea," *Science* 166, 72–76.

† Includes offshore areas of high productivity.

Nutrient concentrations are much higher on the continental shelf and slope and upwelling regions, because vigorous vertical mixing returns nutrients to the photic zone as they are released by decomposition processes in deeper waters. In the Pacific Ocean upwelling occurs along much of the eastern margin, especially the Peruvian–Chilean coast and California, as well as off Japan, Kamchatka (Fig. 2-17), India and western Canada. Similar processes occur in the Atlantic off West Africa, northeast Brazil, and the South American southeast coast. Regions of divergence and upwelling in the Indian Ocean are caused by monsoon winds, and in the open ocean upwelling is associated with divergences along the equator (see Fig. 12-23) and at the border between the Antarctic and mid-oceanic current gyres. Strong vertical water movements also occur in parts of marginal and inland seas such as the Bering Sea, and Arabian Sea.

Relatively unproductive waters are most widespread in the Pacific Ocean, in the Arctic, and at subtropical latitudes. The Indian Ocean has the largest relative proportion of productive areas, particularly in temperate and equatorial latitude belts.

A total of 25 to 30 billion tons of inorganic carbon is apparently fixed by marine plants every year. Allowing the carbon utilized by autotrophic organisms in respiration, net primary productivity of organic carbon is around 15 to 20 billion tons per year. This would appear to be the limit of primary productivity, unless nutrients can somehow be artificially recycled on an enormous scale.

How do these figures relate to the production of food for human consumption? The answer depends on distribution of productivity throughout the world ocean. About 90 percent of the ocean surface, or 326×10^6 square kilometers, is classified as open ocean. If annual productivity here averages about 50 grams of carbon per square meter per year, this is a total of 16.3×10^9 (16.3 billion) tons. Coastal waters, defined as those less than 180 meters deep (but excluding areas of active upwelling) account for another 9.9 percent of the ocean surface, or 36×10^6 square kilometers. These areas are about twice as productive as the open ocean and contribute about 3.6×10^9 tons of organic carbon annually. Coastal upwelling regions are the most productive of all, especially along the subtropical west coasts of continents. Where eastern boundary currents and prevailing winds drive surface waters away from coasts, and during periods of active upwelling, nutrient-rich waters can support productivity of up to 10 grams of carbon per square meter per day. However only about 1 percent of the world ocean (3.6×10^5 square kilometers) experiences these conditions, so even though productivity may average 300 grams of organic carbon annually, these areas only contribute about 0.1×10^9 tons to the annual total of 20×10^9 tons for the hydrosphere as a whole.

Table 12-5 shows that most of the ocean's organic carbon production takes place in deep, open-ocean waters. But most major fisheries (Chapter 15) are located in continental shelf waters, and particularly in areas of upwelling. In terms of food production, the map of productivity distribution (Fig. 12-23) indicates the richest ocean areas. The explanation for this lies in the relative lengths of oceanic and neritic food chains.

Oceanic food chains are typically long and complex. Recall that only the smallest phytoplankters (less than 0.005 millimeter) grow in the nutrient-poor waters of open oceans, and these can only be consumed by very small herbivores, on the order of 0.1 millimeter. Those tiny animals are preyed upon by primary carnivores in the 1 millimeter size range, and they in turn by secondary carnivores that are about ten times as large, or about a centimeter long. (Examples of these organisms will be described in Chapter 13.) In many cases, one or two larger invertebrate animals or fishes form additional links in an open-ocean food chain before it reaches a top carnivore such as a mackerel or tuna.

Energy is lost at each trophic level in the chain, the more so because food is so scarce in open oceans. Predatory animals grow more slowly and expend more energy hunting for food than do coastal predators. It is estimated that ecological efficiencies do not average more than 10 percent at any level of an open-ocean food chain.

In contrast, areas of upwelling are characterized by a large proportion of phytoplankton (diatoms) that can aggregate into clumps or long filaments. These can either be eaten directly by schooling fishes, or there can be a small herbivorous consumer in the food web. In any case, highly productive ocean areas commonly support short food chains of rapidly growing organisms whose ecological efficiencies may be as as high as 20 percent, on the average. Little energy is wasted in hunting, and the yield of food to humans is much higher for the same amount of primary production.

Coastal areas where upwelling is not a common feature are of intermediate productivity. Large phytoplankton are often present, especially during the short intense blooms of temperate zones, but density of even very small phytoplankton is higher in coastal waters than in the open ocean, and long food chains undoubtedly exist along with very short ones, depending on local conditions. Ecological efficiency in coastal waters is assumed to average around 15 percent.

Results of these calculations are summarized in Table 12-6. They indicate that the food potential of an ocean area is typically proportional to the rate at which nutrients are recycled to surface waters.

TABLE 12-6

Estimated Fish Production of Ocean Areas *

PROVINCE	PRIMARY PRODUCTIVITY (10^9 tons of organic carbon)	TROPHIC LEVELS	ECOLOGICAL EFFICIENCY (percent)	FISH PRODUCTION (millions of tons, fresh weight)
Oceanic	16.3	5	10	1.6
Coastal	3.6	3	15	120
Upwelling	0.1	1½	20	120
Total	20			240

* After Ryther, 1969.

SUMMARY OUTLINE

Marine environment
   Relatively homogeneous—contains gases, nutrients, water
   Small temperature fluctuations—organisms mainly cold-blooded
   Plants typically one-celled
   Major classifications of organisms: bacteria, phytoplankton, zooplankton, nekton
   Major environments: intertidal, neritic, pelagic, benthic, abyssal
   Photic zone is region of food production

Limiting factors in marine ecosystems
   Ecosystems self-sustaining: include autotrophs, heterotrophs, decomposers
   Light can limit plant production—winter, Arctic, dense blooms
   Availability of nitrates, phosphates, silicates limits productivity—at low latitudes in open ocean, centers of ocean basins, during heavy blooms in temperate climates
   S/P ratio is small for elements that limit plant growth
   Organic trace materials, e.g. vitamins, toxins, can promote or inhibit growth

Phosphorus and nitrogen cycles
   Nutrients in organic matter sink below photic zone, quickly recycled in shallow water
   Inorganic soluble phosphorus high when particulate phosphorus is low
   Zooplankton release nutrients during metabolism
   Nitrogen cycles more slowly than phosphorus, requires energy to "fix"
      Bacteria oxidize and reduce nitrogen compounds

Organic growth factors
   Vitamins synthesized by some bacteria, phytoplankton
   Chelating agents—make metals, e.g. iron, available to plants

Trophic level and biomass
   Primary producers, primary and secondary consumers form food web
   Position in food web defined as trophic level—"who eats whom"
      Energy in food webs used up, not recycled
   Ecological efficiency a measure of energy transfer between trophic levels
   Biomass—weight of organisms in population

Photosynthesis—solar energy stored as chemical energy in carbohydrates
   Bonds broken, energy released in respiration
   Net production is energy remaining after respiration, cellular metabolic processes
   Light and dark reactions in photosynthesis; light reactions absorb insolation
      NADPH created for reducing power
      ATP created from ADP and phosphate; cyclical energy-transport chain

Measuring primary productivity—a rate function
   Productivity equals average standing crop times rate of generation
   Measured by takeup of $^{14}CO_2$, or by oxygen produced
   At compensation depth, production equals respiration

Factors in plant productivity—recycling of nutrients to surface; stable surface layer
   Coastal zone a productive area
   Critical depth: when mixed layer is no deeper than photic zone, blooms can occur
   Red tides caused by water stratification, rapid multiplication at surface
   Some species photosynthesize best at lower light intensities.

Nonliving organic matter
   Most organic matter nonliving, very small particles

338

Sources of dissolved organic matter
    Decomposition of dead organisms
    Secretion by plants
    Excretion and exudation by animals
Sources of particulate organic matter
    Dead organisms
    Solid wastes excreted
    Adherance of dissolved material to surfaces

Bacteria in food webs
    Have low biomass, but important as decomposers
    Decomposition less effective in very deep water
    A concentrated source of protein for filter-feeders, scavengers
    Bacterial slime accumulates, followed by fouling, on surface-active materials

Distribution of oxygen in the ocean
    Abundant at surface, exchanged with atmosphere
    Least abundant where respiration, decomposition is high

Abundant in deep, open ocean because of sinking polar waters, abyssal circulation
Deficient in bottom waters of basins where surface production is high, resupply of bottom water limited

Productivity of the world ocean
    Ocean surface waters can quickly be depleted of nutrients
    Rate of nutrient renewal, abundance of light, stability of surface zone are primary factors
    Nutrients renewed quickly where deep waters cycle to surface: shallow regions, divergences, upwelling areas
    Food production in the ocean
        Most carbon produced in open ocean areas
        Oceanic food chains long, coastal food chains short
    Ecological efficiency greatest in coastal, upwelling areas, short food chains

SELECTED REFERENCES

BATES, MARSTON. 1960. *The Forest and the Sea*. Random House, New York. 277 pp. Compares life in the ocean and in a tropical rain forest.

PARSONS, T. R. AND M. TAKAHASHI. 1973. *Biological Oceanographic Processes*. Pergamon Press, Oxford. 186 pp. Review of recent research in biological oceanography. Advanced.

RAYMONT, J. E. G. 1963. *Plankton and Productivity in the Oceans*. Macmillan, Pergamon Press, New York. 625 pp. Comprehensive survey of plankton research literature, including aspects of physical oceanography as they relate to marine biology.

RUSSELL-HUNTER, W. D. 1970. *Aquatic Productivity: An Introduction to Some Basic Aspects of Biological Oceanography and Limnology*. Macmillan, London. 306 pp. Modern treatment of plankton and productivity in lakes and in the ocean, emphasizing North Atlantic forms. Good sections on plankton and world food production.

STEELE, J. H., ed. 1970. *Marine Food Chains*. University of California Press, Berkeley and Los Angeles. 552 pp. Collection of papers dealing with relationships between plankton, productivity, and energy transfers in the ocean and estuaries. Intermediate to advanced in difficulty.

WIMPENNY, R. S. 1966. *The Plankton of the Sea*. American Elsevier, New York. 426 pp.

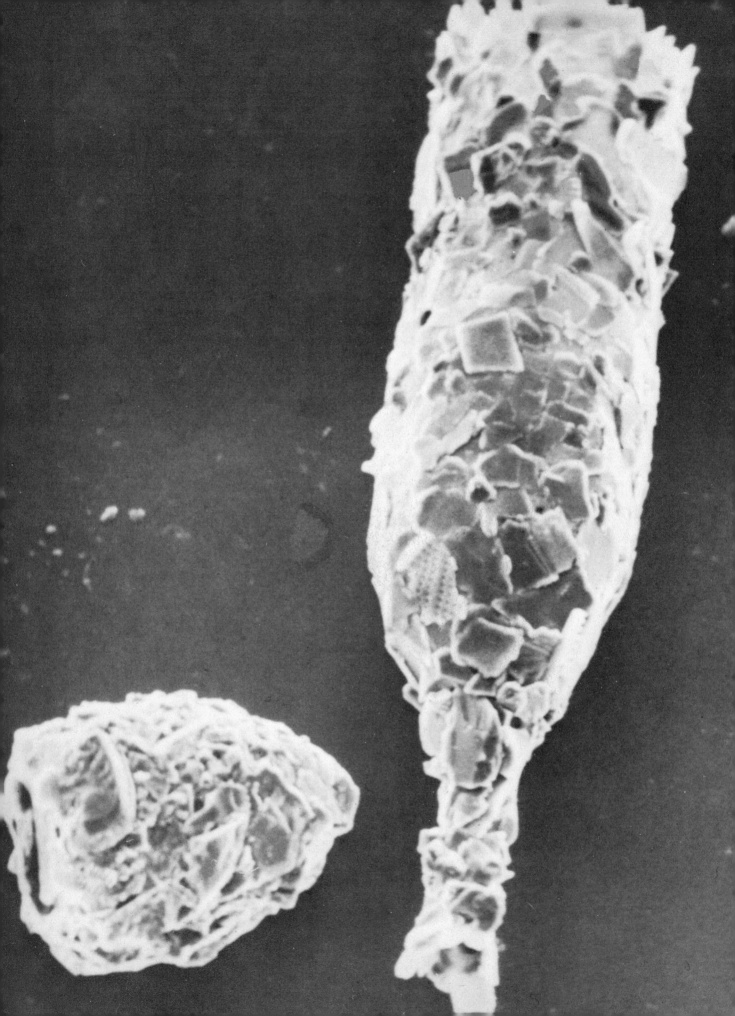

These two tintinnids, taken from Narragansett Bay, R.I., have bell-shaped tests encrusted with debris. The tiny animals sweep phytoplankton into a central cavity by means of a circlet of waving cilia just inside the opening (see Fig. 13-4). The larger of the two is about 0.13 millimeter (130 microns) in length. (Photograph courtesy of Dr. J. M. Sieburth and University Park Press, Baltimore, Md.)

plankton and fish: marine food webs

PHYTOPLANKTON Small size is an advantage to a floating plant, because it depends on molecular diffusion in the water for nutrient supply and waste removal. This is in contrast to the situation on land, or in very shallow water, where air and/or water move with respect to the rooted plant, bringing gases and nutrients and carrying away wastes. A floating object does not move with respect to the water. Each water parcel, with whatever is suspended in it, moves as a unit (Chapter 8). Where nutrient concentrations are low, as in open-ocean waters, availability of nutrient ions within water parcels limits growth. Therefore small size is advantageous because a large ratio of surface area to body mass provides a relatively larger area across which dissolved substances can be exchanged with the water.

Some of the most productive phytoplankton do move a little, relative to the water around them. Many have tiny, whiplike flagellae, with which they propel themselves by lashing motions, while others sink slowly through the water.

An object's rate of sinking in the ocean is determined by two factors: its density relative to the water around it, and the drag or resistance offered by the medium. Phytoplankton protoplasm is generally little heavier than seawater, having a density range of 1.02 to 1.06 grams per cubic centimeter, but still heavy enough to sink. The second sinking-rate determinant, resistance of the medium, is greater for small objects than for large ones.

An object with a large surface-to-volume ratio sinks more slowly than one of the same density but with a smaller ratio of surface area to volume. The smaller the object, the larger the ratio:

$$\text{ratio of surface to volume} = \frac{\text{area}}{\text{volume}} = \frac{4\pi r^2}{\frac{4\pi r^3}{3}} = \frac{3}{r}$$

Thus an object having a radius of 0.1 millimeters has a surface/volume ratio of 30, whereas an object of the same shape and density with a radius of 0.01 millimeters has a surface/volume ratio of 300 and sinks more slowly. A very slow rate of sinking is an advantage, because phytoplankton must remain in the photic zone long enough to grow and reproduce, yet motion through the water column increases nutrient supply. Mixing in the surface zone returns part of a sinking population to the surface. In addition, many flagellated forms exhibit a positive light response which causes them to swim upward.

Major phytoplankton forms are described in Table 13-1. The smallest *ultraplankton* (including bacteria) are less than 0.005 millimeter in diameter. Next largest are the *nannoplankton*, ranging from 0.005 to 0.07 millimeter, followed by *microphytoplankton* (0.07 to 1.0 millimeter), which are in the same size range as many zooplankton.

Nannoplankton, and the very small flagellated ultraplankton, are most important in coastal and open-ocean equatorial waters, where they provide 50 to 80 percent of total standing crop and CO_2 uptake. This may be because very small organisms absorb nutrients more efficiently than larger ones in low-nutrient waters. In the coastal ocean, proportionately more of the larger microphytoplankton are present than in open seas, but nannoplankton may still contribute most of the primary production on an annual basis.

TABLE 13-1
Dominant Forms of Marine Phytoplankton

TYPE AND CHARACTERISTICS	LOCATION	COLOR AND APPEARANCE	METHOD OF REPRODUCTION
Diatoms: silica and pectin "pillbox" cell wall, sculptured designs; of major importance for coastal ocean productivity; has floating and attached forms	Everywhere in surface ocean, especially in colder waters, upwelling areas, even in polar ice; some heterotrophic below photic zone; some form "resting spore" under adverse conditions	Size: 0.01–0.2 mm Yellow-green or brownish; single cells or chains of cells; radial or bilateral symmetry; many have spines or other flotation devices	Division, splitting of nuclear material; average reduction of one cell-wall thickness at each division (Fig. 13-1); when limiting size is reached cell contents escape, form new cell
Dinoflagellates: next to diatoms in productivity; many heterotrophic, injest particulate food; some have cellulose "armor"; very small open-ocean species are naked	In all seas, and below photic zone; some parasitic; warm-water species very diverse; some have resting stage for protection; sometimes abundant in coastal areas as "red tide" (see Chap. 12)	Size: 0.005–0.1 mm Usually brownish, one-celled; have 2 whiplike flagellae for locomotion; many are luminescent	Simple, longitudinal, or oblique divisions; daughter cells achieve size of parent before dividing
Coccolithophores: covered with calcareous plates, embedded in gelatinous sheath; important source of food for filter-feeding animals	Mainly in open seas, tropical and semitropical; sometimes proliferates near coasts; some heterotrophic forms at depths to 3000 meters	Size: 0.005–0.05 mm Many flagellated; often round or oval single cells; when present in great numbers they give the water a milky appearance	Some individuals form cysts, from which spores arise to develop into new individuals
Silicoflagellates: very small, have silica skeleton; some heterotrophic forms	Widespread in colder seas worldwide, especially in upwelling areas	Size: about 0.05 mm Single-celled, 1 or 2 flagellae; starlike or meshlike skeleton	Simple cell division
Blue-Green Algae: small, relatively simple cell structure; cell wall of chitin	Mainly inshore, warmer surface waters, tropics	Size: filaments to 0.1 mm or more Blue-green or red rafts of mottled filaments; can cause a colored "bloom" in the water	Simple division of each cell into two

Microphytoplankton (or *netplankton*) have large standing crops at higher latitudes. They proliferate during spring blooms, when more than 99 percent of plants in an area may belong to a single species. Diatoms (Fig. 13-1) are usually the most important. Rate of division may exceed once a day, and population typically increase 500 to 2000 times the winter "seed crop." Microzooplankton, locally present as consumers of the year-round nannoplankton population, cannot ingest the larger diatoms. But when larger zooplankters begin to grow and reproduce in response to the abundance of food, their grazing quickly reduces phytoplankton biomass (see Fig. 12-12).

Diatoms (Table 13-1) have protruding, gelatinous threads that permit them to form long chains, especially in nutrient-rich coastal and upwelled waters. They grow by division (Fig. 13-1), some species forming *auxospores* at definite seasons, others every two or three years. A particular population may be recognizable over a long time period by its gradually diminishing shell size; some North Sea water masses can be traced by their characteristic diatom population.

Often a single species dominates an area for two or three weeks, or a season, after which it is succeeded by another. Diatoms appear to be most buoyant when they are young and growing rapidly, so that as one population ages it sinks to lower levels and another species which had previously been represented by a small number of individuals begins to bloom in near-surface waters. This process may be repeated several times during spring and summer.

Diatoms can form resting spores, when environmental conditions become less favorable, remaining in this condition for long periods if necessary. After death, their glasslike shells sink to the ocean floor.

Dinoflagellates are second to diatoms in abundance (Fig. 13-2). They are sometimes considered to be one-celled animals, since many are heterotrophic. Some can live on dissolved or particulate organic matter absorbed or ingested from seawater. Many require less light than diatoms, and can tolerate lower nutrient concentrations. Dinoflagellate blooms commonly exceed diatom production in some areas, due in part to the fact that a scarcity of silicon can limit diatom growth, but does not affect dinoflagellates. Changing light intensity also affects the succession of species in a particular area. In the mid-Atlantic coastal ocean, for example, nannoplankton dominates during November and December, but diatoms are the important producers during the January to March blooms.

The so-called "armored" dinoflagellates are covered with small plates of cellulose that form the organism's cell wall. There are usually two flagellae set in grooves at right angles in the cell wall for propulsion. This adaptation, together with a sensitivity to light, permits the organism to remain at its preferred light level by swimming upward. Several species are bioluminescent, which means that they emit chemically produced light, particularly when agitated. In tropical waters, breaking waves are often lit by the phosphorescence of these tiny organisms.

Coccolithophores (Fig. 13-2(c)), another important group of flagellates, are covered with tiny, calcareous plates that are important contributors to marine sediments in certain parts of the deep ocean. Less common are *silicoflagellates* (Fig. 13-2(d)), and numerous other types of flagellated nannoplankton. Near coasts, filamentous *blue-green algae* may be locally abundant, and sometimes green algae, though these are more common in fresh water. Rooted seaweeds of many kinds grow in shallow waters; some of them will be mentioned in the chapter on benthic environments.

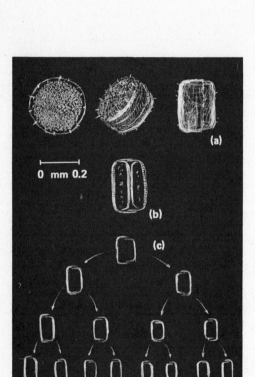

0 mm 0.2

Figure 13-1

Reproduction in diatoms. There is a "pillbox" shell (a), perforated with pores for exchange of metabolic products. When a diatom grows large enough to divide, the "lid" and "box" separate, (b), and each gets one-half of the cell contents. A new "box" is then secreted over the exposed protoplasm, so that the original shell half becomes the new "lid." The daughter cells become progressively smaller in this type of division. When a certain size limit is reached, both old shells are discarded and the resulting bare *auxospore* doubles or triples in size before forming a new set of shells.

Figure 13-2

Variously shaped diatoms (a), collected on a plankton net, include rod-shaped, jointed, spool-shaped and "pillbox" type species. The longest measures about 0.08 millimeter (80 microns). (b) The dinoflagellate *Peridinium* shown here is about 0.4 millimeter (400 microns) long. Its locomotive flagellum (not visible in this photograph) is attached at the groove near the upper left. (Photographs courtesy Dr. J. M. Sieburth and University Park Press, Baltimore, Md.) (c) This living coccolithophore, *Cyclococcolithina leptopora,* was photographed through an electron microscope. It was collected in the North Pacific Ocean and measures a little over 0.01 millimeter (10 microns) in diameter. (Photograph courtesy Dr. Susumu Honjo.) (d) The silicoflagellate *Distephanus,* whose skeleton is shown containing organic debris, is only about 0.03 millimeter (30 microns) in diameter, including the spines. (Photograph courtesy Dr. J. M. Sieburth and University Park Press, Baltimore, Md.)

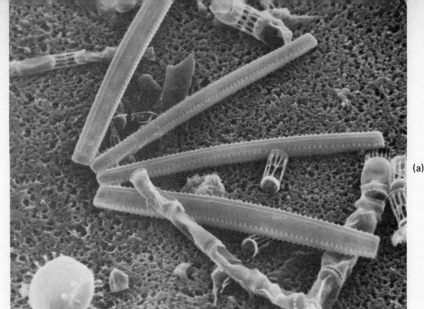

(a)

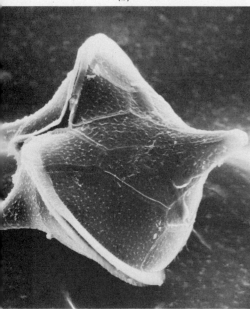

(b)

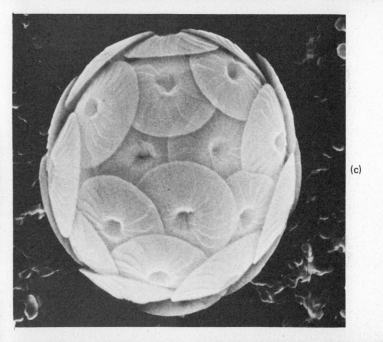

(c)

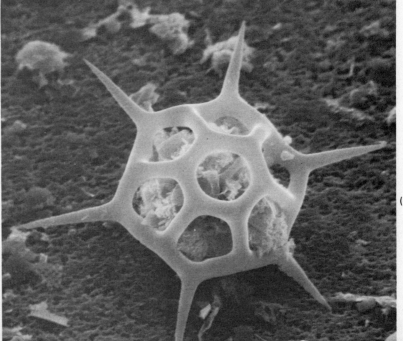

(d)

ZOOPLANKTON　Most of the lower trophic level consumers in marine food webs are zooplankters. As herbivores and primary carnivores, they transfer matter and energy from phytoplankton to the high-level predators consumed by humans. And as conspicuous members of pelagic ecosystems, their life cycles and behavior have been extensively studied.

Zooplankton display a great variety of adaptations for food gathering and reproduction. Some can swim well enough to pursue prey but for the most part they are filter feeders, bearing traps of tiny hairs or mucous surfaces to which floating food particles adhere. Since these animals are usually limited to food particles of a particular size, their distribution depends largely on availability of the organisms that they are adapted to ingest.

Another important factor in zooplankton distribution is the narrow temperature range (generally only a few degrees) at which they can reproduce. Adult populations have greater temperature tolerance and may be borne far out of their breeding range by currents, making them available to predators in a wider area. The greatest number of plankton species breed in tropical waters, with a steady decrease toward higher latitudes. But the number of individuals in a given area is normally a function of its productivity; there may be 500 or 1000 times as many animals per square meter of near-surface water in the North Atlantic coastal ocean, for example, than in the tropical Atlantic open ocean. Again, currents can carry a nutrient-rich water mass through a relatively sterile area, so that abundant zooplankton are sometimes found in a region of normally low productivity.

Not all zooplankton remain free-floating throughout life. Those which do, known as *holoplankton*, are generally more important in marine food webs than the *meroplankton*, which attach to the bottom and become benthic in the adult stage.

HOLOPLANKTON　Many small consumers of the nannoplankton are single-celled *protozoans,* of which *Foraminifera* (Fig. 13-3(a)), *Radiolaria* (Fig. 13-3(b)), and *Tintinnidae* (Fig. 13-4 and frontispiece) are important examples. Foraminiferans live nearly everywhere in the ocean. Many have delicate, porous shells or "tests" which are conspicuous constituents in calcareous sediment (Chapter 4). Thin extrusions of protoplasm called *pseudopodia* (false feet) extend through holes in the shells to capture food particles by surrounding them. Digestive juices are secreted onto the food to dissolve it, so that it can pass directly into the cell. Wastes are excreted from body surfaces. Protoplasmic strands interconnect outside the animal's body to form a branching network around and through the tests as the animals grow. Additional calcium carbonate is laid down around newly extruded pseudopodia so that successive interconnected chambers are added to the shell.

Radiolarian protozoa have an internal capsule of *chitin* (a tough protein substance) and the siliceous skeleton characteristic of cool-water organisms. The skeleton may be arranged as a sphere or as concentric spheres, with radiating pieces. Their protoplasmic strands do not form a net but project in all directions as long, sticky filaments. These trap tiny particles, which are then borne in a protoplasmic stream toward the center of the body to be digested. Individuals range from 0.1 to more than 10 millimeters in diameter. Reproduction is by division into two "daughter" organisms.

Tintinnids ("bell-animals") and other very tiny, ciliated protozoans are common throughout open oceans where plants are small and populations sparse. Tintinnids are enclosed in a goblet-shaped or tubular hard shell of protein a few tenths of a millimeter long. The mouth is surrounded by a circle of hairlike cilia (Fig. 13-4), whose

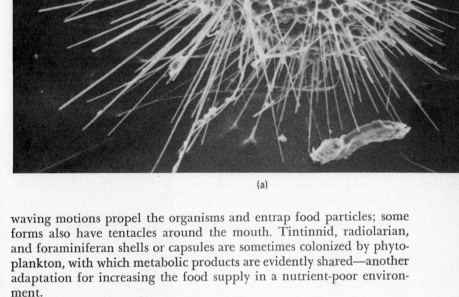

(a)

Figure 13-3

(a) A foraminiferan, *Globigerinoides ruber,* from Bermuda waters. Two apertures, through which food is ingested, are visible just above center. Excluding the spines, this organism is about 0.3 millimeter (300 microns) in diameter. (Photograph courtesy Dr. J. M. Sieburth and University Park Press, Baltimore, Md.) (b) This delicate radiolarian skeleton was drawn by the famous zoologist Ernst Haeckel. (Report of Scientific Results, H.M.S. *Challenger,* Zoology, vol. 18, 1803 pp., 1887.)

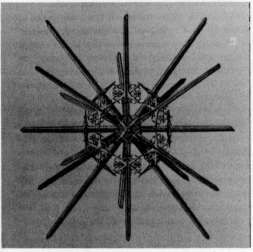

(b)

waving motions propel the organisms and entrap food particles; some forms also have tentacles around the mouth. Tintinnid, radiolarian, and foraminiferan shells or capsules are sometimes colonized by phytoplankton, with which metabolic products are evidently shared—another adaptation for increasing the food supply in a nutrient-poor environment.

Rotifera are not one-celled animals, but tiny roundworms that are seasonally common in coastal and open oceans. A double wheel of beating cilia on the head serves to propel the animal and to sweep microorganisms into its mouth.

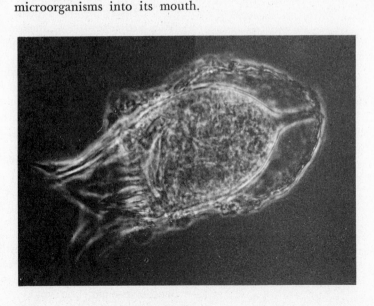

Figure 13-4

This cultivated specimen of *Tintinnopsis lohmani* (on the order of 0.1 millimeter in length) has clearly visible cilia and lorica (outer sheath) surrounding the goblet-shaped body. (Photograph courtesy Kenneth Gold.)

Figure 13-5

Plankton crustacea. Clockwise, beginning at the bottom:

Amphipods hatch in brood-pouches, emerging as tiny adults. Some groups are locally very abundant in the plankton. The *Hyperiid* type shown here has characteristically huge compound eyes. The species illustrated is *Themisto,* an important food of the North Sea herring.

Cumaceans are more common in shallow coastal waters than in the deep ocean. They often remain on the ocean floor during the day, rising to surface layers at night to feed.

Mysids are a shrimplike form. Many species live near continental margins, but a few are found at depths of 4000 meters or more. Metamorphosis, from egg through nauplius stages to adult form, takes place in a brood-pouch under the front legs of the female.

Copepods (meaning "oar-footed") occur at all depths, in all parts of the ocean. They have been termed the "insects of the sea" because of their abundance and variety. The eggs hatch into free-swimming nauplii, which often moult many times before attaining adult form.

Euphausiids are relatives of the edible shrimp, which is a bottom-dweller. All euphausids are planktonic, and are found everywhere in the ocean. Many are filter-feeders, especially in high-latitude waters; others have grinding jaws to accommodate larger food particles.

Many *Isopod* species are land-dwellers. The species *Gnathia,* of which the larva (left) and adult male are shown here, lives as a fish parasite during the larval stages, but is free-swimming in adult life.

In contrast to the very small, sparsely distributed animals of open oceans, the characteristic zooplankters of productive waters are relatively large and complex. *Crustacea* (Fig. 13-5) are the most numerous of these; they constitute 70 percent or more of the zooplankton, both in bulk and numbers. *Copepoda* (Fig. 13-6) and *Euphausiidae* (Fig. 13-7) are most important in marine food webs.

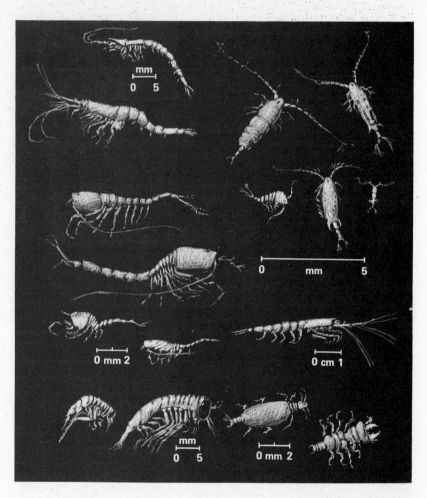

Crustaceans have been called the "insects of the sea." In fact, both are arthropods, meaning "jointed-feet," characterized by segmented bodies and appendages. Each pair is usually specialized for a particular function, such as feeding, movement, sensation, or reproduction. There is a stiff, calcareous outer shell. The characteristic, free-swimming *nauplius* larva commonly molts many times (Fig. 13-7) before assuming the adult shape.

Copepods are present throughout the ocean and may be the most numerous animals in the world; certainly they are the most numerous marine herbivores. Studies in the northwestern Pacific have showed an average of 15,000 individuals per cubic meter in surface waters. In Arctic waters there may be nearly twice as many of a single species in a cubic meter of seawater; even at 500 meters depth well over a thousand individuals may occur in a cubic meter. Depending on temperature and availability of food (usually diatoms) large copepods can double their numbers a few times in a year; smaller species reproduce even more frequently. Daily food consumption for older larvae ranges from 50 percent of body weight to much more when food is plentiful. There are 6000 different species, including a few benthic and parasitic types.

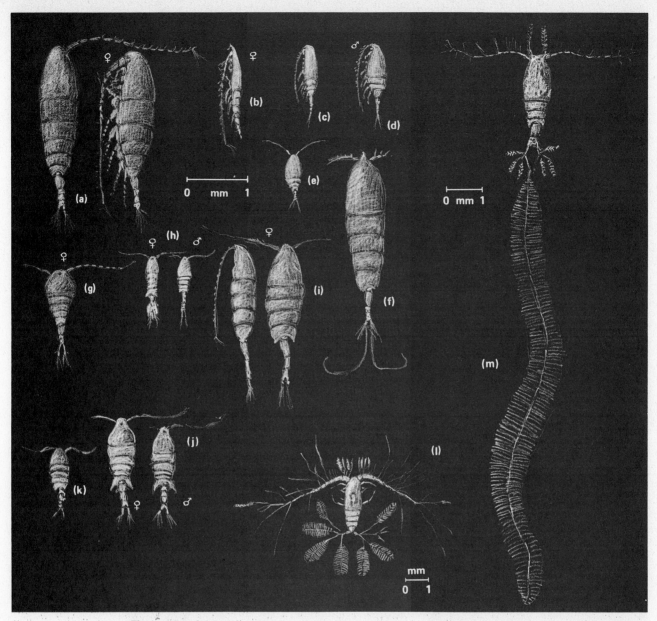

Figure 13-6 (*above*)

Calanoid copepods: (a) *Calanus
finmarchicus,* (b) *Rhincalanus*
(c) *Pseudocalanus,* (d) *Paracalanus,*
(e) *Microcalanus,* (f) *Euchaeia,* (g) *Temora,*
(h) *Eurytemora,* (i) *Metridia,* (j) *Centropages,*
(k) *Isias*—all found in the vicinity of Great
Britain. (Redrawn from Newell, 1967.)
(l) and (m) are tropical varieties of
Calocalanus. (Redrawn from A. Hardy, 1967.
Great Waters, Harper & Row, New York,
541 pp.)

Figure 13-7 (*right*)

Life cycles of *Euphausia superba:* Egg (a),
early and late nauplius stages—(b), (c), (d)—
the intermediate *mysis* stage, and mature
adult are shown.

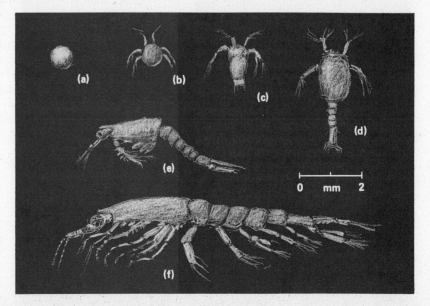

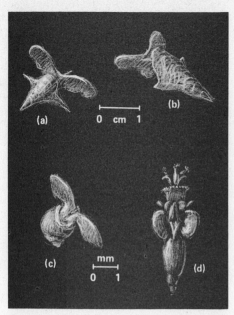

Figure 13-8

Marine pteropods: *Clio*—(a) and (b)—found throughout the world ocean, *Spirialis* (c), and *Dexiobranchaea* (d). (Redrawn in part from Hardy, 1965.)

Figure 13-9

These chaetognath worms include several species of *Sagitta* common to North Atlantic waters. These are sometimes used as tracers for water masses, since different species are found in waters having different nutrient compositions. (Redrawn from G. E. Newell and R. C. Newell, 1967. *Marine Plankton: A Practical Guide*, Hutchinson, London, 221 pp.)

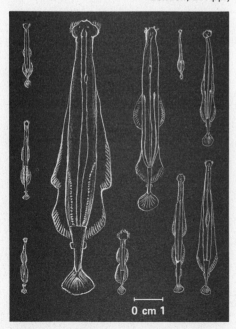

Copepods range in length from about 0.3 millimeter to about 8 millimeters and have feathery, curved bristles which form a filter chamber behind the mouth. By constant movement of the appendages near the head, two opposing spiral currents are set up. One moves the animal forward, the other forces a stream of water into the filter chamber. Tiny plants and fine particles thus trapped are passed along to the mouth. Copepods in northern seas, their bodies rich in proteins and fats, are eaten in great numbers by herring, whales, and many of the other organisms that are important as a source of food for humans.

Shrimplike *euphausiids,* also known as "krill," are another important crustacean group. Dense swarms of these animals feed on diatoms, and themselves comprise the chief food of the large, filter-feeding whales, as well as of many fishes on the high seas. Euphausiids are larger than copepods, being up to 5 centimeters long, and they are found on or near the bottom as well as in surface waters. The group includes herbivorous and, especially in warm waters, carnivorous species.

In Antarctic and colder temperate waters, euphausiids mature slowly and live up to 2 years. It has been estimated that a large baleen whale (see Fig. 13-24) eats an average of 850 liters of euphausiids per day. *Euphausia superba* may produce 10^{11} kilograms annually, nearly twice the world's commercial fishery production for 1974 (0.6×10^{11} kilograms).

Another important type of crustacean, especially in nearshore waters, is *Cladocera* ("water-fleas") such as *Podon* and *Evadne* (see Fig. 13-29). These sometimes occur in tremendous concentrations in coastal ocean (e.g., the North Sea) and in estuaries such as Chesapeake Bay. They are small—around 1 millimeter long—and possess a single large eye formed by fusion of the two larval eyes.

Besides the crustaceans, certain other kinds of zooplankton may be locally very conspicuous. *Pteropoda* ("wing-footed") are small pelagic snails that occur in dense swarms in all seas. Carbonate-shelled species are common in tropical oceans, where their remains can accumulate in sediments forming "pteropod oozes." In all members of this group the characteristic snail foot is modified into fins (Fig. 13-8). Rows of beating cilia set up currents that bring food particles against a mucous secretion, and then bear the food toward the mouth. In surface waters pteropods eat phytoplankton, in deeper currents they are carnivorus. Northern species ordinarily do not have shells; some of these resemble a little slug with wings, 2 centimeters or less in length.

Some small marine carnivores are strong swimmers, though usually classed as plankton. Among the most important are the *arrow-worms* or *Chaetognatha* (Fig. 13-9) whose name means "bristle-jawed." Mature individuals, 2 to 8 centimeters in length, are active and abundant from the surface to great depths in all seas. They are transparent, and possess chitinous spikes embedded in muscle around the mouth. These spikes can open out like a fan and then turn inward as seizing jaws to consume their prey. By fixing their long bands of longitudinal muscles, arrow-worms can dart rapidly through the water in pursuit of small zooplankton. There is no larval stage; small worms hatch directly from fertilized eggs.

Oceanographers can distinguish different North Atlantic water masses by determining the species of arrow-worm that lives in them. *Sagitta setosa* is found only in low-phosphate North Sea water masses, and *Sagitta elegans* in more fertile oceanic waters. The distribution of these water masses changes from year to year, affecting pelagic species as well as the benthic animals that may develop more success-

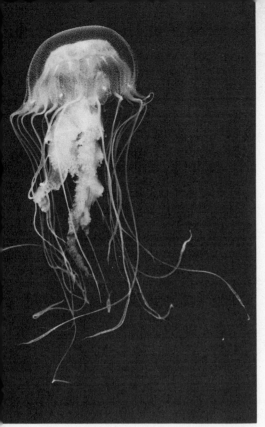

Figure 13-10

The adult jellyfish *Chrysaora* measures about 10 centimeters across the bell and is found worldwide. (Photograph courtesy of Michael J. Reber.)

fully in one or the other type of water. They can be distinguished by chemical analysis, but the quickest and easiest method of determining whether a water sample is *setosa* or *elegans* water is to capture and identify one or two of the arrow-worms that live in it.

The animals discussed so far are all important members of food chains that lead to fish, and from there to human predation. But other zooplankters seem to be dead ends in food webs exploited by man—for example, the well-known jellyfish. They have a continuous, two-layered body wall surrounding a digestive cavity that has only one opening, around which are tentacles bearing stinging cells. The tentacles capture swimming or drifting particles and move them into the mouth, through which wastes are also eliminated. The animal moves by rhythmic pulsations of its bell (Fig. 13-10).

This group includes the spectacular colonial *siphonophores* of the open sea like the Portuguese man-o'-war pictured in Fig. 13-11. Both jellyfish and siphonophores paralyze their prey with stinging cells that consist of a barb attached to a sack of poison. Dangling tentacles entangle the food and sweep it toward the mouth, inside the bell. Some forms swim upward, then sink with bell and tentacles extended, trapping prey beneath. A siphonophore is not a single animal, but a colony of individuals that live together and function as one. Some species (e.g., *Physalia*) have a gas-filled float, but most have one or more swimming bells. Prey-catching, reproductive, and digestive polyps trail below like a fisherman's drift net.

Physalia are common along Mediterranean shores and in nearly all tropical waters, where their stinging cells are feared by swimmers. The poison works instantly and can be fatal to a child. But *Physalia* has enemies of its own. A small, blue-shelled snail, *Janthina*, about 2 to 3 centimeters long, floats at the water surface by means of gas-filled bladders; two of these snails can devour a 10-centimeter long *Physalia* colony in 24 hours.

Figure 13-11

The Portuguese man-o'-war *(Physalia)* has a purple, air-filled bladder that floats at the surface. This animal is colonial in the sense that beneath the float hang clusters of polyp "persons," some of which entrap, paralyze, and engulf their victims, digesting them and absorbing their juices on the spot. Other polyps serve a reproductive function, producing sexual medusae. The man-o'-war fish, immune to *Physalia*'s poison, feeds on discarded scraps and acts as a lure to attract victims. (Photograph courtesy Wometco Miami Seaquarium.)

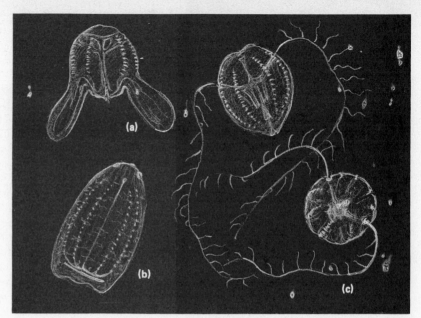

Figure 13-12

"Comb-jellies" (ctenophores) include various forms, three of which—(a) *Bolinopsis,* (b) *Meroë* and (c) *Pleurobrachia*—are shown here. They swim by motions of the fused cilia, of which there are typically eight rows on the body. Medusae, on the other hand, swim by pulsations of the bell.

The delicate *Ctenophora* (Fig. 13-12) look something like coelenterates. Small and jellylike, they are sometimes known as "sea-walnuts" or "comb-jellies." Some have trailing tentacles for capturing prey. Voraciously carnivorous, ctenophores often occur in great numbers, and may substantially reduce populations of crustaceans and young food fishes.

Ctenophores are bilaterally symmetrical, rather than radially symmetrical like the coelenterates, and they lack the stinging cells characteristic of the former group. They swim by moving the bands of fused cilia that forms their "combs." All are luminescent and can sometimes be seen at night in surface waters when disturbed.

Figure 13-13

Tunicates (sea-squirts): these animals pump water through their bodies by contraction of the muscular bands encircling them. *Doliolum* (a) has a life cycle of alternative sexual and asexual generations—the asexual is illustrated here. Reproduction is by budding—little groups of cells are formed on the underside of the mature individual (upper left), whence they migrate along the body of the "parent." They then attach themselves in rows to an appendage at the front end, the *cadophore,* shown enlarged at lower left. As the cadophore grows longer, the buds grow to maturity. Some remain attached as a colony; others break off to live as individuals. (Redrawn from Hardy, 1965.) *Salpa* (b) also exhibits alternation of generations—the asexual stage is shown here, in solitary and aggregate forms (the individual at lower right is an aggregating type). At the top of the group is a "nurse" salp with a cluster of buds on its underside that will break off to form sexually reproducing individuals. (Redrawn from Hardy, 1965.)

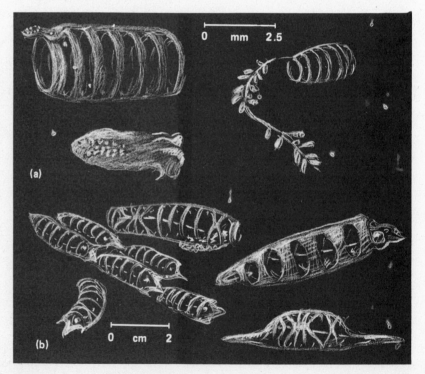

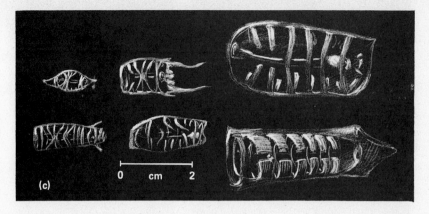

Tunicates are primitive relatives of the *Vertebrata* (animals having backbones). In the plankton they are represented by *sea-squirts* such as the transparent, barrel-shaped *salps* and *doliolids* illustrated in Fig. 13-13. Significant numbers of these occur in tropical waters, where they feed on tiny plankton by pumping water through their bodies. Sometimes these animals form long chains that lie in masses at the surface of quiet waters. Large tunicates seem not to be widely eaten by fishes and, like siphonophores and ctenophores, may represent a dead end in marine food webs.

MEROPLANKTON

Planktonic larval forms of benthic animals are locally important in coastal waters, where they are an important food for deep-water fishes such as cod and flounders. There are more than 125,000 species of benthic animals, most of which have a free-swimming larval stage that lasts a few weeks. Eggs and sperm of benthic animals often are discharged in great clouds to fertilize in the water (Fig. 13-14). These may number in the tens of millions per individual per year, but mortality due to predation and other hazards is high. The number of eggs produced just about balances their loss during development, and the number of benthic adults remains roughly the same over a period of years unless the environment changes.

Figure 13-14

Oysters (male on left, female on right) puffing out sperm and eggs. (Photograph courtesy of Michael J. Reber.)

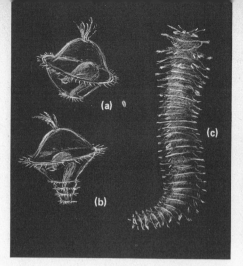

Figure 13-15

Larval stages of generalized polychaete worm. The trochophore stage (a) spins through the water like a top by waving its tuft of cilia; the gut can be seen within. In (b) and (c), we can see how the larva metamorphoses to adult form. (Redrawn from Hardy, 1965.) (d) Photograph of an adult benthic bristleworm. (Photograph courtesy Wometco Miami Seaquarium.)

Figure 13-16

Metamorphosis of a marine snail: (b), (d), and (g), which diagram the internal structure of (a), (c), and (f), respectively, show that the gut, originally straight, twists into a loop so that mouth and anus are both at the bottom of the shell, where the opening is in the adult form. (Redrawn in part from Hardy, 1965.)

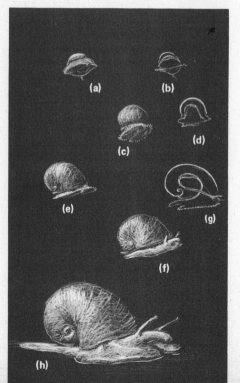

The ability of maturing larvae to find suitable bottom material upon which to settle is an important factor. For example, currents may carry worm larvae into rocky-bottomed areas where there is no sediment in which to burrow, so the larvae die. A barnacle, on the other hand, needs a solid surface to grow on, so that landing on soft, clayey sediment would prevent larvae from completing their metamorphosis into the adult form. Some larvae are adapted to resume a swimming or floating existence if the first attempt at finding a suitable bottom material fails. After a while they seek the bottom again, and some can crawl about in search of an appropriate place to attach themselves.

Predation causes larval mortality of 90 percent or more. Jellyfish, comb jellies, larval fish, arrow-worms and many other carnivorous plankton and nekton consume large numbers of benthic larvae, few of which have any defenses against predators.

Among benthic *invertebrates* (animals without backbones), worms are some of the most widely distributed. *Polychaeta* ("bristle-worms") are segmented animals named for the rows of chitinous bristles along their sides. Some, like *Tomopteris* (see Fig. 13-29) are permanently planktonic but many have free-swimming larvae (Fig. 13-15). At first the young worm is toplike or rounded, as in Fig. 13-15(a). The mouth is at the "equator," below a ciliated girdle, with the anus at the bottom. Waving cilia send the animal spinning through the water, and currents set up by the spinning motion move food particles toward the mouth. This type of larva is called a *trochophore*. As development continues, other ciliated rings and segments appear below the mouth, as shown in Fig. 13-15(b). In some forms the trochophore becomes the head of the adult worm, in others the girdle and other larval structures are suddenly sloughed off when the animal settles on the ocean floor and completes metamorphosis, as shown in Fig. 13-15(c).

The *Mollusca* include snails and slugs (*gastropods*), clams and oysters (*bivalves*), squid and octopus (*cephalopods*), as well as a few less important forms. Most of this group are benthic as adults, with planktonic larvae. The first stage is a trochophore, which usually remains in the original egg capsule, shown in Figs. 13-16(a) and (b). In the later, free-swimming stage, the ciliated girdle is extended at the sides to form large lobes—shown in Fig. 13-16(c)—which not only support the increased weight of the larva (now known as a *veliger*), but also prevent the typical spinning motion of the trochophore, so that the animal swims straight ahead. The lobes extend out beyond the body of the larva, and the internal organs twist into a loop, as shown in Fig. 13-16(g). Mouth and anus use the same shell aperture; when the head and foot are withdrawn for safety inside the shell, a

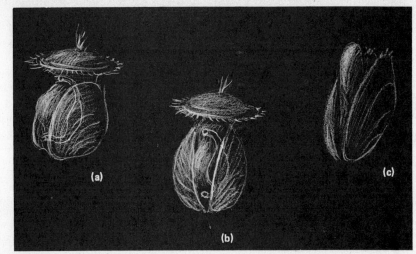

Figure 13-17

Metamorphosis of a generalized bivalve, showing the trochophorelike larva (a) and (b) with a shell forming at the sides. It has been drawn as if transparent, to show formation of the gut. Nearing the adult state (c), the foot is visible where the shell is partially open.

Figure 13-18

Developmental stages of the acorn barnacle *Balanus.* The larval "nauplius" (upper left) swims freely, metamorphosing into the secondary "cypris" stage (upper right and center, showing inner and outer views). Having found an appropriate location for its stationary adult life, the animal forms a hard shell which can be opened for feeding or closed for protection. Cutaway views (bottom) show the animal inside its shell, closed and open.

hard plate (the *operculum*) attached to the foot covers that opening tightly. As the shell develops, the snail sinks to the ocean floor to take up benthic life. The development of the oyster (illustrated in Fig. 13-17), which does not have to twist its body to fit into a single shell, is a variant on the same theme.

Benthic crustaceans such as barnacles develop in much the same way as their pelagic relatives the copepods and euphausiids (see Fig. 13-7). Nauplii of the common barnacle *Balanus* (Fig. 13-18), about 0.5 millimeter in length, may number 13,000 larvae at a time, of which perhaps 30 survive to attachment. Very few live the 2 years to sexual maturity.

Figure 13-19

(a) Development of the Atlantic fish *Sardina pilchardus;* the tiny fish is at first seen developing inside the egg. (b) Larvae of (from the top) mackeral, anchovy, rockfish, croaker, halibut.

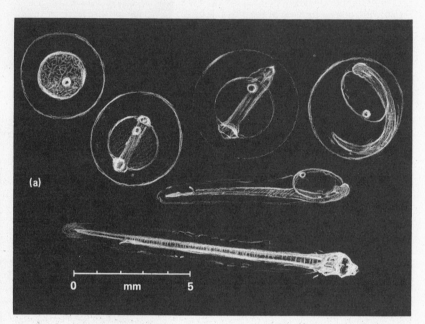

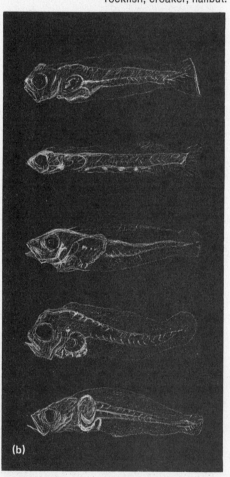

Planktonic fish eggs, larvae, and juveniles (shown in Fig. 13-19) are an important source of food for carnivores in the ocean. Some fishes, such as herring and sand eel, attach their eggs to rocks or vegetation, some lay them in "nests" near the shore, and some lay them in gelatinous masses; but most fish eggs are released and fertilized in deep water near the continental margins. These drift with the plankton, and when sufficiently developed they hatch and begin to feed. Depending on temperature and species, this may be as soon as 1 or 2 days after release. They may then remain in the planktonic stage for weeks. Success of the young fish in a particular *year-class* (that is, those spawned during a single season) is affected by a variety of environmental factors. One of the most important is that the young fish find a good source of food before the yolk sac, shown in Fig. 13-19(a), is completely absorbed. This can depend, for example, on currents carrying the fish into a region where recently hatched invertebrate larvae are the right size for the developing fish to eat.

LIFE IN THE DEEP OCEAN

A varied and distinctive fauna inhabits the *mesopelagic zone,* from 200 to 1000 meters depth. Only about 1 percent of light penetrates to the top of this zone, even in very clear waters. This is close to the lower limit for photosynthesis, but as far down as 1000 meters some animals respond to diurnally changing light levels by migrating vertically every 24 hours.

At mesopelagic depths swarms of planktonic euphausiids and copepods feed on small particles from above and are in turn preyed on by squids and fishes. Food requirements are less in this environment because animal metabolism is slower in colder waters.

This dimly lit zone has relatively more bioluminescent animals than any other part of the ocean. *Stomiatoids* are an abundant group

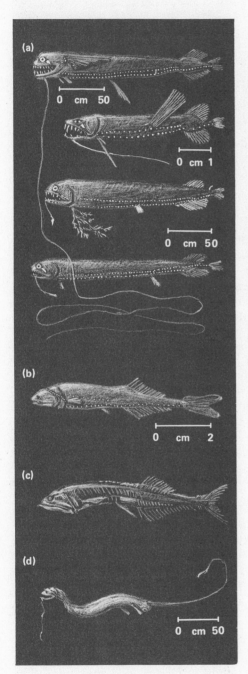

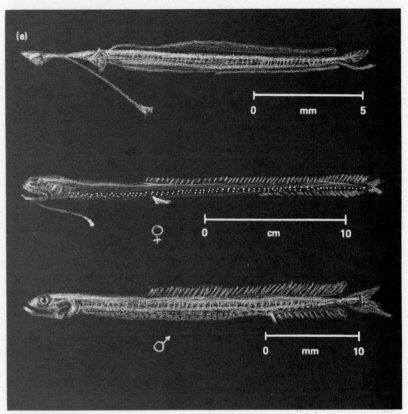

Figure 13-20

Stomiatoid fishes are slim, generally dark or silvery color, and range from a few centimeters to a few tens of centimeters long. Probably the most numerous fishes in the ocean, they are especially common at mid-depths of 500 to 2000 meters. Populations may not be dense in any one place, in the manner of coastal herring or anchovies, but they are common through a vast area of the ocean. *Melanostomiatoids* (a) characteristically bear chin barbels, often associated with a luminous lure. These fleshy organs are sensitive to water movements, which cause the fish to snap in the direction of the disturbance. Stomiatoids have big mouths and sharp, strong teeth, and many can swallow victims larger than themselves. Luminous dots along the body are common. The mid-water *Cyclothone* (bristlemouths) are believed to be the largest single genus (group of species) among marine vertebrates. They are small (2 to 8 centimeters) and rapacious, with mouths that open nearly 180° to capture prey. Illustrated are (b) *C. microdon* and (c) *C. pallida*. *Idiacanthus niger* (d), about 20 centimeters long, is shown after having swallowed a fish longer than its own gut, the prey being doubled up to fit inside the predator. *Idiacanthus fasciola* (e) is commonly found at depths of 900 to 1800 meters, in low-latitude waters. Females are black, reaching a length of 30 centimeters; males are pale and seldom exceed 5 centimeters in length, with lights on their cheeks, but lacking the characteristic barbel of the female. The larvae, whose eyes are on the ends of long stalks, drift at the surface while feeding on plankton. When they reach a length of about 5 centimeters, they sink to 900 meters or lower, and the eyestalks resorb to a normal position.

358
plankton and fish:
marine food webs

whose form and habits illustrate some typical adaptations for life in dark, deep waters (Fig. 13-20). *Lantern fishes* and *hatchet-fishes* (Fig. 13-21), both common from 100 to 500 meters depth, have patterns of light organs that identify particular species. This adaptation permits the fish to recognize potential breeding mates, especially where males and females bear different patterns.

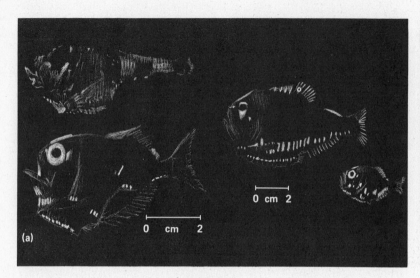

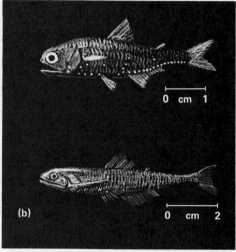

Figure 13-21

Bioluminescent deep-ocean fishes. (a) Four species of hatchet-fishes, all small —the largest species is *Argyropelacus gigas*, at top. These fishes are common at depths of 100–500 meters, although some have been taken at depths to 2000 meters. They are characterized by luminous organs and by their distinctive "hatchet" shape.

(b) Second in numbers only to the stomiatoids among deep-water fishes, there are at least 170 species of lantern fishes, in all but the very coldest seas. Great masses often migrate to the surface at night to feed. Light organs dotted along the sides and a bright light near the tail are characteristic. Although sunlight and even moonlight repel them, lantern fishes often experience a reversal of the normal reaction in the presence of a very bright light, so that they swim toward a searchlight on the water like moths.

Below 1000 meters there is no light at all from the sun, and waters are uniformly cold (Chapter 6). Animals do not vary much in terms of geographical location at these depths; the same kinds may be found in virtually every part of the world ocean. A few filter-feeding crustaceans are supported by what is left of the detrital rain from above, but predation is the rule, and many species are adapted to eat anything that will go into their mouths. This includes animals twice their own size, as shown in Figs. 13-20 and 13-22.

Animals living in absolutely dark waters are usually either black or reddish, either color making them invisible. Lighted lures are common (Fig. 13-22), since many marine animals are attracted by them.

Squid (Fig. 13-23) are high-level predators in the deep ocean as well as in surface waters. These swimming mollusks have no external shell but are stiffened internally by a blade-like chitinous "pen." Encircling their mouths are long prehensile tentacle-bearing suckers. All marine mollusks have a *mantle,* or tough protective membrane through which water circulates, but in squids this is highly developed into a muscular, torpedo-shaped sheath. By powerfully contracting the mantle they force water out of a siphon, propelling themselves violently backward. Some small species can leap out of the water and glide. This is an extremely efficient system of propulsion, making squid one of the fastest animals in the ocean and one of the hardest for humans to catch. They also have very good eyes, an asset in hunting and in evading predators.

It is difficult to estimate the abundance of these elusive and wide-ranging animals, but one indication is their high rate of consumption by *sperm whales* (Fig. 13-24), whose principal food they form. The annual take of squid by sperm whale may equal the total world commercial fish production.

Many zooplankters, as well as squids and fish, migrate vertically on a 24-hour cycle responding to changing light levels. Echo sounders have detected a movable "deep-scattering layer" or dense aggregation

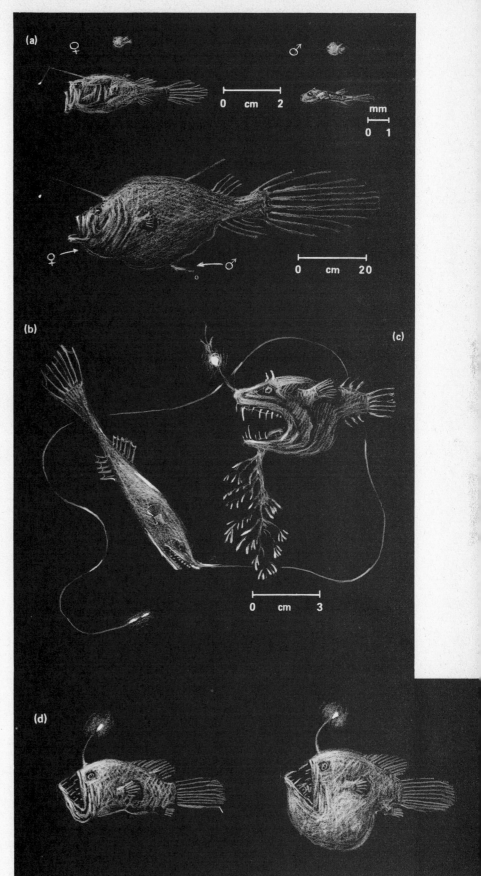

Figure 13-22

Deep-sea angler fishes. The many species of these curious fishes are most abundant at depths of 1500–2000 meters, where the ocean is dark and virtually boundless, and food is extremely scarce. In life cycle and in body structure, some angler fishes illustrate two adaptations for survival at great depths: a movable, lighted fishing lure which dangles in front of the huge mouth, and a foolproof means of guaranteeing mating—the male is parasitically attached to the female.

In the species *Ceratius holboelli* (a), the male swims freely until adolescence, at which time he is only a few millimeters long. Locating a female, he grips her belly skin with his teeth and grows to a length of several centimeters. Their fertilized eggs float to the surface, where the larvae hatch and feed on copepods until adolescence, when they return to deep water. Sexual maturity is not attained until after the fish have joined together permanently. Adult females are about 1 meter long.

Not all angler fishes have the "attached male" adaptation, but all females possess some sort of attractive light organ to lure prey. The female ceratioid angler *Gigantactis* (b), living at depths to 5000 meters, carries a maneuverable "fishing rod" many times the length of her body. Other species, such as *Linophryne* (c), have luminous "barbels" on their chins, or a light organ in their mouths, just behind the teeth.

A common adaptation at these depths is the ability to extend jaws and belly so that prey much larger than the predator can be swallowed.

Melanocetus johnsoni (d) is illustrated before and after having eaten a fish twice its own length.

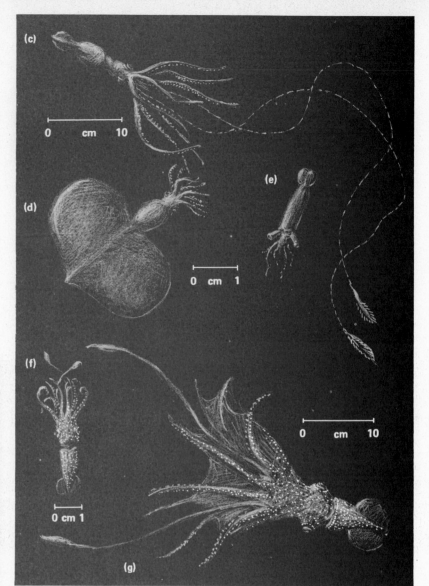

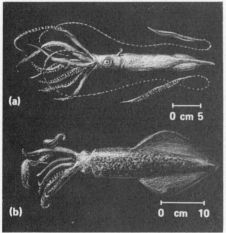

Figure 13-23

Oceanic squids. There are many species of these active, predatory cephalopods in the open ocean, ranging from luminous deep-sea species a few centimeters long to the giant squid that attain a length of over 16 meters. They live in all oceans, sometimes to depths of 3500 meters. Illustrated are: (a and c) two species of the deep-sea octopod *Cirrotheuthis;* (b) the common nearshore squid *Loligo;* two deep-water Atlantic squids— (d) *Octopodotheuthis,* and (e) *Taonidium;* and two mid-water species—(f) *Calliteuthis reversa,* and (g) *Histioteuthis bonnelliana.* *Histioteuthis* has been captured at about 1000 meters depth, and *Calliteuthis* as deep as 1500 meters, but both are sometimes seen at the surface.

of small animals that rise to feed at about 200 meters or less during the night, then return to depths of 500 to 1000 meters during the day. The precise makeup of a deep-scattering layer is difficult to determine. It is known, however, that some copepods, for example, can swim upwards at speeds of 17 to 30 meters per hour and the larger euphausiids three or four time as fast. Fish having gas-filled swim bladders that resonate when sound waves pass through them, may also make up a sizeable fraction of this migratory population.

Besides making the rich variety of foods near the bottom of the photic zone available to deep-water organisms, vertical migration may help zooplankton to maintain their position. Surface currents may move them away from their preferred location while they are feeding at night. By encountering countercurrents that flow below the surface layer, they are carried in the opposite direction.

Vertical migration seems to be an important mechanism for transporting food to the deep ocean. Deep-water animals consume food near the surface, and then carry down food that might otherwise have remained and been recycled in surface waters.

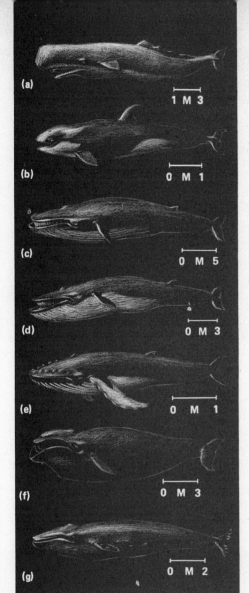

Figure 13-24

Sperm whales (a) are the largest of the toothed whales. Males average 15 meters, females are smaller. Their principal food is squid, which they hunt from the tropics to subarctic latitudes in all seas, diving to depths of 1000 meters or more and remaining below for as long as 1 hour. Giant squids up to 10 meters in length have been found intact in the stomachs of sperm whales. There is a reservoir of *spermaceti oil* in the animal's large, square head. Its function is unknown, although naturalists and scientists have speculated that it might be involved in the exchange of gases during long dives, or perhaps in sound production.
Before a dive, a whale fills its lungs with air; while it is underwater, most of the fresh gases are exchanged for carbon dioxide or other products of its metabolism. Large amounts of oxygen are stored in the myoglobin of its muscle tissue, as well as in blood. The heart beats very slowly while a whale (or other diving animal) is submerged, which helps to reduce oxygen demand. Small lungs and a flexible thoracic cavity permit compression without damage as the volume of air decreases with depth.
Sperm whales are sometimes attacked by packs of *killer whales* (b). These are not really whales at all, but members of the dolphin and porpoise family, most common in cold water. An adult sperm whale can usually resist such attacks, even if surprised while sleeping— probably the only animal in the ocean powerful enough to do so.
Baleen whales (whalebone whales) are the largest animals that have ever lived. They are mammals, and thus breathe air and bear live young. A horny sieve which hangs in plates from the palate is the "whalebone"; it is used to strain copepods and krill from seawater. They have no teeth, and plankton or small fishes are their only food.
Blue whales (c) are the largest of all, reaching lengths of up to 31 meters. A 27-meter blue has been found to weigh 120 tons. Baleen species also include: (d) the *finback,* which attains a length of about 20 meters; (e) the *humpback,* about 15 meters at maturity; (f) *right whales,* 17 meters or more; and (g) the smaller *California gray whale,* about 12 meters.
Baleen whales live in all seas, but particularly in the Arctic and Antarctic, where during the short southern summer (with a peak in February), krill have been reported so thick that a large ship was "slowed to half speed by them." Not only whales but fish, penguins, seals, and huge populations of flying birds gorge themselves on the seasonally abundant food supply. The greatest numbers of baleen whales have been observed in the Antarctic when plankton are dense. During the rest of the year, they range widely throughout the oceans.

FOOD FISHES

Fish are active predators. The typical streamlined shape of most pelagic species testifies to the value of speed for capturing prey and avoiding enemies. They can usually swim very rapidly for short periods, about 10 times their body length in a second. Over extended periods, as during migration, a medium sized fish such as salmon, herring, or cod can travel hundreds of kilometers in a few days. Large oceanic fishes that feed at high trophic levels, such as the *skipjack tuna* (Fig. 13-25) can swim 100 or more kilometers per day for weeks at a time while hunting schools of smaller fish. Estuarine and bottom-dwelling species, however, tend to hide near rocks or in soft sand and their bodies are often adapted for concealment rather than for speed.

Figure 13-25

Skipjack tuna, an inhabitant of tropical and temperate waters, is related to the great bluefin sport fish (which may be up to 5 meters in length), and also to the albacore tuna and the yellowfin. Tunas maintain a body temperature several degrees higher than the surrounding ocean, perhaps due to an unusually high metabolic rate related to their adaptation for high-speed, long-distance swimming.

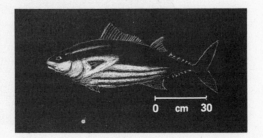

Figure 13-26

Some commercially important demersal fishes. Cod (a) live in cold and temperate seas, chiefly in the Northern Hemisphere, close to rocky or sandy bottoms. A schooling carnivore, its eggs are planktonic and produced in millions to ensure survival. Pollack (b), haddock (c), and silver hake (whiting) (d), are found from Newfoundland to the Caribbean, to 450 meters depth. Related species live in deep waters over continental shelves in both hemispheres, and over a wide temperature range. The young are adapted to eat planktonic crustaceans, while adults prey on small fishes. Like the rest of the cod family, and other less-oily fishes, they can be effectively preserved by drying. The squirrel hake is shown in (e).

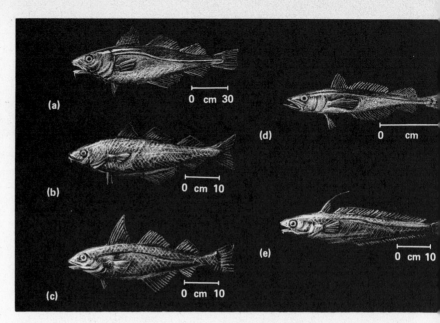

Of the *demersal* fishes, those that live on or near the ocean floor, the most important to man are the *cod* and its relatives (Fig. 13-26). They are mostly first-level carnivores, feeding on invertebrates, small fishes, and larvae. Another demersal group, the *flatfishes* (Fig. 13-27), are modified to lie on the ocean floor. They are elliptically shaped, although their larvae are bilaterally symmetrical. As flatfishes mature the eyes migrate to one or the other side of the head, so that as adults they can conceal themselves on the bottom with only their eyes exposed. The upper side, left or right depending on the species, tends to be darker than the lower. This is a common protective adaptation in pelagic as well as demersal fishes; organisms viewing a fish

Figure 13-27

Some important flatfishes. Atlantic halibut (a) are found from the Arctic Ocean to New Jersey; Pacific species live from the Bering Sea to the latitude of California. At lower latitudes, these species seek deeper waters. Halibut sometimes grow to 3 meters in length and are active predators. Flatfishes live on large invertebrates and small fishes. Northern fluke (summer flounder) (b) range from Cape Cod to Cape Hatteras, and reflected species inhabit all but the coldest seas. This flounder is really a member of the halibut family. Winter flounder (Atlantic sole) (c) are true flounders, as are European plaice (d) and turbot (e).

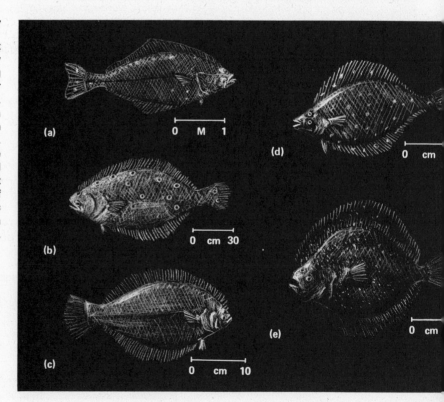

from below will be looking toward the light, so a light-bellied fish is less visible. But looking into deeper water, a dark-backed fish is more difficult to see.

The *herring* family (Fig 13-28) are commercially the most important pelagic fishes (see Table 15-1). Collectively known as *clupeoids,* they are most abundant in nutrient-rich continental shelf waters where diatoms are abundant. Many clupeoid fishes have specially modified gill rakers for straining clumps of phytoplankton out of the water; they also feed on copepods and other zooplankters. Clupeoids and other low-trophic-level feeders have a characteristically high oil content. They are often rendered into oils, fertilizers, or livestock feed, as well as consumed directly as human food.

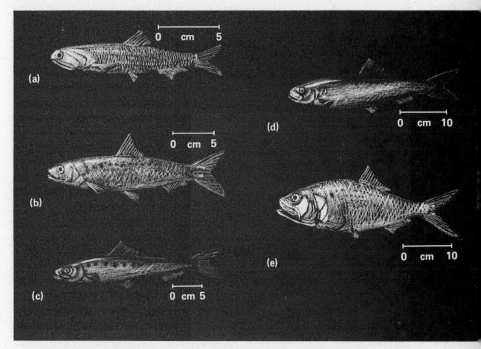

Figure 13-28

The herring family. (a) Anchovies and related species are abundant in warm seas, near shore and in the open ocean. These valuable fish are widely used for food and as fresh and frozen bait. They are tolerant of a wide salinity range, and are a basic food for many larger fishes. (b) Pilchard are found in coastal waters throughout the world, between isotherms 12° and 20°C. They are of major commercial value off Australia and South Africa. (c) Sardine, (d) herring, and (e) menhaden are distributed from Nova Scotia to Brazil and are a staple in the diet of many sea birds and fishes.

All of these relatively small fishes tend to *school,* that is, to aggregate in such a way that they are uniformly spaced and oriented in a common direction. This behavior is typical of fishes that swim unprotected in the surface layer of open seas and are constantly subject to predation. In general, all members of a school are of one species and more or less uniform size.

Schooling fish orient themselves so as to keep their own mirror image constantly in view. When thus positioned they seem to respire more slowly, eat more, learn better, and exhibit less nervous behavior. If a school is startled its members give off an "alarm substance" into the water, then pack more closely together to confront the source of the disturbance. Location of many individuals within a relatively small area decreases the chances of any one individual being attacked.

Smaller predators also school to hunt; they surround the prey school so that its members become confused and less able to escape. But some of the very large predators do not school, nor do those fishes that live below the photic zone. Schools may partially disperse toward nightfall, coincident with a period of intensive feeding. Some, for instance the Atlantic herring, descend toward the floor of the continental shelf during midday, where they may be eaten by demersal fishes such as cod.

North Sea herring have been extensively studied because of their importance as a fishery. Their life cycle and feeding habits are typical of coastal species (Fig. 13-29).

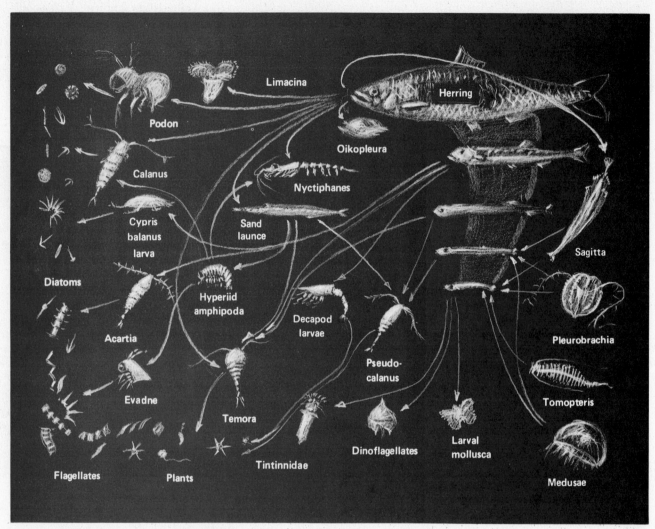

Figure 13-29

Life of the Atlantic herring. (Redrawn from A. Hardy, 1965. *The Open Sea. Part II: Fish and Fisheries,* Houghton Mifflin Co., Boston, p. 62.)

Herring deposit their eggs in thick blankets which may cover 100 square meters or more of the continental shelf. Bottom-living fishes, especially haddock, consume great numbers of these eggs. Herring develop as tiny, wormlike creatures, curled around the yolk of their eggs. After a few weeks, at about 5 millimeters length, the transparent larvae break free of the egg membrane and rise toward the water surface. Their first food is phytoplankton, later nauplii, very small crustaceans, and young stages of copepods.

Others starve when they do not find suitable food. If predatory ctenophore or arrow-worm populations are unusually large, the herring larvae may be eaten in great numbers. At a length of about 30 millimeters, they begin to eat some of the smaller adult copepods, and at about 40 millimeters they grow scales and begin to look like young fish instead of tiny eels. They form enormous school—some of them as much as 5 kilometers long and travel toward shallow water, often passing into estuaries, now eating mainly estuarine crustaceans. After 6 months or so, the young herring disperse throughout the North Sea. When sexually mature, after a period of 3 to 5 years, they

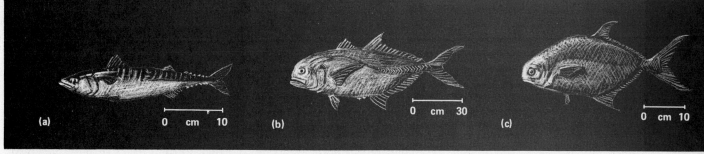

Figure 13-30

(a)   (b)   (c)

Atlantic mackerel (a) are distributed from Labrador to Cape Hatteras, and from Norway to Spain. In the western Atlantic, they migrate northward in schools along the coast in summer, southward offshore in winter. Preyed upon by birds, squids, and other fishes, they themselves eat anything they can swallow. (b) The common jack is related to a number of warm-water species off American shores; jacks range from Cape Cod to Brazil and southward of the Gulf of California. (c) Pompano. Related species include bonitas, tunas, and swordfishes.

join schools of spawning fish, and the cycle is repeated. The same species inhabits the Atlantic coast of Canada and the United States.

The *mackerel* and related species (Fig. 13-30) are all well adapted for strong, continuous swimming in open waters, both near the surface and at depth. Open-ocean species feed at high trophic levels and are voracious carnivores. Mackerel provide an interesting example of migratory patterns in fish. They leave the surface waters about October and aggregate in small areas near the ocean floor. During this period they eat crustaceans and small fish. In January the mackerel move to the surface in schools and migrate to their April spawning-grounds south of Ireland. They spawn near the edge of the continental shelf, gradually moving closer to the land, and during this time they feed on plankton, especially copepods. From June to July they form smaller schools and move close to the shore, changing their diet from plankton to the small fishes that swarm in inshore bays. In the fall, they again seek deeper waters.

Ocean perch and related fishes (Fig. 13-31) are taken in large quantities throughout the world for food (see Table 15-1). *Basses, croakers, snappers, kingfish, walleye pike, porgies,* and *groupers* are all members of this varied *percoid* class, especially common in warm near-shore waters. There are many sport fishes and also many freshwater species included among the percoids.

Another group of valuable food fishes are the *salmons* and *trouts.* Both are active predators, generally confined to northern, fairly cold water, and they typically spend part or all of their lives in fresh water. Salmon invariably spawn in rivers but attain most of their growth in ocean waters (Fig. 13-32).

Figure 13-31 (below)

Ocean perch (rosefish, redfish). These live in the North Atlantic with related species worldwide. Some of the many species in this family are referred to collectively as rockfishes because they prefer near-bottom waters. The flesh tends to be reddish and many groups have long spines, often poisonous in tropical members of the family. North Atlantic redfish and eastern Pacific rockfish are examples of major fisheries opened after World War II to exploit what used to be considered "trash" (noncommercial) fishes in meeting world protein needs.

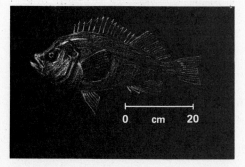

Figure 13-32 (above)

Chinook salmon (king salmon, spring salmon). Ranging from Southern California to northwest Alaska, chinook are especially prized by sport fishermen. They are common along the British Columbia coast. In spring and early summer, mature adults leave coastal waters to "run" up large rivers, where they spawn in fresh water and then die. The young may go to sea in the first year or remain for a year or more in the streams. Maximum length is about 1.5 meters. (Photograph courtesy Washington State Department of Fisheries.)

Most important sport and food fishes are native to coastal waters and more than half are estuarine-dependent during at least part of their lives. Estuaries serve as nursery areas for many species, providing a protected environment as well as abundant food for larvae and juveniles. These areas are highly productive, deriving nutrients from three sources: subsurface coastal ocean waters carried in by the estuarine circulation; dissolved minerals leached from soils in the river's drainage basin; and nutrient-rich by-products of agriculture, industry, and human-waste discharge. Furthermore, nutrients tend to be retained within the estuary because phosphates and nitrates released by decomposition in lower layers are returned to the surface by entrainment of subsurface waters in the estuarine circulation (Chapter 10).

Being shallow as well as rich in nutrients, estuarine tidal flats, marshes, and shallow inlets support dense stands of benthic plants. These include flowering *seagrasses* that grow in mud and sand, benthic algae, and *epiphytes* (plants, often unicellular, such as diatoms, that grow attached to larger plants). Primary productivity of seagrass communities may exceed 600 grams of carbon per square meter annually. For example, in Atlantic estuaries the marsh grass *Spartina* and associated algae produce up to 250 tons of plant material per square kilometer (10 tons per acre) in a year, several times the productivity of an average wheat crop. Shore-based vegetation, for instance in salt marshes, and phytoplankton further increase estuarine productivity.

Fishes, jellyfish, and benthic animals throughout the estuary and coastal ocean are supported by estuarine productivity. Migratory and developing fish populations are major consumers. Food webs in shallow estuaries are based largely on microscopic benthic and epiphytic algae and on decaying vegetation. Thus the primary plant food is consumed directly by herbivorous fishes such as mullet, and by benthic fauna, rather than by zooplankton as is the typical pattern in open waters.

Patterns of spawning and development vary widely among estuarine-dependent salt-water fishes and shellfish. Off the Atlantic coast and the Gulf of Mexico, many coastal fishes and crustaceans spawn in the ocean, but their young soon move into estuaries. Larvae often travel many tens of kilometers, moved passively by strong coastal currents, and enter small inlets where currents are quite strong. Larvae that swim at the surface (such as mullet) may come in on the tides. Bottom species such as shrimp, whose life cycle is illustrated in Fig. 13-33, are aided by inshore currents associated with estuarine circulation systems. Once inside, the larvae are distributed within the bays by currents; possibly some are sensitive to salinity gradients and swim toward less saline water. In general, lower salinities are preferred by fish larvae where waters are warmer. Tolerance for low salinities declines toward higher latitudes. North of Cape Cod, Massachusetts, for example, most fisheries depend on deep-water fishes, such as cod and haddock, which do not enter estuaries.

In the Northern Hemisphere in spring and summer, fish migrations are typically directed inshore and northward; in late summer and fall, the pattern reverses. The bottom-dwelling fluke (see Fig. 13-27), for example, winter offshore along the middle-Atlantic coast, sometimes as far as 150 kilometers from land. In spring they move toward the coast to feed, spending late spring and summer in nearshore waters. In early fall, fluke start back toward their wintering grounds. They spawn during October and November, 15 to 100 killometers offshore on the continental shelf, and surface currents carry the buoyant eggs (later the larvae) southward until early spring. They swim toward the coast and spend the summer in shallow estuarine waters; in autumn

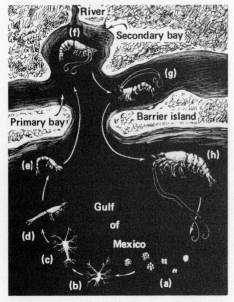

Figure 13-33

Life history of the Gulf Coast shrimp: (a) shrimp eggs; (b) nauplius larva; (c) late nauplius; (d) mysis; (e) postmysis; (f) juvenile; (g) adolescent; (h) adult.

they reenter the ocean. This type of breeding cycle, with variations, is common among Atlantic and Gulf coast species.

Some migratory fishes enter Atlantic estuaries to spawn. Striped bass, for instance, lay their eggs in the fresh waters of Chesapeake Bay and Albemarle Sound tributaries, and the young drift downstream into the estuary. At maturity (2 or more years) some of them enter coastal waters and migrate northward as far as New England during the summer. Winters are spent in deep holes or channels, and in spring they again swim upstream to spawn.

Productivity in the North Sea (see Figs. 3-20 and 12-23) has been studied in order to determine whether fish yields in the area might be increased. But judging from known levels of primary production, it seems unlikely that present yields can be greatly increased. For example, it has been calculated that net primary productivity in the North Sea averages about 90 grams of carbon per square meter per year. Production of pelagic fish has ranged as high as 2×10^6 tons during peak fishing years in the 1960's and demersal species were caught at a rate of 0.9×10^6 tons annually. Assuming that "natural mortality," or death from causes unrelated to man, claimed at least half of the pelagic fishes and more than half of demersal production, it has been calculated that 4.0×10^6 tons of pelagic and 1.3×10^6 tons of demersal fish represent a maximum potential production for the North Sea as a whole. This averages out to 8.0 grams of carbon per square meter (wet weight) of pelagic fish and 2.6 grams of carbon per square meter of demersal fish for each 90 grams of carbon per square meter per year of net phytoplankton produced in the region.

Figure 13-34 is a representation of the North Sea as an ecosystem; it illustrates the food web through which energy passes from phytoplankton to man. Each gram of organic carbon is considered equivalent to 10 kilocalories (kcal) of potential energy. Each organism in the food web has been assigned a value in kilocalories, calculated either from its biomass or on the basis of its trophic level and presumed ecological efficiency. For instance copepods are assumed to be about 20 percent efficient in transforming ingested phytoplankton into animal biomass ($\frac{170}{900} = 19$ percent). But only half the herbivores are eaten by pelagic fish in the direct food chain to man, and pelagic fish are only considered to be about 10 percent efficient at building fish tissue from the copepods they eat. Furthermore, even if all chaetognaths (arrowworms) are ultimately eaten by herring, the additional trophic level would use virtually all the energy potentially transferable from arrowworm to fish.

On the decomposer–demersal fish side of the model, it is postulated that benthic forms are supported by detrital material from above, which may go at least in part through bacterial transformation before being consumed by bottom-dwellers. Bacteria apparently have high ecological efficiency and a very low ratio of biomass to productivity because they live only a short time but reproduce often. They evidently provide substantial food for both the *macrobenthic epifauna* (large animals living on the bottom) and the *meiobenthos* (minute animals that live in bottom sediments). These organisms probably also have a low ratio of biomass to productivity, especially the very small species. But if efficiencies of around 10 percent are typical of fish and benthic animals, the model indicates that present populations utilize virtually all primary production in the North Sea. No large increase in fish yields is likely under these circumstances without drastically disturbing energy flows in the food webs.

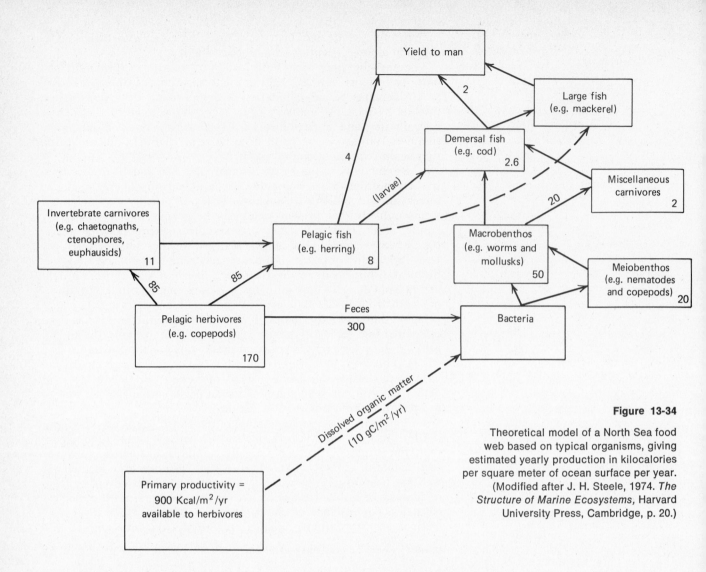

Figure 13-34

Theoretical model of a North Sea food web based on typical organisms, giving estimated yearly production in kilocalories per square meter of ocean surface per year. (Modified after J. H. Steele, 1974. *The Structure of Marine Ecosystems,* Harvard University Press, Cambridge, p. 20.)

SUMMARY OUTLINE

Phytoplankton
  Small size results in large ratio of surface to volume
    Presents relatively larger surface for nutrient, gas diffusion
    Retards sinking
  Major classifications—ultraplankton, nannoplankton, microplankton
    Small forms dominate in open ocean
  Diatoms—dominant producer in coastal, mid- and high latitudes
    Characterized by heavy spring blooms, clumping in high-nutrient areas, siliceous shells
  Dinoflagellates—next most productive; some heterotrophic
  Coccolithophores, silicoflagellates also important; blue-green algae in warm coastal areas

Zooplankton—transfer matter and energy from primary producers to predators

Variety of feeding adaptations for trapping small particles
Currents carry them away from narrow temperature ranges of breeding grounds

Holoplankton—permanent members of plankton
  Protozoa—consume nannoplankton
    Foraminifera—calcium carbonate shell; *Radiolaria*—siliceous; *Tintinnidae;* and *Rotifera* important in open oceans
  Crustaceans—abundant in productive ocean areas
    Copepoda—most numerous
    Euphausiidae—"krill," abundant in Antarctic, eat diatoms, eaten by whales
  Pteropods—"snail with wings"; herbivorous or carnivorous
  Chaetognaths—active, predatory worms
  "Dead-ends" in food chains—Coelenterates (jellyfish); Ctenophores ("comb-jellies"); Tunicates (salps, doliolids)

Meroplankton—larval stages only in plankton
    Eggs, larvae planktonic, metamorphose to adult form at attachment to bottom
    Worms, mollusks—have trochophore larvae
    Crustaceans—similar to planktonic forms
    Fish larvae—success depends on availability of food, other environmental factors

Life in the deep ocean—mesopelagic zone—200 to 1000 meters
    Less than 1 percent of light penetrates
    Many organisms bear patterned lights, lighted lures
    Squid—powerful swimmers, high-level predators
    Vertical migration by many fishes, invertebrates
        Deep-scattering layer: 200 meters deep at night; 500–1000 meters by day
        Deep-water organisms get food from surface zone

Food fishes
    High-trophic-level feeders, e.g., tuna, swim powerfully

Demersal fishes (cods, flatfishes)—coastal ocean, deep-water, low-trophic-level feeders
Herring family—clupeoids; herbivorous, tend to school for protection
    Life-cycle typical of coastal fishes
Mackerel—migrate throughout wide coastal-ocean area

Fish and coastal-ocean productivity
    Most food and sport fishes depend on coastal, estuarine productivity
    Estuaries—high productivity nursery areas; benthic vegetation productive
    Different species have various migratory patterns; enter nearshore and estuarine areas at different times of year, different stages of life-cycle
    North Sea: cannot yield much more than at present
        Primary productivity approximately 90 grams of carbon per square meter per year; fish yield approximately 3×10^6 tons
        Present population evidently utilizes all energy from primary productivity

SELECTED REFERENCES

FRASER, JAMES. 1962. *Nature Adrift: the Story of Marine Plankton*. G. T. Foulis, London. 178 pp. Concise, readable survey of marine-plankton forms, their behavior and ecology.

HARDY, ALISTER. 1965. *The Open Sea: Its Natural History*. One-volume ed. Houghton Mifflin, Boston, 355 pp. A classic in the field, by a well-known British naturalist; nicely illustrated.

HOLT, S. J. 1969. "The Food Resources of the Ocean." *Scientific American*, 221 (3) (Sept., 1969), 178–94. Survey of contemporary technology and outlook for the future of marine food resources.

IDYLL, C. P. 1964. *Abyss: The Deep Sea and The Creatures That Live In It*. Thomas Y. Crowell, New York. 396 pp. Elementary discussion of marine life with particular emphasis on deep-ocean organisms; also discusses various curiosities such as sea monsters.

MARSHALL, N. B. 1958. *Aspects of Deep-Sea Biology*. Hutchinson, London. 380 pp. Authoritative treatment of the deep-ocean environment and its ecology.

RAY, CARLETON, AND ELGIN CIAMPI. 1956. *The Underwater Guide to Marine Life*. A. S. Barnes, New York. 337 pp. Classifies and briefly describes thousands of marine organisms, with line drawings.

WIMPENNY, R. S. 1966. *The Plankton of the Sea*. American Elsevier, New York. 426 pp. Emphasis on the taxonomy of animal plankton; well illustrated.

Figure 14-1

On this rocky Maine coast, brown algae grow thickly along a steep rock face, then cease abruptly except near the bottom, where growth extends further out toward the water.

the benthos

Benthic organisms live in a two-dimensional world, in contrast to the three-dimensional, nearly boundless world experienced by plankton and nekton. There are three main strategies: attachment to a firm surface, movement on the bottom, or burrowing in sediment. These generally correspond to three ways of obtaining food: it can be filtered from seawater; caught by predation; or ingested by swallowing soft sediment, removing its organic contents, and either regurgitating or excreting the indigestible portions.

The three lifestyles can be combined in a variety of ways. A crab or worm, for instance, may live in a sand or rock burrow but emerge to hunt actively for prey or scavenge for detritus. Slow-moving animals with heavy shells, such as snails or sea urchins, feed on attached organism or detrital particles. But some fixed animals, such as sea anemones, can be predators; they capture organisms that swim or float past them.

Figure 14-2

Sargassum (sometimes called "gulfweed") is the only large nonattached seaweed. It is found in the south-central North Atlantic, where surface currents converge in an area of calm sea. There, it puts out shoots which break off as new plants. Patches of this algae form "floating islands"; they harbor communities of smaller plants and animals which can live nowhere else in the world. These include many kinds of specialized crabs, snails, copepods, worms, fish, sea horses, and small octopus, as well as the *sargassum fish* itself.

Patches of sargassum collect in windrows along lines of convergence formed by wind-driven Langmuir cells and other circulation systems. Surface films, plankton, and particles of detritus are caught there and so become available to decomposers and consumers in the sargassum community. Large numbers of eggs and larvae of open-ocean fishes become trapped in the dense growth, which offers protection from predators as well as a rich supply of food particles. Some of these eggs and young fishes, an energy source that originated outside the sargassum community, are consumed within the community, thus augmenting the local nutrient supply.

Seven million tons of "gulfweed" is estimated to float in the approximately 5000 square kilometers of the Sargasso Sea.

Attached plants can only grow on about two percent of the ocean floor, shallower than 30 meters depth, and many kinds require a firm reef or rock substrate (Fig. 14-1). Fixed seaweeds thus contribute little to total productivity in the open ocean, although they are highly productive in shallow, coastal areas. A notable exception is *sargassum*, a leafy brown *alga* (seaweed) which forms a unique floating "benthic" ecosystem in the Sargasso Sea (Fig. 14-2).

The makeup of a benthic community is controlled mainly by light, temperature, salinity, and bottom material, and by whether these parameters are constant or variable. Environmental stability seems to favor the evolution of highly diverse communities, where many kinds of plants and animals coexist. For example, in surf zones and other nearshore areas where waves keep bottom materials in constant motion, benthic life may be extremely scarce. And in parts of shallow polar oceans, a short growing season for plants, together with marked salinity changes due to melting and freezing of sea ice, create a rigorous climate where relatively few species thrive. Quiet waters in nearshore tropical areas, on the other hand, usually have highly diverse bottom communities. Thousands of different species can make up a single ecosystem, such as a coral reef community.

Shallow-water ecosystems in temperate climates tend to have relatively few different species, but more individuals of each kind (Fig. 14-3). This lack of diversity may be due in part to the stress of pronounced temperature changes that have occurred at mid-latitudes over the past several hundred thousand years. Tropical and deep-ocean environments, on the other hand, have evidently experienced stable temperatures for millions of years. They are also less rigorous, being relatively free from seasonal climate changes. Such environments include a greater variety of organisms than do communities where temperature, salinity (see Fig. 12-2), or other conditions are subject to change.

Development of juvenile stages also varies with environment. Benthic organisms commonly have pelagic larvae in tropical and subtropical zones, but in rigorous polar and subpolar climates, nonpelagic larvae predominate (Fig. 14-4).

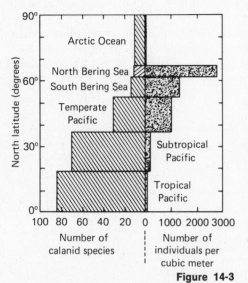

Figure 14-3

Diversity gradient for species of calanoid copepods (planktonic crustaceans) from the upper 50 meters indicate that the number of species is greatest in tropical climates. Numbers of individuals are greatest at subpolar latitudes. This relationship also holds true for benthic organisms. (After A. G. Fischer, 1960. "Latitudinal Variations in Organic Diversity," *Evolution*, 14, 84–81.)

Figure 14-4

Moving south from polar to subtropical waters, the percentage of snail species having pelagic (free swimming) larvae increases with increasing temperatures. (After Raymont, 1963.)

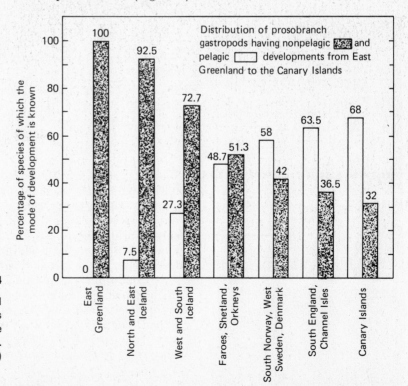

The abundance of benthic organisms is governed by productivity of surface waters and by water depth. It is greatest on continental shelves and below areas of upwelling or divergence. Number of organisms decreases steadily with increasing water depth and distance from land.

Over the short term, benthic populations vary according to fluctuations in year-class size. Variations in temperature, food supply, or number of predators determine the number of larvae that can grow to maturity. This in turn affects population size, particularly for organisms whose life cycle is not much longer than their generation period. This is illustrated by the change in biomass of 5 species of clam over a 4-year period, as shown in Fig. 14-5. Among relatively long-lived species, where many generations coexist, populations are less subject to fluctuations in size.

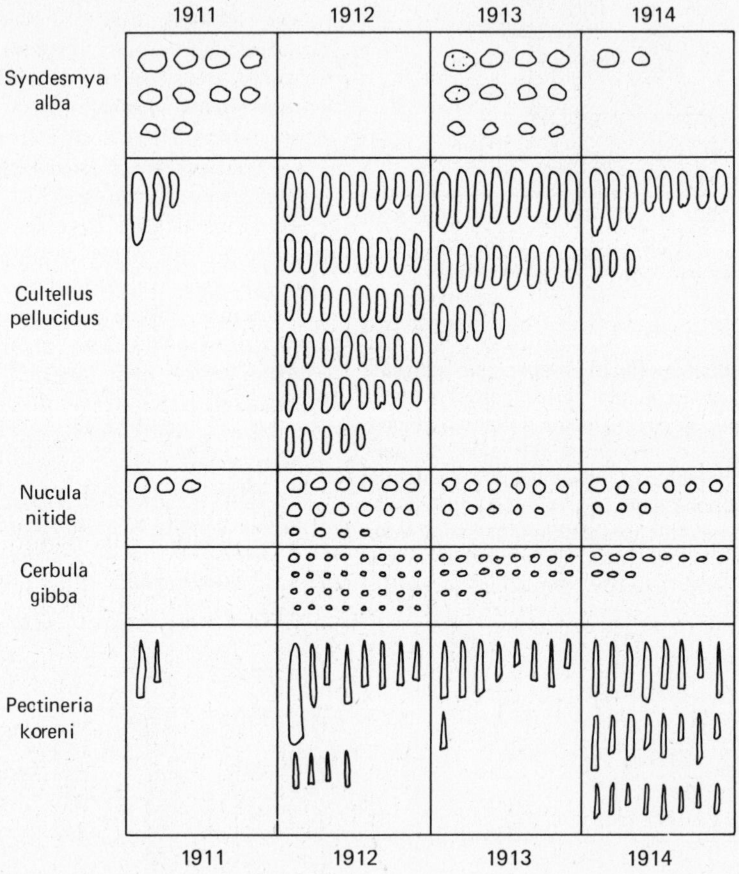

Figure 14-5

Annual variation in relative abundance of five inverterbrate species studied over a 4-year period. (After J. E. G. Raymont, 1963. *Plankton and Productivity in the Ocean*, Macmillan, Pergamon Press, New York, 625 pp.)

ROCKY SHORE COMMUNITIES

Intertidal areas provide a vertical sequence of habitats. Each *zone* or horizontal band is occupied by a particular assemblage of plants and animals. This is easily seen on a rocky beach, where organisms are attached at the surface, though many animals hide in moist, protected crevices when the tide is out. Zonation is also present on sandy and muddy shores, but there many zones occur within the sediment and are less obvious to the casual observer.

Conditions in intertidal areas range from nearly always dry (highest high-tide level) to nearly always submerged (lowest low-tide level). Population zones are often sharply divided at depth contours rather than grading into one another. This is in keeping with the intense competition for living space in intertidal regions. At the line where the two populations meet, neither enjoys marked advantage.

Above and below the line, species assemblages are determined by differences in the organisms' ability to endure air exposure, and often by the survival of their predators and diseases under the same conditions. Zonation among barnacles on the Maine coast can be seen in Fig. 14-6.

Where winds, sun or waves create a severe climate, or where the rock face is steep, attached seaweeds may be sparse or absent. In more protected areas a variety of plants attach to bare rocks along the seashore. Above the high-tide mark, seawater only reaches the rocks at higher spring tides and during storms, but salt spray usually wets the area. Resistance to drying is the prime requirement for plants and animals in this region. Several varieties of blue-green algae, lichens (algae growing symbiotically among fungus filaments), and certain small snails are common above the high-tide mark. Filamentous green algae known as sea-hair sometimes grow in moist, protected spots.

Between the high-tide and low-tide marks, especially where heavy surf is a factor or where current scouring is intense, firm attachment to the rock is a necessity. Barnacles (Fig. 14-6) often dominate the upper, more exposed regions, where the rock is covered by water less than half of each tidal day. In shady, protected areas, the barnacle zone extends farther up the rock face than it does on dry, sunny surfaces. These animals feed while covered by water at high tide, filtering small particles from the water (Fig. 13-18). The barnacle shell has a complex, four-part lid which shuts tightly, protecting the animal from exposure.

Figure 14-6

Zonation among barnacles, Marine coast. A small species dominates the upper layer, separated by a sharply defined boundary from the larger species below.

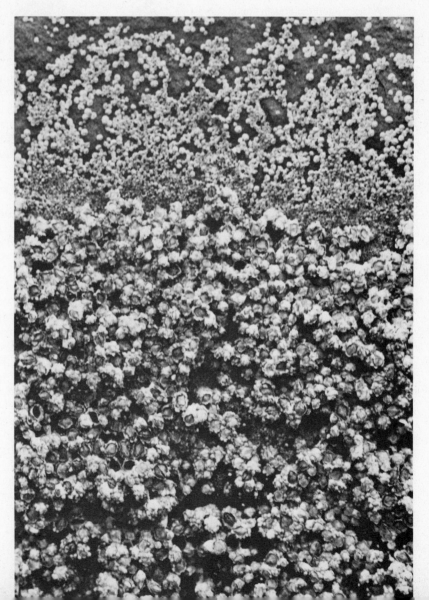

Figure 14-7

A rocky intertidal area in Maine shows zonation of attached animals and plants: barnacles above, brown algae below, and at the bottom, underwater even at low tide, green and red algae grow in a shady pool. Snails, worms, sea stars, and tiny crustaceans live among the plants.

Limpets, cone-shaped snails, also live in rocky intertidal areas. They feed by scraping off accumulated algae and detritus. During low tide or when subjected to wave action, some limpets return each to its own depression on the rock face, which closely fits the rim of the animal's shell. The seal is so good that it is nearly impossible to dislodge a limpet by hand. Some species favor attachment on top of another limpet. Lower in the intertidal zone, tube-worms and sea anemones compete with the barnacles; closer to the low-tide mark, dense mussel beds are common, and various kinds of attached algae as seen along the Maine coast in Fig. 14-7.

Preference for a specific light intensity governs distribution of plants. Green algae may be found in areas ranging from somewhat above the low-tide mark to a depth of perhaps 10 meters in temperate and tropical zones. Varieties range from lush-looking bright green sea-lettuce, up to 1 meter in length, to delicate, mossy types only a few centimeters long. Red algae grow worldwide, but most abundantly in temperate and tropical seas. They prefer the dim light of very deep water or well-shaded pools. Brown algae flourish in colder water, although some kinds are found on rocky coasts throughout the world. An abundant form in North America is brown rockweed, with its cluster of berrylike floats. Marine algae have no true roots, but absorb nutrients from seawater directly into their fronds. Some benthic algae attach to the bottom by root-like structures called *"hold-fasts."*

Tide pools contain rather specialized plants and animals. The protected environment afforded by the rocks often permits more delicate organism to live in these pools, and a large variety of plants and animals may live in a small area (Fig. 14-8). Tiny, shrimplike crus-

Figure 14-8

Many varieties of red and green algae shade one another in this tide pool on the Maine coast, providing shelter for small shellfish and worms. The area photographed is about 1 meter across.

taceans (amphipods), swimming worms, and many kinds of snails are common in tide pools, especially where deep crevices or beds of seaweed retain water during low tide. Evaporation, overheating, and oxygen depletion occur on warm, sunny days, whereas heavy rains can cause dilution within a few minutes. Survival for a tide pool organism requires a tolerance for sudden changes in temperature, salinity, and dissolved-oxygen content of the water.

The most populous zone on a rocky beach is generally located around and below low-tide level. Starfish and crabs are common, usually hidden in rocky crevices. Small scavenging snails inhabit protected niches containing stagnant water and decaying debris. Sea anemones (Fig. 14-9), sea urchins, and sea cucumbers may be locally abundant below the low-water mark. Hydroid colonies (Fig. 14-10) grow in quiet but not stagnant waters, as do the nudibranchs (a type of shell-less snail, see Fig. 14-11) which feed on them.

Figure 14-9

Warm-water sea anemones with sea stars at the New York Aquarium. The animals at left center and far right are open; the mouth opening is at the center of the "flower." When startled, they close instantly, giving the appearance of a wrinkled stump (right center).

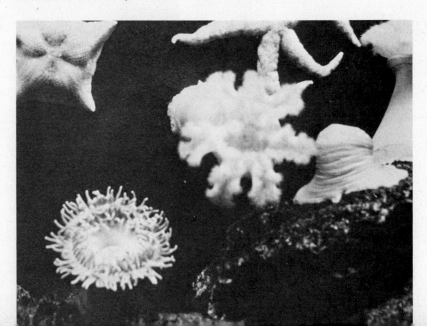

(a)

(b)

(c)

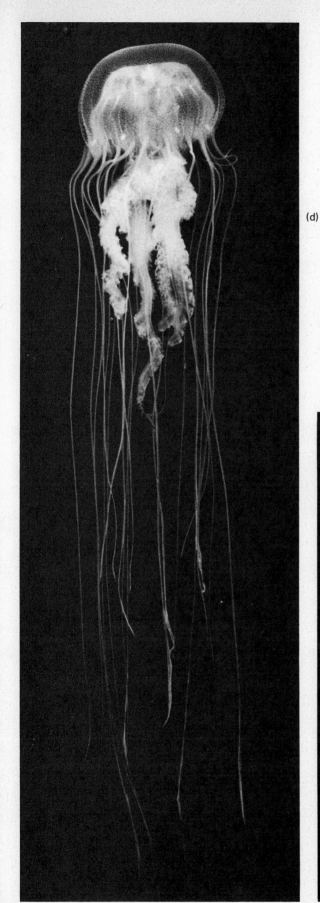

(d)

Figure 14-10

The coelenterate *Chrysaora quinquecirrha*, a native of Chesapeake Bay: (a) hydroid polyps, growing on an oyster shell; (b) detail of a typical feeding polyp, about 1 millimeter tall—the mouth, surrounded by sixteen tentacles, is at the top of the polyp; (c) reproductive polyp shows budding of ephyrae, and may produce a new stack full-grown, free-swimming *medusae* in 2 to 4 weeks; polyp may give rise to two to twelve ephyrae, and may produce a new stack up to four times in one summer; (d) full-grown medusa of *Chrysaora*, about 10 centimeters across the bell; their unpleasant sting sometimes closes the bay to swimmers. (Photographs courtesy Michael J. Reber.) Medusae reproduce sexually to form a new, sessile animal. The hydroid *Obelia*, showing feeding polyps and reproduction cycle, is illustrated in the diagram (e).

(e)

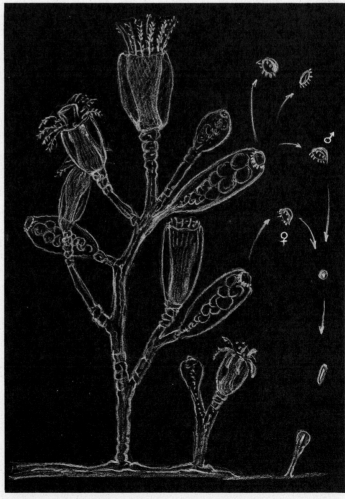

Figure 14-11

The nudibranch *Craten pilata,* length about 20 millimeters, which feeds actively on hydroid polyps in Chesapeake Bay. This shell-less snail breathes through the gill-like projections along the sides. Because of their noxious taste, they have few natural enemies. (Photograph courtesy Michael J. Reber.)

Figure 14-12

American lobster, which lives off the East Coast of the United States. (Photograph of a model in the U.S. National Museum, Washington, D.C.)

Figure 14-13

Spiny lobster (crayfish), similar to a true lobster but lacking the large claws, is found in the warm waters of the West and Gulf Coasts. (Photograph courtesy Marineland of Florida.)

Figure 14-14

Octopus. The eyes are on top of the head; below is a siphon through which water is ejected after passing over the gills. There are eight tapered tentacles, or arms, all equipped with double rows of suckers. Maximum size is around 50 kilograms with a 9-meter arm-spread, but many common species are much smaller. (Photograph courtesy Marineland of Florida.)

Lobsters (Fig. 14-12) and crayfish ("spiny lobsters," Fig. 14-13) scavenge on subtidal hard bottoms, both near shore and far out on the continental shelf, around most of the North American continent. They walk about at night, searching for worms, mollusks, and organic debris, and usually seek shelter during daylight under rocks or among seaweed. In autumn, Maine lobsters move offshore to breed, probably seeking warmer deeper waters and thereby avoiding the cold shelf waters in winter time.

The octopus (Fig. 14-14) is another common resident of hard-bottom and continental shelves. It prefers the protection of crevasses, and usually emerges only at night to feed.

SOFT-BOTTOM COASTAL COMMUNITIES

Tidal marshes, beaches, and estuarine shorelines offer a variety of habitats for benthic flora and fauna. Marshes received fine-grained sediments from fresh-water runoff and tidal influx, and abundant rooted plants add to a high organic content, besides protecting the marsh from erosion by tidal currents. On open coastlines, where wave action or longshore currents inhibit rooted plant growth and also carry away silt and detritus, coarse, sandy bottoms with low organic content are the rule. In beach and marsh environments, plankton, rooted plants, benthic invertebrates, fish, birds, and mammals interact in a complex variety of ecosystems (Fig. 14-15).

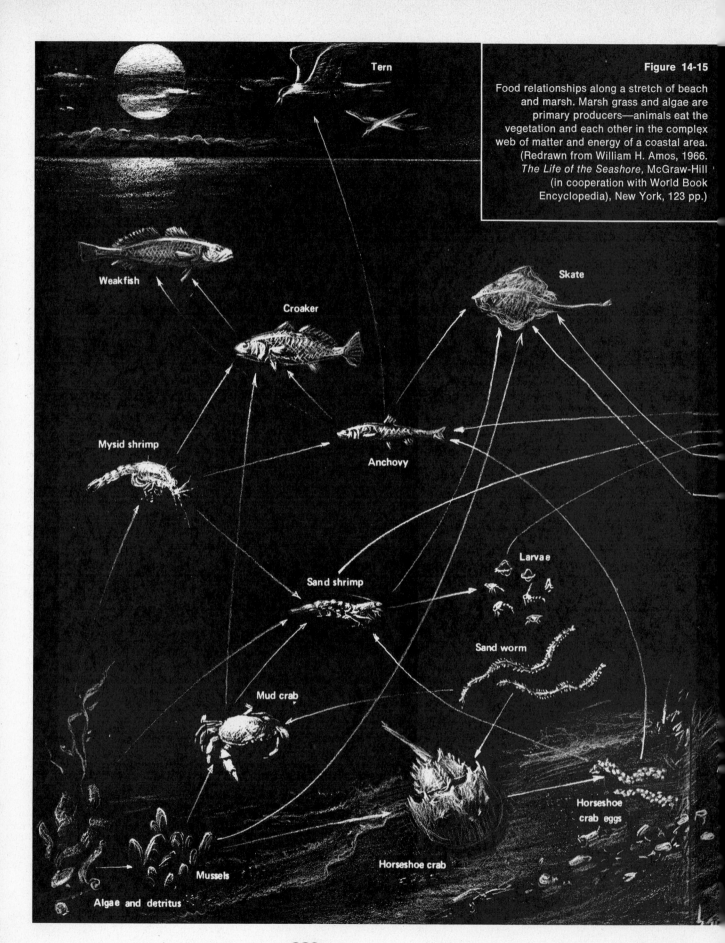

Tern

Figure 14-15

Food relationships along a stretch of beach and marsh. Marsh grass and algae are primary producers—animals eat the vegetation and each other in the complex web of matter and energy of a coastal area. (Redrawn from William H. Amos, 1966. *The Life of the Seashore,* McGraw-Hill (in cooperation with World Book Encyclopedia), New York, 123 pp.)

Weakfish

Croaker

Skate

Mysid shrimp

Anchovy

Larvae

Sand shrimp

Sand worm

Mud crab

Horseshoe crab eggs

Mussels

Horseshoe crab

Algae and detritus

Gull

Bat

Swallow

Diatoms

Grackle

Dead
horseshoe crab

Sandpiper

Insects

Zooplankton

Rat

Dead bird

Mole crab

Ghost shrimp

Ghost
crab

Beach hoppers

Amphipods
and
isopods

Skunk

Sand fauna

Beach wrack

Infauna—animals that live buried in sediment—are much less conspicuous than surface-dwelling animals, or *epifauna*. Many of the former are *selective deposit-feeders* (Fig. 14-16(a)), lifting and sucking organic particles out of the mud, while others pump large quantities of water across a filtering device to remove the edible material. Still others feed *unselectively* on sediment deposits, eating their way through the substrate and absorbing their food through the gut; these include some nudibranchs, sea-cucumbers, and many worms, such as the one in Fig. 14-16(b).

Distribution of infauna is governed in part by grain size of the sediments. Animals which feed by filtering plankton and detritus from seawater predominate in sandy deposits below low tide. They require relatively clear, nonturbid water, which will not clog delicate, mucus-coated filtration devices; fine muds, easily suspended by bottom currents, are not usually a satisfactory substrate for filter-feeders. Muds and clays, however, are well suited to organisms which feed unselectively by ingesting sediments, because the smaller particles, with their higher ratio of surface to volume, generally adsorb more dissolved and suspended organic matter. Such detritus supports a population of bacteria and very small animals which are food for deposit-feeders. These microscopic members of the infauna will be discussed more extensively in the following section.

Figure 14-16

(a) Benthic bivalves: the cockle is a filter-feeder; the clam *Tellina* sucks edible particles from the sediment surface. (After Hedgpeth, 1957.)
(b) Sipunculid worms can retract the tentacle-bearing head into the body. Many live in soft ocean sediments and feed by swallowing great quantities of mud and sand, from which the organic component is extracted in the gut. (Photograph of a model in the U.S. National Museum, Washington, D.C.)

(a)

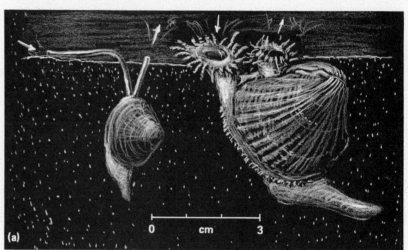

(b)

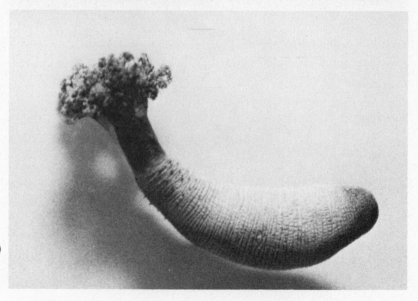

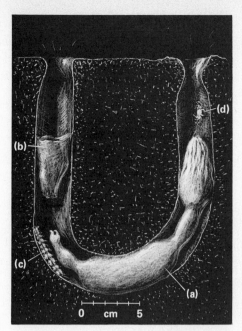

Figure 14-17

The pink echiuroid worm *Urechis caupo* (a) is a common inhabitant of Pacific mud flats, at or below the low-tide mark. This so-called "innkeeper worm" constructs a deep, U-shaped burrow with narrow openings, by scraping away mud with bristles at each end of its body. Inside, it secretes a funnel-shaped, fine-meshed mucus net that fits over its neck like a collar (b). As *Urechis* pumps water through the burrow, food particles are trapped in the net. When the funnel becomes clogged, the worm eats it and constructs a fresh one. Wastes and debris are ejected by a blast of water pumped from the gut, but enough remains in the burrow to support other animals. A worm *Hesperonoë* (c) lives behind the "host," along with tiny pea crabs *(Scleroplax),* and sometimes small fish called *gobies* shelter in the burrows.

Even for deposit-feeders, the most desirable sediments are not necessarily entirely silt- or clay-sized particles, especially where waters contain large amounts of dissolved or suspended organic matter. Excess organic material in the sediment may result in a degree of oxygen depletion in the near-bottom waters which is intolerable to most benthic animals. About 3 percent organic matter appears to be optimal for deposit-feeding bivalves.

Mud-burrowing animals require specialized breathing structures that are not clogged by fine silt. Breathing through the skin—as sea stars do, for example—is a poor adaptation for a muddy environment. One widely used mechanism involves setting up a flow of water through the burrow (Fig. 14-17) or in and out of the shell of a bivalve. In this way, oxygen is supplied for respiration, fine particles are swept off the gills and a constant food supply is maintained.

For filter-feeders, fine sands are more suitable than coarse. Particle size does not affect the animal directly, but coarse-grained deposits are usually associated with strong bottom currents. Fine sands occur where currents are weaker. In intertidal areas, deposit-feeders generally predominate whatever the substrate, because an unstable sediment surface prevents the accumulation of phytoplankton or other particulate food utilized by filter-feeders.

Many deposit-feeders make fecal pellets that are quite durable (Fig. 14-18). This tends to reduce turbidity at the sediment-water interface, and is thus beneficial to certain benthic organisms. In deep, quiet water, the bottom may be nearly covered by fecal pellets, which often exhibit a form or texture characteristic of the species which made them.

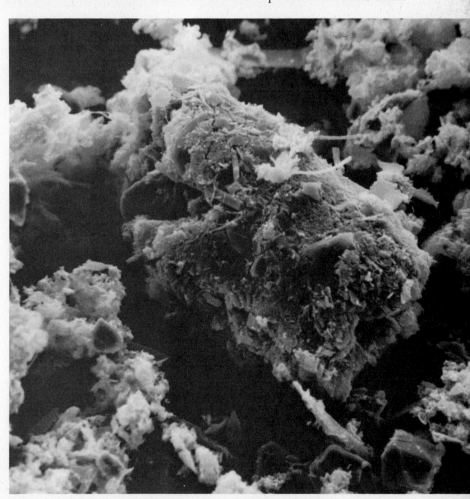

Figure 14-18

Fluffy, light-colored particles or organic debris surround this fecal pellet. Darker mineral grains appear at the bottom of the photograph, which shows an area about 0.14 millimeters square. (Photograph courtesy Dr. J. M. Sieburth and University Park Press, Baltimore, Md.)

Figure 14-19

A Long Island, New York, tidal marsh at high tide (above) and at low tide (below), when the surface is wet mud rather than water.

In the absence of strong wave or current action, intertidal sediment deposits may be stabilized by highly productive marsh grasses, forming *salt marshes* (Fig. 14-19). Such sediments are rich in detritus. Walking through the area at low tide one sees little evidence of life, but the ground is full of burrows which testify to an active, if unseen, population. Mussels live buried in the wet mud, and so do burrowing worms.

Around the high tide mark fiddler crabs (Fig. 14-20) dig their burrows. These burrows may be up to one meter in length, but they are dug obliquely to be shallow so that the crab can remain dry. While the tide is low, a crab picks through the mud, selecting bits of plant or animal detritus. He enters his burrow when flooding is imminent, sealing the doorway against the rising tide, and remains inside until the water subsides. Fiddler crabs require only enough contact with the sea to keep their gill chambers moist, although they can survive for weeks immersed in seawater.

Figure 14-20

Fiddler crab and his burrow: (a) fiddler posed on the sand; (b) about to dart into his burrow, fighting claw held protectively over his head; (c) a freshly made burrow surrounded by lumps of compacted sediment excavated by the crab and deposited outside.

(a)

(c)

(b)

(a)

Figure 14-21

(a) Model of an eelgrass community, Woods Hole, Mass. Scallops, shrimps, and crabs are shown in the eelgrass; several kinds of worms live in the soft bottom. The burrow at left contains the polychaete *Chaetopterus,* which maintains a current through the tunnel by flapping leaflike *parapods* ("side-feet"), seen along the middle of its body. The burrow lining is a tough, parchmentlike material, shown in the W-shaped burrow at center. At right, the tube-dwelling polychaete *Diopatra,* with gills emerging from the mouth of the tube. (Photograph courtesy American Museum of Natural History.) (b) Bacteria and several species of diatoms (magnified about 200 times), grow as a crust on eelgrass. (Photograph courtesy of Dr. J. M. Sieburth and University Park Press, Baltimore, Md.)

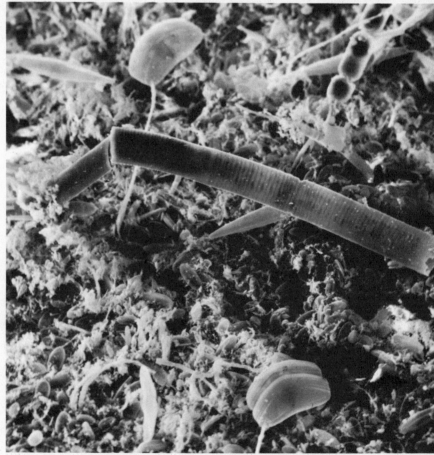

(b)

The burrowing land crab *Cardisoma* and the beach scavenger *Ocypode* (ghost crab) seem to maintain even less contact with seawater than does the fiddler. These animals have enlarged gill chamber linings which function as lungs. Temperate to tropical environments seem to favor this adaptation to a semiterrestrial way of life; relatively higher humidities at these latitudes keep the animal's gill membranes moist while it is away from the water.

Dense stands of *eelgrass*, a seed-bearing marine plant, grow in shallow-water sediments, as shown in Fig. 14-21(a), along both coasts of the United States. Eelgrass has an underground stem and ribbonlike leaves as much as 1 meter long. Specialized communities of plants and animals find food and shelter in eelgrass beds; its leaves are coated with epiphytic diatoms, blue-green algae, protozoans, organic detritus, and small hydroids (Fig. 14-21(b)). Tube-building crustaceans and worms attach to the leaves, and small grazing shrimps and snails scrape off accumulated detritus. This regular cleaning of the eelgrass by animals is essential to the continued growth of the epiphytes, especially the diatoms, for if they become coated with detritus they die for lack of sunlight. Other common residents of eelgrass communities include tunicates and scallops.

The composition of an eelgrass community depends in part on local conditions such as temperature, depth, and current speed. Where currents are strong, for instance, small crustaceans living in tubes (which they themselves secrete and cement to the blades of eelgrass) are likely to survive, while heavier, crawling species are swept away by the water. Infauna grow profusely in the sediments, because the grass has a stabilizing effect and turbidity is reduced. The eelgrass and epiphytes are food for many animals in the community, and detritus for bottom-dwellers is abundant. In addition, sediments are usually firm enough to permit construction of burrows; holes dug in coarse sands tend to collapse, although some worms and other animals are able to build tubes which prevent such accidents.

Below low-tide level, in the muddy sediments of the continental shelf, there are usually large numbers of deposit-feeding bivalves (such as the clam in Fig. 14-22). Where water is too deep for algae and eelgrass, sediments are often soft and easily eroded, due in part to constant reworking by burrowing polychaetes and mollusks. A few tens of clams per square meter can rework all newly deposited sediment once or twice each year, to a depth of perhaps 2 centimeters. Attached animals are unable to colonize the semiliquid substrate, and filter-feeders are choked by the turbidity.

Figure 14-22

Feeding orientation of *Yoldia limatula*. Half-buried in mud, the clam uses its feeding palps to collect sediment and bring it into the mantle cavity. The large inedible fraction is forcibly ejected as a cloud of loose sediment, creating mounds of laminated, reworked sediment as the clam feeds. When resting, *Yoldia* commonly burrows a few centimeters below the surface. (After Donald C. Rhoads, 1963. "Rates of Sediment Reworking by *Yoldia limatula* in Buzzards Bay, Massachusetts, and Long Island Sound," *Journal of Sedimentary Petrology*, 33 (3), 723–27.)

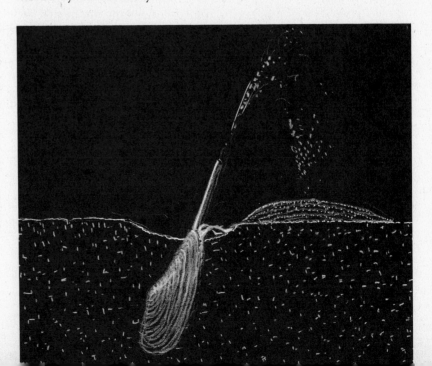

The deposit-feeding clam *Yoldia* is equipped to take advantage of this habitat. Its low-density body is often swept up by strong bottom currents, and suspended along with clouds of silt from the soft bottom. Heavier species, such as an occasional filter-feeding bivalve that tries to settle there, tend to be buried and suffocated by these turbid flows. If *Yoldia* happens to be buried by sediment it tunnels upward toward the new mud–water interface, where oxygen is available. If exposed at the surface by wave action, it just as quickly digs in again.

Where deposit-feeders are abundant, filter-feeders are scarce. This may happen where the activities of the former make the area less easily habitable by the latter. In this case, one organism forces another out of the area—not by competing for food or even for living space, but by creating an undesirable habitat. On the other hand, oxygen introduced by biological reworking of sediment (*bioturbation*) permits bacteria, protozoans, and other small benthic animals to live deeper below the sediment-water interface than is possible in compacted, oxygen-deficient muds.

MICROORGANISMS IN MARINE SEDIMENTS

Marine sediments contain numerous one-celled and small multicellular animals called *meiobenthos* (from the Greek *meion,* meaning "smaller"), about 1 millimeter or less in length. A relatively small number are larvae of larger animals. Conspicuous groups of permanent meiobenthos include *harpaticoids* (very small copepods), *turbellarians* (flat-worms), *ostracods* (small bivalved crustaceans), and several kinds of roundworms, especially *nematodes.* Some common types do not fit into classifications we have previously described, for example the *tardigrade* or "water-bear" illustrated in Fig. 14-23. Benthic foraminiferans and ciliate protozoans are especially abundant in sediments all over the world.

Meiobenthic organisms crawl, climb, and swim between sand grains, feeding on bacteria, protozoans, diatoms, and other epiphytes or on minute particles coated with organic matter. Average biomass of meiobenthic organisms in coastal marine sediments ranges from 0.2 to 2.0 grams wet weight of silty sand. This is only 3 percent of total benthic faunal biomass in such regions, although numbers of individuals may range from 50,000 to more than a million per square meter. Relative biomass is often greater than this in brackish water regions and on beaches. Shallow-water and intertidal muds usually support denser populations than do purely sandy deposits or the deeper, finer sediments offshore.

On the continental shelf, four thousand to 32 million meiobenthic individuals live beneath each square meter of ocean bottom. In coastal waters, for example in the Mediterranean Sea and nearshore waters off South India, meiobenthic biomass may be 10 to 30 percent or more of the total fauna, depending in part on whether larger (e.g., harpaticoids) or smaller (e.g., nematodes) organisms predominate. In the Norwegian Sea at 1500 to 1800 meters depth, where there are large populations of nematodes, biomasses of 2 grams of carbon (wet weight) per square meter have been recorded. In continental shelf and slope waters deposit-feeding macrobenthos preying on the microfauna may contribute to the diet of demersal fishes. In this way meiobenthos probably enter pelagic food webs that include large predators such as fishes, squid, and eventually man.

Deep-sea sediments often contain as much meiofaunal as macrofaunal biomass, although there are generally fewer individual organisms than in coastal waters. For example, at 3000 to 5000 meters depth, Indian Ocean sediments may contain from 20,000 to 110,000 individuals per square meter. This may only represent a few tenths of a gram,

Figure 14-23

Batillipes Mirus, one of the 340 known species of marine tardigrades. It is found in shallow coastal waters. (After S. Hinton.)

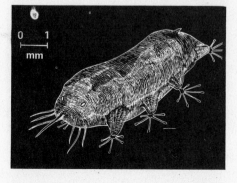

391

role of bacteria
in the chemistry
of marine sediments

or less, per square meter, but macrofauna have extremely low density in the deep ocean and average only 0.002 to 0.2 grams per square meter, so that meiobenthos here are proportionately quite important.

Meiobenthic activity governs availability of dissolved oxygen in the uppermost 1 or 2 centimeters of marine sediments. In loose sediment deposits, where water and oxygen are available due to stirring by the meiobenthos, oxygen-consuming animals occur at depths of a few tens of centimeters or more. Anaerobic species (those that do not require dissolved oxygen) inhabit deeper, oxygen-deficient sediments. In intertidal environments many organisms move up and down continuously with the tide, so as to remain near the air-water interface in flooded sand.

Minute animals have higher metabolic rates and consume relatively more food and oxygen per unit of body weight than larger ones. In general they also have shorter life cycles and consequently populations exhibit more rapid turnover. Oxygen consumption in meiofauna ranges from 200 to 2000 cubic millimeters of oxygen per hour per gram wet weight. This is 20 times higher than the figure for large macrobenthos, with medium-sized animals utilizing intermediate amounts. On the average, the meiofaunal metabolic rate is estimated at five times that of larger animals.

Generation rates are quite variable. Some groups reproduce every few days, others as infrequently as once a year, and the majority a few times a year. Standing crop represents perhaps a third of gross production, due mainly to regular predation and death from other causes. Using three generations per year as an average, meiofaunal production is estimated to be about 5 times the rate for macrobenthos. In subtidal silty sand, populations representing only a few percent of the total standing crop may provide some 15 percent of the annual contribution to food webs at low trophic levels.

Microbenthos (including bacteria and fungi, ciliate and flagellate protozoans, amoebae, benthic diatoms, and filamentous algae) are smaller than 100 microns and range in weight from 10^{-4} to 10^{-9} milligrams. Autotrophs live at the surface and heterotrophs mainly within the sediment. In near-shore deposits, the microbenthos often constitute an even smaller mass than do the meiobenthos, but at the same time constitute a hundredfold larger number of individuals. Most of their food is used in respiration and reproduction; relatively little contributes to increased biomass.

When large quantities of organic matter are present, a dense bacterial film often forms at the sediment surface. Biomass in the topmost layer of silt may then reach several tens of grams per square meter and numbers of individuals may reach 10,000 to 100 million per gram of mud.

Bacteria in marine sediments contain abundant proteins, fats, and oils. Thus though they may supply little energy as carbohydrates, they are an important food for deposit-feeding macro- and meiofauna, with which they also compete for detritus in bottom deposits.

ROLE OF BACTERIA IN THE CHEMISTRY OF MARINE SEDIMENTS

The chemistry of marine sediments, including their state of oxidation or reduction and their acidity, is largely controlled by bacterial activity. For instance bacteria reduce sediments in several ways. Some bacteria utilize compounds such as ammonia (NH_4), methane (CH_4), and nitrate (NO_3^-), or elements such as hydrogen, iron, and sulfur, as primary energy sources. By this process, known as *chemosynthesis*, they fix CO_2 and form new organic matter, just as plants do by utilizing solar energy. The process (compare with photosynthesis) can be shown as follows:

$$(1) \quad \underset{\substack{\text{inorganic} \\ \text{compound} \\ \text{or element}}}{AH_2} + \underset{\text{water}}{H_2O} \xrightarrow[\text{dehydrogenase}]{\text{enzyme}} \underset{\substack{\text{oxidized end} \\ \text{product}}}{AO} + \underset{\substack{\text{reducing} \\ \text{power} \\ \text{(hydrogen ions} \\ \text{and electrons)}}}{4(H^+ + e^-)}$$

Some of the reducing power ($H^+ + e^-$) is used for energy production through ATP synthesis. Note that some form of oxygen is required for this process:

$$(2) \quad 4(H^+ + e^-) + ADP + \underset{\substack{\text{inorganic} \\ \text{phosphate}}}{P_i} \longrightarrow O_2 + H_2O + ATP$$

Reducing power is also used to form reduced nicotinamide adenine dinucleotide (NAD):

$$(3) \quad 2\,(H + e^-) + NAD \longrightarrow NADH_2$$

which is then used in conjunction with ATP for carbon dioxide assimilation:

$$(4) \quad 12NADH_2 + 18ATP + 6CO_2 \longrightarrow$$
$$C_6H_{12}O_6 + 6H_2O + 18ADP + 18P_i + 12NAD$$

In chemosynthesis the reduced inorganic compound or element (such as ammonia, methane, or sulfur) used to create reducing power is also the sole source of the organism's energy. Such reduced substances are mainly produced through metabolic processes carried out in the absence of oxygen (anaerobically). This happens in some deep fjords, estuaries, and silled basins where oxygen used for respiration or decomposition is not replaced by circulation of oxygenated waters. Anaerobic bacteria obtain oxygen for chemosynthesis (see step 2 above) and respiration from carbon dioxide, nitrite, nitrate (NO_3^-), or sulfate (SO_4^{2-}). In so doing they reduce these compounds, forming the nitrites, ammonia, methane, and sulfur that are involved in chemosynthesis. For this reason, oxygen-deficient environments such as occur at the bottom of the Black Sea, in the Cariaco Trench (Caribbean Sea), and in some oxygen-deficient sediments are favorable for chemosynthetic activity. Reduced sulfur combines with hydrogen forming hydrogen sulfide gas (H_2S) or with iron to form the iron sulfide that gives anoxic subsurface muds their characteristic black color. Breakdown of organic matter releases phosphates (PO_4^{2-}), resulting in precipitation of inorganic calcium and iron phosphates in sediments.

We have seen that biological processes govern the nonconservative properties of seawater, such as distribution of dissolved oxygen and nitrates. Bacteria also play a role in the distribution of elements not directly involved in plant or animal metabolism, for example manganese. Manganese eroded from the earth's crust and carried to the ocean in fresh water runoff is not concentrated in seawater or by marine organisms to any great extent. Rather it precipitates as iron–manganese nodules (see Chapter 4) over large areas of the deep-ocean floor. Heterotrophic bacteria attach themselves to these nodules and the enzymes they produce seem to accelerate manganese oxidation and precipitation. Energy derived from this process apparently enables the bacteria to form ATP and assimilate organic matter more efficiently, so that manganese serves as an energy source for these oxidizing organisms.

Several types of marine bacteria are active and abundant at very low temperatures, even below 0°C. Aerobic bacteria live in the top few centimeters of sediment, while anaerobic types occur to 40 or more centimeters below the surface; both types, however, are most abundant just beneath the surface and decrease in numbers with depth. The principal determinant of abundance seems to be the amount of organic material present.

Some organic material reaching the ocean bottom is extremely resistant to decomposition. Bacteria in sediments break down many resistant organic substances, making them available to deposit-feeders, but much organic material resists even bacterial decomposition. Squid beaks, sharks' teeth, and fecal matter from deposit-feeders, for instance, resist physical and chemical degradation. Sediment below the zone of bacterial action seems to contain as much organic material as do surface layers in deep-ocean deposits. The oxygen deficit that generally occurs below the surface may be responsible for this, in that anaerobic decomposition processes seem to be less efficient than aerobic ones. Chitin, for example, contains a large proportion of the nitrogen in marine sediments, but this is attacked slowly in anoxic environments. Fermentation of cellulose, an organic substance common in land plant fibers, is known to occur in deep-ocean sediments where turbidity currents have deposited land-derived organic matter.

REEF COMMUNITIES

A reef community begins to develop when a dominant organism builds a rigid, wave-resistant structure on the bottom, thereby creating environments favorable to other organisms. The reef provides a shallow environment, where nutrients are readily recycled.

Barnacles, algae, mussels and other sessile organisms colonize a reef, as they do any firm surface in shallow water. The process by which such organisms invade a previously uninhabited area is known as *ecological succession,* or "fouling," and it begins with an accumulation of bacterial slime (see Chapter 12). Benthic diatoms and protozoans appear next on an intertidal reef. They multiply rapidly, utilizing adsorbed organic compounds and products of bacterial decomposition. Hydroids (Fig. 13-10) and multicellular algae follow, and then come the planktonic larvae of barnacles (Fig. 13-18), mussels, and snails. Eventually the ecosystem reaches a balanced state or *climax community* in which no further colonization occurs, and ecological succession ceases unless a disturbance of the system causes the process to start afresh.

Oysters are sessile, bivalved mollusks, abundant in coastal areas worldwide, and an important source of food in many regions (see Chapter 15). Each mature female produces many millions of eggs in a year (see Fig. 13-14). Most of these are fertilized, but only a small fraction of the larvae (Fig. 13-17) survive the planktonic stage (which lasts a few weeks, more or less, depending on latitude) and settle to the bottom to develop as adults. Larvae select a suitable spot, preferring oyster shells to any other substrate, and apparently favoring live oysters over dead shells. One to 5 years' growth is required to mature an oyster, during which time most of those which settled together as larvae have been killed by crowding, competition for food, or predation.

Oyster reefs consist of enormous numbers of individuals, their shells cemented to rocks and to one another. The flats of partially enclosed bays and river mouths often provide advantageous conditions, especially where moving water brings fresh supplies of plankton and oxygen and there is a minimum of silt. (An exception to this is the Virginia oyster *Crassostria virginica*, which is well adapted to life in a silty environment.)

An oyster pumps many gallons of water each hour. Plankton or other food particles in the water are caught on a mucus net that is

Figure 14-24

Sabellarian polychaete worms (a) secrete a tube of mucus that hardens into a parchmentlike substance. Feathery gills, often brightly colored, are the only part of the worm that emerges from the end of the tube. The gills, which absorb oxygen and also trap plankton and detritus in a mucus coating, have eyespots that are sensitive to changes in light intensity; the passing of a shadow causes them to retract into the tube. Sabellarian worms can build reefs of their own. Colonial species live in surf zones, where the shoreline is rocky enough to support their tubes. Currents supply filterable food particles and sand grains, which they bond with mucus secretion to make wave-resistant structures. As a sabellarian colony grows, new worms settle around and over old ones, and a reef is formed which may extend for several meters along a beach or breakwater. This photograph was made at the Wometco Miami Seaquarium. The calcareous tube-worm *Hychoides dianthus* (b) is commonly found encrusting oysters or other mollusks.

(a)

(b)

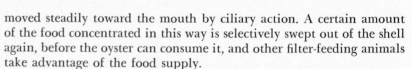

moved steadily toward the mouth by ciliary action. A certain amount of the food concentrated in this way is selectively swept out of the shell again, before the oyster can consume it, and other filter-feeding animals take advantage of the food supply.

Attached organisms such as barnacles, mussels, and tube-worms (Fig. 14-24) add to the bonding that holds oysters together in heavy clusters, thus increasing the stability of the structure. Crevices between shells provide homes for small filter-feeders. Microorganisms, flat-worms, and tiny crabs feed parasitically on the fixed animals, while certain fishes (gobies and killifish, for example) are especially well adapted for life in the shallow tidal waters over the beds. The latter prey on small crustaceans and soft-bodied animals among the rocks and crevices. Some animals, like the pea crab, inhabit the shells of live oysters.

Sea stars, snails, crabs, and various fishes prey directly upon oysters. The oyster-drill snail chemically dissolves and physically bores a hole through the heavy shell to eat the oyster. Sea stars exert tremendous suction with their pneumatically powered tube-feet to pull open the valves, then consume the meat by everting their stomachs to surround the oyster tissues.

Oysters thrive in various habitats. In the Chesapeake Bay area, for example, there are large natural beds in the intermediate salinities of the estuary (7–18 parts per thousand), where oysters can survive but their principal enemies and diseases cannot. Experiments have indicated that, in the absence of predation, oysters grow faster and more prolifically near the mouth of the estuary than they do upstream. In the high-salinity water of the seaside along the Ocean shore, they can survive only in the intertidal zone.

394

Coral reefs occur mainly in subtropical oceans (see Chapter 3), where carbonates precipitate readily. Reef-building coral is a coelenterate polyp that builds a cup-shaped calcareous skeleton as shown in Fig. 14-25. New individuals bud from the side of the parent. As animals die, consumed by predators or scavengers, subsequent generations build over and around the disused skeletons. Unicellular algae, called *zooxanthellae,* live in coral tissues. Theirs is a symbiotic relationship in which the algae make photosynthetic products directly available to the host coral and in return receive nutrients and CO_2. Zooxanthella photosynthesis apparently alters CO_2 concentrations in coral animal tissue, greatly increasing the animal's ability to extract $CaCO_3$ from seawater for production of the carbonate skeleton. Calcification rates of reef-building corals are strongly influenced by the amount of insolation received by the algae and are highest in sunlit waters less than 20 meters deep.

But corals are only the most conspicuous contributors to the framework of a reef. Benthic encrusting plants known as *crustose red algae* are important cement depositors at shallow depths. In addition, red and green *calcareous algae* produce much of the loose carbonate sediment incorporated in the reef mass.

Coral reefs create a complex shallow-water benthic environment. The ecosystems that flourish on and around them are some of the most productive in the ocean, supporting extensive benthic and pelagic communities. They are also highly diversified, both in the kinds of living space available and in the kinds of organisms that inhabit them. It has been estimated that more than 3000 animal species may coexist on a large reef tract. Competition for firm surface space, rather than lack of food, probably limits the population of filter-feeding benthos, at least above 200 meters depth.

On the reef crest, where tides and waves periodically expose parts of the surface, hardy, soft-bodied algae, barnacles, and coralline algae form rimmed pools that project above sea level perhaps 10 centimeters or more. Water splashed into these pools drains or dries out slowly, keeping plants and animals in the pool moist. Coralline algae may form a purplish-red ridge, the *algal* or *Lithothamnion ridge,* at the water line. Crabs, small fishes, worms, and benthic algae live in the shallow pools.

Figure 14-25

Encrusting coral growing under experimental conditions. (Photo courtesy Michael J. Reber.)

In the photic zone, reef-building corals form the predominant cover, but attached plants may greatly exceed the biomass of animal material. Besides the coralline algae that add calcareous material to the reef framework, filamentous green algae embedded all over its surface manufacture food during the day, using nutrients released by animals and bacteria. At night, coral polyps extend their filter-feeding apparatus (Fig. 14-26) to capture plankton and detritus from the water.

(a)

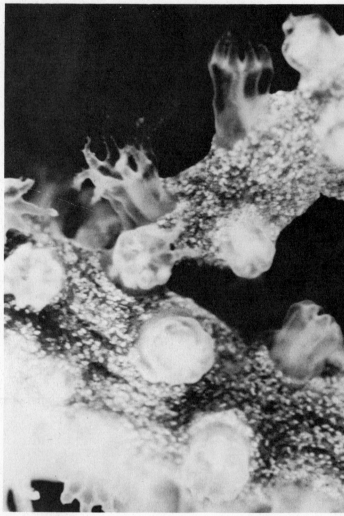

(b)

Figure 14-26

Branching whip coral, *Leptogorgia:* (a) branches of coral with a redbeard sponge in the background; (b) close-up of a feeding polyp showing tentacles extended. (Photo courtesy Michael J. Reber.)

Many kinds of invertebrates and fishes (Fig. 14-27) live in coral reef communities, though there are relatively few individuals of any one kind. Some, like the sea star in Fig. 14-28, consume coral polyps and algae; others feed on detritus. Still others—for example, the moray eel and sea anemone—are predators. Parrot fish and other browsers actually graze on the reef itself to get at the algae and animals living in it. They excrete clouds of calcareous silt; this material, along with mollusk shells and calcareous debris from benthic foraminifera, corals, and algae, accumulates in interstices on the forereef.

Zonation of plants and animals on the forereef is largely controlled by light and temperature. In shaded environments such as caves and tunnels and on the undersides of ledges, *sclerosponges* deposit a skeletal material made of slender silica needles and calcium carbonate, the only living organisms known to deposit both of these minerals.

Figure 14-27 (above)

Fish swimming over a coral reef, Midway Island. (Photo courtesy U.S. Geological Survey.)

Figure 14-28 (right)

Crown-of-thorns sea star *(Acanthaster planci)* feeds on coral reef at Guam. In the lower photo, the same area of coral is shown as it appeared a few months later. The coral is dead and its skeleton has been overgrown by algae. (Photo courtesy Westinghouse.)

(a)

(b)

Figure 14-29

These photographs of artificial fishing reefs were taken at Artificial Shoals in Hawaii: (a) coral growth on a reef of junk car bodies; (b) a school of *Chromis verator* in a car body; (c) a reef made of damaged concrete pipe. (Photographs courtesy State of Hawaii, Department of Fish and Game.)

(c)

They evidently contribute substantially to some reefs, especially below 70 meters depth where reef-building corals are rare. Sediments trapped behind sclerosponges and corals apparently undergo rapid cementation, contributing significantly to the solidification of the reef framework.

Many attached organisms of the deep forereef are restricted to protected areas where they will not be covered by falling sediment. Nonreef-building corals, soft sponges, hydroids, crustose red and filamentous green algae, ascidians (benthic tunicates), *crinoids* (see Fig. 14-31) bivalve mollusks, tube-dwelling worms, and sea anemones are among the many kinds of attached organisms below 70 meters depth. Some, such as tube-worms and ascidians contribute their skeletal materials to the reef framework. Small lobsters, crabs, sea stars, and sea urchins move over the surface, grazing on attached organisms. Boring and cavity-dwelling sponges and many kinds of small and large worms live within the reef itself, and fish prey on the benthic fauna. Thick sediment deposits on flat surfaces contain animal burrows and trails.

Below about 200 meters, the sediment layer is thicker. Shrimps, sea stars, small fishes, worms, sea anemones, and sea cucumbers are among the animals that inhabit holes and mounds in the bottom. Stalked crinoids are also abundant at these depths. Scavengers and grazers include hermit crabs, brittle stars (see Fig. 14-32), and sea urchins. Corals grow where sediment layers are thin, as do unstalked crinoids and encrusting sponges. Attached animals can support mobile organisms; for example, brittle stars are often found clinging to corals. Among the pelagic fauna found at these depths are jellyfish, chaetognaths, mysids (see Fig. 13-5), and many kinds of crustacean nauplii.

An artificial reef can form the basis for an ecosystem, just as a naturally occurring reef can. Most of the U.S. Atlantic and Gulf Coast continental shelf is flat and sandy, with little rock or natural reef on which communities can build or pelagic populations find shelter. But migratory bluefish, tuna, mackerel, flounder, pollock, and jacks often gather around wrecked ships because of the protection they offer, particularly for juveniles. In addition, formation of benthic communities provides a source of food.

To construct artificial reefs, heaps of quarry rock, construction debris, lengths of culvert pipe, discarded automobile tires, and specially built concrete blocks with holes have been deposited in several U.S. coastal areas for the benefit of sport fishermen. Junk car bodies have also been tried, but they tend to break up after a few years and the metal thus released may locally be a nuisance, catching fishing gear, for example (Fig. 14-29). In the Gulf of Mexico, offshore drilling platforms function as artificial reefs for sport fishes.

The ideal artificial reef would be high enough to cause upwelling of nutrient-rich bottom waters, and it would have a variety of surfaces and openings to provide diverse habitats for benthic organisms and juvenile fishes. It may be that a well-designed reef system could help solve solid waste disposal problems for coastal communities.

DEEP-OCEAN BENTHOS

The deep-ocean benthos exhibits great diversity, due in part to the stability of the environment. Below about 2000 meters depth (see Figs. 6-18 and 6-19) temperature and salinity vary little over most of the world ocean. Deep waters were probably 3° to 8°C warmer several million years ago, before the Ice Ages began, but since then bottom conditions have been nearly constant. Immigration and evolution under these colder conditions have resulted in diversity comparable to that of tropical soft-bottomed communities; for example below the Central North Pacific gyre diversity can be as high as 54 species per 176 individuals.

Figure 14-30

Deep-ocean sea-cucumbers: (a, right) *Scotoplanes;* (b, below) *Psychropotes* (drawn from the underside, as if crawling on a sheet of glass). These cylindrical, muscular echinoderms creep along the bottom by contractions of the body. At the anterior end the mouth sweeps up sediment, detritus, and small animals. (From H. Theel, 1882 and 1886. "Report on the Holothurioidea of the *Challenger* Expedition," *Report of the Scientific Results* H.M.S. *Challenger, Zoology,* vol. 14, pt. 1, 176 pp., 1882; pt. 2, 290 pp., 1886.)

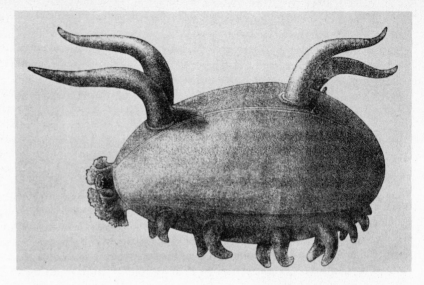

(a)

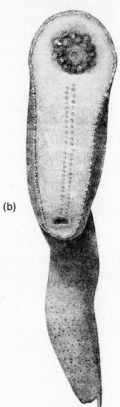

(b)

Figure 14-31 (*right*)

Sea lilies have cup-shaped bodies and long arms. The skeleton is made of calcareous plates, absent in the vicinity of the mouth. (Photograph of a model in the U.S. National Museum, Washington, D.C.)

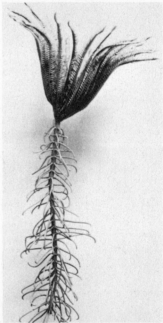

Figure 14-32 (*right*)

Brittle-star and sea-cucumber, photographed at a depth of about 1500 meters in Hydrographer's Canyon, western North Atlantic. (Photograph courtesy Woods Hole Oceanographic Institution, David Owen, photographer.)

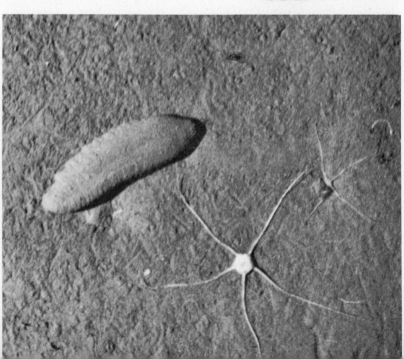

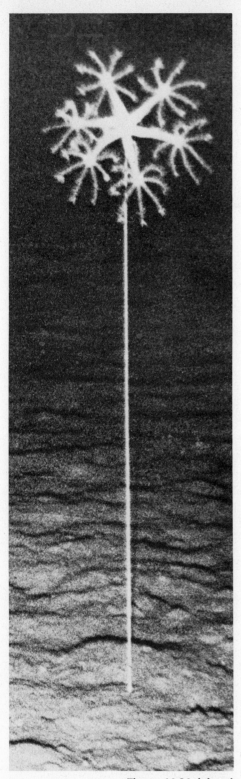

Soft sediments dominate the deep-ocean bottom, and deposit-feeders (Fig. 14-30) predominate, especially in central ocean basins far from shore. In areas covered by red-clay deposits, where sediment accumulation is slow and deposits contain less than 0.25 percent of organic carbon, filter-feeders may be conspicuous. The crinoid in Fig. 14-31, for example, stands on a long stalk above easily eroded oozes. Predatory forms, such as *brittle-stars* (Fig. 14-32), are more abundant where organic carbon content is higher, moving about on long legs that support their bodies above the sediment surface.

Filter-feeding sponges (as in Fig. 14-33), coelenterates (Fig. 14-34), segmented and unsegmented worms (Fig. 14-35), bivalved mollusks and crustaceans—in fact all the major groups we have encountered among shallow-water benthos—occur in deep-water habitats. Uniform coloration (gray or black among the fishes, often reddish in crustaceans) and delicacy of structure are typical in these quiet, dark waters. Smaller animals in deep-ocean sediments include polychaete worms, isopods and amphipods (see Fig. 13-5), bivalves, and *tanaids,* deep-water crustaceans similar in appearance to isopods.

(a)

(b)

Figure 14-33

Deep-ocean sponges: (a, left) *Toegeria;* (b, right) *Lefroyella.* (From John Murray, and A. F. Renard, 1891. "Report on Deep-Sea Deposits Based on the Specimens Collected During the Voyage of H.M.S. *Challenger* in the Years 1872 to 1876," *Report of the Scientific Results H.M.S. Challenger,* vol. 5, 525 pp.)

Figure 14-34 (*above*)

The stalked coelenterate *Umbellula,* whose pencil-thick stem appears to be about 1 meter long, was photographed at a depth of more than 5000 meters off the Atlantic coast of Africa by a Naval Oceanographic Office crew aboard the research vessel *Kane.* (Photograph courtesy U.S. Navy.)

Figure 14-35 (*right*)

The unsegmented worm *Priapus* swallows its prey whole. It is found in bottom sediments of polar regions in both hemispheres. (Photograph of a model at the U.S. National Museum, Washington, D.C.)

Deep-sea organisms are smaller than their shallow-water counterparts. Ten times as many individuals are caught if a screen of 0.06 millimeter mesh is used to filter the sediment, instead of one that catches only organisms longer than 0.3 millimeter. Meiofauna taken in this way include benthic foraminiferans, nematodes, harpaticoid copepods and ostracods (Fig. 14-36).

Most of the detritus that falls from surface or mid-depth zones is probably consumed or decomposed before reaching the deep zone, so that little biodegradable material reaches the deep-ocean floor. Inedible remains of large animals such as shark's teeth, squid beaks, and whale's earbones are common but they are not a source of food for deep-ocean organisms. Eggs of some benthic forms float to the surface zone where food is more plentiful, and hatch there before descending to the bottom.

Biomass is very low at abyssal depths. On outer continental shelves and upper slopes, 5000 to 22,000 individuals may live in each square meter of sediment. The lower slope and upper rise below 500 meters depth average around 1000 individuals per square meter, but only around 30 animals inhabit the same sized area on lower continental rises and abyssal plains. Below the central ocean gyres, biomass of the 30 to 200 organisms in each square meter is only a few hundredths of a gram, wet weight, or a few percent of the infaunal biomass in nearshore sediments.

Large fishes, sharks, or whales may fall to the ocean floor after death. The carcasses attract large numbers of swimming scavengers, but such events are rather rare so that the ability to sense food from a distance is an important adaptation in a dark, relatively barren environment. Near continents, where turbidity currents flowing through abyssal channels bring plant material to the deep ocean, a specialized population exists. A certain bivalve, for example, bores holes in the coconut husks, bamboo stems, and parts of trees that are carried to the deep ocean. Abandoned holes may then be occupied by small mussels, worms, or amphipod crustaceans. Plant debris is eventually decomposed by bacteria, which then make an important contribution to the diet of deposit-feeders. But at abyssal depths, as we know, bacterial activity proceeds very slowly and may not provide much food for infauna.

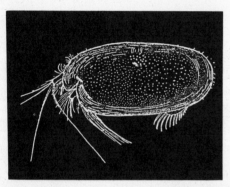

Figure 14-36

A generalized ostracod, about 1.5 millimeters in length. (Adapted from M. E. Johnson and H. J. Snook, 1967. *Seashore Animals of the Pacific Coast,* Dover, New York.)

SUMMARY OUTLINE

The benthos: two-dimensional world
Life strategies for benthic animals and feeding behaviors:
Attachment to a firm surface—mainly filter-feeders
Free movement on the bottom—mainly active predators
Burrowing in sediment—mainly deposit-feeders

Distribution and diversity of benthic life
Attached plants rare except in shallow coastal waters
Benthic community structure controlled by light, temperature, salinity, type of bottom, environmental stability
Diversity—governed by stability, absence of rigorus conditions
Greatest in deep-ocean and tropical, near-shore waters
Less in temperate climate communities
Benthic biomass governed by local productivity

Zonation on rocky beaches
Zonation based on duration of submergence
Attached organisms dominate in intertidal zone
Tide pools—specialized environment
Zone below low-tide level most densely populated
Mobile crustaceans common on hard-bottomed continental shelves

Soft-bottomed coastal communities
Infauna—filter-feeders, selective and unselective deposit feeders, in sediment
Epifauna—surface-dwellers

Nature of deposits governs community composition
  Filter-feeders—coarser deposits, less turbid waters
  Deposit-feeders—fine-grained deposits, rich in organic matter
Fecal pellets can stabilize deposits
Salt marshes—highly productive; support diverse communities
Eelgrass community—specialized population; grass roots stabilize bottom
Infauna of soft, continental-shelf deposits tolerate turbidity and avoid burial by sediments

Microorganisms in marine sediments
  Meiobenthos—sediment-dwelling animals, smaller than 1 millimeter; numerous; high energy flow; short lives
    Relative biomass of meiofauna increases with distance from continental shelf
    Organisms stir, oxygenate uppermost sediment layers
  Microbenthos (less than 100 microns)—bacteria, fungi, protozoans
    Bacteria synthesize protein

Role of bacteria in the chemistry of marine sediments
  Chemosynthesis—bacteria reduce inorganic compounds to fix CO_2 as an energy source; can happen in anaerobic environments

Bacterial activity promotes formation of manganese nodules
Some organic materials resist decomposition in anoxic sediments

Reef communities—occur on, around a rigid, wave-resistant structure; nutrients recycled
  Ecological succession (fouling) gives rise to climax community
  Oyster reefs—oysters pump water that contains particles for filter-feeders
  Coral reefs—subtropical
  Coelenterate polyps, zooxanthella algae form symbiotic relationship
    Crustose algae contribute to reef structure
    Many species coexist
    Attached plants highly productive
    Sclerosponges contribute siliceous material
    Carbonate sediment incorporated in reef
  Artificial reefs—on U.S. mid-Atlantic and Gulf Coasts; attract sport fishes

Deep-ocean benthos
  Great diversity due to stable environment
  Soft sediment bottom—deposit-feeders and stationary organisms
  Delicate structures, small size, low biomass characteristic
  Scavengers sense presence of food from a great distance

SELECTED REFERENCES

HEDGPETH, JOEL W. (ed.). 1957. *Treatise on Marine Ecology and Paleoecology, Vol. I: Ecology.* Geological Society of America, New York. 1296 pp. Detailed, scholarly monographs on habitats and biological conditions in the ocean, with emphasis on biological communities; extensive bibliography.

JOHNSON, M. E. AND H. J. SNOOK. 1967. *Seashore Animals of the Pacific Coast.* Dover, New York. 659 pp. Nontechnical account of marine biology, stressing Pacific Coast organisms.

NICOL, J. A. COLIN. 1960. *The Biology of Marine Animals.* Interscience, New York. 707 pp. A standard reference text, with emphasis on physiology.

RICKETTS, EDWARD F., AND JACK CALVIN. 1962. *Between Pacific Tides,* 3rd ed. Stanford University Press, Stanford. 516 pp. Excellent field guide and reference book, well illustrated and including detailed bibliography.

THORSON, GUNNAR. 1971. *Life in the Sea.* World University Library, McGraw-Hill, New York. 256 pp. Elementary biological oceanography, emphasizing benthic environments.

15

Offshore drilling rig exploring for petroleum and natural gas in the United Kingdom sector of the North Sea. (Photograph courtesy Continental Oil Company.)

ocean resources

T ransportation, defense, and fishing are traditional uses of the ocean. But increasing demand for fuels, metals, and construction materials has in many instances outstripped the production capabilities of land resources, and increasingly the ocean is being exploited for needed materials. This has been particularly obvious in the case of petroleum and natural gas, for which the continental shelf is expected to become a prime source in the next few decades.

International conferences on the Law of the Sea have answered some questions about who owns ocean resources, and thus removed a major stumbling block for their development. The pace of ocean resource development will likely expand greatly in consequence. The new legal regime will also affect management and protection of such traditional marine resources as fisheries. Under the ancient system of free entry, where anyone with the money to buy a boat could fish as much as desired, fish stocks have been systematically overfished and in many instances essentially destroyed as a resource. The new legal boundaries will permit better control and perhaps management of the fishing industry.

Despite the long use made of the ocean, it is still a frontier area for development. We are only slowly developing techniques that permit us to mine salt and sulfur from the sea bed or drill down to recover gas and oil. There is still no capability to work for long periods at the bottom of ocean basins except with expensive, specialized devices built primarily for military purposes or petroleum production.

As in any frontier area, there is much uncertainty and a great deal of inflated expectation. Conversely, resources that have not been identified are likely to become important within a few decades. In this chapter we explore some traditional uses of the ocean and consider possibilities for some less-developed techniques. But first we need to consider the nature of resources based on the rate of their formation and extraction. We distinguish between *renewable* resources, which are replenished through growth or other processes at rates that equal or exceed our rate of consumption, and *nonrenewable* resources. Fresh water and forests are examples of renewable resources; petroleum and metal ores are nonrenewable.

FISHERIES AND AQUACULTURE   Fisheries are the best example of an exploited renewable marine resource. Man is now a major predator (Fig. 15-1), harvesting a large fraction of the yearly production of several groups of fishes (Table 15-1(a)), most of which are coastal ocean species.

Figure 15-1

Today's commercial fishing methods are not very different from those used for centuries. Bottom-dwelling fish such as halibut are scooped up in a trawl net (a) that is dragged over the sea floor. The photograph shows a trawl catch before it is dumped on the deck of a boat of the National Marine Fisheries Service in the North Atlantic. Fish that feed in schools near the surface are often trapped or entangled in a vertically suspended net, or in a purse-seine as shown in (b) fishing for menhaden at Sealevel, North Carolina.

(a)

(b)

Purse-seines are useful for catching densely schooling fishes in protected waters, but they are also used on the high seas in some part of the world to catch tunas. The third common fishing method uses hook and line (c); here, fishermen land a 30-kilogram yellowfin tuna, using a three-pole, single-line rig in the Gulf of Guinea, off Africa. (Photographs courtesy National Marine Fisheries Service.)

(c)

TABLE 15-1(a)

World Commercial Catch of Marine Fish, Crustaceans, and Mollusks, by Species Groups, 1974* (excluding whales and seals)

SPECIES GROUP	THOUSANDS OF METRIC TONS (live weight)	PERCENT OF TOTAL
Herring, sardines, anchovies, et al.	13,731	23.2
Cods, hakes, haddocks, et al.	12,697	21.4
Miscellaneous marine and migratory (fresh-salt) fishes	9,112	15.4
Redfish, basses, perches, et al.	4,587	7.7
Mackerels, snoeks, cutlassfishes, et al.	3,621	6.1
Mollusks	3,437	5.8
Jacks, mullets, sauries, et al.	3,312	5.6
Salmon, trouts, smelts, et al.	2,449	4.1
Crustaceans	1,937	3.3
Tunas, bonitos, billfishes, et al.	1,875	3.2
Flounders, halibuts, soles, et al.	1,178	2.0
Shads, milkfishes, et al.	749	1.3
Sharks, rays, chimaeras, et al.	558	0.9
Total	59,245	100.0

* Data from Food and Agricultural Organization of the United Nations, *Yearbook of Fishery Statistics, 1974*. Vol. 39.

Figure 15-2

Some lesser-known fishes of potential commercial importance. (a) Sauries (garfish) are distributed from Alaska to lower California and south of Japan to the Asian mainland. There are coastal and open-ocean species. These schooling fishes feed near the surface at low trophic levels. Related species, such as the halfbeak, are some of the bullet-shaped "flying" fishes reported skittering along the surface in open waters. Pacific sauries have been fished commercially offshore, but to a limited extent. (b) Grenadiers, often called "rattails," are found worldwide along lower continental slopes at 200–800 meters depth. These are perhaps the most numerous of all fishes on the edges of the deep sea. They feed on benthic invertebrates, also on fishes and plankton at mid-depths. Like many deep-water organisms, they are fragile and easily damaged in fishing. The tapering "rat-tailed" body is characteristic of many deep-water species. (c) Sand lances swim in large schools near northern shores. Their habit of diving into sand and sometimes remaining there after the tide is out gives them their name.

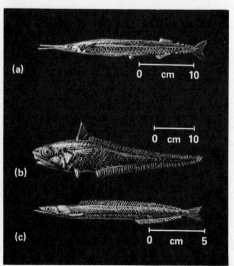

Fish provide about 3 percent of human protein consumption, and about 10 percent of all animal protein consumed. Between 1950 and the mid-1970's, worldwide marine fish production increased steadily from about 20 million to about 60 million tons. It was hoped that with proper management the oceans would ultimately provide much of the protein needed to feed rapidly increasing populations. But the collapse of the California sardine fishery in the early 1950's showed that unlimited fish production from the ocean is not possible.

In 1974 the total world production of marine fish and shellfish was about 59 million metric tons (live weight). Another 9 million metric tons of various freshwater fishes were harvested. About 60 percent of the catch was used directly for food, while most of the remainder was made into either fish oil or meal for use as livestock feed. Major fish-producing countries were (in order of tonnage): Japan, U.S.S.R., China, Peru, United States, and Norway. Until the collapse of the anchovetta fishery in 1972, Peru was the world's top fish producer.

In 1975 the 1780 metric tons of fish landed in the United States were valued at about $971 million. Menhaden, a fish used for industrial purposes, accounted for 46 percent of the volume but only about 10 percent of the value. Luxury seafoods, including shrimp, salmon, tuna, crabs, and lobsters were the most valuable products, accounting for 61 percent of the total value of the U.S. catch (Table 15-1(b)).

Recreational fisheries in the United States take a large and steadily increasing amount of marine fish each year. The amount of money spent on recreational fishing every year exceeds the value of the U. S. commercial fish catch (Table 15-1(b)).

The maximum yield of fish that can be harvested from the ocean is estimated by many fishery scientists at about 200 million tons a year, about four times the catches of the 1970's. Achieving this level of production would require harvesting many fishes not now taken (Fig. 15-2). It might also lead to serious reduction—perhaps extinction—of certain species. Increasingly heavy fishing has severely depleted several fish stocks; combined with natural climatic fluctuations, such pressure can deplete a stock to the point where the fishery never recovers. An overview of fishery dynamics will point up some of the factors that control fish production in the ocean and demonstrate the effects of overfishing.

TABLE 15-1(b)

U.S. Recreational and Commercial Fisheries Catch †

TYPE OF FISHERY	THOUSANDS OF METRIC TONS	MILLIONS OF DOLLARS
Recreational fisheries (1970)	72	1,225‡
Commercial fisheries (1975)		
Finfish		
Tuna	177	108
Salmon	91	116
Menhaden	818	49
Flounders	71	43
Other	623	168
Subtotal	1,780	484
Shellfish		
Shrimp	156	226
Crabs	136	84
Lobsters	17	59
Oysters	24	43
Clams	50	41
Other	29	34
Subtotal	412	487‡
Total	2,192	971

† Data from National Marine Fisheries Service, 1976. *Fisheries of the United States, 1975.* Current Fishery Statistics No. 6900. National Oceanic and Atmospheric Administration, Washington, D.C.

‡ Expenditures by individuals for recreational fishing activities.

Figure 15-3

The percentage of assimilated energy used by the Pacific sardine for growth and reproduction. Ecological efficiency, that is, ratio of growth to food intake, may achieve a high of 20 percent or more during its early life. (After J. H. Steele, 1974. *The Structure of Marine Ecosystems,* Harvard University Press, Cambridge.)

A stock of fish is largest in a virgin, unfished state. Growth of the young is balanced by death and predation, and there is a relatively large proportion of mature and older individuals. As a fishery is developed, the number of older individuals is reduced first, because the largest fish are trapped most easily. Thus the heaviest burden on a fully exploited fishery falls on young adult individuals that have recently experienced their peak growth efficiency (Fig. 15-3). A sardine, for example, grows 121 millimeters in its first year but only 30 millimeters in the third year of life.

When a fish population has grown as large as possible, it is in a steady state and does not increase or decrease. But if the population is reduced, for instance by heavy predation as in a fishery, the number of young fish to reach maturity increases due to the reduction of competition for food and/or living space. Rate of population increase is highest at some intermediate number of individuals, but it approaches zero if the population is reduced to a minimum. In a young fishery, therefore, the number of young fish may increase rapidly for a few years so that the population as a whole grows larger.

Figure 15-4 shows the relationship of yield to population size. When the harvest reaches about 70 percent of the maximum potential, rate of increased yield per unit of fishing effort begins to drop rapidly. *Maximum sustainable yield* for the fishery is attained by slightly more than doubling the effort required to reach the 70 percent mark. Additional effort may increase yields for a while, but eventually the population shrinks and its rate of natural increase slows to zero. Thus

Figure 15-4

Yield curve for a fishery. (After R. Edwards and R. Hennemuth, 1975. "Maximum Yield: Assessment and Attainment," *Oceanus,* 18(2).)

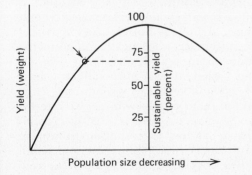

overfishing, or fishing past the maximum sustainable yield, causes future yields to decline and damages the stock's capacity to recover, even if fishing is later restricted. The growth rate of fishes is also a factor. A shorter time is required to correct overfishing in the case of fast-growing fishes like the anchovy, than for a fishery in which individuals mature slowly, such as halibut.

A mature fishery producing the maximum sustainable yield is usually about half the size of the original population. But even if this take is not exceeded, other changes can threaten a fishery. For example recruitment from a new *year-class* (the individuals hatched in a single spawning season) varies with changes in food supply, predation, and oceanic conditions such as temperature and current patterns. If fishermen take the same number of fish year after year, or increase their efforts in a poor year to compensate for smaller stocks, overfishing can result. A "bad year," combined with heavy fishing, may reduce a stock to such an extent that a different species takes over the living space and food supply, replacing the commercial species. If there is no market for the successor species, the fishery collapses.

The Pacific sardine fishery off central California illustrates this sequence of events. Once the largest in the Western Hemisphere, it has produced virtually nothing since the mid-1960's. Beginning in 1915, this fishery expanded rapidly. Reaching its peak in 1936, it maintained an average production exceeding 500,000 tons until 1944 and then fell off sharply. But even after the near disappearance of sardines, there is no evidence of a reduction in fish biomass of California waters. Instead it seems that the sardine has been replaced by other fishes, especially anchovy. The anchovy standing crop has increased at least twenty-fold while the sardine dropped to about 5 percent of its former biomass. But restrictions on commercial fishing off California made it impossible to develop a market for fish meal and oil from the anchovy. Thus this fish population has not been commercially exploited.

The decline of the Peruvian anchovetta fishery is more recent. Located in the highly productive upwelling system off Peru, this was once the world's largest fishery. It expanded rapidly from a modest level in the 1950's, and during its 1968–1969 peak season yielded over 10 million metric tons of fish, one-fifth of the world's total production. At that time this was estimated to be the maximum sustainable yield of the fishery. A harvest of 9.3 tons was taken by fishermen, with the remainder being consumed by about 4.5 million guano birds whose droppings supported a local fertilizer industry. But in addition to exploitation by man and birds, changes in environment had a serious impact on the fishery. Every so often an anomalous weather pattern occurs, bringing changes in the Pacific Trade Winds and causing sudden surface warming with inhibition of upwelling. This so-called El Niño condition sharply reduces the productivity of coastal waters, so that anchovetta are no longer available to either birds or fishermen. Following El Niño conditions during the 1957–58 and 1965–66 seasons (Fig. 15-5), the population of guano birds was greatly reduced; it recovered from the earlier setback though not from the later one, possibly due to heavier commercial fishing. But in 1972–73, severe El Niño conditions reduced the anchovetta catch to 1.8 million metric tons. This depressed the fishery to the extent that Peru dropped from the largest fish-producing nation in 1971 (10.6 million metric tons) to sixth place in 1974, only 2.3 million metric tons.

Whaling is an example of an industry that will require international regulation if stocks are to be preserved, because whaling grounds extend through international and national waters. Antarctic whales, for example, breed in winter off the African, Australian, and South American coasts. In summer they migrate toward the Antarctic continent to feed on the abundant krill.

Figure 15-5

Commercial catch of anchovetta by calendar years, and population of guano birds from censuses. (After M. B. Schaeffer, 1970. "Men, Birds and Anchovies in the Peru Current—Dynamic Interactions," *Transactions of the American Fisheries Society,* 99 (3), 461–467.)

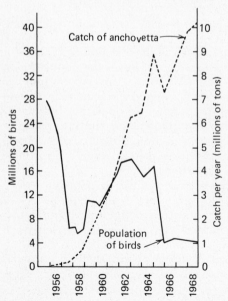

Whaling as an industry dates from at least the twelfth century. At first whales were harpooned from small boats launched at the shore; later they were hunted far out at sea by ships carrying small boats, a few of which would be launched at once to attack a whale from all sides. In the 18th and 19th centuries, when whalebone for dresses and oil for lamps were in demand, the seas near Greenland were a great whale fishery. Sperm whales (see Fig. 13-24) were hunted out of New England for spermaceti, used for cosmetics and candles. This fishery was extended to baleen (whalebone) whales and became worldwide, especially in tropical waters and in the Southern Hemisphere. As each new hunting-ground was located, it was exploited to near-depletion of the whale stock, and many stocks became dangerously low. By the mid-1970's, right whales had been reduced to about 5000 individuals, total. There were about 7000 humpbacks, 11,000 grays and 12,000 blue whales. Fin whales numbered about 100,000 and sperm whales over 600,000. Whales mature slowly and may live to be 80 years old, so that when a stock becomes depleted it takes decades to recover.

Mariculture, or *aquaculture,* the marine equivalent of agriculture, represents a means of extracting more food from the ocean while avoiding some of the uncertainties inherent in harvesting wild stocks. The simplest form of aquaculture is to take young animals from their native spawning grounds and relocate them in another area where they can fatten for the market. This is done, for example, with oysters (Fig. 15-6), which are taken from spawning areas in Connecticut to fatten in bays on Long Island before harvesting. More complicated forms of aquaculture involve enclosures (fish pens) where feeding is controlled. Lack of domesticated marine species and ignorance of fish diseases have limited the growth of aquaculture in the United States. Also the legal status of many marine areas is unclear, so that it is often impossible to control access for prevention of loss or damage to a crop.

Figure 15-6

A commercial oyster dredging operation. (Photograph courtesy Michael J. Reber.)

Figure 15-7

In Japan, a bamboo raft is used to suspend strings of oysters off the bottom. This method offers greater protection from predatory starfish and snails. (Photograph courtesy Consulate General of Japan, N.Y.)

None of these problems is technically insurmountable, and mariculture is likely to become increasingly familiar as a food source, as it already has in Asia (Fig. 15-7).

RESOURCES FROM SEAWATER

Large quantities of salt, bromine, and magnesium are extracted from seawater. The ocean is an attractive source of these raw materials for several reasons. First, there is the abundance of seawater; at most conceivable rates of use, extraction of metals and salt from seawater is not likely to deplete any significant fraction of the ocean in the predictable future. Furthermore, the annual supply of several substances from the erosion of continents is equal to or greater than present levels of extraction from the ocean. In the late 1960's, 29 percent of the world's salt came from the ocean, as well as 70 percent of its bromine, 61 percent of its magnesium metal, 6 percent of its magnesium compounds, and 59 percent of its "manufactured" water. These materials were produced in about 60 countries and were worth more than $400 million.

Isolated evaporating basins, located in relatively dry coastal regions, use solar energy to evaporate seawater. Evaporation of brine is carefully controlled so that only sodium chloride or another desirable component of sea salt is recovered at a single stage. Salts recovered from seawater must otherwise be treated to remove magnesium sulfate (Epsom salt) and calcium carbonate, a gritty impurity. Evaporating ponds for extraction of sea salts operate around southern San Francisco Bay (Fig. 15-8). Sea salt is used by the chemical industry, as is magnesium, a lightweight metal. Bromine extracted from seawater is used as a component of antiknock compounds in gasoline.

Figure 15-8

About 5 percent of the U.S. production of salt comes from evaporating seawater in southern San Francisco Bay. Brine from large evaporating ponds is transferred to "pickle ponds" and finally to a crystallization pond, where the salt forms. The photo shows a harvesting operation after the last of the liquid, known as "bittern," has been removed and the layer of crystallized salt has reached a depth of about 10 centimeters. (Photograph courtesy Leslie Salt Company.)

Despite the wide variety and large amount of valuable materials dissolved in ocean waters, the extremely dilute nature of seawater makes it prohibitively expensive to produce most of them. Gold is an example; its concentration in seawater is extremely low, about 4 grams per million tons (10^{12} grams) of seawater. This amounts to about 5 million tons of gold in the ocean, but the cost of energy for pumping seawater, added to the cost of chemical treatment, far exceeds the value of the gold recovered. High energy costs also discourage recovery of metals dissolved in seawater. Another problem in many areas is the difficulty of obtaining adequate supplies of undiluted, uncontaminated seawater.

The development of cities in arid coastal regions makes water from the ocean an increasingly important resource. In obtaining it we effectively duplicate the natural cycling of water from the ocean through evaporation (or freezing), using fossil or nuclear fuels instead of solar energy, in order to have water when and where it is needed.

The simplest way to obtain water from seawater is to build a glass shelter—a greenhouse, so to speak. Evaporation from pans of seawater occurs during the day when the sun is shining, and condensation takes place at night on the cool surfaces of the shelter. This creates a humid, nearly tropical, environment where certain plants can be grown at the same time for local marketing. Such man-made oases are useful for small production of high-value products such as fresh vegetables or fruit, but seem to have little potential for large-scale agricultural development.

The desalination method of fresh-water recovery is easily done using a simple still (Fig. 15-9). Seawater is fed into a closed container and heated. The water vapor, free of salt, is removed and condensed. The salt remaining behind and most of the original seawater (typically 90 percent) are discharged as a hot brine, the disposal of which constitutes a potential pollution problem for desalination projects.

Figure 15-9

Simple single-stage evaporation–condensation apparatus.

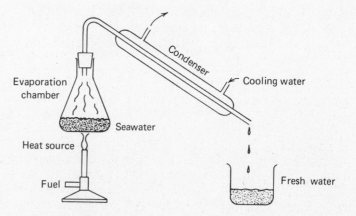

Although easily visualized, recovery of fresh water from a single-stage evaporation process like that in Fig. 15-9 is impracticable. Too much heat is lost in the cooling water, and high energy costs restrict the process to small-scale or emergency uses.

One way to increase the efficiency is to use the seawater for cooling before it is fed into the boiler (see Fig. 15-10). In this way the energy consumption is reduced. It is also possible to reduce the boiling point by carrying out the process under a slight vacuum, a technique known as "flash evaporation." Such a multistage flash distillation system has been built for Elat, Israel.

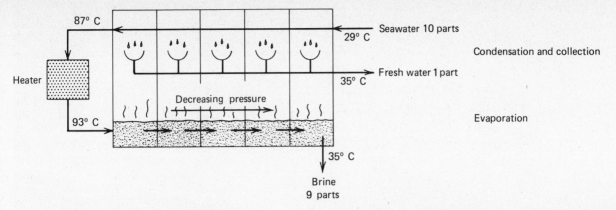

87° C

29° C ← Seawater 10 parts

Condensation and collection

Heater

35° C → Fresh water 1 part

Decreasing pressure

Evaporation

93° C

35° C

Brine
9 parts

Figure 15-10

Multi-stage flash distillation used at Elat, Israel, where water temperatures are relatively high. (After B. J. Skinner and K. K. Turekian, 1973. *Man and the Ocean,* Prentice-Hall, Englewood Cliffs, N.J.)

Another way of desalinating seawater is to duplicate the natural freezing cycle in which water and seasalts are separated as discussed in Chapter 5. In this process, a coolant is added to seawater so that when it evaporates it removes heat from the liquid surface, causing it to freeze. If the ice crystals are removed and washed free of brine and salt, they can then be melted to obtain fresh water. The coolant is recycled and brines are discharged with approximately twice the salinity of the original seawater. A plant using these principles was designed for Ipswich, England, to produce water at a cost of approximately 50 cents per 100 gallons. Such a process is particularly suited to application in northern latitudes where ocean waters are initially cold, thereby reducing costs.

PETROLEUM AND NATURAL GAS

Petroleum and natural gas are familiar examples of nonrenewable resources from the ocean. Some petroleum is forming today, but at rates far slower than it is being used.

Chapter 2 mentioned the derivation of fossil fuels from altered organic matter, usually phytoplankton. In areas where productivity is exceptionally high and bottom water circulation sluggish, dissolved-oxygen concentrations may be depleted due to decomposition of uneaten phytoplankton. Growth of consumer organisms is thereby inhibited, and organic matter deposited. Eventually bacteria break down some of this material, forming gas and oily residues in areas where rapid sediment accumulation prevents complete decomposition of organic matter.

Sediments gradually compact under the weight of overlying deposits. Water and oil are expelled and the sediments gradually become consolidated into rock. Liquids move through these rocks unless they are trapped in porous material, typically sandstone. If petroleum accumulates in this way, it can be extracted commercially. Thus the factors required to form major petroleum deposits are (1) thick accumulations of sedimentary rock rich in organic matter and (2) permeable, porous rocks to hold the petroleum in an extractable form after its movement has been stopped by an impermeable layer.

Petroleum and natural gas have been produced off California, in the Gulf of Mexico and in the Persian Gulf from extensions of well-known oil fields on land (Fig. 15-11). Elsewhere, as in the North Sea between England and Holland and in Bass Strait between Australia and Tasmania, there were no corresponding land deposits. Known major petroleum sources lie on the continental shelf under highly productive coastal ocean waters. Promising areas include the continental shelves north of Alaska, around Sumatra and Borneo in Indonesia, and on the Atlantic and Pacific margins of North America.

Figure 15-11

Production facilities in the Fateh Field, Persian Gulf. Equipment on the platforms is used to control the production of the wells, to pump the crude oil, and to flare excess natural gas. The tanker in the background stores the oil until it is transhipped for transportation to market. (Photograph courtesy Continental Oil Company.)

Salt domes are common on continental shelves and often serve as petroleum traps (Chapter 3). Fossil fuels accumulate under uplifted layers of rock around the domes and are readily recovered from them (see Fig. 15-15). Ninety percent of known gas and oil reserves occur in continental shelf domes, and petroleum has also been detected in cores from salt domes on the Sigsbee Abyssal Plain. This suggests that fossil fuel accumulations are not limited to shallow-water areas but may occur in the extensive sediment deposits of marginal basins.

It is difficult—perhaps impossible—to predict the petroleum possibilities of the continental slope and rise. Thick sediment accumulations and relatively high concentrations of organic matter suggest that deposits occur there too. But difficult engineering and technical problems must be solved before these deep-water deposits can be successfully exploited.

Estimates of undiscovered petroleum and natural gas resources are shown in Table 15-2. About 70 percent of the undiscovered resources are expected to be found on continental shelves and in shallow marginal ocean basins, much of which can be exploited using available techniques. About 23 percent is expected to come from the continental slope, which cannot be exploited using present production techniques. Relatively little is expected to come from the continental rise and deep-ocean floor.

TABLE 15-2

Estimated Undiscovered Total Offshore Petroleum Resources (oil plus equivalent amount of natural gas) *

OCEAN AREA	RESOURCES	
	(billions of metric tons)	(billions of barrels)
Undiscovered reserves		
Continental shelves, shallow seas	183	1,370
Continental slopes	61	460
Continental rises	12	90
Deep-sea trenches and ridges	3.5	26
	260	1,950
Proved reserves	19	143

* From National Academy of Sciences (COMRATE), 1975. *Mineral Resources and the Environment*, Washington, D.C., 348 pp.

OTHER POTENTIAL ENERGY SOURCES IN THE SEA

The ocean itself is a source of renewable energy (Table 15-3). Tides are now used for power, and concepts have been formulated for generating power from waves and from the temperature differences between warm surface waters and the much colder deep-ocean waters. Geothermal power, now used in Iceland, exploits the heat flows from mid-ocean ridges.

TABLE 15-3

Energy from Fossil Fuels and from other Earth Sources *

SOURCE	ENERGY	
	($\times 10^{20}$ cal/yr)	($\times 10^{12}$ watts)
Combustion by man	0.6	8.0
Dissipated by tides	0.2	2.7
Geothermal losses by earth	3.2	42
Solar energy at earth's surface	6500	86,000

* From National Academy of Sciences (COMRATE), 1975. *Mineral Resources and the Environment*, Washington, D.C., 348 pp.

Tides have been used for centuries to power small mills in coastal regions; for example, in 1650 Boston had a tidally powered mill to grind corn. Tidal power is now used to generate electricity for use far from the coast. In the early 1970's two tidal power plants were operating, one on the Rance Estuary in Brittany (France), the other on a bay in the U.S.S.R. near Murmansk, in the White Sea. Some estuaries in England, China, and Korea are possibilities, as is the Bay of Fundy.

Three factors limit use of tidal power; the necessity for large tidal ranges, suitable topography, and the timing of power generation. The largest tidal ranges in the ocean are around 15 meters, for example in the Bay of Fundy (Chapter 9). Even using the best of large, modern turbines designed to work on the ebb and flood tides, the tidal range must exceed 5 meters to be useful. Such ranges are relatively rare (Fig. 15-12), since the tidal range for most coasts is only about 2 meters (see Chapter 9). Furthermore, many of the potential sites are in remote areas, from which power transmission to urban and industrial centers would be expensive.

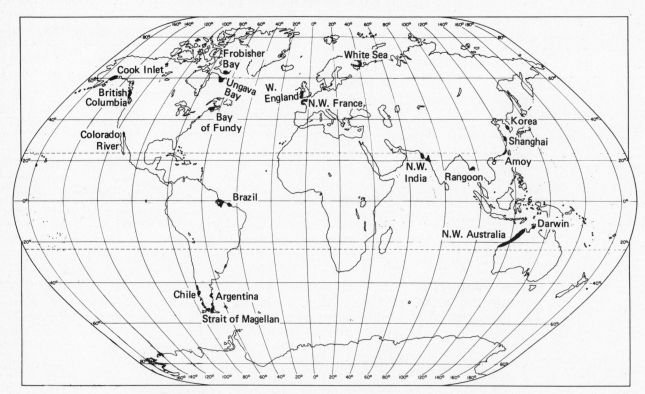

Figure 15-12

Areas whose tidal range exceeds 5 meters.

The second limiting factor for a tidal-power generating station is topography. Most tidal power schemes involve one or more dams. In a typical system, the gates through the dam are opened when the tide is high and then closed, keeping the water behind the dam at the level of the high tide. When the tide has dropped sufficiently, the water is allowed to flow out through turbines, generating power. The larger and wider the opening into the bay or estuary, the more expensive the dam. Many of the most attractive sites are at high latitudes where glaciers have cut deep, narrow embayments and scoured the landscape down to bedrock, providing good foundations for dams.

Thirdly, there is the timing problem. Tidal power generation is tied to the tidal cycle, which does not usually coincide with periods of peak power demand. For example the tidal power plant on the Rance Estuary produces about four times as much power during spring tides as it does during neap tides. Several ideas have been advanced to solve this problem. One is to use a vast network of power lines so that the electricity can be used somewhere regardless of when it is generated. Another uses several dams to store water at high levels, so that one basin serves as reservoir and another as collector. Finally, there is the option of generating electricity and storing it in some way for later use. A fuel such as hydrogen could be made and stored in the same way.

Waves are another potential energy source. One wave 1.8 meters high in water 9 meters deep releases about 10 kilowatts of power in each meter of wave front (Chapter 8). The power dissipated on a beach during a single storm is enormous. Small amounts of power generated by the waves have been used to power whistles or gongs on navigational buoys for many years. The problem is to develop schemes for large-scale extraction of this energy at reasonable costs.

Temperature differences between water masses represent another potential energy source. Since surface waters are heated by the sun to about 25°C over much of the ocean, while waters a few hundred meters below are 5°C or less, power can be generated by transfer of heat between deep and surface layers. This has been done on an experimental basis. Warm surface waters are used to evaporate specially chosen liquids much as an oil-fired boiler is used to evaporate water in a steam electrical power plant, and waters from below the thermocline can be used to cool the system. Schemes have been proposed to generate power on tropical islands while using the cold, nutrient-rich deep waters to support aquaculture. Still other schemes envision large floating power plants in subtropical waters, "grazing the thermocline" to produce power.

In areas of recent volcanic activity, the water trapped underground in porous or fractured rocks becomes heated through contact with the heated rocks (Chapter 2). Wells drilled into such reservoirs tap either hot water or superheated steam. This can be used as an energy source for a conventional steam electric power plant as has been done in Tuscany (Italy), New Zealand, northern California, Mexico, Japan, and the U.S.S.R. Steam from a large geothermal field is used for heating greenhouses in the town of Hveragerdi, Iceland, and development is underway to harness the power for generating electricity. Similar plants are possible on other volcanic islands and exposed portions of mid-oceanic ridges.

MINERALS FROM THE OCEAN BOTTOM

Despite the attention given to petroleum and natural gas from the world's continental shelves, sand, gravel, and shell account for the bulk of the material taken from the ocean, and for much of the value of ocean resources produced in the past. Sand, gravel, and shell are in constant demand in urban areas for paving roads, constructing buildings, and filling low-lying areas. As an urbanized region expands it typically uses up local sand and gravel quarries and often builds over potential supplies. Thus construction firms must go long distances to acquire needed building materials. For coastal cities, nearby bays and harbors have been dredged to provide sand and gravel, especially for landfill operations where composition and size requirements are less stringent than in making concrete. With increased concern about the environmental impact of such operations, these sources have also become more difficult to exploit, so that interest has increased for use of offshore deposits.

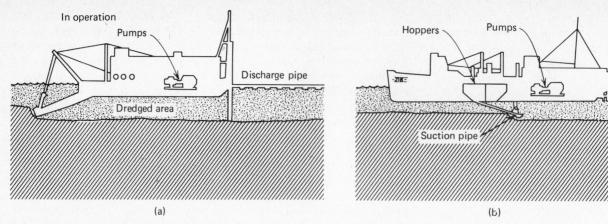

(a)   (b)

Figure 15-13 (above)

Schematic representation of simple hydraulic and hopper dredge operations. In the former, the cutter head is moved across the bottom, stirring and suspending the sediment. This slurry is then pumped into the vessel and out through a discharge pipe to a disposal or storage site where the sand and gravel settle out. The water and fine sediment particles run off. In the hopper dredge, suspended sediment and water are pumped into tanks aboard ship where the sand and gravel settle out and the water is discharged overboard.

Over the next 10 to 20 years, tens of millions of cubic meters of sand and gravel will be produced from U.S. offshore waters. Such deposits have long been exploited around the North Sea. Carbonate sands have been dredged in island areas where limestone for making cement is scarce on land, as in the Hawaiian Islands and Iceland.

Sand is also used in large quantities to repair and restore beaches damaged by normal currents or by storms. One resort community on the U.S. Atlantic coast uses more than one million cubic meters of sand yearly for such purposes.

Figure 15-14

A simple hydraulic dredge, with the cutter head shown out of the water, at left. The discharge pipe can be seen at right.
(Photograph courtesy Ellicott Machine Corporation.)

Material for construction purposes or landfill is commonly obtained by dredging. In open-ocean waters, dredging is carried out by seagoing ships which drag a special pipe behind the vessel and pump up large volumes of water and sand off the bottom (see Fig. 15-13). The water is discharged but the sand settles out in the large tanks on the vessel, called a hopper dredge. The hoppers are dumped or pumped out at the designated site. Another common type is the hydraulic dredge, which is basically a set of large pumps mounted on a barge. The sand is pumped and then discharged through large pipes to the shore site (or other location) being filled, or where the material can be stockpiled (see Fig. 15-14).

The rivers that cut their way across continental shelves during periods of lower sea level, the movements of the shoreline across the shelf as sea level rose, and the constant scouring due to storms, together with tidal current action, have left deposits of valuable minerals on many continental shelves (Table 15-4). These so-called heavy minerals,

TABLE 15-4

Materials Commonly Recovered
from Surficial Continental Shelf Sediment Deposits *

MATERIAL	Mineral		
	NAME	CHEMICAL COMPOSITION	DENSITY (g/cm^3)
Gold	Gold	Au	19.3
Tin	Cassiterite	SnO_2	7.0
Chromium	Chromite	$FeCr_2O_4$	4.5
Titanium	Rutile	TiO_2	4.3
Titanium	Ilmenite	$FeTiO_3$	4.7
Rare earths	Monazite	Rare-earth phosphates	5.0
Diamond	Diamond	C	3.5
Sand and gravel	—	—	2.5

* After B. J. Skinner and K. K. Turekian, 1973. *Man and the Ocean,* Prentice-Hall, Englewood Cliffs, N.J., 149 pp.

or *detrital deposits,* are much denser than normal sands and gravels (1.5 to 2.4 grams per cubic centimeter), and they are often concentrated and left behind when other sediments are moved out. Channels formed by rivers that cut across the continental shelf during periods of glaciation, when sea level was lower, are often sites of heavy mineral concentration. The rivers that flowed through them eroded their granite walls and carved deep valleys. Light and soluble materials were washed away, but heavier grains, containing for instance cassiterite, were left in the river valley. Some detrital deposits formed near beaches, where waves and longshore currents separated particles containing titanium and rare earths from the lighter sand grains, leaving concentrations of the heavy minerals as beach deposits. Where such minerals are sufficiently valuable, they are recovered for processing as ores. Gold, tin, chromium, and titanium have been produced from continental shelves in many areas of the world (Fig. 15-15).

Manganese nodules are another mineral resource of unknown but potentially great importance for world production of copper, cobalt, and nickel. There are large deposits in central basin areas, especially in the Pacific Ocean where cobalt (an estimated 5 billion tons) and nickel (15 billion tons) contents of the nodules are especially high in a large area of about one million square kilometers, at around 5 kilometers depth (Fig. 15-15). Special recovery techniques and specially built recovery vessels will be required before these deposits can be commercially worked. Present planning calls for special ships to recover and partially concentrate the nodules at sea. The concentrates will then be brought ashore for chemical processing to recover the metals in them.

Among nonmetal ocean-bottom resources (excluding fuels and building materials), sulfur and phosphorite are particularly important. Salt domes are rich sources of sulfur, in addition to petroleum as previously mentioned (Fig. 15-16). Sulfur accumulates around a salt plug, because an insoluble sulfur-containing compound, anhydrite ($CaSO_4$), is commonly present in the original salt deposit. As the plug is forced upward due to pressure from surrounding and overlying denser rock layers, it often comes in contact with groundwater that dissolves some of the salt. Anhydrite, commonly altered to gypsum, and other insoluble impurities such as clay, form a "cap rock" mantle over the salt core. If anaerobic bacteria are also present, in association with petroleum deposits, they act to reduce anhydrite as follows:

$$CaSO_4 \quad + \quad CH_4 \quad \longrightarrow \quad CaCO_3 \quad + \quad H_2S \quad + \quad H_2O$$

anhydrite   hydrocarbon   calcium carbonate   hydrogen   water
or gypsum   (petroleum)   (limestone)   sulfide

Hydrogen sulfide gas is released, and later oxidized in the presence of aerated groundwater to yield elemental sulfur:

Figure 15-15

Areas of present petroleum, natural gas, and mineral extraction from the ocean bottom. Areas of potential manganese nodule production are also shown.

$$2H_2S \quad + \quad O_2 \quad \longrightarrow \quad 2H_2O + \quad S$$

hydrogen   oxygen   water   sulfur
sulfide

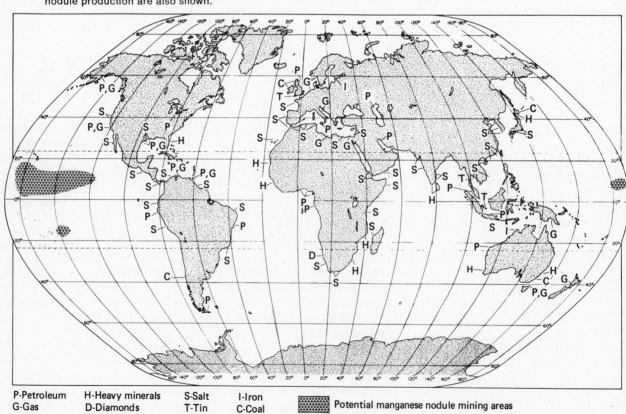

P-Petroleum   H-Heavy minerals   S-Salt   I-Iron   Potential manganese nodule mining areas
G-Gas   D-Diamonds   T-Tin   C-Coal

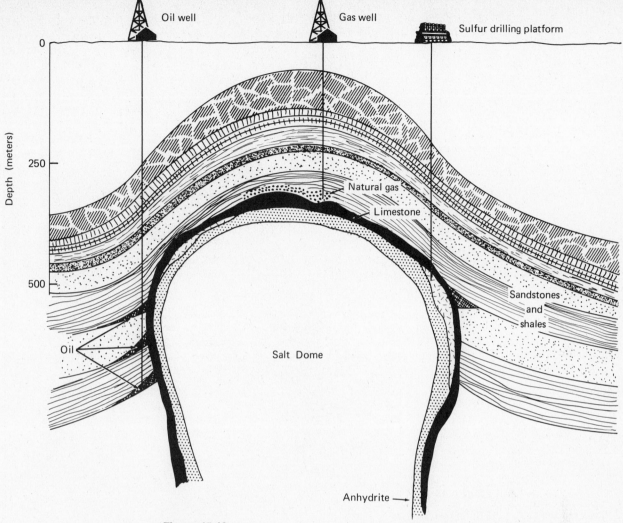

Oil well

Gas well

Sulfur drilling platform

Depth (meters)

0

250

500

Natural gas

Limestone

Sandstones and shales

Oil

Salt Dome

Anhydrite →

Figure 15-16

Schematic representation of a salt dome, showing deposits of petroleum, natural gas, and sulfur.

Sulfur is recovered by forcing hot air and water into a salt dome, to melt the low-density sulfur and cause it to rise through pipes to the surface.

Phosphorites, relatively insoluble deposits from which fertilizer is made, occur as nodules over the ocean floor, particularly on continental shelves. They can easily be dredged, and may become quite valuable in coastal areas short of phosphates. Phosphate content (P_2O_5) of the nodules tends to be relatively low, on the order of 20 percent, but many billions of tons are thought to be available. Present world production of phosphate is about 100 million tons per year.

MAN-MADE ISLANDS AND OFFSHORE PORTS

The shallow coastal ocean provides space for urban and industrial growth. In many urban regions there is little space remaining for expansion, so municipalities, utilities, and government agencies are considering construction of facilities on continental shelves.

This seaward thrust is nothing new to the Netherlands. Between A.D. 1200 and 1950, the Dutch reclaimed about 1.6 million acres (6300 square kilometers), and reclamation of new areas continues still (Fig. 15-17). Land is reclaimed from the shallow sea by dikes which enclose fields, called *polders*. Some projects have been carried out on a monumental scale. In the 1930's, the Zuider Zee, a large, shallow embayment,

Figure 15-17

Areas reclaimed from the North Sea, in the Netherlands. (After J. Van Veen, 1962. *Dredge, Drain, Reclaim! The Art of a Nation,* 5th ed., Martinus Nijhoff, The Hague.)

was cut off by a dike, changing it from a brackish estuary to the fresh-water Ijssel Lake in a few years. Agricultural land was reclaimed from the former bay bottom. Land reclamation has also been done on a smaller scale in low-lying coastal areas around the North Sea. Parts of the lower Rhine estuary were shut off from the North Sea in the 1950's and 1960's to prevent a repetition of the disastrous flooding accompanying the 1953 storm surge (Chapter 8).

Offshore port facilities in various parts of the world have been used for loading and unloading supertankers, tankers carrying hundreds of thousands of tons of petroleum and drawing up to 30 or 35 meters of water. Before the mid-1960's, tankers drew less than 15 meters, so that they could move through the Suez Canal. Closures of the Canal, and the savings in transportation cost provided by larger vessels, made deep-draft ships more attractive. Unfortunately, such ships cannot operate in most estuaries without extensive and expensive dredging. But by locating loading and unloading facilities offshore, dredging is avoided. Offshore port facilities have been used for years in the Persian Gulf and promise to be widely used in U.S. waters along the Atlantic and Gulf coasts.

424

425
man-made islands
and offshore ports

Difficulties in obtaining plant sites on land and problems of access to adequate cooling waters have made offshore sites for large nuclear plants especially attractive. Distance from shore reduces residents' objections to the disruptive appearance of a power plant on the coastal landscape and also reduces the exposure of the coastal population to potential radiation hazards caused by accidents at the plant (Fig. 15-18).

Figure 15-18

Artist's model of proposed offshore nuclear power plant. In this scheme the nuclear reactors float on large barges. A semicircular breakwater protects the plant from wave action.

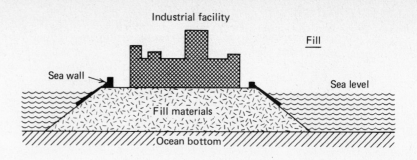

Industrial facility

Fill

Sea wall

Sea level

Fill materials

Ocean bottom

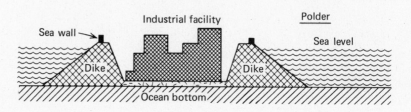

Sea wall

Industrial facility

Polder

Sea level

Dike

Dike

Ocean bottom

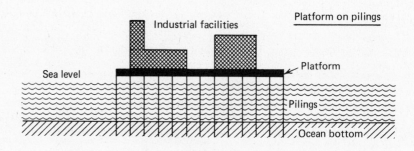

Industrial facilities

Platform on pilings

Sea level

Platform

Pilings

Ocean bottom

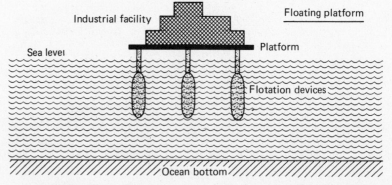

Industrial facility

Floating platform

Sea level

Platform

Flotation devices

Ocean bottom

Figure 15-19

Methods of constructing offshore facilities
for industrial or other purposes.

Offshore facilites can be constructed in several ways (Fig. 15-19). Simplest are tanker mooring and offloading facilities consisting of little more than a few pilings and moorings for the vessel to tie to. Oil is pumped ashore through a system of pipes to be stored and refined on land.

Other construction techniques suitable for building large offshore structures include:

1. Dike and polder construction similar to techniques used in the Netherlands; costs range between $2000 and $30,000 per acre (1970 dollars);

2. Conventional fill (including breakwater construction) in which the ocean bottom is built up by dumping fill materials; costs range between $8000 and $200,000 per acre;

3. Pile-supported deck or platform built on pilings; costs vary between $400,000 and $2.2 million per acre;

4. Floating platform using flotation devices to support a platform; costs range between $1.3 million and $4.3 million per acre.

Regardless of which construction approach is used, costs of artificial islands are certain to be large. An offshore structure near New York City to be used as an airport had estimated costs of $1 billion to $7 billion in 1971. Smaller structures will be less expensive, but most will cost hundreds of millions. Despite their high costs, islands are likely to be built on the continental shelf. Many are likely to be used for several purposes to reduce the individual costs of a single use.

WASTE DISPOSAL   The ocean serves many little-known but vital functions. One of the most controversial is waste disposal. Near coastal cities and industries, the ocean is used to dilute and disperse a great variety of wastes, both liquid and solid (see Table 15-5). Urban wastes are often discharged by sewers into bays and harbors or into the coastal ocean directly from large pipes that take the wastes offshore for disposal. The waters around discharge pipes dilute the wastes and currents disperse them into the surrounding coastal ocean.

TABLE 15-5
Wastes Discharged into Estuaries and the Coastal Ocean *

SOURCES	WASTES DISCHARGED
Municipal storm sewers	Waste oils Street washings Raw sewage Suspended sediment
Municipal sewage treatment plants	Nutrients (phosphates and nitrates) Sewage sludges (solids from treatment)
Industrial wastes	Waste chemicals, e.g., acids, petrochemicals Waste oil
Runoff from agricultural lands	Nutrients from fertilizers Pesticides and herbicides Animal wastes
Electrical power plants	Waste heat Ash (from coal) Chemicals (corrosion-inhibiting, foam-suppressing, biocides)
Dredging operations and construction activities	Suspended sediment Nutrients from sediments
Petroleum production and exploration	Suspended sediment (drilling muds) Crude oil
Ships (commercial and recreational)	Untreated sewage Garbage Waste oil

* After M. G. Gross, 1976. *Oceanography*, 3rd ed., Charles E. Merrill, Columbus, Ohio, 138 pp.

Coastal and estuarine circulation patterns (Chapter 10) do not always work well for dispersing wastes. Coastal currents tend to move materials along the coast and thereby inhibit movement of the wastes out into the open ocean where they could be more effectively diluted. Wastes that become associated with particles, such as any constitutent taken up by organisms, tend to be carried landward by bottom-waters in estuarine-like circulation. Thus nutrients such as phosphates from detergent and nitrogen compounds from sewage, when discharged to bays and harbors tend to remain in the area. This can stimulate the growth of phytoplankton, often of types not directly usable by organisms that live in the area. Problems arise if the plants are not eaten; they die and are decomposed in the waters, locally causing depletions of dissolved oxygen and fish kills, especially in late summer.

Waste disposal in coastal waters interferes with shellfish production and sometimes with fisheries, especially those dependent on near-bottom species. This is increasingly a problem on heavily populated, industrialized coasts such as the U.S. Atlantic seaboard. Benthic organisms may be poisoned by sewage and chemicals (including pesticides), or suffocated by silt deposits. If not killed, they may be weakened and thus less able to resist disease and predation by natural enemies. A different problem arises when chemical wastes containing a high proportion of trace elements are taken up by shellfish. Silver (Ag), cadmium (Cd), chromium (Cr), and lead (Pb), for instance, have no known function in biological systems. In high concentrations some of these trace elements are toxic to organisms. In the absence of any mechanism for excretion of such substances, they accumulate in plant and animal fats and proteins (Table 15-6). Such materials tend to be conserved in ecosystems, because they are passed along in food chains.

TABLE 15-6
Enrichment of Various Elements in Shellfish *

ELEMENT	Enrichment Factors		
	SCALLOP	OYSTER	MUSSEL
Ag	2,300	18,700	330
Cd	2,260,000	318,000	100,000
Cr	200,000	60,000	320,000
Cu	3,000	13,700	3,000
Fe	291,500	68,200	196,000
Mn	55,500	4,000	13,500
Mo	90	30	60
Ni	12,000	4,000	14,000
Pb	5,300	3,300	4,000
V	4,500	1,500	2,500
Zn	28,000	110,300	9,100

* After Horne, 1969.

Some may substitute for elements essential to metabolic processes; cadmium, for example, substitutes for zinc in humans, resulting in impairment of fat metabolism. Elements incorporated in animal or plant skeletons are likely to have shorter lifetimes in an ecosystem and to be more quickly incorporated in sediments instead.

A more serious problem arises when filter-feeding clams and oysters ingest sewage-borne viruses and bacteria; since these organisms are often eaten raw and whole, including the digestive tract, they constitute a significant potential for disseminating hepatitis and gastro-intestinal diseases. Waste disposal areas are closed to commercial shellfish production when judged unsafe by government inspectors, disrupting the livelihood of oystermen and clam diggers.

Bathing beaches are also closely monitored for evidence of sewage, and closed for swimming if sewage-associated bacteria counts run too high. Thus many of the beaches close to cities in the United States are no longer available for water-contact sports by city residents (Fig. 15-20).

Figure 15-20

Warnings are posted along the Potomac River waterfront, in Alexandria, Virginia. (Photograph courtesy of U.S. Environmental Protection Agency, EPA-DOCUMERICA. Erik Calonius, photographer.)

There may also be other, less obvious, changes associated with disposal of wastes on continental shelves. For example, the presence of sewage solids in the New York Bight and off Southern California has caused an increased incidence of fin rot in bottom-dwelling fishes. In this condition, the fins of the affected fish are eaten away by some unknown disease, leaving the fish subject to easy capture by predators. Somewhat similar conditions are also observed to affect lobster and crabs taken from the floor of a waste-disposal area.

Although sewage solids are a ubiquitous problem in urban coastal areas, they are by no means the only wastes discharged at sea. Dredging operations to build and maintain navigation facilities, extraction of ores such as iron, titanium, and aluminum, and the washing of sand

and gravel are major sources of waste solids. Some of these sources rival or exceed natural processes as sources of sediment particles to coastal waters. Waste disposal operations in the New York metropolitan region are the largest single source of sedimentary materials to the entire Mid-Atlantic area. And such disposal operations are quite widespread. In 1974, dredging produced about 100 million tons of sediment discharged on the continental shelf, about half of this from the Mississippi River alone. Virtually every major port requires dredging, and much of the dredged material is disposed of at sea because of the lack of suitable sites on land.

Other wastes discharged at sea include those carried in liquids. For example, the desalination processes discussed earlier will produce brines that must be piped back to the ocean and with them the waste heat that was lost during the evaporation (or freezing) process. Steam electric power plants and refineries also discharge heated waters that can locally warm a bay or coastal area enough to cause changes in biological communities.

SUMMARY OUTLINE

Marine resources—fishing, defense, transportation; also fuels, metals, construction materials
  Control and management through international laws a positive factor
  A frontier area—techniques being developed
  Renewable resources—replenished naturally
  Nonrenewable resources—e.g., petroleum and metal ores

Fisheries and aquaculture—a renewable resource
  Man an important predator—60 or more metric tons annual production
  Maximum yield probably 200 million tons, including species not now used
  Dynamics of fishery—aids in prediction of yields
    Fishing increases population growth rate by reducing population size
    Overfishing results if maximum sustainable yield is exceeded
    Size of year class depends on environmental conditions, food supply
  Pacific sardine fishery—replaced by noncommercial anchovy off California
  Peruvian anchovetta fishery—El Niño conditions cause upwelling failures
  Whaling—international regulation necessary
  Mariculture—important in Orient; e.g., relocation of year class for fattening, enclosures to control conditions

Resources from seawater—salt, magnesium, bromine, water
  Salt—by evaporation using solar energy
  Metals—production costs usually exceed value of resource
  Water—important in coastal areas; evaporation, distillation, freezing

Petroleum and natural gas—in porous sedimentary deposits, with organic matter
  Major sources on continental shelf, in salt domes

Potential energy sources—renewable
  Tidal power—requires large tidal range, suitable topography, energy storage
  Waves—now used for buoys, whistles
  Temperature differences—a use of solar energy
  Geothermal power—steam or hot water from heat in ocean floor; used in Iceland

Minerals from the ocean bottom
  Sand, gravel, shell—demand from construction industry; for landfill
  Heavy minerals (e.g., tin, titanium)—from erosion of ancient rocks, in valleys and beach deposits on continental shelf
  Manganese nodules—valuable for copper, nickel, cobalt
  Sulfur—produced from salt domes
  Phosphorites—phosphate for fertilizer, not yet produced commercially

Man-made islands and offshore ports—pressure for land in coastal areas.
  Polders reclaimed as farmland in the Netherlands
  Offshore port facilities—for deep-draft vessels, e.g., tankers
  Sites for industry—power production, storage of materials
    Four types: dike and polder, fill, piles supporting deck, floating platforms

Waste disposal—coastal, estuarine circulation retains wastes in coastal waters
  Sewage and chemical wastes—damaging to fish, shellfish, beaches
  Dredging—may be the major sediment source in a coastal area
  Waste heat—carried in brines, coolants

SELECTED REFERENCES

CUSHING, D. H. 1975. *Marine Ecology and Fisheries*. Cambridge University Press, Cambridge, England. 278 pp. Advanced treatment of fisheries, fish production, and marine food webs.

MERO, J. L. 1965. *The Mineral Resources of the Sea*. Elsevier, Amsterdam. 312 pp. Broad coverage of mineral resources.

SKINNER, B. J. AND K. K. TUREKIAN. 1973. *Man and the Ocean*. Prentice-Hall, Englewood Cliffs, N.J. 149 pp.

Elementary discussion of ocean resources in context of oceanographic processes.

VAN VEEN, J. 1962. *Dredge, Drain, Reclaim! The Art of a Nation*. 5th ed. Martinus Nijhoff, The Hague, Netherlands. 200 pp. Elementary discussion of Dutch polder building.

WENK, EDWARD. 1969. "The Physical Resources of the Ocean. *Scientific American* 221 (3), 166–76.

conversion factors

EXPONENTIAL NOTATION

It is often necessary to use very large or very small numbers to describe the ocean or to make calculations about its processes. To simplify writing such numbers, scientists commonly indicate the number of zeros by *exponential notation*, indicating the powers of ten. Some examples:

$1,000,000,000 = 10^9$	(one billion)
$1,000,000 = 10^6$	(one million)
$1,000 = 10^3$	(one thousand)
$100 = 10^2$	(one hundred)
$10 = 10^1$	(ten)
$1 = 10^0$	(one)
$0.1 = 10^{-1}$	(one tenth)
$0.01 = 10^{-2}$	(one hundredth)
$0.001 = 10^{-3}$	(one thousandth)
$0.000\ 001 = 10^{-6}$	(one millionth)
$0.000\ 000\ 001 = 10^{-9}$	(one billionth)

Multiplication: To multiply exponential numbers (powers of ten), the exponents are added. For example $10 \times 100 = 1000$ which is written exponentially as $10^1 \times 10^2 = 10^3$

Division: To divide exponential numbers, the exponent of the divisor is subtracted from the exponent of the dividend. For example, $100/10 = 10$ or written exponentially as $10^2/10^1 = 10^1$

UNITS OF MEASURE

length

1 *kilometer* (km) = 10^3 meters = 0.621 statute mile = 0.540 nautical mile

1 *meter* (m) = 10^2 centimeters = 39.4 inches = 3.28 feet = 1.09 yards = 0.547 fathom

1 *centimeter* (cm) = 10 millimeters = 0.394 inch = 10^4 microns

1 *micron* (μ) = 10^{-3} millimeter = 0.0000394 inch

area

1 *square centimeter* (cm²) = 0.155 square inch

1 *square meter* (m²) = 10.7 square feet

1 *square kilometer* (km²) = 0.386 square statute mile = 0.292 square nautical mile

volume

1 *cubic kilometer* (km³) = 10^9 cubic meters = 10^{15} cubic centimeters = 0.24 cubic statute mile

1 *cubic meter* (m³) = 10^6 cubic centimeters = 10^3 liters = 35.3 cubic feet = 264 U.S. gallons

1 *liter* (l) = 10^3 cubic centimeters = 1.06 quarts = 0.264 U.S. gallon

1 *cubic centimeter* (cm³) = 0.061 cubic inch

mass

1 *metric ton* = 10^6 grams = 2205 pounds

1 *kilogram* (kg) = 10^3 grams = 2.205 pounds

1 *gram* (g) = 0.035 ounce

time

1 day = 8.64×10^4 seconds (mean solar day)

1 year = 8765.8 hours = 3.156×10^7 seconds (mean solar year)

speed

1 *knot* (nautical mile per hour) = 1.15 statute miles per hour = 0.51 meter per second

1 *meter per second* (m/sec) = 2.24 statute miles per hour = 1.94 knots

1 *centimeter per second* (cm/sec) = 1.97 feet per minute = 0.033 feet per second

temperature

Conversion formulas	Conversion table	
	°C	°F
$°C = \dfrac{°F - 32}{1.8}$	0	32
	10	50
	20	68
$°F = (1.8 \times °C) + 32$	30	86
	40	104
	100	212

energy

1 *gram-calorie (cal)* = $\frac{1}{860}$ watt-hour = $\frac{1}{252}$ British thermal unit (Btu)

APPENDIX

data on earth and ocean

Size and Shape of the Earth

DIMENSION	MILES	KILOMETERS
Equatorial radius	3,963	6,378
Polar radius	3,950	6,357
Average radius	3,956	6,371
Equatorial circumference	24,902	40,077

Areas of the Earth, Land, and Ocean

PART OF EARTH	MILLIONS OF SQUARE MILES	MILLIONS OF SQUARE KILOMETERS
Land (29.22 percent)	57.5	149
Ice sheets and glaciers	6	15.6
Oceans and seas (70.78 percent)	139.4	361
Land plus continental shelf	68.5	177.4
Oceans and seas minus continental shelf	128.4	332.6
Total area of the earth	196.9	510.0

Volume, Density, and Mass of the Earth and Its Parts *

PART OF EARTH	AVERAGE THICKNESS OR RADIUS (km)	VOLUME ($\times 10^6$ km^3)	MEAN DENSITY (g/cm^3)	MASS ($\times 10^{24}$ g)	RELATIVE ABUNDANCE (percent)
Atmosphere	—	—	—	0.005	0.00008
Oceans and seas	3.8	1,370	1.03	1.41	0.023
Ice sheets and glaciers	1.6	25	0.90	0.023	0.0004
Continental crust †	35	6,210	2.8	17.39	0.29
Oceanic crust ‡	8	2,660	2.9	7.71	0.13
Mantle	2,881	898,000	4.53	4,068	68.1
Core	3,473	175,500	10.72	1,881	31.5
Whole Earth	6,371	1,083,230	5.517	5,976	

* From Holmes, 1965.
† Including continental shelves.
‡ Excluding continental shelves.

Heights and Depths of the Earth's Surface

Land			Oceans and Seas		
HEIGHT	FEET	METERS	DEPTH	FEET	METERS
Greatest known height:			Greatest known depth:		
Mt. Everest	29,028	8,848	Marianas Trench	36,204	11,035
Average height	2,757	840	Average depth	12,460	3,808

APPENDIX

graphs, charts, and maps

In any scientific discipline, data gathered from many sources are usually at some point compiled and presented in graphic form. Oceanography is no exception, although the wide variety of material obtained from studying the ocean poses some special problems. In this section we shall review some of the common means of graphically presenting data—graphs, profiles, maps, and diagrams—in an effort to show the uses as well as the limitations of each technique.

GRAPHS

Of the various techniques used to portray scientific data, *graphs* are perhaps the most widely used. A graph permits general aspects of the relationship between two properties to be understood at a glance. Furthermore, values of various properties can be measured from a graph.

In constructing a graph, data are often first organized into tables. For example, if we are studying the increase in boiling point of seawater with salinity changes, we may arrange our data as follows:

SALINITY (parts per thousand)	BOILING POINT INCREASE (°C)
4	0.06
12	0.19
20	0.31
28	0.44
36	0.57

Figure A3-1 (*right*)

Graph of the increase in temperature of boiling and water salinity. The dots represent the experimental data from the table.

From this table it is apparent that the boiling point of seawater increases with increased salinity. To see this more clearly, a graph can be drawn as in Fig. A3-1. Note that salinity is plotted on the *x* (horizontal) axis with values increasing to the right. Boiling point increase, in degree Celsius (°C), is plotted on the *y* (vertical) axis, with values increasing upward. We plot our experimental points, and then draw a line through these points.

Generally, a straight line is the best first estimate. In this case, it is a reasonably good approximation. For many other graphs, we may need to use more complicated curves; but even for complicated curves, a straight line is a reasonable estimate for small portions of the curve.

Note that the graph shows that the boiling point is raised as salinity increases. Also, you can estimate boiling point increases for salinities we did not study. For instance, for seawater with a salinity of 24 parts per thousand, we can interpolate a boiling point increase of 0.375°C; for a salinity of 40 parts per thousand, we extrapolate (or extend) the curve to indicate a boiling point increase of 0.63°C.

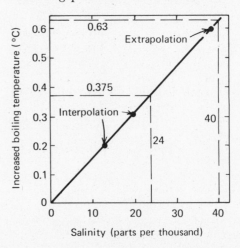

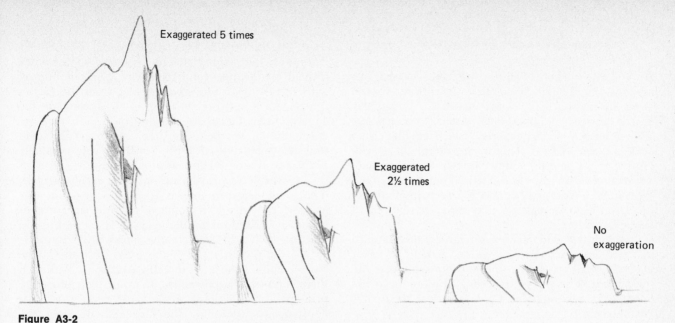

Exaggerated 5 times

Exaggerated 2½ times

No exaggeration

Figure A3-2

Distortions in a profile of a human face resulting from use of different standard exaggerations.

In reading a graph, always determine which property is plotted on each axis. Also, check both the scale intervals and the values at the origins. The appearance of the graph can be changed drastically by changing either; advertisers, for instance, often make use of graphs where the scales and origins are chosen to present their data in the most favorable light.

PROFILES

Profiles are used to show topography, either of ocean bottom or of land. An ocean-bottom profile can be considered as a vertical slice through the earth's surface. Such a profile can be drawn with no distortion—distances are equal vertically and horizontally. But imagine the problems involved in drawing a profile 10 centimeters long of the Atlantic Ocean between New York and London, a distance of 5500 kilometers, but only 3.4 kilometers deep at the deepest point. A pencil line would be too thick to portray accurately the maximum relief; thus such a profile conveys no useful information.

To get around this problem, profiles—including those in this book—are usually distorted. Profiles showing oceanic features are typically distorted by factors of several hundred or several thousand. This causes even gently rolling hills to look like impossibly rugged mountains. The effect of profile distortion can be seen rather dramatically when it is applied to the human profile, as in Fig. A3-2.

CONTOURS AND CONTOUR MAPS

Various means have been used to portray land forms or typography; the most useful employ contours in con-

tour maps. Figure 2-21 is an example of a contour map. *Contour lines* connect points that are at equal elevations or at equal depths. Obviously, not all elevations (or depths) can be connected by contour lines. Only certain ones at selected intervals are shown; otherwise, the map would be solid black. The vertical interval represented by successive contour lines is called the *contour interval.*

To interpret a contour map, imagine the shoreline of a lake. The still-water surface is a horizontal plane, touching points of equal elevation along the shore. The shoreline is thus a contour line. If the water surface were controlled to fall by regular intervals, it would trace a series of contour lines, forming a contour map on the lake bottom (or hillside). Note that the shorelines formed at different lake levels do not cross one another; neither do contours on maps.

Contours reveal topography. For example, contours that are closed on a map (do not intersect at a boundary) indicate either a hill or a depression. To find out which, look to see if the elevation increases toward the closed contour(s). If so, you are looking at a hill. If the elevation decreases toward the closed contour, it is a depression. Contours around depressions are often marked by *hachures*—short lines on the contour pointing toward the depression. When a contour line crosses a valley or canyon, the contour line forms a V, pointing upstream.

Contours can also be used to depict properties other than elevation (or depth). For example, several maps in this book use contours to show distribution of properties such as surface-ocean temperatures (see Figs. 6-6 and 6-7) and salinities (see Fig. 6-14). We would consider them to be temperature (or salinity) hills and valleys. High temperature corresponds to a hill, low temperature to a valley. Contours may also be used to show the distribution of temperature or salinity in a vertical section of the ocean.

COORDINATES—LOCATIONS ON A MAP

A *coordinate* is a sort of address—a means of designating location. The most familiar coordinate system, found in many cities, is the network of regularly spaced, lettered or numbered streets, crossing one another usually at right angles, which enables us to locate a given address. This is an example of a *grid*. As soon as we determine how the streets and avenues are arranged, we can find Fifth Avenue and 42nd Street in New York, or 8th and K Streets in Washington, D.C., even though we may never have been in those cities before.

A printed grid is used on large-scale maps to designate locations of features. In making maps of relatively small areas, it is simplest to assume that the world is flat. This works for areas extending up to 100 miles from a starting point. For larger areas, the earth's curvature must be considered.

Since the earth is round, we must use *spherical coordinates*—a grid fitted to a sphere. A small town or even a state has rather definite starting points for a grid—the edges or the center. But a sphere has no edges or corners, so we must designate those points where our numbering system is to begin.

Distances north or south are easiest to deal with. We can easily identify the earth's geographic poles, where the axis of rotation intersects the earth's surface. Using these points, it is fairly easy to draw a line circling the earth and equally distant from the North and South Poles. This is the *equator*, which serves as our starting point to measure distances north and south. Going from the equator, the distance to the pole is divided into 90 equal parts (*degrees*). The series of grid lines which circle the earth and connect points that are the same distance from the nearest pole are known as *parallels of latitude*.

To see how this works, imagine the earth with a section cut out, as in Fig. A3-3. Now look at the angle formed by the line connecting any point of interest with the earth's center and the line from the earth's center to a point directly south of that point, on the equator. This angle is a measure of the distance between the chosen point and the equator. The North Pole has a latitude of 90°N, Seattle is approximately 47°N, and Rio de Janeiro, approximately 23°S.

Latitude was easily measured by early mariners. The angle between Polaris (the pole star) and the horizon provides a reasonably accurate measure of latitude. At the equator, the pole star is on the horizon (latitude 0°N). Midway to the North Pole (latitude 45°N), the pole star is 45° above the horizon. At the North Pole, Polaris is directly overhead. Although there is no star directly above the South Pole, the same principle holds, except that a correction would be necessary to allow for the displacement from the South Pole of the star used.

Measuring east–west distances on the earth poses the problem—where do we start? The answer has been to establish an arbitrary starting point—the *prime meridian*—and to indicate distances as east or west of that meridian. Several prime meridians have been used by different nations, but today the Greenwich prime meridian is most commonly used. It passes through the famous observatory at Greenwich (a suburb southeast of London).

Longitude—distance east or west of the prime meridian—is indicated on a map by north–south lines, connecting points with equal angular separation from the prime meridian. They are called *meridians of longitude,* and converge at the North and South Poles. Longitude in degrees is measured by the size of the angle between the prime meridian and the meridian of longitude passing through the given point, as in

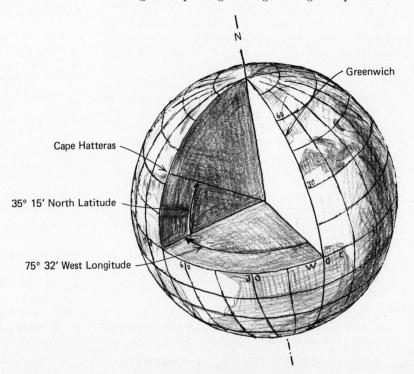

Figure A3-3

Latitude and longitude on the earth shown for Cape Hatteras, North Carolina.

Fig. A3-3. Going eastward from the prime meridian, longitude increases until we reach the middle of the Pacific Ocean, when we come to the 180° meridian. Going westward from Greenwich, longitude also increases until we reach the 180° meridian, which represents the juncture between the Eastern and Western Hemispheres. Through much of the Pacific Ocean, the 180° meridian is also the location of the *international dateline*. This designation of the 180th meridian as the international dateline—where the "new day" begins—is no accident. Its position in the midst of the Pacific avoids the problem of adjacent cities being one day apart in time. This also explains, in part, the choice of the Greenwich Meridian as the Prime Meridian.

Longitude and time are intimately related. The earth turns on its axis once every 24 hours. Since it takes the earth 24 hours to make one complete turn (360°), we calculate that the earth turns 15° per hour. We use this relationship to find our relative position east or west of the prime meridian.

Each meridian of longitude is a *great circle*. If we sliced through the earth along one of the meridians of longitude, our cut would go through the center of the earth. Of the parallels of latitude, only the equator is a great circle. All the other parallels are *small circles*, since a plane (or slice) passing through them would not go through the center of the earth. Great circles are favored routes for ships or aircraft because a *great circle* route is the shortest distance between two points on a globe.

To study time and longitude, let us begin at local noon on the prime meridian, when the sun is directly overhead. One hour later, the sun is directly over the meridian of a point 15° west of the prime meridian; two hours later, it is over a meridian 30° west of the prime meridian. Twelve hours later (midnight), it is over the 180th meridian and the new day begins.

If we have an accurate clock keeping "Greenwich time" (the time on the Prime Meridian), we can determine our approximate longitude from the time of local noon, when the sun is highest in the sky. Assume that our clock read 2:00 P.M. Greenwich time at local noon. The 2-hour difference indicates that our position is 30° from the prime meridian. Since local noon is later than Greenwich, we know that we are west of Greenwich and our longitude is therefore 30°W. Another example—if our local noon occurs at 9:30 A.M. Greenwich time, we are 2.5 hours × 15° per hour = 37.5° east of the prime meridian; our longitude is thus 37.5°E.

Degrees—like hours—are divided into 60 parts known as *minutes*. Each minute is further divided into 60 *seconds*. Consequently, in the last example we would give our position as 37°30′E.

To determine longitude, therefore, a ship need only have accurate time. With modern electronic communications, this poses no problems. For centuries, however, seafarers had no means of keeping accurate time at sea. Not until the 1760's, when the first practical chronometers—accurate clocks for use aboard ship—were designed, was it possible for most ships' pilots to determine longitude. Even the Greek astronomer Ptolemy (A.D. 90–168) made maps with relatively accurate positions north and south. But he overestimated the length of the Mediterranean by 50 percent, an error that was not corrected until 1700.

Each map in this book shows latitude and longitude (usually at 20° intervals) to indicate the positions of the map features (see Fig. 2-6). The parallels of latitude may also be used to determine approximate distance on a map. Each degree of latitude equals approximately 60 nautical miles (69 statute miles or 111 kilometers). Each minute of latitude is approximately 1 nautical mile, or 1.85 kilometers. At the equator, each minute of longitude is 1 nautical mile, but decreases so that 1 minute of longitude is only 0.5 nautical miles at 60° north or south latitude and vanishes at the poles.

MAPS AND MAP PROJECTIONS

A *map* is a flat representation of the earth's surface. Symbols are used to depict surface features. Since the earth is a sphere, making a flat map distorts the shape or size of surface features. The only distortion-free map is a globe, but a globe is not practical for the study of relatively small areas, so maps are used almost exclusively in science.

In making a map, we would like to make the final product as useful as possible. In general, we would like a map to preserve the following properties of the earth's surface:

Equal area—each area on the map should be proportional to the area of the earth's surface it represents.
Shape—the general outlines of a large area shown on a map should approximate as nearly as possible the shape of the region portrayed. A map which preserves shape is said to be *conformal*.
Distance—a perfect map would permit distance to be measured accurately between any two points anywhere on the map. Many common maps, such as the Mercator projection, do not accurately portray distances in a simple way
Direction—ideally, it would be possible to measure directions accurately anywhere on a map.

No map has all these properties; only a globe preserves size, shape, direction, and distance simultaneously.

A *map projection* takes the grid of latitude and longitude lines from a sphere and converts them into a grid on a flat surface. Sometimes, the resulting grid is a simple rectangular one where longitude and latitude lines intersect at right angles. In other projections, latitude and longitude lines are complex curves which intersect at various angles.

TABLE A3-1

Characteristics of Various Map Projections

NAME OF PROJECTION (type)	DISTINCTIVE FEATURES	DESIRABLE FEATURES (UNDESIRABLE FEATURES)	USES
Mercator (cylindrical)	Horizontal parallels Vertical meridians	Compass directions are straight lines (Extreme distortion at high latitudes)	Navigation
Goode homolosine projection	Horizontal parallels Characteristic interruptions of outline	Equal area Little distortion of shape (Interruption of either continents or oceans)	Data presentation
Hoelzel's planisphere (modified)	Horizontal parallels Characteristic nearly oval outline	Shapes easily recognizable (Moderate scale distortion at high latitudes)	Index map Data presentation

With the network formed from the grid lines, the map is drawn by plotting points in the appropriate spot on the new projection. In this way, the various types of maps are prepared. We shall consider only a few of the many map projections that have been developed to serve specific functions (listed in Table A3-1).

Without a doubt, the *Mercator projection* is most familiar. Parallels of latitude and meridians of longitude are all straight lines, and cross at right angles. The outline shape is a square or rectangle, as shown in Fig. A3-4.

Figure A3-4

A mercator projection. Compare the shape of Greenland shown on this map with that in Figs. A3-5 and A3-6 to see the distortion at high latitudes.

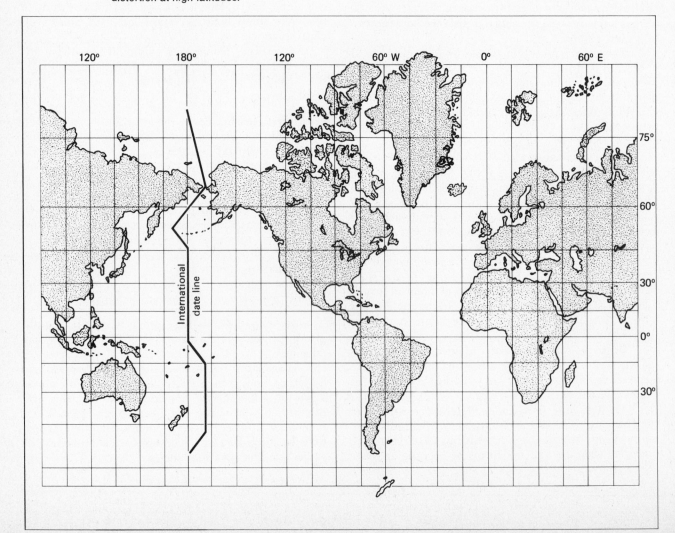

In its simplest form, the Mercator projection can be visualized as being made by a light inside a translucent globe projecting latitude and longitude onto a cylinder surrounding the globe. Even though the cylinder used for the projection is curved, it is easily made into the flat map desired.

The common Mercator projection in our example is most accurate within 15° of the equator and least accurate at the poles. Although shapes are well preserved by this projection, area is distorted, especially near the poles. For example, South America is in reality 9 times the size of Greenland, but this is not obvious from common Mercator projections. The scale of a Mercator projection changes going away from the equator. If the reader uses the length of a degree of latitude as his scale, he can avoid serious error.

Another property of the Mercator projection useful to mariners is that a course of constant compass direction (a *rhumb line*) is a straight line on this projection. While a rhumb course is not a great circle, and thus not the shortest distance between any two points on the earth's surface, it is useful for navigation because a great circle course requires a constant changing of direction. A rhumb line is slightly longer but is easier to navigate.

For world maps, the Mercator projection has distinct limitations; but for relatively small areas, such as navigation charts, the Mercator projection is without equal. Nearly all navigation charts used at sea are Mercator projections.

For use in this book, a *Hoelzel planisphere* (shown in Fig. A3-5) is used. This modified cylindrical projection shows the continents well, permitting most of the word's coastal regions to be easily recognized. Like the Mercator, this projection distorts areas near the poles. Instead of converging to a point at the poles, the merdians converge to a line which is only a fraction of the length of the equator. Because the projection used in this book splits the Pacific Ocean in the middle, it is not overly convenient for showing properties in the ocean.

A special projection (the *interrupted homolosine*, shown in Fig. A3-6) was developed by J. P. Goode in 1923 to show the ocean basins without any interruptions. In addition, the projection shows area equally. The continents are interrupted to show the ocean basins intact.

SELECTED REFERENCES

ALDRIDGE, B. G. 1968. *Mathematics for Physical Science*. Merrill Publishing Company, Columbus, Ohio. 137 pp. Elementary mathematical and data-plotting procedures.

COHEN, P. M. 1970. *Bathymetric Navigation and Charting*. United States Naval Institute, Annapolis, Md. 138 pp. Navigation through use of bathymetric data; nontechnical.

GREENWOOD, DAVID. 1964. *Mapping*. University of Chicago Press, Chicago. 289 pp. Elementary discussion of coordinates, maps, and map projections.

WILLIAMS, J. E., ed. 1963. *World Atlas*, 2nd ed. Prentice-Hall, Inc., Englewood Cliffs, N.J. 96, 41 pp. Includes graphical summary of different map projects.

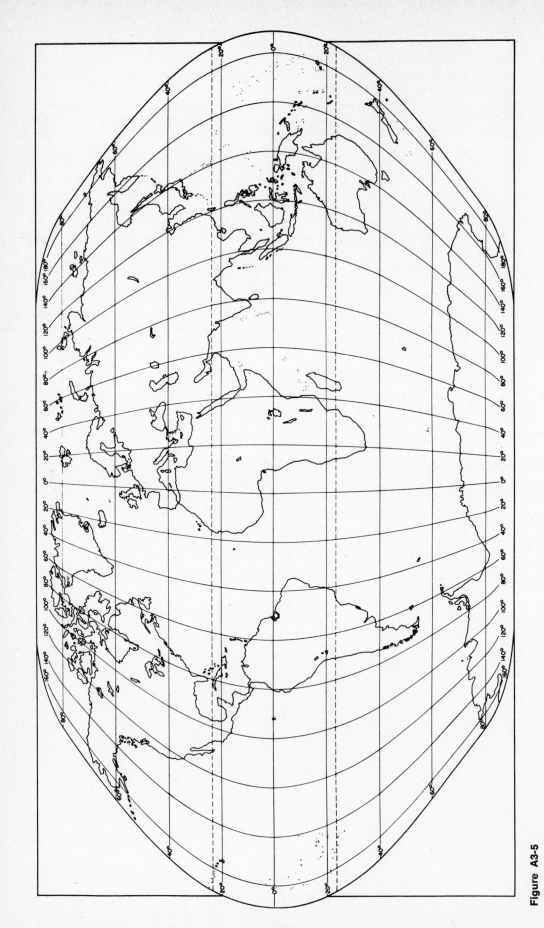

Figure A3-5

A Hoelzel planisphere. Note that the meridians of longitude converge to a line shorter than the equator but still not a point.

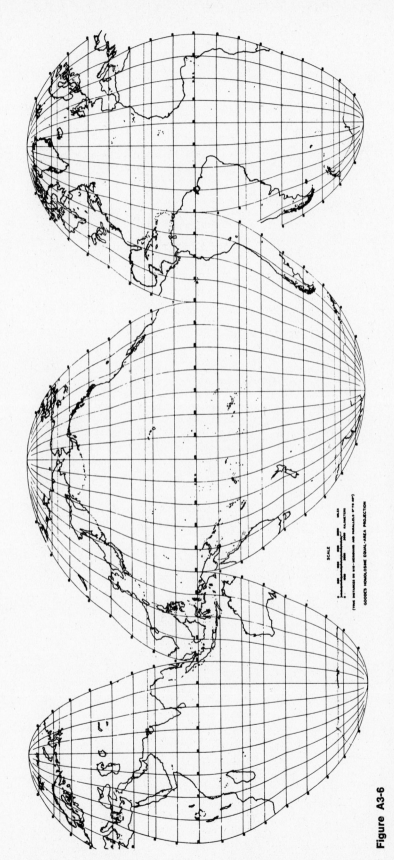

Figure A3-6

Goode's homolosine projection interrupted to show the oceans to best advantage.

APPENDIX

temperature-salinity diagrams

To study the origin, movement and mixing of ocean waters, oceanographers use temperature and salinity to identify individual water masses. Each water mass forms at the surface of the ocean and acquires a temperature and salinity that (usually) permits the source of the water to be determined. After it sinks below the surface zone, and as long as it remains isolated from the atmosphere, neither temperature nor salinity of a water mass can change appreciably unless it mixes with waters whose properties are different.

To study such mixing and to identify water masses, a *temperature-salinity (T-S) diagram* is constructed. Points are plotted to correspond to the measured temperatures and salinities of individual water samples. If conditions are constant and a water mass is homogeneous, the entire mass is represented by a single point on a *T-S* diagram. More commonly, there is some variation within the mass, as well as mixing with adjacent masses, resulting in a *T-S curve* that indicates a continuum of properties within the water column at each station (Fig. A4-1). This *T-S* curve remains the same (or within a small range of difference) for each location at any point in time, because the water masses present have formed in a constant-source area and have the same properties when they reach any given station. Similar curves are often characteristic of large ocean areas.

Consider the two water masses in Fig. A4-1(a), each with its characteristic temperature and salinity. As they mix, the newly formed waters will have temperatures and salinities that plot on the line connect-

ing the two original points. Furthermore, a particular position on this line indicates the relative proportions of the waters that have been mixed. For example, a mixture of equal amounts of two different waters will fall midway between them on the line connecting the two points on the *T-S* diagram.

Adding a third water mass results in a more complicated diagram, but the same general relationship holds. From analyses of *T-S* curves it is possible to determine the sources of the water masses present at that station. Figure A4-2 shows diagrammatically how mixing of subsurface water layers might appear on a *T-S* diagram. There is a characteristic reversal point associated with the core (central depth) of water mass *B*. This can be distinguished until the mass has completely lost its identity, although it becomes less and less obvious as mixing progresses.

Each major ocean area has a set of distinctive *T-S* curves. A set of curves is shown for the North Atlantic in Fig. A4-3.

The *T-S* diagram serves other functions besides the tracing of water masses. It permits ready identification of errors in a set of observations. A plot of temperature and salinity observations for a given area should show water density increasing with depth, in a regular progression or sequential order. Points that fall far out of line on the diagram indicate a possible inaccuracy in equipment function or data-reduction technique. Internal-consistency criteria among curves permit a simple graphic check on the accuracy of oceanographic observations over a wide area.

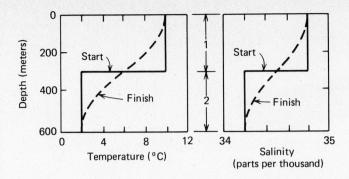

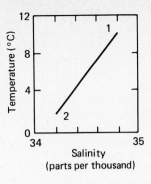

(a) Mixing of two homogeneous water bodies

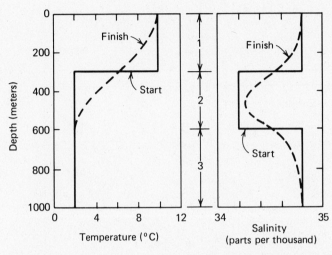

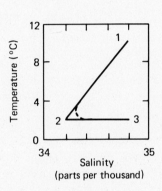

(b) Mixing of three homogeneous water bodies

Figure A4-1

Temperature–salinity relationships resulting from mixing of water masses:
(a) Water mass 1 (0 to 300 meters of depth; $T = 10°C$, $S = 34.8 \, ^0/_{00}$)
mixes with water mass 2 (300 to 600 meters depth; $T = 2°C$, $S = 34.2 \, ^0/_{00}$).
Initially, the boundary between the two masses is sharp and represented by sudden,
large changes in temperature and salinity with depth. After mixing, the boundary is
more diffuse and the changes in temperature and salinity occur over a depth range of
nearly 600 meters. On the T–S diagram, mixing of these two water masses is shown by the
straight line connecting the T–S points characteristic of the two original water masses.
(b) Mixing of three water masses can be represented in a similar manner.
Water mass 3 (600 to 1500 meters depth; $T = 2°C$, $S = 34.8 \, ^0/_{00}$) mixes with
water mass 2 but not with water mass 1. The initial sharp boundaries between masses are
gradually obliterated. On the T–S diagram, the mixing process is shown by two straight lines
indicating that water masses 1 and 2 mix, and that water masses 2 and 3 mix.
The lines meet at a point representing water mass 2, now altered by mixing with water
above and below so that its original identity has been lost and
it can no longer be represented by a single point on the T–S diagram.

Use of a *T–S* diagram to study subsurface water movements. Water mass *B* (intermediate density) flows beneath surface layer *A* and above the denser water mass *C*. A series of oceanographic stations taken at three locations show different but related *T–S* curves. At station 1, closest to the source of *B*, its characteristic temperature and salinity plot at point *B* on the *T–S* diagram. Farther from the source, the water mass mixes with those above and below, so it no longer has its original value although the general shape of the *T–S* diagram is still quite discernible. At station 3, farthest from the source, water mass *B* is no longer recognizable and masses *A* and *C* mix together, resulting in a continuum of *T–S* values intermediate between the two.

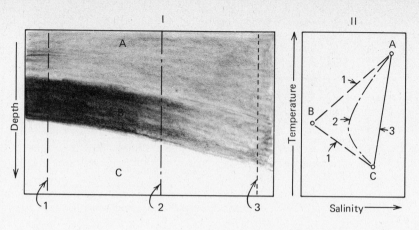

Figure A4-3

Temperature–salinity curves for major water masses in the Atlantic. (Data from Sverdrup et al., 1942.)

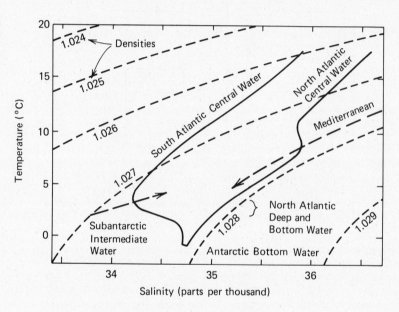

APPENDIX

glossary

* In part after B. B. Baker, W. R. Deebel, and R. D. Geisenderfer (eds.), *Glossary of Oceanographic Terms,* 2nd ed., U.S. Naval Oceanographic Office, Washington, D.C. (1966); and American Geological Institute, *Glossary of Geology and Related Sciences* and supplement, 2nd ed., Washington, D.C. (1960).

Abyssal pertaining to the great depths of the ocean, generally below 3700 meters.

Albedo the ratio of radiation reflected by a body to the amount incident upon it, commonly expressed as a percentage.

Algae marine or freshwater plants, including phytoplankton and seaweeds.

Algal ridge the elevated margin of a windward reef built by actively growing calcareous algae.

Alternation of generations mode of development, characteristic of many coelenterates, in which a sexually reproducing generation gives rise (by union of egg and sperm) to an asexually reproducing form, from which new individuals arise by budding or simple division of the 'parent' animal.

Amphidromic point the center of an amphidromic system; a nodal or no-tide point around which the crest of a standing wave rotates once in each tidal period.

Anaerobic condition in which oxygen is excluded, with the result that organisms which depend on the presence of oxygen cannot survive. Anaerobic bacteria can live under these conditions.

Andesite a type of volcanic rock intermediate in composition between granite and basalt, associated with partial melting of crust and mantle during subduction in the presence of some water. Andesitic mountain ranges are often associated with subduction zones.

Anion a negatively charged atom or radical.

Anoxic devoid of dissolved oxygen. See *Anaerobic*.

Antinode that part of a standing wave where the vertical motion is greatest and the horizontal velocities are least.

Aphotic zone that portion of the ocean where light is insufficient for plants to carry on photosynthesis.

Aquaculture the cultivation or propagation of water-dwelling organisms.

Archipelagic plain a gently sloping sea floor with a generally smooth surface, particularly found among groups of islands or seamounts.

Arrow-worm see *Chaetognath*.

Arthropods animals with a segmented external skeleton of chitin or plates of calcium carbonate, and with jointed appendages—for example, a crab or an insect.

Asthenosphere upper zone of the earth's mantle, extending from the base of the lithosphere to depths of about 250 kilometers beneath continents and ocean basins; relatively weak, probably partially molten.

Atoll a ring-shaped organic reef that encloses a lagoon in which there is no pre-existing land and which is surrounded by the open sea. Low sand islands may occur on the reef.

ATP adenosine triphosphate, a compound involved in energy transfer in metabolism, mainly formed during cell respiration.

Autotrophic nutrition that process by which an organism manufactures its own food from inorganic compounds.

Azoic devoid of life.

Backshore that part of a beach which is usually dry, being reached only by the highest tides; by extension, a narrow strip of relatively flat coast bordering the sea.

Baleen (whalebone) horny material growing down from the upper jaw of large plankton-feeding (baleen) whales, which forms a strainer or filtering organ consisting of numerous plates with fringed edges.

Bank an elevation of the sea floor of a large area; a submerged plateau.

Bar an offshore ridge or mound which is submerged (at least at high tide), especially at the mouth of a river or estuary.

Barrier beach a bar parallel to the shore, whose crest rises above high water.

Barrier island a detached portion of a barrier beach between two inlets.

Barrier reef a reef that is separated from a land mass by a lagoon, usually connected to the sea through passes (openings) in the reef.

Basalt a fine-grained igneous rock, black or greenish-black, rich in iron, magnesium, and calcium.

Basin a depression of the sea floor more or less equidimensional in form and of variable extent.

Bathythermograph a device for obtaining a record of temperature at various depths in the ocean, usually within a few hundred meters of the ocean surface.

Baymouth bar a bar extending partially or entirely across the mouth of a bay.

Beach seaward limit of the shore (limits are marked approximately by the highest and lowest water levels).

Beach cusp one of a series of low mounds of beach material separated by troughs spaced at more or less regular intervals along the beach face.

Beach drift see *Littoral drift.*

Beach face this term is sometimes used to describe the highest portion of the foreshore, above the low-tide terrace and exposed only to the action of wave uprush, not to the rise and fall of the tide.

Bed load see *Load.*

Benioff zone see *Subduction zone.*

Benthic that portion of the marine environment inhabited by marine organisms which live permanently in or on the bottom.

Benthos bottom-dwelling marine organisms.

Berm the low, nearly horizontal portion of a beach (backshore), having an abrupt fall and formed by the deposit of material by wave action. It marks the limit of ordinary high tides and waves.

Berm crest the seaward limit of a berm.

Bight a concavity in the coastline which forms a large, open bay.

Biogenous sediment sediment consisting of at least 30 percent by volume of particles derived from skeletal remains of organisms, which may contribute calcium carbonate, silica, or phosphate minerals to the sediment.

Biogeochemical cycles the paths by which elements essential to life circulate from the nonliving environment to living organisms and back again to the environment.

Bioluminescence the production of light without sensible heat by living organisms as a result of a chemical reaction either within certain cells or organs or in some form of secretion.

Biomass the amount of living matter per unit of water surface or volume expressed in weight units.

Bird-foot delta a delta formed by the outgrowth of pairs of natural levees, formed by the distributaries, making the digitate or bird-foot form.

Bivalves mollusks, generally sessile or burrowing into soft sediment, rock, wood, or other materials. Individuals possess a hinged shell and a hatchet-shaped foot, which is sometimes used in digging. Clams, oysters, and mussels belong to this group.

Black bottle (dark bottle) a container used in measuring respiratory activity of primary producers. The container is covered or coated to exclude light and thereby prevent photosynthetic activity.

Black mud (hydrogen sulfide mud) a dark, fine-grained sediment formed in poorly aerated bays, lagoons, and fjords. This sediment usually contains large quantities of decaying organic matter and iron sulfides and may exude hydrogen sulfide gas.

Bloom see *Plankton bloom.*

Bore see *Tidal bore.*

Bottom water the water mass at the deepest part of the water column. It is densest water that is permitted to occupy that position by the regional bottom topography.

Boundary currents northward- or southward-directed surface currents which flow parallel and close to the continental margins. They are caused by deflection of the prevailing eastward- and westward-flowing currents by the continental land masses.

Brackish water water in which salinity values range from approximately 0.05 to 17.00 parts per thousand.

Breaker a wave breaking on the shore, over a reef, etc. Breakers may be roughly classified into four kinds, although the categories may overlap: *spilling breakers* break gradually over a considerable distance; *plunging breakers* tend to curl over and break with a crash; *surging breakers* peak up, but then instead of spilling or plunging they surge up on the beach face; *collapsing breakers* break in the middle or near the bottom of the wave, rather than at the top.

Breakwater a structure, usually rock or concrete, protecting a shore area, harbor, anchorage, or basin from waves.

Brine water containing a higher concentration of dissolved salt than that of the ordinary ocean. Brine can be produced by the evaporation or freezing of seawater.

Calcareous algae marine plants which form a hard external covering of calcium compounds.

Calorie the amount of heat required to raise the temperature of 1 gram of water by 1°C (defined on basis of water's *specific heat*).

Canyon (submarine) see *Submarine canyon.*

Capillary wave a wave in which the primary restoring force is surface tension. A water wave in which the wavelength is less than 1.7 centimeters is considered to be a capillary wave.

Cation a positively charged ion, tending to be deposited at the cathode in electrolysis or in a battery.

Celsius temperature temperatures based on a scale in which water freezes at 0° and boils at 100° (at standard atmospheric pressure); also called *centigrade temperature.*

Cephalopods benthic or swimming mollusks possessing a large head, large eyes, and a circle of arms or tentacles around the mouth; the shell is external, internal, or absent, and an ink sac usually is present. They include squids and octopus.

Chaetognath small, elongate, transparent, wormlike animals, pelagic in all seas from the surface to great depths. They are abundant and may multiply rapidly into vast swarms. Some, especially *Sagitta,* have been identified as indicator species.

Change of state a change in the physical form of a substance, as when a liquid changes to a solid due to cooling.

Chemosynthesis carbon dioxide fixation (primary production) by certain bacteria in the absence of sunlight, using inorganic compounds such as ammonia or methane, or elements such as reduced iron, hydrogen, or sulfur.

Chitin a nitrogenous carbohydrate derivative forming the skeletal substance in arthropods.

Chloride an atom of chlorine in solution, bearing a single negative charge.

Chlorinity a measure of the chloride content, by mass, of seawater (grams per kilogram, or per mille, of seawater).

Chlorophyll a group of green pigments which occur chiefly in bodies called chloroplasts and which are active in photosynthesis.

Cilia hairlike processes of cells, which beat rhythmically and cause locomotion of the cells or produce currents in water.

Clay soil consisting of inorganic material, the grains of which have diameters smaller than 0.005 millimeter (or 5 microns).

Climate the prevalent or characteristic meteorological conditions of a place or region—in contrast with *weather,* which is the state of the atmosphere at a particular time.

Climax community end-product of ecological succession in which there is a balance of production and consumption in the community and further change in species composition takes place only as a result of changes in environmental conditions or outside forces influencing the community.

Clupeoids fish of the herring family, including sardine, anchovy, pilchard, and menhaden.

Coastal currents currents flowing roughly parallel to the shore and constituting a relatively uniform drift in the water just seaward of the surf zone. These currents may be caused by tides or winds, or associated with the distribution of mass in coastal waters.

Coastal ocean shallow portion of the ocean (in general, overlying the continental shelf), where the circulation pattern is strongly influenced by the bordering land.

Coastal plain low-relief continental plain, adjacent to the ocean and extending inward to the first major change in terrain features.

Cobble a rock fragment between 64 and 256 millimeters in diameter, larger than a pebble and smaller than a boulder; usually rounded or otherwise abraded.

Coccolithophores microscopic, often abundant planktonic algae, the cells of which are surrounded by an envelope on which numerous small calcareous discs or rings (coccoliths) are embedded.

Coelenterates a large, diverse group of simple animals possessing two cell layers and a digestive cavity with only one opening. This opening is surrounded by tentacles containing stinging cells. Some are sessile, some pelagic (medusae), and some undergo alternation of generations.

Compaction the decrease in volume or thickness of a sediment under load through closer crowding of constituent particles and accompanied by decrease in porosity, increase in density, and squeezing out of water.

Compensation depth   the depth at which oxygen production by photosynthesis equals that consumed by plant respiration during a 24-hour period.

Compressional wave   a wave which causes alternate compression and expansion of the medium through which it passes.

Conduction   transfer of energy through matter by internal particle or molecular motion without any external motion.

Conformal projection   a map projection in which the angles around any point are correctly represented —for example, a Mercator projection.

Conservative property   a property whose values do not change in the course of a particular, specified series of events or processes—for example, those properties of seawater, such as salinity, the concentrations of which are not affected by the presence or activity of living organisms, but which are affected by diffusion and currents.

Continental block   a large land mass rising abruptly from the deep ocean floor, including marginal regions that are shallowly submerged.

Continental climate   a climate characterized by cold winters and hot summers in areas where the prevailing winds have traveled across large land areas.

Continental crust   the thickened part of the crust forming continental blocks; typically, about 35 kilometers thick, consisting primarily of granitic rocks.

Continental margin   a zone separating the emergent continents from the deep-sea bottom; generally consists of the continental shelf, slope, and rise.

Continental rise   a gentle slope with a generally smooth surface, rising toward the foot of the continental slope.

Continental shelf   the sea floor adjacent to a continent, extending from the low-water line to the change-in-slope, usually about 180 meters depth, where continental shelf and continental slope join.

Continental slope   a declivity from the outer edge of of a continental shelf, extending from the break in slope to the deep sea floor.

Contour line   a line on a chart connecting points of equal value above or below a reference value; can portray elevation, temperature, salinity or other values.

Convection   vertical circulation resulting from density differences within a fluid, resulting in transport and mixing of the properties of that fluid.

Convergence   area or zone where flow regimes come together or converge, usually resulting in sinking of surface water.

Convergence zone   a band along which crustal area is lost. Colliding edges of crustal plates may be thickened, folded, or underthrust, or one plate may be subducted and destroyed.

Copepods   minute shrimplike crustaceans, most species of which range between about 0.5 and 10 millimeters in length.

Coral   the hard, calcareous skeleton of various sessile, colonial coelenterate animals, or the stony solidified mass of a number of such skeletons; also the entire animal, a compound polyp which produces the skeleton.

Coral reef   a complex ecological association of bottom-living and attached calcareous, shelled marine invertebrates forming fringing reefs, barrier reefs, or atolls.

Coralline algae   red algae, having either a bushy or encrusting form and deposits of calcium carbonate either on the branches or as a crust on the substrate. Certain of them develop massive encrustations on coral reefs.

Core   a vertical, cylindrical sample of sediments from which the nature and stratification of the bottom may be determined; also, the central zone of the earth.

Coriolis effect   an apparent force acting on moving particles resulting from the earth's rotation. It causes moving particles to be deflected to the right in the Northern Hemisphere and to the left in the Southern Hemisphere; the deflection is proportional to the speed and latitude of the moving particle. The speed of the particle is unchanged by the apparent deflection.

Crinoids   a group of echinoderms, most of which are attached by a long stalk to the bottom either permanently or when immature; species without stalks either swim or creep slowly about. Crinoids (such as sea-lily, feather star) occur in shallow water as well as at great depths.

Critical depth   the depth above which occurs the net effective plant production in the water column as a whole.

Crust of the earth   the outer shell of the solid earth. Beneath the oceans, the outermost layer of crust is composed of unconsolidated sediment, the second layer of consolidated sediment and weathered lavas, and the inside layer probably of basaltic rocks. The crust varies in thickness from approximately 5 to 7 kilometers under the ocean basins to 35 kilometers under the continents. Its lower limit is defined to be the Mohorovičić discontinuity.

Crustaceans   arthropods which breathe by means of gills or similar structures. The body is commonly covered by a hard shell or crust. The group includes barnacles, crabs, shrimps, and lobsters.

Ctenophores spherical, pear-shaped, or cylindrical animals of jellylike consistency ranging from less than 2 centimeters to about 1 meter in length. The outer surface of the body bears 8 rows of comblike structures.

Current ellipse a graphic representation of a rotary current in which the speed and direction of the current at different hours of the tide cycle are represented by vectors joined at one point. A line joining the extremities of the radius vectors forms a curve roughly approximating an ellipse.

Current meter any device for measuring and indicating speed or direction (usually both) of flowing water.

Current rose a graphic representation of currents for specified areas, utilizing arrows to show the direction toward which the prevailing current flows and the frequency (expressed as a percentage) of any given direction of flow.

Current tables tables which give daily predictions of the times, speeds, and directions of the currents.

Cypris larva the stage at which the young of barnacles attach to the substrate.

Daily (diurnal) inequality the difference in heights and durations of the two successive high waters or of the two successive low waters of each day; also, the difference in speed and direction of the two flood currents or the two ebb currents of each day.

Daily (diurnal) tide a tide having only one high water and one low water each tidal day.

Debris line a line near the limit of storm-wave uprush marking the landward limit of debris deposits.

Decomposers heterotrophic and chemo-autotrophic organisms (chiefly bacteria and fungi) which break down nonliving matter, absorb some of the decomposition products, and release compounds usable in primary production.

Deep scattering layer a stratified population of organisms in ocean waters which causes scattering of sound as recorded on an echo sounder. Such layers may be from 50 to 200 meters thick. They occur less than 200 meters below the surface at night and several hundred meters below the surface during the day.

Deep-water waves waves in water the depth of which is greater than one-half the average wavelength.

Delta an alluvial deposit, roughly triangular or digitate in shape, formed at the mouth of a stream of tidal inlet or a river.

Demersal fishes those living near the bottom.

Density the mass per unit volume of a substance, usually expressed in grams per cubic centimeter.

In centimeter–gram–second system, density is numerically equivalent to *specific gravity*.

Density current the flow (caused by density differences or gravity) of one current through, under, or over another; it retains its unmixed identity from the surrounding water because of density differences.

Deposit-feeding removal of edible material from sediment or detritus, either by ingesting material unselectively and excreting the unusable portion, or selectively by ingesting discrete particles.

Depth of no motion the depth at which water is assumed to be motionless, used as a reference surface for computing geostrophic currents.

Desalination removal of water from seawater or brine.

Detritus any loose material produced directly from rock disintegration. *Organic detritus* consists of decomposition or disintegration products or dead organisms, including fecal material.

Diatoms microscopic phytoplankton organisms, possessing a wall of overlapping halves (valves) impregnated with silica.

Diffusion transfer of material (e.g., salt) or of a property (e.g., temperature) by eddies or molecular movement. Diffusion causes spreading or scattering of matter under the influence of a concentration gradient, with movement from the stronger to the weaker solution.

Dinoflagellates microscopic or minute organisms, which may possess characteristics of both plants (such as chlorophyll and cellulose plates) and animals (such as ingestion of food).

Discontinuity an abrupt change in a property, such as salinity or temperature, at a line or surface.

Dispersion the separation of a complex, surface gravity-wave disturbance into its component parts. Longer component parts of the wave travel faster than shorter ones and arrive at their destination first, as swell.

Distributary an outflowing branch of a river, which occurs characteristically on a delta.

Diurnal daily, especially pertaining to actions which are completed within approximately 24 hours and which recur every 24 hours.

Divergence horizontal flow of water in different directions from a common center or zone; upwelling is a common example of divergence.

Divergence zone a region along which crustal plates move apart and new lithospheric material solidifies from rising volcanic magma.

Doldrums a nautical term describing a belt of light, variable winds near the equator; an area of low atmospheric pressure.

Drift bottle   a bottle released at sea for use in studying surface currents. It contains a card identifying the date and place of release and requesting the finder to return it, noting the date and place of recovery.

Drift current   slow-moving current, principally caused by wind, usually restricted to the waters above the pycnocline.

Drift net   fishing net suspended in the water vertically so that drifting or swimming animals may be trapped or entangled in the mesh.

Dynamic topography   the configuration formed by the geopotential difference (dynamic height) between a given surface and a reference surface, usually the layer of no motion. A contour map of dynamic "topography" may be used to estimate geostrophic currents.

Ebb current   the movement of a tidal current away from shore or down a tidal stream.

Echinoderms   principally benthic marine animals having calcareous plates with projecting spines forming a rigid or articulated skeleton or plates and spines embedded in the skin. They have radially symmetrical, usually five-rayed, bodies. They include the starfish, sea urchins, crinoids, and sea-cucumbers.

Echiuroids   unsegmented, burrowing marine worms.

Echo sounding   determination of the depth of water by measuring the time interval between emission of a sonic or ultrasonic signal and the return of its echo from the bottom. The instrument used for this purpose is called an *echo sounder*.

Ecological efficiency   ratio of the efficiency with which energy is transferred from one trophic level to to next.

Ecosystem   ecological unit including both organisms and the nonliving environment, each influencing the properties of the other and both necessary for the maintenance of life as it exists on the earth.

Eddy   a current of air, water, or any fluid, often on the side of a main current; especially one moving in a circle, in extreme cases, a whirlpool.

Eddy viscosity   the turbulent transfer of momentum by eddies giving rise to an internal fluid friction, in a manner analogous to the action of molecular viscosity in laminar flow, but taking place on a much larger scale.

Eelgrass   a seed-bearing, grasslike marine plant which grows chiefly in sand or mud-sand bottoms and most abundantly in temperate water less than 10 meters deep.

Ekman spiral   a theoretical representation of the currents resulting from a steady wind blowing across an ocean having unlimited depth and extent and uniform viscosity. Specifically, it would cause the surface layer to drift on an angle of 45° to the right of the wind direction in the Northern Hemisphere; water at successive depths would drift in directions more to the right until, at some depth, the water would move in the direction opposite to the wind. Speed decreases with depth throughout the spiral. The net water transport is 90° to the right of the wind in the Northern Hemisphere.

Encrusting algae   see *Coralline algae, Red algae*.

Epifauna   animals which live at the water–substrate interface, either attached to the bottom or moving freely over it.

Epiphytes   plants which grow attached to other plants.

Equal-area projection   a map projection in which equal areas on the earth's surface are represented by equal areas on the map.

Equilibrium tide   the hypothetical semidaily tide caused by gravitational attraction of the sun and moon on a frictionless, nonrotating entirely water-covered earth.

Estuary   a semienclosed, tidal coastal body of saline water with free connection to the sea, commonly the lower end of a river.

Euphausids   shrimplike, planktonic crustaceans, widely distributed in oceanic and coastal waters, and especially in colder waters. They grow to 8 centimeters in length and many possess luminous organs.

Euphotic zone   see *Photic zone*.

Fan   a gently sloping, fan-shaped feature normally located near the lower termination of a canyon.

Fault   a fracture or fracture zone in rock along which one side has been displaced relative to the other.

Fauna   the animal population of a particular location, region, or period.

Fermentation   a type of respiration in which partial oxidation of organic compounds takes place in the absence of free oxygen, for example the oxidation of glucose to alcohol by yeast.

Fetch   an area of the sea surface over which seas are generated by a wind having a constant direction and speed, also called *generating area*; also, the length of the fetch area, measured in the direction of the wind in which the seas are generated.

Filter-feeding   filtering or trapping edible particles from seawater; a feeding mode typical of many zooplankters and other marine organisms of limited mobility.

Fjord a narrow, deep, steep-walled inlet of the ocean, formed either by the submergence of a mountainous coast or by entrance of the ocean into a deeply excavated glacial trough after the melting away of the glacier.

Flagellum a whiplike bit (process) of protoplasm which provides locomotion for a motile cell.

Flocculation process of aggregation into small lumps, especially with regard to soils and colloids.

Floe sea ice, either as a single unbroken piece or as many individual pieces covering an area of water.

Flood current the tidal current associated with the increase in the height of a tide. Flood currents generally set toward the shore, or in the direction of the tide progression; sometimes erroneously called *flood tide*.

Flora the plant population of a particular location, region, or period.

Food web the interrelated food relationships in an ecosystem, including its production, consumption, and decomposition, and the energy relationships among the organisms involved in the cycle.

Foraminifera benthic or planktonic protozoans possessing shells, usually of calcium carbonate.

Forced wave a wave generated and maintained by a continuous force, in contrast to a free wave that continues to exist after the generating force has ceased to act.

Fore-reef the upper seaward face of a reef, extending above the lowest point of abundant living coral and coralline algae to the reef crest. This zone commonly includes a shelf, bench, or terrace that slopes to 15–30 meters, as well as the living, wave-breaking face of the reef. The terrace is an eroded surface or is veneered with organic growth. The living reef front above the terrace in some places is smooth and steep; in other places it is cut up by grooves separated by ridges that together have been called *spur-and-groove systems*, forming comb-tooth patterns.

Forerunner low, long-period swell which commonly precedes the main swell from a distant storm, especially a tropical cyclone.

Foreshore see *Low-tide terrace*.

Foul to attach to or come to lie on the surface of submerged man-made or introduced objects, usually in large numbers or amounts, as barnacles on the hull of a ship or silt on a stationary object.

Fracture zone an extensive linear zone of unusually irregular topography of the ocean floor characterized by large seamounts, steepsided or asymmetrical ridges, troughs, or long, steep slopes.

Free wave any wave not acted upon by any external force except for the initial force that created it.

Fringing reef a reef attached directly to the shore of an island or continental land mass. Its outer margin is submerged and often consists of algal limestone, coral rock, and living coral.

Fully developed sea the maximum height to which ocean waves can be generated by a given wind force blowing over sufficient fetch, regardless of duration, as a result of all possible wave components in the spectrum being present with their maximum amount of spectral energy.

Gastropods mollusks that possess a distinct head, generally with eyes and tentacles, and a broad, flat foot, and which are usually enclosed in a spiral shell.

Geostrophic current a current resulting from the balance between gravitational forces and the Coriolis effect.

Geothermal power power derived from heat energy that flows from the earth's crust, especially in volcanic areas.

Giant squids large cephalopods, (length may be 15 meters or more) which inhabit mid-depths in oceanic regions but may come to the surface at night. They are eaten by sperm whales.

Gill a delicate, thin-walled structure, often an extension of the body wall, used for the exchange of gases in a water environment, and sometimes for excretion of wastes.

Glass-worm see *Chaetognath*.

Glacial marine sediment high-latitude, deep-ocean sediments that have been transported from the land to the ocean by means of glaciers or icebergs.

Glacier a mass of land ice, formed by recrystallization of old, compacted snow, flowing slowly from an area of accumulation to an area of sublimation, melting, or evaporation, where snow or ice are removed from the glacier surface.

Grab an instrument possessing jaws that seize a portion of the ocean bottom for retrieval and study.

Graded bedding a type of stratification in which each stratum displays a gradation in grain size from coarse below to fine above.

Gradient the rate of decrease (or increase) of one quantity with respect to another—for example, the rate of decrease of temperature with depth.

Granite a crystalline rock consisting essentially of alkali feldspar and quartz; in seismology, a rock in which the compressional-wave velocity varies approximately between 5.5 and 6.2 kilometers per second. *Granitic* is a textural term applied to coarse and medium-grained igneous rocks.

Gravel loose material ranging in size from 2 to 256 millimeters.

Gravity anomaly a disturbance in the earth's normal gravity field.

Gravity wave a wave whose velocity of propagation is controlled primarily by gravity. Water waves of length greater than 1.7 centimeters are considered gravity waves.

Great circle the intersection of the surface of a sphere and a plane through its center—for example, meridians of longtiude and the equator are great circles on the earth's surface.

Greenhouse effect the result of the penetration of the atmosphere by comparatively short-wavelength solar radiation which is largely absorbed near and at the earth's surface, whereas the relatively long-wavelength radiation emitted by the earth is partially absorbed by water vapor, carbon dioxide, and dust in the atmosphere, thus warming the atmosphere.

Groin a low artificial damlike structure of durable material placed so that it extends seaward from the land; used to slow littoral drift.

Group velocity the velocity with which a wave group travels. In deep water, it is equal to one-half the individual wave velocity.

Guyot a flat-topped submarine mountain or seamount.

Gyre a circular or spiral form, usually applied to a semiclosed current system.

Hadley cell a semiclosed system of vertical motions in the earth's atmosphere. Warm, moist air rises in equatorial regions, flows to mid-latitudes (30°N, 30°S), where it sinks and returns to the equatorial zone as the trade winds.

Half-life the time required for the decay of one-half the atoms of a sample of a radioactive substance. Each radionuclide has an unique half-life, related to its rate of disintegration.

Halocline water layer with large vertical changes in salinity.

Headland a high, steep-faced point of land extending into the sea.

Heat budget the accounting for the total amount of the sun's heat received on the earth during any one year as being exactly equal to the total amount which is lost from the earth to radiation and reflection.

Heat capacity amount of heat required to raise the temperature of a substance by a given amount.

Herbivore an animal that feeds only upon plant matter.

Heterotrophic nutrition that process by which an organism utilizes preformed organic compounds for its food.

High water the upper limit of the surface water level reached by the rising tide; also called high tide.

Higher high water the higher of the two high waters of any tidal day. The single high water occurring daily in a diurnal tide is considered to be a higher high water.

Higher low water the higher of two low waters occurring during a tidal day.

Holoplankton organisms whose life-cycle is completely passed in the floating state.

Hurricane a cyclonic storm, usually of tropical origin, covering an extensive area, and containing winds of 120 kilometers per hour or higher.

Hydrogen bond the relatively weak bond formed between adjacent molecules in liquid water, resulting from the mutual attraction of hydrogen and nearby atoms.

Hydrogenous sediment particles precipitated from solution in water.

Hydrography the measurement, description, or mapping of surface waters.

Hydroid the polyp form of coelenterate animals which exhibit alternation of generations. It is attached, often branching, and gives rise to the pelagic, medusa form by asexual budding.

Hydrologic cycle the composite cycle of water exchange, including change of state and vertical and horizontal transport, of the interchange of water among earth, atmosphere, and ocean.

Hydrophilic having the property of attracting water.

Hydrophobic not attractive to water; unwettable.

Hydrosphere the water portion of the earth, as distinguished from the solid part and the gaseous outer atmosphere. It consists of liquid water in sedimentary rocks, rivers, lakes, and oceans, as well as ice in sea ice and continental ice sheets.

Hydrothermal associated with high temperature ground waters, for example the alteration or precipitation of minerals and mineral ores.

Hypsometric graph a representation of the respective elevations and depths of points on the earth's surface with reference to sea level.

Iceberg a large mass of detached land ice floating in the sea or stranded in shallow water.

Ice shelf a thick ice formation with a fairly level surface, formed along a polar coast and in shallow bays and inlets, where it is fastened to the shore and often reaches bottom.

Igneous rock formed by solidification of magma.

Infauna animals who live buried in soft sediment.

Insolation in general, solar radiation received at the earth's surface; also, the rate at which direct solar radiation is incident upon a unit horizontal surface at any point on or above the surface of the earth.

Instability a property of a system such that any disturbance grows larger, rather than diminishing, so that the system never returns to the original steady state; in oceanography, usually refers to the vertical displacement of a water parcel.

Interface surface separating two substances of different properties (such as different densities, salinities, or temperatures)—for example, the air–sea interface or the water–sediment interface.

Internal wave a wave that occurs within a fluid whose density changes with depth, either abruptly at a sharp surface of discontinuity (an interface) or gradually.

Interstitial water water contained in the pore spaces between the grains in rocks and sediments.

Intertidal zone (littoral zone) generally considered to be the zone between mean high-water and mean low-water levels.

Invertebrate animal lacking a backbone.

Ionic bond a linkage between two atoms, with a separation of electric charge on the two atoms; a linkage formed by the transfer or shift of electrons from one atom to another.

Iron-manganese nodules see *manganese nodules.*

Island-arc system a group of islands usually having a curving archlike pattern, generally convex toward the open ocean, with a deep trench or trough on the convex side and usually enclosing a deep-sea basin on the concave side. Volcanoes usually associated.

Isopods generally flattened marine crustaceans, mostly scavengers.

Isostasy the balance of portions of earth's crust, which rise or subside until their masses are in equilibrium relationship, "floating" on the denser mantle below. Isostatic theory implies that large continental masses are supported by "roots" of low-density crustal material.

Isotherm line connecting points of equal temperature.

Isotope one of several nuclides having the same number of protons in their nuclei, and hence belonging to the same element, but differing in the number of neutrons and therefore in mass number or energy content; also, a radionuclide or a preparation of an element with special isotopic composition, used principally as an isotopic tracer.

Jellyfish see *Medusa.*

Jet stream high-altitude swift air current.

Jetty in United States terminology, a structure built to influence tidal currents, to maintain channel depths, or protect the entrance to a harbor or river.

Knoll an elevation rising less than 1 kilometer from the ocean floor and of limited extent across the summit.

Lagoon a shallow sound, pond, or lake, generally separated from the open ocean.

Lamina a sediment or sedimentary-rock layer less than 1 centimeter thick visually separable from the material above and below. *Lamination* refers to the alternation of such layers differing in grain size or composition.

Laminar flow a flow in which the fluid moves smoothly in streamlines, in parallel layers, or sheets; a nonturbulent flow.

Langmuir circulation cellular circulation, with alternate left- and right-hand helical vortices, having axes in the direction of the wind, which is set up in the surface layer of a water body by winds exceeding 7 knots.

Lantern-fishes (myctophids) small oceanic fishes which normally live at depths between a few hundred and a few thousand meters depth and characteristically have numerous small light organs on the sides of the body. Many undergo diurnal vertical migration.

Latent heat the heat released or absorbed per unit mass by a system undergoing a reversible change of state at a constant temperature and pressure.

Lava fluid rock, issuing from a volcano or fissure in the earth's surface, or the same material solidified by cooling.

Layer of no motion see *Depth of no motion.*

Levee an embankment bordering one or both sides of a seachannel or delta distributary.

Lithogenous sediment sediment composed primarily of mineral grains transported into the oceans from the continents.

Lithosphere the outer, solid portion of the earth; includes the crust of the earth.

Littoral see *Intertidal zone.*

Littoral drift sand moved parallel to the shore by wave and current action.

Load the quantity of sediment transported by a current. It includes the *suspended load* of small particles which float in suspension distributed through the whole body of the current, and the bottom load or *bed load* of large particles which move along the bottom by rolling and sliding.

Longshore bar see *Bar.*

Longshore current a current located in the surf zone moving generally parallel to the shoreline; usually generated by waves breaking at an angle with the shoreline.

Low-tide terrace the zone that lies between the ordinary high- and low-water marks and is daily traversed by the oscillating water line as the tides rise and fall. This area, together wth the vertical scarp that often occurs at its upper limit, is sometimes called the *foreshore,* which ends at the highest point of normal wave uprush.

Low water the lowest limit of the surface-water level reached by the lowering tide; also called *low tide.*

Lower high water the lower of two high tides occurring during a tidal day.

Lower low water the lower of two low tides occurring during a tidal day.

Lunar tide that part of the tide caused solely by the gravitational attraction of the moon, as distinguished from that part caused by the gravitational attraction of the sun.

Magma mobile rock material, generated within the earth, capable of intrusion and extrusion, from which igneous rock solidifies.

Manganese nodules concretionary lumps of manganese and iron, containing copper and nickel; found widely scattered on the deep ocean floor.

Mantle the bulk of the earth, between crust and core, from about 40 to 3500 kilometers depth. Phase changes caused by increasing pressure with depth divide the mantle into concentric layers. Also the tough, protective membrane possessed by all mollusks, within which water circulates.

Map projection a method of representing part or all of the surface of a sphere, such as the earth, on a plane surface.

Marginal sea semienclosed body of water adjacent to, widely open to, and connected with the ocean at the water surface but bounded at depth by submarine ridges—for example, the Yellow Sea.

Mariculture see *Aquaculture.*

Marine humus partially or entirely undecomposed organic matter in ocean sediment.

Marine snow particles of organic detritus and living forms. The downward drift of these particles and living forms, especially in dense concentration, appears similar to a snowfall when viewed by underwater investigators.

Maritime climate climate characterized by relatively little seasonal change, warm moist winters, cool summers: result of prevailing winds blowing from ocean to land.

Marsh an area of soft, wet land. Flat land periodically flooded by salt water is called a *salt marsh.*

Maximum sustainable yield the maximum yield of a fishery that can be sustained for many years without steady depletion of the stock.

Meander a turn or winding of a current that may become detached from the main stream.

Mean sea level the average height of the surface of the sea for all stages of the tide over a 19-year period, usually determined from hourly readings of tidal height.

Mean tidal range the difference in height between mean high water and mean low water, measured in feet or meters.

Medusa (jellyfish) any of various free-swimming coelenterates having a disk- or bell-shaped body of jellylike consistency. Many have long tentacles with stinging cells.

Meiobenthos animals in the size range 100 to 500 microns that live between the grains in nearshore sediments, including many kinds of small invertebrates and larger protozoans.

Mélange a large-scale formation containing diverse rock materials that were originally mixed and consolidated at great pressure, characteristic of compression or subduction zones.

Meridian (of longitude) a great circle passing through the North and South Poles. It connects points with an equal angular separation from the prime meridian.

Meroplankton mainly organisms whose early developmental stages occur in the floating state; adults are benthic.

Metabolite substance necessary to survival and growth of an organism; this includes vitamins and other materials required only in trace amounts, as well as the nutrients required for synthesis of living tissue.

Metamorphic rocks formed in the solid state, below the earth's surface, in response to changes of temperature, pressure, or chemical environment.

Microbenthos one-celled plants, animals, and bacteria living in or on the surface of bottom sediments.

Mid-ocean ridge a great median arch or sea-bottom swell extending the length of an ocean basin and roughly paralleling the continental margins.

Mid-ocean rise see *Oceanic rise.*

Mixed tide type of tide in which a diurnal wave produces large inequalities in heights and/or durations of successive high and/or low waters. This term applies to the tides intermediate to those predominantly semidaily and those predominantly daily.

Mohorovičić discontinuity the sharp discontinuity in composition between the outer layer of the earth (crust) and the inner layer (mantle).

Molecular diffusion see *Diffusion.*

Mole formula weight of a substance in grams—for example, 1 mole of H_2O weighs 18 grams (oxygen = 16; 2 hydrogen = 2).

Molecular viscosity an internal resistance to flow in liquids arising from the attractive interactions between molecules.

Monsoons a name for seasonal winds (derived from the Arabic *mausim,* meaning "season"), first applied to the winds over the Arabian Sea, which flow for 6 months from the northeast and the remaining 6 months from the southwest, but subsequently extended to similar winds in other parts of the world.

Mud detrital material consisting mostly of silt and clay-sized particles (less than 0.06 millimeter) but often containing varying amounts of sand and/or organic materials. It is also a general term applied to any fine-grained sediment whose exact size classification has not been determined.

Mysids elongate crustaceans which usually are transparent (or nearly so) and benthic or deep-living.

Nannoplankton plankton whose length is less than 50 microns. Individuals will pass through most nets and usually are collected by centrifuging water samples.

Nansen bottle a device used by oceanographers to obtain subsurface samples of seawater. The "bottle" is lowered by wire; its valves are open at both ends. It is then closed *in situ* by allowing a weight ("messenger") to slide down the wire and strike the reversing mechanism. This causes the bottle to turn upside down, closing the valves and reversing the thermometers. Usually a series of bottles is lowered, and then the reversal of each bottle releases another messenger to actuate the bottle beneath it.

Natural frequency the characteristic frequency (number of vibrations or oscillations per unit time) of a body controlled by its physical characteristics (dimensions, density, etc.).

Nauplius a limb-bearing early larval stage of many crustaceans.

Neap tide lowest range of the tide, occurring near the times of the first and last quarter of the moon.

Nekton those pelagic animals that are active swimmers, such as most of the adult squids, fishes, and marine mammals.

Net primary production the total amount of organic matter produced by photosynthesis minus the amount consumed by the photosynthetic organisms in their own respiratory processes.

Nitrogen fixation conversion of atmospheric nitrogen to oxides usable in primary food production.

Nodal line a line in an oscillating area along which there is little or no rise and fall of the tide.

Nodal point the no-tide point in an amphidromic region.

Node that part of a standing wave where the vertical motion is least and the horizontal velocities are greatest.

Nonconservative property a property whose values change in the course of a particular specified series of events or processes—for example, those properties of seawater, such as nutrient or dissolved oxygen concentrations, which are affected by biological or chemical processes.

Nonrenewable resource one which is not replenished at a rate comparable to its rate of consumption.

Nuclide a species of atom characterized by the constitution of its nucleus. The nuclear constitution is specified by the number of protons, number of neutrons, and energy content—or alternatively, by the atomic number, mass number, and atomic mass.

Nudibranchs (sea slugs) gastropods in which the shell is entirely absent in the adult. The body bears projections which vary in color and complexity among the species.

Nutrient in the ocean, any one of a number of inorganic or organic compounds or ions used primarily in the nutrition of primary producers. Nitrogen and phosphorus compounds are examples of *essential nutrients.*

Ocean basin that part of the floor of the ocean that is more than about 2000 meters below sea level.

Oceanic crust a mass of igneous material, approximately 7 kilometers thick, which lies under the ocean bottom.

Oceanic (mid-ocean) rise general term for the discontinuous ocean-bottom province which rises above the deep ocean floor; area of crustal generation.

Ooze a fine-grained deep-ocean sediment containing at least 30 percent (by volume) undissolved sand- or silt-sized, calcareous or siliceous skeletal remains of small marine organisms, the remainder usually being clay-sized material.

Open ocean that part of the ocean which is seaward of the approximate edges of the continental shelves, usually more than 2 kilometers deep.

Ophiolitic suite an assemblage of rocks on land, containing deep-sea sediments, submarine lavas, and oceanic crust, apparently thrust upward during crustal plate subduction.

Oxidation loss of hydrogen or electrons; opposite of reduction.

Pack ice a rough, solid mass of broken ice floes forming a heavy obstruction, preventing navigation.

Pangaea single continent thought to have existed prior to Permian times. Fragments of it split apart about 200 million years ago to form the present continents.

Partial tide one of the harmonic components comprising the tide at any point. The periods of the partial tides are derived from various combinations of the angular velocities of earth, sun, and moon, relative to each other.

Patch reef a term for all isolated coral growths in lagoons of barrier and atolls. They vary in extent from expanses measuring several kilometers across to coral pillars or even mushroom-shaped growths consisting of a single colony.

Pelagic a primary division of the ocean which includes the whole mass of water. The division is made up of the coastal ocean (which includes the water shallower than 200 meters) and the open ocean (which includes that water deeper than 200 meters).

Pelagic deposits deep-ocean sediments that have accumulated by the settling out of the ocean on a particle-by-particle basis.

Photic zone (euphotic zone) the layer of a body of water which receives ample sunlight for the photosynthetic processes of plants; it is usually 80 meters deep, or more, depending on angle of incidence of sunlight, length of day, and cloudiness.

Photosynthesis the manufacture of carbohydrate food from carbon dioxide and water in the presence of chlorophyll, by utilizing light energy and releasing oxygen.

Phytoplankton the plant forms of plankton.

Planetary winds see *Prevailing wind systems*.

Plankton passively drifting or weakly swimming organisms.

Plankton bloom an unusually high concentration of plankton (usually phytoplankton) in an area, caused either by an explosive or gradual multiplication of organisms and usually producing a discoloration of the sea surface.

Plate tectonics the theory that accounts for seismicity, mountain building, volcanism, and other manifestations of crustal plate movement with sea-floor spreading.

Polder a land area that has been reclaimed from the ocean bottom; often separated from the ocean by dikes, then drained.

Polychaetes segmented marine worms, some of which are tubeworms; others are free-swimming worms.

Polyp an individual, sessile coelenterate.

Prevailing wind systems (planetary winds) large, relatively constant wind systems which result from the shape, inclination, revolution, and rotation of the planet. They are the northeast and southeast trade winds, the westerlies, and the polar easterlies.

Primary coastline a coastline shaped primarily by terrestrial forces rather than by wave action or other marine processes.

Primary productivity the amount of organic matter synthesized by organisms from inorganic substances in unit time in a unit volume of water or in a column of water of unit area cross-section and extending from the surface to the bottom; also called *gross primary production*.

Prime meridian the meridian of longitude (0°), used as the reference for measurements of longitude, internationally recognized as the meridian of Greenwich, England.

Productivity see *Primary productivity*.

Profile a drawing showing a vertical section along a surveyed line.

Progressive wave a wave which is manifested by the progressive movement of the wave form.

Protozoa mostly microscopic, one-celled animals, constituting one of the largest populations in the ocean.

Pseudopod an extension of protoplasm that has no permanent shape but can be projected or withdrawn by the animal, for capturing food or for locomotion; characteristic of some protozoans.

Pteropods free-swimming gastropods in which the foot is modified into fins; both shelled and nonshelled forms exist. In some shallow areas, accumulated shells of these organisms form a type of bottom sediment called *pteropod ooze*.

Pycnocline a vertical density gradient in some layer of a body of water, positive with respect to depth and appreciably greater than the gradients above and below it; also, a layer in which such a gradient occurs.

Radioisotope any radioactive isotope of an element; also, a word loosely used as a synonym for *radionuclide*.

Radiolarians single-celled planktonic protozoans possessing a skeleton of siliceous spicules and radiating threadlike pseudopodia.

Radionuclide a synonym for *radioactive nuclide*.

Red algae reddish, filamentous, membranous, encrusting or complexly branched plants in which the color is imparted by the predominance of a red pigment over the other pigments present. Some are included among the coralline algae.

Red clay a more or less brown to red deep-sea deposit. It is the most finely divided clay material that is derived from the land and transported by ocean currents and winds, accumulating far from land and at great depths.

Red tide a red or reddish-brown discoloration of surface waters most frequently in coastal regions, caused by concentrations of certain microscopic organisms, particularly dinoflagellates.

Reduction gain in electrons or in hydrogen; opposite of oxidation.

Reef an offshore consolidated rock hazard to navigation with a least depth of 20 meters or less.

Reef flat (of a coral reef) a flat expanse of dead reef rock which is partly or entirely dry at low tide. Shallow pools, potholes, gullies, and patches of coral debris and sand are features of the reef flat. It is divisible into inner and outer portions.

Reflection the process whereby a surface or discontinuity turns back a portion of the incident radiation into the medium through which the radiation approached.

Refraction of water waves the process by which the direction of a wave moving in shallow water at an angle to the contours is changed, causing the wave crest to bend toward alignment with the underwater contours; also, the bending of wave crests by currents.

Relict sediment sediment having been deposited on the continental shelf by processes no longer active.

Renewable resource one which may be replenished at a rate comparable to its rate of consumption, either as a result of natural growth, or by careful management of the resource.

Residence time time required for a flow of material to replace the amount of that material originally present in a given area. Assuming a steady flow, replacement time can be calculated for any substance, such as salt or water.

Respiration an oxidation-reduction process by which chemically bound energy in food is transformed into other kinds of energy upon which certain processes in all living cells are dependent.

Reversing thermometer a mercury-in-glass thermometer that records temperature upon being inverted and thereafter retains its reading until returned to the first position.

Reversing tidal current a tidal current that flows alternately in approximately opposite directions, with a period of slack water at each reversal of direction. Reversing currents usually occur in rivers and straits where the flow is restricted. When the flow is toward the shore, the current is flooding; when in the opposite direction, it is ebbing.

Rift valley a trough formed by faulting in a divergence area.

Rill a small groove, furrow, or channel made in mud or sand on a beach by tiny streams following an outflowing tide.

Rip agitation of water caused by the meeting of currents or by a rapid current setting over an irregular bottom—for example, a *tide rip*.

Rip current a strong current usually of short duration, flowing seaward from the shore. It usually appears as a visible band of agitated water and is the return movement of water piled up on the shore by incoming waves and wind.

Ripple a wave controlled to a significant degree by both surface tension and gravity.

Rise a long, broad elevation that rises gently and generally smoothly from the ocean bottom.

River-induced upwelling the upward movement of deeper water which occurs when seawater mixes with fresh water from a river, becoming less dense, and moves toward the surface.

Rotary current, tidal a tidally induced current which flows continually with the direction of flow, changing through all points of the compass during the tidal period. Rotary currents are usually found in the ocean where the direction of flow is not restricted by any barriers.

Sabellid see *Tube-worm*.

Salinity a measure of the quantity of dissolved salts in seawater. Formally defined as the total amount of dissolved solids in seawater in parts per thousand $(^0/_{00})$ by weight when all the carbonate has been converted to oxide, the bromide and iodide to chloride, and all organic matter is completely oxidized.

Salps transparent pelagic tunicates. The body is more or less cylindrical and possesses conspicuous ringlike muscle bands, the contraction of which propels the animal through the water.

Salt dome a generally cylindrical mass of salt, a kilometer or more in diameter, that rose to its present position through surrounding sediments that were originally deposited above it. Reservoirs in rocks around a dome may contain gas and oil.

Salt marsh see *Marsh*.

Salt-water wedge an intrusion in a tidal estuary of seawater in the form of a wedge characterized by a pronounced increase in salinity from surface to bottom.

Sand loose material which consists of grains ranging between 0.0625 and 2.0 millimeters in diameter.

Sargassum a brown alga characterized by a bushy form, substantial "holdfast" (rootlike structure) when attached, and a yellowish brown, greenish yellow, or orange color.

Scarp an elongated and comparatively steep slope of the ocean floor separating flat or gently sloping areas.

Scattering the dispersion of light when a beam strikes very small particles suspended in air or water. In light scattering there is, theoretically, no loss of intensity but only a redirection of light.

School a large number of one kind of fish or other aquatic animal swimming or feeding together.

Scour the downward and sideward erosion of a sediment bed by wave or current action.

Scyphozoans coelenterates in which the polyp or hydroid stage is minimized or insignificant and the medusoid stage is well developed. The true jellyfish belong to this group.

Sea waves generated or sustained by winds within their fetch, as opposed to *swell;* also, a subdivision of an ocean.

Seabed drifter a plastic float designed to drift in near-bottom currents; movements of the drifter indicate speed and direction of the current.

Sea breeze a light wind blowing toward the land caused by unequal heating of land and water masses.

Seachannel a long, narrow, U- or V-shaped shallow depression of ocean floor, usually occurring on a gently sloping plain or fan.

Sea-floor spreading the mechanism by which oceanic crust is generated at divergence zones such as mid-ocean ridges, and adjacent crustal plates are moved apart as new material forms.

Seamount an elevation rising 900 meters or more from the ocean bottom.

Sea state (state of the sea) the numerical or written description of ocean-surface roughness.

Seawall a man-made structure of rock or concrete built along a portion of coast to prevent wave erosion of the beach.

Secondary coastline a coastline shaped primarily by marine forces, such as wave action, or by marine organisms.

Sediment particulate organic and inorganic matter which accumulates in a loose, unconsolidated form. It may be chemically precipitated from solution, secreted by organisms, or transported from land by air, ice, wind, or water and deposited.

Sedimentation the process of breakup and separation of particles from the parent rock, their transportation, deposition, and consolidation into another rock.

Seiche a standing-wave oscillation of an enclosed or semienclosed water body that continues, pendulum-fashion, after the cessation of the originating force, which may have been seismic, atmospheric, or wave-induced; also an oscillation of a fluid body in response to a disturbing force having the same frequency as the natural frequency of the fluid system.

Seismic pertaining to, characteristic of, or produced by earthquakes or earth vibrations.

Seismic reflection the measurements, and recording in wave form, of the travel time of acoustic energy reflected back to detectors from rock or sediment layers which have different elastic-wave velocities.

Seismic sea wave see *Tsunami.*

Semidaily (semidiurnal) tide a tide having a period or cycle of approximately one-half a tidal day; the predominating type of tide throughout the world is semidiurnal, with two high waters and two low waters each tidal day.

Semipermeable membrane a membrane through which a solvent, but not certain dissolved or colloidal substances, may pass.

Sensible heat the portion of energy exchanged between ocean and atmosphere which is utilized in changing the temperature of the medium into which it penetrates.

Sessile permanently attached, by a base or stalk; not free to move about.

Set (current direction) the direction toward which the current flows.

Shear wave a wave which causes particles in a medium to vibrate back and forth at right angles to the direction of propagation.

Shoreline the boundary between a body of water and the land at high tide (usually mean high water).

Significant wave height (characterisic wave height) the average height of the highest one-third of waves of a given wave group.

Sill shallow portion of the ocean floor which partially restricts water flow; may be either at the mouth of an inlet, fjord, etc., or at the edge of an ocean basin—for example, the Bering Sill separates the Pacific and Arctic portions of the Atlantic Ocean.

Silt particles between sands and clays in size.

Sinking (downwelling) a downward movement of surface water generally caused by converging currents or as a result of a water mass becoming more dense than the surrounding water.

Siphonophores medusoid coelenterates, many of which are luminescent and some venomous. Some possess a gas-filled float. Some are colonial, so that polyp and medusoid individuals function as a single individual.

Sipunculids wormlike marine animals, unsegmented, with the mouth surrounded by tentacles. The anterior (head) end can be withdrawn into the body. They are deposit-feeders.

Slack water the state of a tidal current when its velocity is near zero, usually occurs when a reversing current changes its direction.

Slick area of quiescent water surface, usually elongated. Slicks may form patches or weblike nets where ripple activity is greatly reduced.

Slump the slippage or sliding of a mass of unconsolidated sediment down a submarine or subaqueous slope. Slumps occur frequently at the heads or along the sides of submarine canyons; triggered by any small or large earth shock, the sediment usually moves as a unit mass initially, but often becomes a turbidity flow.

Solar tide the (partial) tide caused solely by the tide-producing forces of the sun.

Sounding the measurement of the depth of water beneath a ship.

Specific gravity the ratio of the density of a substance relative to the density of pure water at 4°C; in the centimeter–gram–second system, *density* and *specific gravity* may be used interchangeably.

Specific heat the quantity of heat required to raise the temperature of 1 gram of any substance by 1°C. The common unit is calories per gram per degree centigrade.

Sperm whale see *Toothed whales.*

Spicules crystals of newly formed sea ice; also, a minute, needlelike or multiradiate calcareous or siliceous body in sponges, radiolarians, some gastropods, and echinoderms.

Spit a small point of land projecting into a body of water from the shore.

Spring bloom the sudden proliferation of phytoplankton which occurs from the temperate zone to the poles whenever the critical depth (as determined by penetration of sunlight) exceeds the depth of the mixed, stable, surface layer (as determined by the establishment of a pycnocline).

Spring tide tide of increased range which occurs about every two weeks when the moon is new or full.

Spur-and-groove structure see *Fore-reef.*

Stability the resistance to overturning or mixing in the water column, resulting from the presence of a positive density gradient.

Stand of the tide the interval at high or low water when there is no appreciable change in the height of the tide; its duration will depend on the range of the tide, being longer when the tidal range is small and shorter when the tidal range is large.

Standing crop the biomass of a population present at any given moment.

Standing wave a type of wave in which the surface of the water oscillates vertically between fixed points, called *nodes,* without progression. The points of maximum vertical rise and fall are called *antinodes.* At the nodes, the underlying water particles exhibit no vertical motion, but maximum horizontal motion.

Steady state the absence of change with time.

Still-water level the level that the sea surface would assume in the absence of wind waves; not to be confused with mean sea level or half-level tide.

Storm surge (storm wave, storm tide, tidal wave) a rise or piling-up of water against shore, produced by wind-stress and atmospheric-pressure differences in a storm.

Stratosphere that part of the earth's atmosphere between the troposphere and the upper layer (ionosphere).

Subduction zone an inclined plane descending away from a trench, separating a sinking oceanic plate from an overriding plate. A region of high seismicity, known as a Benioff zone, coincides with this plane.

Sublimation the transition of the solid phase of certain substances into the gaseous—and vice versa—without passing through the liquid phase. Water possesses this property; thus ice can change directly into water vapor or water vapor into ice.

Submarine canyon a V-shaped submarine depression of valley form with relatively steep slope and progressive deepening in a direction away from shore.

Subsurface current a current usually flowing below the pycnocline, generally at slower speeds and frequently in a different direction from the currents near the surface.

Subtropical high one of the semipermanent highs of the subtropical high pressure belt.

Succession, ecological the orderly process of community change, whereby communities replace one another in sequence.

Surf collective term for breakers; also, the wave activity in the area between the shoreline and the outermost limit of breakers.

Surf zone the area between the outermost breaker and the limit of wave uprush.

Surface-active agent substance, usually in solution, which can markedly change the surface or interfacial properties of the liquid fraction, even when present in trace amounts only.

Surface tension (surface energy, capillary forces, interfacial tension) a phenomenon peculiar to the surface of liquids, caused by a strong attraction toward the interior of the liquid acting on the liquid molecules in or near the surface in such a way to reduce the surface area. An actual tension results, and is usually expressed in dynes per centimeter or ergs per square centimeter.

Surface zone (mixed zone) water layer above the pycnocline where wind waves and convection cause mixing of the water, resulting in more or less uniform temperature and salinity with depth.

Surge horizontal oscillation of water with comparatively short period accompanying a seiche (see also *Storm surge*).

Swallow float a tubular buoy, usually made of aluminum, which can be adjusted to remain at a selected density level to drift with the motion of that water mass. The float is tracked by shipboard listening devices, to determine current velocities.

Swash the rush of water up onto the beach following the breaking of a wave.

Swell ocean waves which have traveled out of their generating area.

Swimbladder a membranous sac of gases, lying in the body cavity between the vertebral column and the alimentary tract of certain fishes. It serves a hydrostatic function in most fishes that possess it; in some, it participates in sound production.

Symbiosis a relationship between two species in which one or both members are benefitted and neither is harmed.

Tablemount see *Guyot*.

Tectonic estuary an estuary occupying a basin formed as a result of mountain building.

Thermocline a vertical temperature gradient in some layer of a body of water, negative with respect to depth and appreciably greater than the gradients above and below it; also, a layer in which such a gradient occurs.

Thermohaline circulation vertical circulation induced by surface cooling, which causes convective overturning and consequent mixing.

Tidal bore a very rapid rise of the tide in which the advancing water forms an abrupt front, sometimes of considerable height, occurring in certain shallow estuaries where there is a large tidal range.

Tidal bulge (tidal crest) a long-period wave associated with the tide-producing forces of the moon and sun; identified with the rising and falling of the tide. The trough located between the two tidal bulges present at any given time on the earth is known as the *tidal trough*.

Tidal constituent one of the harmonic elements in a mathematical expression for the tide-generating force and in corresponding formulas for the tide or tidal current. Each constituent represents a periodic change or variation in the relative positions of the earth, moon, and sun.

Tidal current the alternating horizontal movement of water associated with the rise and fall of the tide caused by the astronomical tide-producing forces.

Tidal day the interval between two successive upper transits of the moon over a local meridian. The period of the *mean tidal day*, sometimes called a *lunar day*, is 24 hours, 50 mintes.

Tidal flats marshy or muddy areas which are covered and uncovered by the rise and fall of the tide; also called *tidal marshes*.

Tidal period the elapsed time between successive high or low waters.

Tidal range the difference in height between consecutive high and low waters.

Tidal trough see *Tidal bulge*.

Tide the periodic rise and fall of the earth's ocean and atmosphere that results from tide-producing forces of moon and sun acting on the rotating earth.

Tide curve a graphic presentation of the rise and fall of tide; time (in hours or days) is plotted against height of the tide.

Tide-producing forces the slight local difference between the gravitational attraction of two astronomical bodies and the centrifugal force that holds them apart. Gravitational attraction predominates at the surface point nearest to the other body, while centrifugal "repulsion" predominates at the surface point farthest from the other body.

Tide pool depression in a rock within the intertidal zone, alternately submerged and exposed with water remaining inside, by the rise and fall of the tide.

Tide rip see *Rip*.

Tide tables tables which give daily predictions, usually a year in advance, of the times and heights of the times and heights of the tide.

Tide wave a long-period gravity wave that has its origin in the tide-producing force and which manifests itself in the rising and falling of the tide.

Tintinnids microscopic planktonic protozoans which possess a tubular or vase-shaped outer shell.

Tombolo an area of unconsolidated material, deposited by wave or current action, which connects a rock, island, etc., to the main shore or other body of land.

Toothed whales this group includes dolphins, porpoises, killer whales, and sperm whales.

Trade winds the wind system, occupying most of the tropics, which blows from the subtropical highs toward the equatorial trough; a major component of the general circulation of the atmosphere. The winds are northeasterly in the Northern Hemisphere and southeasterly in the Southern Hemisphere.

Transform fault a fault connecting the offset portions of a mid-ocean ridge, along which crustal plates slide past each other.

Trawl a bag or funnel-shaped net for catching bottom fish by dragging along the bottom; also a large net for catching zooplankton and fishes by towing in intermediate depths.

Trochophore the free-swimming pelagic stage of some segmented worms and mollusks.

Trophic level a successive stage of nourishment as represented by links of the food chain. Primary producers (phytoplankton) constitute the first trophic level, herbivorous zooplankton the second trophic level, and carnivorous organisms the third and higher trophic levels.

Tropics equatorial region between Tropic of Cancer and Tropic of Capricorn; climate found in the belt close to the equator; characteristics are daily variation in temperature exceeding seasonal variation and generally high rainfall.

Tropopause the upper limit of the troposphere.

Troposphere that portion of the atmosphere next to the earth's surface in which temperature generally rapidly decreases with altitude, clouds form, and convection is active. At middle latitudes, the troposphere generally includes the first 10 to 12 kilometers above the earth's surface.

Tsunami (seismic sea wave) a long-period sea wave produced by a submarine eathquake or volcanic eruption. It may travel unnoticed across the ocean for thousands of miles from its point of origin, and builds up to great heights over shallow water.

Tube-worm any polychaete, chiefly the sabellids and related groups, that builds a calcareous or leathery tube on a submerged surface. Tube-worms are notable fouling organisms.

Tunicates globular or cylindical, often saclike animals, many of which are covered by a tough flexible material. Some are sessile, others are pelagic.

Turbidite turbidity-current deposit characterized by both vertically and horizontally graded bedding.

Turbidity reduced water clarity resulting from the presence of suspended matter. Water is considered turbid when its load of suspended matter is visibly conspicuous, but all waters contain some suspended matter and therefore are turbid to some degree.

Turbidity current a gravity current resulting from a density increase by suspended material.

Turbulent flow a flow characterized by irregular, random-velocity fluctuations.

Upwelling the process by which water rises from a lower to a higher depth, usually as a result of divergence and offshore currents.

Van der Waals forces weak attractive forces between molecules which arise from interactions between the atomic nuclei of one molecule and the electrons of another molecule.

Veliger the planktonic larval second stage of many gastropods.

Viscosity internal resistance-to-flow property of fluids which enable them to support certain stresses and thus resist deformation for a finite time; arises from molecular and eddy effects.

Volcanic center ("hot spot") a locus of continuing volcanism due to long-term activity in the mantle. Sea-floor spreading moves the earth's crust with respect to such a center, so that as a volcanic ridge or island chain is built at the surface, its orientation is governed by the direction of sea-floor motion.

Volcanic island island formed by the top of a volcano or solidified volcanic material, rising above the sea surface.

Water budget the accounting for the interchange of water substance among the earth, the atmosphere, and the ocean.

Water mass a body of water usually identified by its T–S curve or chemical content.

Water parcel water mass with a certain temperature and salinity, separated from surrounding waters by fairly sharp boundaries across which mixing occurs.

Wave a disturbance which moves through or over the surface of the ocean (or earth); wave speed depends on the properties of the medium.

Wave age the state of development of a wind-generated sea-surface wave, conveniently expressed by the ratio of wave speed to wind speed. Wind speed is usually measured at about 8 meters above still-water level.

Wave energy the capacity of a wave to do work. In a deep-water wave, about half the energy is kinetic energy, associated with water movement, and about half is potential energy, associated with the elevation of water above the still-water level in the crest or its depression below still-water level in the trough.

Wave group a series of waves in which the wave direction, wavelength, and wave height vary only slightly.

Wave height vertical distance between crest and preceding trough.

Wavelength horizontal distance between successive wave crests measured perpendicular to the crests.

Wave period the time required for two successive wave crests to pass a fixed point such as a rock or an anchored buoy.

Wave spectrum a concept used to describe by mathematical function the distribution of wave energy (square of wave height) with wave frequency (1/period). The square of the wave height is related to the potential energy of the sea surface so that the spectrum can also be called the *energy spectrum*.

Wave steepness the ratio of the wave height to wave length.

Wave train a series of waves from the same direction.

Wave velocity speed at which the individual wave form advances; also, a vector quantity that specifies the speed and direction with which a wave travels through a medium.

Weathering process of destruction or partial destruction of rock by thermal, chemical, and mechanical processes.

Wetland see *Marsh*.

Whitecap the white froth on crests of waves in a wind, caused by wind blowing the crest forward and over.

Wind-driven circulation surface-current system driven by the force of the winds.

Wind mixing mechanical stirring of water due to motion, induced by the surface wind.

"Wind tide" vertical rise in the water level on the leeward side of a body of water caused by wind stress on the surface of the water.

Wind waves waves formed and growing in height under the influence of wind; loosely, any wave generated by wind.

Windrows rows of floating debris, aligned in the wind direction, formed on the surface of a lake or ocean by Langmuir circulation in the upper water layer.

Year-class organisms of a particular species spawned during a single year or breeding season.

Zonation organization of a habitat into more or less parallel bands of distinctive plant and animal associations where conditions for survival are optimal.

Zone a layer which encompasses some defined feature, structure, or property in the ocean.

Zooplankton the animal forms of plankton.

index

Bacteria (cont.)
  in the coastal ocean, 330
  and coral reefs, 396
  as decomposers, 95, 115, 316, 321,
    330–331, 333, 334, 367, 402, 415
    (see also Organic detritus, and
    bacteria)
  in the deep ocean, 133, 331, 392, 402
  epiphytic, 388
  and food web, 318, 321, 322, 330–333,
    367, 368, 384, 396, 402
  and fossil fuel production, 415
  and fouling, 333, 393
  as heterotrophs, 323
  and manganese nodules, 96, 392
  and nitrogen cycle, 132, 319–320
  in the open ocean, 330–331
  and organic detritus (see Organic
    detritus, and bacteria)
  and organic growth factors, 317, 321
  and oxygen distribution, 334
  and phosphorus cycle, 318–319, 322
  and pressure (see Pressure, and bac-
    terial activity)
  in sediments, 331, 390–393
  and waste disposal, 428–429
Baffin Bay, 82, 273, 274
Baffin Island, 83
Bahama Banks, 109, 134, 195
Bahama Islands, 257
Baird, Spencer Fullerton, 8, 10
Baja California, 26, 39, 41, 43
Balanus, 355, 356, 364
Balboa, Vasco Núñez de, 4
Baleen whales, 350, 361, 411
Baltic Sea, 96, 165
Barbados, 71
Barents Sea, 14, 80, 83, 165
Barite, 91
Barnacles:
  and larval development, 354, 355,
    356, 393
  and reefs, 393, 394, 395
  and zonation, 375, 376
Barnegat Inlet, N.J., 294
Barnegat Light, N.J., 218
Barrier beaches, 264, 265, 283, 294, 308
Barrier islands, 256, 264, 273, 283, 286,
    292, 294, 299, 300
Barrier reefs, 75
Basalt, 29, 32, 34, 42, 47, 49, 50, 52, 71
Bass, 365, 367, 408
Bass Strait, 415
Bat, 382–383
Bathebius, 7
Bathythermograph, 17, 142–143
Baymouth bars, 286, 294, 295
Bay of Bengal, 226, 237, 270, 288, 289
Bay of Fundy, 233, 242–243, 260, 418
Beach cusps, 302
Beach drift (see Littoral drift)
Beaches, 286, 292–297 (see also Barrier
    beaches)
  elevated, 283
  and erosion (see Erosion, of coastal
    features)

Beaches (cont.)
  general features, 282
  and marine organisms, 283, 301, 381–
    383, 390
  preservation and modification of,
    306, 420
  and sedimentary processes, 108, 285,
    287, 293–294, 297, 306, 421
  and tides, 283, 299, 300
  U.S., 108, 429
  and waste disposal, 429
  and waves, 204, 210, 216, 217, 219,
    222, 282, 283, 297, 300, 419
Beach grasses, 296
Beach hopper, 382–383
Beach processes, 297–302 (see also
    Beaches, and sedimentary pro-
    cesses)
Beach sand, 92, 96, 293, 296 (see also
    Sand, on beaches)
Beach slope, 293
Beach wrack, 382–383
Beaufort numbers, 211
Beaufort Sea, 83
Bed load, 100
Belgium, 249, 250
Bellany Island, 41
Benguela Current, 178
Benioff zone, 46, 48, 52
Benthic algae, 366, 376, 393, 395, 396,
    397 (see also Seaweeds)
Benthic colonies, development of, 333,
    399 (see also Fouling)
Benthic community, environmental fac-
    tors, 373, 374
Benthic life, strategies of, 372
Benthic plants (see Plants)
Benthos, 314, 350, 428 (see also names
    of groups and individuals, e.g.
    Crabs; Sabellarian worms; Um-
    bellula)
  estuarine, 366
  planktonic larvae of (see Meroplank-
    ton; Pelagic larvae, of benthos)
Bering Sea, 17, 80, 336, 362, 373
Bering Sill, 80, 81, 196
Bering Strait, 81, 83
Berm, 282, 295–296, 297
Bermuda, 65, 67, 69, 70–71, 109, 293,
    347
Bicarbonate ion, 126, 131, 133, 134
Bigelow, Henry Bryant, 12
Bikini Atoll, 18
Billfishes, 408
Biogenous sediment (see Sediment,
    biogenous constituents)
Bioluminescence, 343, 344, 352, 357–
    359
Biomass, 322 (see also Standing crop)
  of bacteria, 330, 391
  and coral reef communities, 396
  of deep-ocean organisms, 390, 402
  of fish in California waters, 410
  of meiobenthos, 390
  of microbenthos, 391

Biomass (cont.)
  of phytoplankton reduced by graz-
    ing, 324, 344
  and productivity, 324–325, 367
  and trophic level, 367
  and year class, 374
Biotin, 321
Bioturbation, 390 (see also Sediment,
    reworked by burrowing orga-
    nisms)
Bird-foot delta, 287, 292
Birds, 169, 312, 361, 363, 365, 381, 382–
    383, 410
Biscayne Bay, Florida, 215, 305
Bivalves, 354–355, 384, 385, 389, 390,
    399, 401, 402 (see also Clams;
    Mussels; Oysters)
Black Sea, 80, 85, 109, 336, 392
Blake Plateau, 96, 111
Bloom, 315, 317–318, 325–327, 329, 337,
    343, 344 (see also Spring bloom)
Bluefin tuna, 361
Bluefish, 399
Blue whale, 361, 411
Boiling point (see Water, boiling point)
Bolinopsis, 352
Bones, in sediment, 90, 91, 95, 102
Bonin Trench, 45, 69
Bonita, 365, 408
Bore (see Tidal bore)
Boreal forest, 169
Borneo (see Kalimantan)
Boron, 127, 130, 135
Boston, Mass., 65, 184, 306, 418
Bottom currents (see Near-bottom cur-
    rents)
Bottom topography:
  and barrier bars, 294
  and distribution of marine orga-
    nisms, 354, 373 (see also Sedi-
    ment, and benthic organisms)
  and sediments (see Sediment, and
    bottom topography; Sediment
    transport, and bottom topog-
    raphy)
  and tides, 239, 240
  and water movements, 44, 46, 64, 158,
    159, 182–183, 196, 198, 222, 262,
    268
  and waves, 217–218, 219, 221, 225
Boulders, 91, 92, 108, 262
Boundary currents, 180–183, 200, 256,
    260 (see also Eastern boundary
    currents; Western boundary cur-
    rents)
Bouvet Island, 41
Boyle, Sir Robert, 4
Brahmaputra River, 80, 84, 97, 98, 284,
    288, 289
Brazil, 56–57, 336, 363, 365
Brazil basin, 45
Brazil Current, 178
Breakers, 205, 206, 219, 222, 223, 282
  and coral reefs, 77
  energy released by, 216
  formation of, 205, 217, 294

Eastern boundary currents, 181, 183, 261, 275, 276, 337
Eastern South Pacific Water, 163
East Greenland Current, 178
East Pacific Rise, 37, 40, 41, 43, 45, 63, 64, 66, 105
Ebb current (ebb tide), 244–245, 418
  (see also Tidal currents)
Echinoderms, 400
Echiuroid worms, 385
Echo sounding, 17, 19, 68, 139, 294, 358
Ecological efficiency, 322, 337, 338, 367–368, 409
Ecological succession (see Succession, of species)
Ecosystems, 23, 315, 317, 322, 333, 334, 367, 373, 381, 393, 395, 399
  trace elements conserved in, 428
Eddies, 177, 181, 182, 183, 260
Edinburgh, Scotland, 250
Eel, 13 (see also Moray eel)
Eelgrass communities, 388–389
Eggs (see also Fish eggs):
  of benthic larvae in surface waters, 402
Eifel (volcanic center), 41
Egyptian culture, 289
Ekman, V. W., 11, 12, 14
Ekman, Walfrid, 192
Ekman current meter, 175
Ekman spiral, 191, 192
Ekman transport, 192
Elat, Israel, 414, 415
Electrolytes, 131, 132
Electromagnetic spectrum, 144, 145, 147
Electron cloud, and water molecule, 116–117, 118
Ellesmere Island, Canada, 83, 263
El Niño, 410
El Segundo, Calif., 302
Emperor Seamounts, 41, 42
Energy:
  and chemosynthesis, 392
  and ecological efficiency, 367–368, 409
  and photosynthesis, 323, 324
  the ocean as a source of, 417–419
Energy flow, and food web, 315, 322, 323, 337, 346, 367, 368, 382
England, 3, 9, 12, 64, 85, 101, 214, 249, 250, 313, 415, 418
English Channel, 64
  and amphidrome system, 241
  and nutrients, 319, 320
  plankton in, 314, 324
  and tidal currents, 246, 249
English coast:
  and long-period swell, 214
  and tidal phenomena, 230
Eniwetok Atoll, 18
Enzymes, 318, 323, 329, 392
Ephyrae, 378, 379
Epifauna, 367, 384
Epiphytes, 366, 389, 390
Epsom salt (see Magnesium sulfate)

Ecuador, 78
Equator, 190, 191
  and determination of latitude, 441
  radius and circumference of, 436
Equatorial Countercurrent, 170, 177, 178, 180
Equatorial currents, 105, 170, 180, 200
  (see also North Equatorial Current; South Equatorial Current)
Equatorial region, 153, 169, 170, 177
  and heat budget, 146, 148, 149, 151–152, 183, 200
  and plankton, 343
  and productivity, 102, 111, 336
  and salinity, 170
  and tidal phenomena, 235, 236, 237, 239
and upwelling, 336
Equilibrium tide, 234–238
Erebus (ship), 5
Erosion:
  of coastal features, 217, 270, 283, 285, 294, 295, 297, 299–300, 303, 305, 306 (see also Glaciers, and erosion; Waves, and erosion)
  and marshes, 307, 381
  and mineral resources in the ocean, 413
  and ocean-bottom topography, 66
  and sedimentation, 32, 47, 50, 83, 98–101, 105, 217, 285, 286, 292–294, 303
  of volcanic islands, 42, 46, 70–71, 75, 285–286
  wave-induced (see Waves, and erosion)
Estuaries:
  and anaerobic processes, 392
  formation of, 262, 263, 264, 283 (see also River valleys, drowned)
  and marine life, 305, 313, 361, 364, 366–367
  modification of, 305–307, 424
  North America (see North America, and estuaries)
  oxygen in, 334
  and productivity, 334, 335, 366
  and salinity distribution, 265–268, 269, 305, 313, 394
  and sedimentation, 100, 262, 264, 269, 287, 289, 292, 294, 305, 427 (see also names of specific estuaries e.g. Chesapeake Bay; Columbia River estuary; Laguna Madre)
  submerged, 58, 59, 264 (see also River valleys, drowned)
  and tidal phenomena, 248, 251–252, 265–269, 283, 308
  and tidal power, 418
  and waste discharge (see Wastes, in estuaries and harbors)
Estuarine circulation, 265–268 (see also names of specific estuaries, e.g. Chesapeake Bay; Columbia River estuary; Laguna Madre)

Estuarine circulation (cont.)
  and coastal ocean areas, 275, 276–278, 325
  and migration of larvae, 366
  and nutrients, 325, 366, 428
  and sediment transport, 100
  and wastes, 428
Estuarine environment, and habitats, 381
Estuarinelike circulation, 428 (see also Coastal circulation)
Euchaeia, 349
Euphausia superba, 349, 350
Euphausiids, 348–350, 356, 368 (see also Krill)
Eurasia, 29, 37, 54, 171
Eurasian Plate, 36, 41
Europe, 26, 37, 40, 49 50, 52, 80, 85, 97, 152, 167, 199
Eurytemora, 349
Evadne, 350, 364
Evaporating ponds, 413
Evaporation:
  and coastal ocean, 258, 263
  and conservative properties of seawater, 217
  distribution of, 79, 82, 84, 85, 153–155, 169, 182, 198, 273–275
  and heat, 119, 122, 123, 156, 158, 159, 188, 259, 430
  and desalination, 414, 415, 430
  and residence time of water, 129
  and salt deposits, 128
  and salt production, 413, 430
  in tide-pools, 377
  and water budget, 153, 155, 156, 159
Evolution, 7, 13, 26, 53
Ewing, Maurice, 18, 19
E.W. Scripps (ship), 17
Excretion:
  and release of nutrients, 318, 321, 333, 336 (see also Nitrogen cycle; Phosphorus cycle)
  and release of organic matter, 321, 329, 333
Exponential notation, 434

F

Faeroe Islands, 46, 65, 373
Falkland Current, 178
Fall line, 279
Falmouth, Mass., 8
Fateh Field, Persian Gulf, 416
Fats, 323, 329, 350, 391, 428
Fatty acids, 329
Faults:
  at boundaries of crustal plates, 36, 37, 38
  as earthquake zones, 46
  and hydrothermal ore deposits, 48
  and mid-oceanic ridge system, 61, 63, 64, 66

Garfish, 408
Gas (see Natural gas)
Gases:
   in the atmosphere, relative abundance of, 133
   dissolved (see Dissolved gases)
   ionized, and heat budget, 151
Gastropods, 354, 373
Gemini XI, 84
Geomorphologist, 285
Georges Banks, 9, 326
Georgia–Carolina coast, 186
Geostrophic currents, 188, 193–196, 198, 247, 260
Geothermal power, 34, 417, 419
German Antarctic Expedition of 1911–12, 13
Germany, 12, 249, 305
Ghost crab, 382–383, 389
Gibbsite, 91, 105
Gigantactis, 359
Gill rakers, 363
Gills, of marine organisms, 381, 385, 387, 388, 389, 394
Glacial marine sediment, 92, 104, 108, 110, 111 (see also Sediment transport, and glaciers)
Glacial moraine (glacial deposits), 285, 293
Glacial period (see Ice Age; Pleistocene glacial period)
Glaciers, 68 (see also Continental ice sheets)
   area, density, mass, volume of, 436
   and coastline formation, 262, 264, 285, 418
   distribution of, 169, 284 (see also Pleistocene glacial period)
   and erosion, 66, 83, 92, 99, 107, 262, 264, 285, 292, 418
   and fjords, 59, 262, 263, 268, 285
   and icebergs, 167
   and isostatic adjustment of continental blocks, 33, 59, 283
   and sea level, 59, 108, 264, 283, 420 (see also Ice Age, and sea level)
   and sediment transport, 90, 96, 105, 108, 262, 285
   water contained in, 30, 31, 152
Glass worms (see Arrow-worms)
Glauconite, 114
Global Atmospheric Research Programme (GARP), 22
Globigerina, 93, 106
Globigerinoides ruber, 347
Glomar Challenger (ship), 19, 40
Glucose, 333
Gnathia, 348
Gobies, 385, 394
Godirari River, 80
Goethite, 91
Gold, 48, 127, 130, 414, 421
Gondwana, 39
Goode, J. P., 423 (see also Homolosine projection)
Gough Island, 44

Grackle, 382–383
Graded bedding, 101
Grain size (see Sediment size)
Grand Banks, 101, 167, 168, 181, 183, 198
Grand Canyon (of the Colorado River), 59
Granite, 29, 32, 34, 47, 60, 421
Granules (sediment), and size of, 91
Graphs, how to use, 438–439
Grasses, 325, 335 (see also Beach grass; Seagrasses)
Gravel, 59, 96, 99, 100, 285, 286, 293, 300, 304
   as a marine resource, 20, 419–421, 430
Gravitational field of the earth, 19
Gravity:
   and currents, 190, 195 (see also Geostrophic currents)
   and tides, 235, 236, 238
   and waves, 207, 209, 223, 225
Gravity anomalies, 33, 68
Gravity corer, 88–89
Gravity waves, 209
Gray whale, 361, 411
Great Barrier Reef, 74, 75, 109
Great Britain (see England)
Great circle, 441
Great Lakes, 225, 257, 292
Greece, 50
Greenhouse effect, 147, 151
Greenland, 14, 65, 334, 373, 443
   continental composition of, 82
   and currents, 174, 199
   and icebergs, 15, 167, 168
   and ice cover, 80, 98, 167, 283
   and submarine ridges, 46, 158
   and water masses, 196
   and whaling, 411
Greenland Sea, 83
Greenwich, Conn., 250
Greenwich, England, and meridian of longitude, 241, 440, 441
Grenadiers, 408
Groins, 283, 300, 306
Gross productivity, 325
Groundwater, 152, 291, 422
Groupers, 365
Group speed, 209
Grunt (fish), 139
Guam, 397
Guano birds, 410
Guinea Current, 178
Gulf Coast of the U.S.:
   and artificial reefs, 399
   and estuarine systems, 264, 265, 273
   and marine organisms, 366, 367, 380
   and offshore facilities, 424
   and shoreline features, 292
   and storm surges, 226
   and tides, 233
Gulf of Aden, 40, 54
Gulf of Alaska, 17, 42, 64, 70
Gulf of Aqaba, 86
Gulf of Bothnia, 165

Gulf of California, 17, 39, 80
   evolution of, 26, 43, 85
   and marine life, 365
   and mid-oceanic ridge system, 64, 85
   and sedimentation, 109
   and tidal phenomena, 233, 243
Gulf of Guinea, 407
Gulf of Maine, 12, 59, 245, 256, 257, 327
Gulf of Martaban, 270
   and artificial reefs, 399
   and continental rifting, 68
   and fossil fuel production, 415
   and freshwater input, 82
   as a marginal ocean basin, 68, 82, 85
   and marine organisms, 366
   and salt domes, 68
   and sedimentation, 100, 109
   as a source of moisture, 170
   and tidal phenomena, 232, 233, 243, 273, 289, 308
Gulf of Mexico coastal-ocean region (U.S.), 256, 257
Gulf of Panama, 155
Gulf of St. Lawrence, 233
Gulf of Siam, 72
Gulf of Suez, 86
Gulf Stream system, 8, 9, 10, 111, 177, 178, 181–186, 195, 196, 254–256, 330
Gulf-weed (see Sargassum)
Gull, 382–383
Guyots, 19, 46, 71
Gymnodinium, 321
Gypsum, 422
Gyres (see Current gyres)

H

Habitats, 374, 390, 399
Hachures, 439
Haddock, 362, 364, 366, 408
Hadley cell, 152
Hake, 362, 408
Halfbeak, 408
Halibut, 356, 362, 407, 408, 410
Halocline, 160, 223, 276, 277
Harpaticoids, 390, 402
Harrison, John, 3
Hatchet-fishes, 358
Hawaiian Islands, 41, 420
   and artificial reefs, 398
   and beaches, 285, 293
   and seismic sea waves, 223
   and volcanic coastal features, 285
   as a volcanic island chain, 42, 67, 69
Headlands, 217, 218, 294, 299, 303
Heat budget, 22, 144–147, 151, 156, 158, 159, 200
Heat capacity, 119, 122, 151, 170, 216, 313
Heat flow, from the earth, 19, 33, 34, 35, 66, 146

Heat transport, global (*see* Heat budget)

Helium, 127, 130, 133

Helland-Hansen, Bjorn, 10, 14, 15

Hematite, 91

Henson, Victor, 12

Hepatitis, 429

Heptane, 119

Herbicides, 427

Herbivores, and food webs, 315, 321, 322, 324, 337, 346, 348, 366, 367, 368

Hermit crabs, 399

Herring, 348, 350, 356, 361, 363, 364–365, 367, 368, 408

Hesperonoë, 385

Hess, Harry H., 19, 20, 46

Heterotrophic organisms, 315, 321, 323, 391 (*see also* Phytoplankton, heterotrophic)

Heterotrophic uptake, 333

Higher high water (high tide), 232, 233, 374

Higher low water, 232, 233

High Plains (of mid-continental U.S.), 67

High-pressure cells, 169

High tide (*see* Higher high water; Lower high water)

Himalaya Mountains, 26, 39, 41, 49, 50, 171

Histioteuthis bonnelliana, 360

Hjort, Dr. Johann, 7, 13

Hoelzel's planisphere, 442, 443, 444

Hold-fasts, 376

Holland (*see* Netherlands)

Holoplankton, 346–353

Homolosine projection, 442, 443, 445

Homotrema rubrum, 293

Honolulu, Hawaii, 233, 234

Honshu Island, Japan, 48

Hook of Holland, 226

Hoover Dam, 216

Hopper dredge, 420–421

Hormones, 317

Horse latitudes, 152

Horseshoe crab, 382–383

"Hot spots" (*see* Volcanic centers)

Housatonic River, 263

Hudson Bay, 65, 85

Hudson River, 254
    as an estuary, 248, 263
    and submarine canyon, 60, 217, 218, 264

Human wastes, 366

Humpback whale, 361, 411

Hurricanes, 211, 214, 215, 225, 226, 273, 282, 283, 299

Huxley, Thomas H., 7

Hveragerdi, Iceland, 419

Hwang Ho River, 80, 97, 98, 284

Hychoides dianthus, 394

Hydration atmosphere (sheath), 131, 132

Hydraulic dredge, 420–421

Hydraulic models, 308

Hydrogen:
    in ancient atmosphere and ocean, 53
    and chemosynthesis, 391, 392
    as a fuel, 419
    and reducing power, 323, 392
    in seawater, 127, 130, 134
    in water molecule, 118, 119

Hydrogen bond, 116–117, 119, 120, 121, 123, 124, 136

Hydrogen chloride, 53

Hydrogen ion, 134, 316

Hydrogen sulfide, 133, 392, 422

Hydroids, 377, 378–379, 380, 389, 393, 399

Hydrological cycle, 153, 154

Hydrophilic substances, 136, 137

Hydrophobic substances, 137

Hydrophone, 34

Hydrosphere, 29, 30, 337

Hydrostatic pressure (*see* Pressure)

Hydrothermal ore deposits, 48, 49

Hydroxides, in marine sediments, 91

Hydroxy acids, 321

Hyperiidae, 348, 364

Hypersaline lagoons, 273

Hypsographic curve, 28, 29

I

Ice, 31, 129 (*see also* Floe ice; Glaciers; Pack ice; Sea ice; Shelf ice)
    density of, 120–121, 124, 125
    effect of heat on, 122, 123, 124
    structure of, 116–117, 20–121

Ice Age, 22, 99 (*see also* Pleistocene glacial period)
    and glacial-marine sediments, 111
    and marine organisms, 399
    and sea level, 29, 262

Icebergs, 15, 108–109, 166, 167, 168, 169, 263

Icecap (terrestrial climate zone), 169

Iceland, 34, 36, 38, 41, 43, 46, 64, 65, 83, 199, 373, 417, 419, 420

Ideal waves 204, 207

Idiacanthus, 357

Igneous rocks, 32

Ijssel Lake, Netherlands, 424

Illite, 91, 105, 107

Ilmenite, 421

Incoming solar radiation (*see* Insolation)

India, 37, 43, 84, 85, 336, 390
    and monsoon circulation, 170
    and sediment discharge, 97, 99, 289
    and tectonic processes, 26, 39, 49, 50
    and upwelling, 261

Indian–Antarctic basin, 45

Indian basin, 45, 58, 61, 64
    dimensions of, 79, 84
    evolution of, 40, 49, 51
    as a gulf, 28

Indian basin (*cont.*)
    and sediment deposits, 84, 104, 106, 107, 111 (*see also* Sedimentation, of Indian basin)
    and topography, 67, 68, 72
    and trenches, 69, 72

Indian Central Water, 163

Indian Equatorial Water, 163

Indian Ocean, 7, 54, 78
    and barriers to water exchange, 46, 65, 334
    boundaries of, 80, 81, 84
    and coral reefs, 74
    and currents, 170, 177, 178, 179–180, 197
    fresh water discharged to, 79, 80, 85, 270
    and islands, 85
    and marine organisms, 390
    and nutrients, 317, 336
    and oxygen distribution, 334
    and productivity, 336
    and salinity, 79, 157, 160, 162
    and sea-surface topography, 194
    surface and drainage areas, and average depth, 79
    and temperature, 79, 149–150, 160, 161, 197
    and tidal phenomena, 230, 243
    and upwelling, 336
    volume of, 31, 81
    and waves, 214, 216

Indonesia, 47, 71, 72, 78, 84, 85, 109, 415

Indus River, 80, 84, 98, 300

Industrial Revolution, 97

Industrial wastes (*see* Wastes, industrial)

Infauna, 384, 389, 402 (*see also* Deposit-feeding; Meiobenthos; Microbenthos, *and names of particular groups or individuals, e.g.* Tube-worms; *Urechis caupo*)

Infra-gravity waves, 209

Inlets (*see* Tidal inlets)

Insects, and food webs, 382–383

Insolation (*see also* Light, in marine environments; Light in the ocean; Solar energy):
    distribution of, 146–147, 159–160, 169, 335
    and food webs, 315, 358
    and heat budget, 122, 144, 146–147, 151, 156, 159, 419
    and reef-building corals, 395

Intergovernmental Oceanographic Commission (IOC), 20

Internal waves, 223, 224, 266

International Council for the Exploration of the Sea (ICES), 11, 12

International Council of Scientific Unions (ICSU), 18

International dateline, 441

International Decade of Ocean Exploration, 22

International Geophysical Year (IGY), 18

International Ice Patrol, 15, 168

International Indian Ocean Expedition, 20

International Law of the Sea (see Law of the Sea)

Intertidal region, 258
   and marine organisms, 230, 313, 314, 374–376, 385, 390, 391, 393, 394

Intra-Coastal Canal, Laguna Madre, 275

Invertebrates, 337, 354, 356, 362, 374, 381, 396, 408 (see also names of particular groups and individuals, e.g. Barnacles; Umbellula)

Iodine, 126, 127, 130, 136, 316

Ionic bond, 119, 120, 129

Ion pairs, 131, 132

Ipswich, England, 414

Ireland, 365

Iron:
   in earth's core, 29, 52
   and marine organisms, 316, 321, 392, 428
   as a mineral resource (see Iron ore)
   in rocks, 32, 35, 47
   in seawater, 53, 127, 130, 131, 316
   in sedimentary deposits, 91, 97, 108, 111, 392

Iron–manganese nodules (see Manganese nodules)

Iron ore, 48, 49, 53, 429

Iron oxide, 91

Iron phosphates, 392

Iron sulfide, 392

Irrawaddy River, 80, 98, 271

Isias, 349

Island arc-trench systems, 47, 48, 52, 62, 70–71, 72, 85, 109 (see also Volcanic islands, formation of)

Isohalines, 162, 258, 276

Isopods, 348, 401

Isostasy, 33, 40, 42, 283

Isotherms, 149, 160, 161, 186, 193, 259

Isthmus of Panama, 26

Italy, 50, 419

J

Jacks, 365, 399, 408

Jamaica, W.I., 3, 77

James River, 263, 270, 271

Janthina, 351

Japan, 47, 48, 71, 81, 101, 223
   and aquaculture, 412
   and fish production, 408
   and geothermal power, 419

Japan Sea, 48, 80

Japan Trench, 41, 45, 69

Jaramilla event, 38, 114

Java Sea, 72

Java Trench, 41, 45, 69, 71, 72

Jeanette (ship), 174

Jefferson, Thomas, 8

Jellyfish, 351, 354, 366, 399

Jet stream, 107

Jetties, 204, 273, 275, 283, 306

John F. Kennedy International Airport, New York, 306

Joint Oceanographic Institution for Deep Earth Sampling (JOIDES), 19, 20

Julius Caesar, 230

K

Kalimantan (Borneo), 72, 415

Kamchatka Peninsula, 41, 71, 101, 336

Kane (ship), 401

Kaolinite, 91, 105, 107

Kermadec Trench, 69, 71

Key West, Fla., 292

Killer whale, 361

Killifish, 394

Kingfish, 365

Knorr (ship), 21

Kodiak Island, 42

Kolyma River, 80, 83

Korea, 418

Kovachi, 73

Krill, 361, 410 (see also Euphausiids)

Krypton, 52, 127, 130, 133

Kurile-Kamchatka Trench, 41, 45, 69

Kuroshio, 178, 179, 181

L

Labrador, 168, 365

Labrador Current, 167, 177, 256

Labrador Sea, 80

Lagoons, 283, 294
   and coral reefs, 75, 76, 77
   as estuarine systems, 262–264
   and sea ice, 258
   and sedimentation, 293, 294, 300
   submerged, 59
   U.S., 256, 262, 263, 292, 294, 299

LaGuardia Airport, N.Y., 306

Laguna Madre, 257, 263, 273–275

Lakes, 295
   and deltas, 291
   and density of water, 124–125
   and lagoons, 264–265, 283
   and productivity, 335
   as sediment traps, 99, 292
   and standing waves, 225
   and water contained in, 30, 31, 152, 153

Lake Tanganyika, 64

Laminations (beach features), 300–301

Lamont-Doherty Geological Observatory, Columbia University, 19

Landfill operations, 419, 421

Langmuir, Irving, 187

Langmuir circulation, 187–188, 372

Lantern-fishes, 358

La Plata River, 80, 284

Larvae:
   of benthos, 373, 374 (see also names of specific organisms, e.g. Balanus)
   planktonic (see Fish larvae; Meroplankton; Pelagic larvae, of benthos; and names of groups and individuals, e.g. Balanus; Polychaetes; Worms, larvae of)

Latent heat, 123, 147, 151

Latitude, parallels of, 4, 440, 441

Laurasia, 39

Lava, 34, 35, 37, 43, 48, 50, 53, 61, 69

Law of the Sea, 406

Lead, 48, 127, 130, 428

Lefroyella, 401

Lena River, 80, 83, 284

Lepidocrocite, 91

Leptogorgia, 396

Levees, 60, 68, 291

Lichen, 375

Lighthouse Point, Fla., 296

Light, in marine environments, and distribution of benthic organisms, 373, 375, 376, 396

Light in the ocean (see also Bioluminescence; Insolation):
   and adaptations of marine organisms, 343, 344, 356, 358–360, 363, 394, 402 (see also Color, of marine organisms, and environmental factors)
   extinction of, 314, 356, 358
   as a limiting factor, 315–316, 319, 325, 334–335
   measurement of, 21
   and photosynthesis, 102, 159, 315–317, 325–326, 327, 328, 334–335, 389
   and release of photo-assimilated carbon, 329
   and succession of species, 344

Light reaction, 323

Limacina, 364

Limestone, 70–71, 75, 420, 422, 423 (see also Calcium carbonate)

Limiting factors, 315–317, 319, 321

Limpets, 376

Line Island chain, 41, 42

Linophryne, 359

Lithogenous sediment (see Sediment, lithogenous constituents)

Lithosphere, 29, 30, 32, 33, 47

Lithospheric plates, 36–38, 40–42, 46, 48, 52 (see also names of individual plates, e.g. Pacific Plate; South American Plate)

Lithothamnion ridge, 395

Littoral currents (see Longshore currents)

Littoral drift, 299, 300

P

Pacific–Antarctic Ridge, 41
Pacific basin, 58, 61, 78–79
  and continental mragins, 59, 60
  dimensions of, 79, 216
  evolution of, 37, 39, 40, 43
  as a gulf, 28
  and islands, 43, 71, 73, 76, 81 (see also Volcanic islands, Pacific Ocean)
  and sediment deposits, 66, 95, 97, 102, 104–107, 109, 111, 112, 421 (see also Sedimentation, of Pacific basin)
  and subsidence, 69, 71
  and topography, 19, 67–70, 72
  and trenches, 31, 34, 37, 47, 69
Pacific Coast of the United States, 216, 222, 256
  and beaches and dunes, 108, 293–294, 297
  and coastal ocean, 258, 260, 262, 276
  and estuarine systems, 262–265
  and marine organisms, 380, 385
  and shoreline features, 262, 264, 283
  and tidal phenomena, 233
Pacific Equatorial Water, 163
Pacific Northwest (U.S.), 264, 276
  and coastal ocean region (see Northeast Pacific coastal ocean region)
Pacific Ocean, 37, 43, 59, 171 (see also Northeast Pacific coastal ocean; North Pacific Ocean; South Pacific Ocean)
  and barriers to water exchange, 196, 334
  boundaries of, 80, 81
  and currents, 170, 174, 178, 197, 200, 399
  and freshwater discharge, 79, 80, 269
  and icebergs, 167
  and marine organisms, 348, 362, 365, 373, 408, 409, 410
  and nutrients, 317, 336
  and organic carbon, 330
  and oxygen distribution, 334
  and productivity, 336
  and salinity, 79, 155, 157, 160, 162, 198, 276
  and sea-surface topography, 194
  and surface and drainage areas and average depth, 79
  and temperature, 79, 149–150, 160, 161, 197
  and tidal phenomena, 233, 243
  and upwelling, 336
  volume of, 31, 81
  and water masses, 197, 198, 200
  and waves, 214, 216, 223
Pacific Plate, 26, 36, 39, 41, 42, 43
Pacific Southwest (U.S.), and coastal ocean region, 256–257
Pacific Subarctic Water, 163
Pack ice, 14, 83, 165, 167, 168, 169

Padre Island, Texas, 273
Pago Pago, American Samoa, 286
Pakhoi, China, 234
Pakistan, 48
Palau Trench, 45
Paleozoic era, 26, 40
Palmer Peninsula, Antarctica, 81
Pamlico River, N.C., 263
Pamlico Sound, N.C., 254, 257, 263, 264
Pancake ice, 166, 167
Pangaea, 26, 39, 49, 51
Paracalanus, 349
Parachute drogue, 176
Parapods, 388
Parasites, 343, 348, 394
Parrot-fish, 396
Partial tide, 238–239, 243 (see also Lunar partial tide)
Patch reefs, 76, 77
Pea crab, 385, 394
Peat, 291
Pebbles, 91, 293
Pectin, 343
Pectinaria koreni, 374
Pelagic environment, and ecosystems, 314, 322, 346, 350
Pelagic fauna, and reef communities, 399
Pelagic fishes, 362, 363, 367, 368, 390, 399 (see also names of groups and individuals, e.g. Cod; Sardina pilchardus)
Pelagic larvae, of benthos, 373 (see also Meroplankton)
Pelagic sediment deposits, 109, 110
Penguins, 361
Pensacola, Fla., 232
Peptides, 329
Perch, 365, 408
Percoid fishes, 365
Peridinium, 345
Period, of a basin, 240, 243, 248 (see also Wave period, of standing waves)
Periodic table of the elements, 127
Persian Gulf, 84, 85, 156, 258, 415, 416, 424
Peru, 22, 78, 237, 336
  and fish production, 22, 408, 410
Peru Basin, 45
Peru–Chile Trench, 37, 41, 45, 47, 60, 69
Peru Current, 177, 178
Pesticides, 275, 427, 428
Petitcodiac River, N.B., 242-243
Petroleum, 17, 60, 187, 404–405, 406, 415–417, 419, 422, 424, 427
Pettersson, Otto, 11, 12
pH, of seawater, 134, 321
Philadelphia, Pa., 184
Philippine Basin, 45
Philippine Islands, 48
Philippine Plate, 41
Philippine Sea, 80
Philippine Trench, 45, 69

Phillipsite, 106
Phosphates (see also Phosphorus cycle):
  and chemosynthesis, 392
  and depletion of, in surface waters, 262, 315
  in the English Channel, 320
  in estuaries, 366
  as a mineral resource, 7, 423
  in the North Sea, 350
  and photosynthesis, 192, 262, 323, 324 (see also Phosphorus cycle)
  and phytoplankton (see Phosphorus cycle)
  and sediments, 95, 102, 105 (see also Sediment, phosphatic constituents)
  and S/P ratio, 316
  as wastes, 428
Phosphorite, 422, 423
Phosphorus, 127, 130, 193, 316, 317
Phosphorus cycle, 317–319, 329
Phosphorylation, 323
Photic zone, 342, 343
  and coral reefs, 396
  depth of, 314
  as food source for deep-water organisms, 321, 357–358, 360, 402
  mixed layer extending below, 319, 325–326
  and nutrient cycles, 315, 317, 319, 335 (see also Nutrients, recycled to surface waters)
  organic matter below, 327, 329
  oxygen in, 133, 325, 334
  and productivity, 325, 342
  and schooling of fishes, 363
Photosynthesis, 323–324
  and biogenous sediment, 102
  and carbon dioxide (see Carbon dioxide, and photosynthesis)
  and dissolved organic matter, 329, 333
  and food webs, 315, 321, 333
  and insolation, 144, 159, 315, 356
  and oxygen production, 53, 54, 133, 325
  and productivity, 325
  and temperature, 313
  and zooxanthellae, 395
Photosynthetic zone (see Photic zone)
Physalia, 351
Phytoplankton, 314
  characteristics of, 342–344
  and chelating agents, 321
  in the coastal ocean, 181, 262, 314, 343, 344
  colonial, 347
  composition of, 317
  density of, 342
  and dissolved organic matter, 318–320, 329
  in the English Channel (see English Channel, and plankton in)
  in estuaries, 366
  and food webs, 319–322, 337, 341, 346, 350, 363, 364, 367, 385

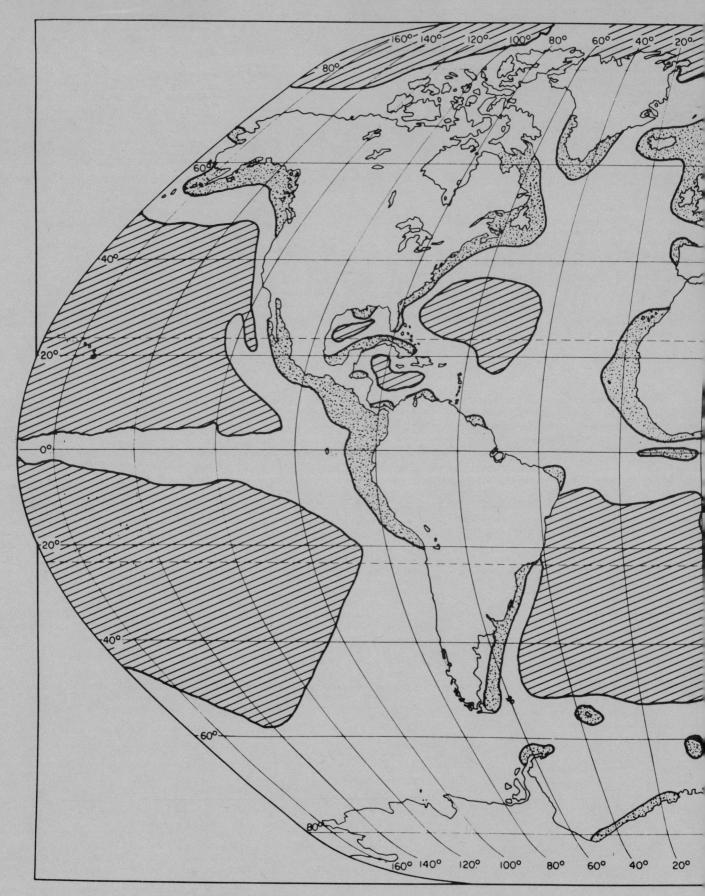

Distribution of primary productivity in surface waters
of the world ocean. (Redrawn from O. J. Koblentz-Mishke,
V. V. Volkovinsky and J. G. Kabanova, 1970.
"Plankton Primary Production of the World Ocean,"
in *Symposium on Scientific Exploration of the South Pacific*,
National Academy of Sciences, Washington, D.C., pp. 183-93.)